S0-BRS-596

THE MEDITERRANEAN AIR WAR

THE MEDITERRANEAN AIR WAR

Airpower and Allied Victory in World War II

Robert S. Ehlers, Jr.

University Press of Kansas

For Leisa, Emily, Bailey, and Kina
Your Love Sustains Me
Your Patience Astounds Me

Published by the University Press of Kansas (Lawrence, Kansas 66045), which was
organized by the Kansas Board of Regents and is operated and funded by
Emporia State University, Fort Hays State University, Kansas State University,
Pittsburg State University, the University of Kansas, and Wichita State University

Library of Congress Cataloging-in-Publication Data
Ehlers, Robert.
The Mediterranean air war : airpower and Allied victory in World War II /
Robert S. Ehlers, Jr.
pages cm. — (Modern war studies)
Includes bibliographical references and index.
ISBN 978-0-7006-2075-3 (cloth : alk. paper) — ISBN 978-0-7006-2076-0 (ebook)
1. World War, 1939–1945—Aerial operations. 2. World War, 1939–1945—
Campaigns—Mediterranean Region. 3. Air power—Mediterranean Region—History—
20th century. 4. Combined operations (Military science)—Case studies.
5. Mediterranean Region—History, Military—20th century. I. Title.
II. Title: Airpower and Allied victory in World War II.
D785.E54 2015
940.54′21—dc23
2014040567

British Library Cataloguing-in-Publication Data is available.

Printed in the United States of America

10 9 8 7 6 5 4 3 2 1

The paper used in this publication is recycled and contains 30 percent postconsumer
waste. It is acid free and meets the minimum requirements of the American National
Standard for Permanence of Paper for Printed Library Materials Z39.48-1992.

CONTENTS

Preface and Acknowledgments   vii

Abbreviations   xiii

PART I: PRELUDE

1 The Approach to War   3

PART II: CONTEST

2 War Comes to the Mediterranean   25

3 Triumph and Tragedy: Operation Compass and the German Attack on Greece   46

4 The RAF Holds—and Learns   86

5 The Back Door Secured and the Front Door Strengthened   114

6 Preparations for and the Conduct of Operation Crusader   150

7 Rommel Strikes Again   177

8 Axis High Tide   200

9 The Tide Turns   223

10 El Alamein and Operation Torch   242

11 Building a New Air Command and Clearing North Africa   265

PART III: EXPLOITATION

12 Operations Husky, Avalanche, and Baytown   291

13 The Italian Campaign and the Invasion of Southern France   322

14 The Second Aerial Front   356

15 The Balkan Air Force   386

PART IV: RETROSPECTIVE

16 Taking Stock   397

Notes   407

Selected Bibliography   475

Index   491

PREFACE AND ACKNOWLEDGMENTS

THE ROMANS CALLED IT *Mare Nostrum* (Our Sea)—the body of water stretching from the Straits of Gibraltar to the shores of the Holy Land. In their day, when warfare existed in two dimensions, they discovered, initially at great cost, that their sea must remain so without fail. The dangers to the republic were too great to consider what might happen should they lose control of the waters surrounding and connecting their territorial possessions. The Carthaginians came as close as anyone to bringing about Rome's undoing, by sea and by land. The Romans never forgot that.

For the next two millennia, the Mediterranean Sea's grand-strategic importance waxed and waned. At times, it was more a trade thoroughfare than a battleground; at others, it was the domain in which the destinies of the Great Powers were decided. Even as the age of sail replaced that of galley warfare in the Mediterranean, the Middle Sea's importance remained great as an artery for economic exchange and as a strategic pivot connecting the many states, communities, and peoples ringing its shoreline. Control of this body of water, with its central location and the rapid movement this allowed, gave it continuing strategic importance even as the Great-Power rivalries of the eighteenth and nineteenth centuries shifted the focus to the north. The completion of the Suez Canal in 1868 and the United Kingdom's ultimate control over it, along with Gibraltar at the other end of the Middle Sea and Malta in the middle, conferred strategic advantages. Among these was the means to move goods and military forces around the world with unprecedented speed. Even as the Great War (World War I) made the Mediterranean a relative backwater, the British were not about to risk losing control of the Suez Canal. In fact, quite the opposite: Allenby's campaign against the Ottomans was designed not only to destroy their armies and create fissures in the Turks' already shaky facade, but in the process to push this menace away from Suez.

During the period between the world wars, at a time when the Mediterranean might easily have receded further into its widely perceived backwater status, several things happened that made it important once again in fundamental ways. The development of the internal-combustion engine and the British discovery and exploitation of oil reserves in Iraq and Iran made access to this commodity—and the ability to protect it from hostile powers—a prominent part of British strategy. Benito Mussolini's adventurism in

Abyssinia, as he proclaimed a "New Rome," further concerned the British and made clear the vulnerability of the Suez Canal and oil resources. The world economic crisis, which led directly to the collapse of a fledgling democracy in Japan and set that country on a decidedly warlike course by the early 1930s, further alarmed policy makers in London, who now recognized the importance of Egypt and the Suez Canal as the means for moving military forces to defend their holdings in the Far East and, most importantly, India. Finally, the Spanish Civil War, and the German and Italian roles in it, concerned Great Britain and France about these countries' wider aims in the Mediterranean.

That said, Adolf Hitler soon revealed his real focus: the countries of Central and Eastern Europe and those of the West as stage setters for his ultimate prize—the Soviet Union. However, he maintained a strong interest in controlling the western Mediterranean, and the approaches to it, in order to keep both British and US forces away from Europe and Africa. His failures in the subsequent diplomatic engagements with Vichy France, Spain, and Italy played an important role in Germany's defeat both in the Mediterranean and in the larger war. While Hitler focused predominantly on the Continent, British and Italian leaders set their sights on the protection and expansion, respectively, of their Mediterranean holdings. And so, once again, the Middle Sea would become a place of battle. Yet several things had changed fundamentally by 1939, particularly in the realm of air warfare.

The Romans had been able to conquer the lands surrounding the Mediterranean through a deft and ruthless form of combined-arms warfare in two dimensions. With the advent of modern aircraft, a third dimension entered the combined-arms equation, and the countries that employed all three branches of their armed services—army, navy, and air force—with the highest degree of coordination, creativity, and operational effectiveness would have a major advantage in the coming war. This advantage was magnified in the Mediterranean theater from Italy's declaration of war on 10 June 1940 until Victory in Europe Day precisely because of its unique geography and central location. It was an arena ideally suited for effective combined-arms operations and a fatal one for those who proved wanting in this respect. Those who learned quickly had a great advantage. During the five-year campaign, the Allies ultimately proved more adept than their Axis adversaries at this kind of warfare. This occurred both in spite of and because of early German victories and Allied defeats. Pushed back on their heels, the British had to learn or lose. Conversely, the Germans believed, with some justification, that their way of war had proven itself quite convincingly in the conflict's first year. However, the Battle of Britain gave them their first taste of defeat at the hands of an adversary that employed aircraft and the various assets and capabilities facilitating their use with greater skill. The Germans faced the British Royal Air Force (RAF) again in the Mediterranean, and despite some impressive but costly victories in Greece and on Crete, the Luftwaffe (German Air Force) again proved unable to prevail. This defeat in the air over North Africa and the Mediterranean, though a gradual and complex process, and one in which the Luftwaffe often performed effectively at the tactical and operational levels, was a major boon for the Allies as they gained control of the Mediterranean and put their now formidable air assets to work attacking high-value Axis targets in Europe.

This work traces the course of the air war in the Mediterranean theater from its earliest phase in 1940, when a makeshift RAF command defeated the Regia Aeronautica, or Italian Air Force (IAF), in the Western Desert and Italian East Africa (IEA) through the seesaw combined-arms and logistics-dependent campaigns for control of North Africa and then during the period of Allied ascendancy in the air from the conclusion of the North African campaign to Victory in Europe Day. It takes at face value Richard Overy's and Air Chief Marshal Arthur Tedder's assertions that the Allies won in the air—and thus more quickly and with fewer casualties on the ground and at sea—because they treated airpower as a key part of a larger, coordinated strategy that involved all three services (air, ground, and sea), and in which all played key roles.[1]

It would, however, be a mistake to think of this as a study of airpower alone. Any such inquiry would, because of the geographical and combined-arms realities already mentioned, be shallow and therefore of little use to historians, military professionals, or policy makers. The focus of this study is in fact on how the Allies succeeded in achieving their policy and military-strategic objectives in the Mediterranean theater precisely because they became more expert at waging combined-arms operations than did their adversaries. Therefore, this work is about the proper role of airpower *within a given historical, geographical, and grand-strategic context* and the ways in which it came together with ground and sea power to bring the Allies a resounding series of victories—and the Axis a series of catastrophes from which, in conjunction with the disasters on the Eastern Front, they never recovered.

One of the most crucial aspects of the campaigns in this theater was the extraordinary level of interdependence between the services. Interestingly, the historical record is full of army and navy—and even air force—references to "supporting air forces." However, research for this work has uncovered only a handful of instances when senior officers referred to the army or navy as "supporting forces," even though victory or defeat ultimately hinged as much on control of airfields and the effective use of land-based air forces within the distinctive context of the theater as it did on armies or navies. Although airpower may simply have been too new for that kind of nuanced insight, it is worth pondering what this kind of approach might mean for armed conflicts today, in which a continuing use of the terms "supporting" and "supported" stunts our thinking about how combined-arms operations should work. During the war, Air Marshal John Slessor, one of the best RAF commanders and among the foremost thinkers in any uniform, said, "The fact is that 'Army Air Support' is really an obsolete term, as is the conception that the Air is a 'supporting arm' just like artillery."[2] He said that although the soldier had to take the ground and was best suited to determine how to do that, he was not well suited by training or experience to tell airmen how to help. In fact, Slessor continued,

> It's not only a question of the Air supporting the Army but of the Army supporting the Air. It is a question of seeing how the Air and the Army . . . can best collaborate and play into each other's hands—and the Air factor *may* have a preponderating influence on the whole plan. It may—in fact already has—determined *where* an attack can be made. And the primary essential principle underlying the whole thing must be the old principle of concentration of decisive force at the decisive time and place—i.e., flexibility. . . . These are my own views stated frankly. I . . . should like to think that the people who are thinking about this subject do understand, rather more clearly than they now appear to, what the Air Staff are driving at.[3]

Similarly, Air Chief Marshal Charles Portal, the RAF chief of Air Staff (CAS), felt compelled to state once again at the 260th Chiefs of Staff (CoS) meeting the RAF's position regarding command of air forces in the Middle East. "I am of course aware," he said, "that the C.I.G.S.' [chief of the Imperial General Staff Field Marshal Alan Brooke's] conception of the correct state of affairs is that the Air Force should always be subordinate to the Army in any theatre of war in which the Army and Air Force are together engaged. This conception I regret I am quite unable to accept."[4]

This question of who "supports" and is "supported" in armed conflicts—each of which is unique and brings about vastly different requirements relating to this concept—should prompt us to ask whether we might be able to approach such questions with greater insight by thinking of the services in terms of their cooperation, their interdependence, and the effects and effectiveness of their combined-arms efforts. To do otherwise when studying the war in the Mediterranean would lead to the gravest misunderstandings about how each side fought the war there and why the Allies won. To make the same error in the future, in armed conflicts with unique contexts but with similar levels of complexity, would be to invite defeat.

Locating sources for this work proved challenging for two reasons. The first is that the vast majority of secondary sources published on the Mediterranean theater focus on the ground and naval wars, in that order, with works about airpower in a very distant and almost nonexistent third place. The second is that, with the partial exception of the 1941–1943 campaigns, the war in the Mediterranean often comes across as something of a minor sidelight, not just to the war on the Eastern Front but to the Pacific theater and to the Grand Alliance's effort in general. The result in the case of this work has been a heavy reliance on primary sources, official histories, and the few secondary works dedicated to the air war in the Mediterranean and Middle East theater. This problem is compounded by the fact that there are relatively few surviving German and Italian records about the air war in the Mediterranean. By default, Allied records, the many German documents and field reports captured and translated during and after the war, and specialized secondary sources such as the German official history of the war and Karl Gundelach's two-volume study of the Luftwaffe in the Mediterranean theater, represent the majority of readily available sources. Nonetheless, these allow us to assess how effectively the Allied and Axis air forces operated in this key theater and the ways in which the air efforts were linked in some meaningful way to policy and military-strategic objectives relating not just to the Mediterranean but to the larger war.

There are many people to thank for their roles in making this book a reality. Mike Briggs and his team at the University Press of Kansas were, as always, incredibly helpful and patient as I struggled to balance a heavy workload in my day job with the completion of this work. Sebastian Cox, head of the Royal Air Force's Air Historical Branch, has been a superb mentor and gave me access to the many crucial primary sources residing in the hidden gem at RAF Northolt. Joe Guilmartin, my Ph.D. adviser at Ohio State, the man to whom I dedicated my first book and a combat veteran and scholar for whom I have the utmost respect and admiration, kept after me to get this one done as well. Tami Davis Biddle, another outstanding historian and mentor, provided in-depth comments that helped me to make the work much stronger than it would otherwise have been. I also owe the staffs of the Air Force Historical Research Agency, Air Force Academy Special Collections Branch, National Archives and Records Administration, Library of Congress, Air Historical Branch, and most of all the National Archives in Kew, London, a huge debt of gratitude for helping me to find the primary-source documents upon which this work depends. Joe Panza of the

Air University Foundation very generously provided funds for the indexing work. James Corum pointed the way to German and Italian sources that contributed substantially to this effort. Richard Muller, my former comrade from the School of Advanced Air and Space Studies (SAASS) and longtime mentor, was kind enough to listen patiently as my idea for this work took shape, to offer superb advice and insights, and to inspire me with his tremendous level of scholarship. In fact, I owe all of my old SAASS colleagues a great debt of gratitude for their unstinting support as I began my research for this book. Although it is both proper and a pleasure to acknowledge these outstanding people and institutions, I of course take full responsibility for the contents of this work. I hope it will give scholars, military personnel, policy makers, and interested citizens an opportunity to engage with and think deeply about one of the least-studied aspects of World War II—the role of airpower in the vital Allied victory in the Mediterranean and Middle East. Perhaps, in the process, they will contemplate the ways in which events from that time can help us understand how best to assess and engage with even modestly analogous current and future developments.

I would be particularly remiss if I did not thank the most important people in my life—my family—for their patience and forbearance as I spent very full days at work followed by virtually every evening and weekend at the computer writing this book. With the project now behind me, I will give them more of the attention and affection they so richly deserve. All of them— Leisa, Emily, Bailey, and Kina—are a blessing to me. The same is true of my parents, Robert and Carol, who gave me a love of history and both the moral support and the means to study it. I am in their debt for that great gift.

ABBREVIATIONS

A2	director of US Intelligence
A3	director of US Operations
AAA	antiaircraft artillery
AASC	army air support control
ABC 1	American-British-Canadian talks (first meeting)
ACTS	air corps tactical school
AEFP	air explosives and fuels park
AFHQ	Allied Forces Headquarters
AHQ	Air Headquarters
AI	air-intercept radar
AIL	air intelligence liaison
ALO	air liaison officer
ALS	air liaison section (comprised of ALOs attached to an army formation)
AMU	aircraft maintenance unit
AO	Airfields Organization
AOA	Air Officer Administration
AOC	air officer commanding
AOC in C	air officer commanding in chief
AOP	air observation post
ASC	air support control (RAF counterpart collocated with AASCs)
ASGC	Army Support Group Command
ASV	air-to-surface surveillance radar
ASW	antisubmarine warfare
ASP	air stores park
AWM	Australian War Memorial
AWPD 1	Air War Plans Division Plan 1
BAF	Balkan Air Force
BDA	bomb-damage assessment
C2	command and control (includes communications and intelligence)
CAS	chief of Air Staff
CBO	combined bomber offensive
CCS	Combined Chiefs of Staff
CIA	Central Intelligence Agency

C in C(s)	commander(s) in chief
CMO	chief maintenance officer
CoS	Chiefs of Staff (British)
CSDIC	Combined Services Detailed Interrogation Center
CSO	chief signal officer
CSTC	Combined Strategic Targets Committee
DAF	Desert Air Force (succeeded WDAF)
DAOC in C	deputy air officer commanding in chief
DLMM	Die deutsche Luftwaffe im Mittelmeer
DSCAEF	deputy supreme commander Allied Expeditionary Force
EAC	Eastern Air Command
EOU	enemy objectives unit
FAA	Fleet Air Arm
FAC	forward air controller
FASL	forward air support liaison
FCNA	Führer Conferences on Naval Affairs
GCCS	Government Code and Cipher School (at Bletchley Park)
GHQME	General Headquarters Middle East
GOC	general officer commanding
H2X	radar bombing aid (USAAF version of RAF's H2S)
HE	high explosive
HF	high-frequency radio
HF/DF	height-finding/direction-finding radar
HMS	his majesty's ship
HMSO	his/her majesty's stationery office
HQ	headquarters
IAF	Italian Air Force
IEA	Italian East Africa
IFF	interrogation friend or foe
IWM	Imperial War Museum
JCS	Joint Chiefs of Staff
JG	*Jagdgeschwader* (Luftwaffe fighter wing)
JIC	Joint Intelligence Centre
JPRC	Joint Photographic Reconnaissance Center
JPS	Joint Planning Staff
KG	*Kampfgeschwader* (Luftwaffe bomber wing)
LOC	line of communication
LRDG	long-range desert group
MAAF	Mediterranean Allied Air Forces
MAC	Mediterranean Air Command
MACAF	Mediterranean Allied Coastal Air Force

MAPRW	Mediterranean Allied Photographic Reconnaissance Wing Unit
MASAF	Mediterranean Allied Strategic Air Forces
MATAF	Mediterranean Allied Tactical Air Forces
MEAF	Middle East Air Forces (USAAF command under RAFME)
MPIC	Mediterranean Photographic Interpretation Center
MTLRU	mechanical transport light repair unit
MV	merchant vessel
NAAF	Northwest African Air Force
NACAF	Northwest African Coastal Air Force
NAPRW	North African Photographic Reconnaissance Wing
NASAF	Northwest African Strategic Air Force
NATAF	Northwest African Tactical Air Force
OC	officer commanding
OKH	Oberkommando des Heeres (German Army High Command)
OKL	Oberkommando der Luftwaffe (German Air Force High Command)
OKW	Oberkommando der Wehrmacht (German Armed Forces High Command)
OP	observation post
OSS	Office of Special Services
OTU	operational training unit (RAF aircrew-training unit)
PFF	Pathfinder Force (Fifteenth Air Force term for H2X-equipped lead bombers)
PI	photographic intelligence
POL	petroleum, oil, and lubricants
POW	prisoner of war
PR	photographic reconnaissance
PSP	pierced-steel planking
QMG	quartermaster general
RAF	Royal Air Force
RASL	rear air support liaison
RAFDEL	Royal Air Force delegation (to the United States)
RAFME	Royal Air Force Middle East
RSU	repair-and-salvage unit
R/T	radio telephony
SAAF	South African Air Force
SAS	Special Air Service
SASO	senior air staff officer
SCAEF	supreme commander Allied Expeditionary Force

SIS	Signal Intelligence Service (German)
SLOC	sea line of communication
SOE	Special Operations Executive
SOS	Special Operations Service
SS	*Schutzstaffel* (*Waffen SS* ["Weapon SS"]—comprised SS combat units)
Tac R	tactical reconnaissance
TAF	Tactical Air Force
USAAC	US Army Air Corps
USAAF	US Army Air Forces
USAF	US Air Force
USSAFE	US Strategic Air Forces in Europe
USS	US ship
USSR	Union of Soviet Socialist Republics
VCAS	vice chief of air staff
VE Day	Victory in Europe Day (8 May 1945)
VHF	very high-frequency radio
WAC	Western Air Command
WDAF	Western Desert Air Force
WT	wireless telephony

PART I

Prelude

The Mediterranean and Middle East theater (this was the British title; the Americans referred to it as the Mediterranean Theater of War) covered an area of more than 3 million square miles. Its geography demanded a capable and complex interplay between land, sea, and air forces. The side that learned more effectively to employ combined-arms operations within the theater, based on sound grand and military strategies, would have a major advantage. (Maps Department of the US Military Academy, West Point)

I

The Approach to War
Events to June 1940

We will fight to the last inch and ounce for Egypt.
—Winston Churchill

The Mediterranean will be turned into an Italian lake.
—Benito Mussolini

I go the way that Providence dictates with the assurance of a sleepwalker.
—Adolf Hitler

The Grand-Strategic Context

ANY EXAMINATION OF THE ROLE of airpower in the Mediterranean theater of war must begin with a wider look at the theater's grand-strategic importance. Because of the Mediterranean's geographical makeup, combat in any of its many subregions was bound to feature a major air presence given this relatively new weapon's ability to strike deeply from land bases. Land-based airpower in many ways set the tone and direction of the conflict in the Mediterranean, although always in conjunction with, and correspondingly dependent upon, land and naval forces. How each of the warring powers chose to employ air assets was tied in large measure to its grand-strategic views of the Mediterranean.

Unfortunately, most existing studies mischaracterize the Mediterranean as anything from a subsidiary (if important) theater to an irrelevant one. This "irrelevance" was, as a number of historians have asserted, the result of a misconceived Allied approach to the theater. As one scholar noted, "The consequence of this Allied stumble into a poorly thought-out and 'opportunistic' Mediterranean strategy was a dreadful slogging match in which the British and subsequently the Americans were outgeneraled and outfought around the shores of a sea of trifling importance."[1] Another was nearly as categorical when he argued that the Mediterranean was a "cul-de-sac . . . a mere byplay in the conclusion of a war won in mass battles on the Eastern and Western Fronts."[2] One particularly negative assessment holds that "the reality was that Great Britain sacrificed vital interests, such as the home front and Singapore, and paid an exorbitant cost in shipping to maintain

3

for three years a small army in a peripheral campaign far from the German jugular. Britain fought in the Mediterranean because from 1940 to 1943 there was no other place where it could fight without the prospect of total defeat."[3] Finally, another asserts that "Britain was ambivalent about the Mediterranean. While it was neither vital to its own survival nor to that of the British Empire, Britain was incapable of releasing it. It repeatedly devoted to it significant resources which could have more effectively been used elsewhere."[4] Other scholars opined that "from the strategic point of view the Middle East offered [Hitler] few possibilities."[5]

Others have hedged their bets. One said of the Mediterranean theater, "Whether the expenditure in resources was worth the results remains a matter of controversy."[6] He concluded that "although a secondary theater of war . . . the Mediterranean campaigns nevertheless are a vital part of the history of the Second World War."[7] Given the importance of the Middle Sea to the British as a conduit to their Far East holdings and the means of protecting their access to oil supplies, these assertions ring hollow, especially when one considers the major strategic benefits conferred by victory in the Mediterranean, including the destruction of an entire Axis field army, air supremacy over southern Europe, Italy's surrender, and heavy-bomber missions from the Foggia airfields against key economic and transportation targets in occupied Europe, among others. The war may have been won elsewhere, but the Allies could have lost it here. One of the reasons they did not, aside from poor Axis grand strategy and relatively ineffective employment of airpower, was the Allies' more effective use of airpower, within a combined-arms context, driven by a clear grand strategy.

Douglas Porch places the Mediterranean theater within the larger context of a global conflict for the survival of the democratic powers and then portrays it as a combined-arms effort in which the Allies ultimately prevailed. Further, he argues that "while the Mediterranean was not the *decisive* theater of the war, it was the *pivotal* theater, a requirement for Allied success," one in which the Allies were able to "acquire fighting skills, audition leaders and staffs, and evolve the technical, operational, tactical, and intelligence systems required to invade Normandy successfully in June 1944."[8] A Commonwealth loss in the Mediterranean in 1941 or 1942 would have been disastrous for the Allied war effort. The Axis would have seized enormous oil resources, passage for U-boats through the Suez Canal to the Indian Ocean, a back door into the Soviet Union and Turkey, and a possible linkup with the Japanese that would have destroyed the British position in India.[9]

We must view all of this within the context of the Mediterranean's geography, which was a "strategist's nightmare," particularly with the advent of land-based airpower.[10] Because of the geography of the Mediterranean and adjoining regions such as the Middle East and Italian East Africa (IEA), it was a sea-air-land theater more so than any other in the war. The British understood this, mostly because of their imperial interests, and they planned to fight for control of the region. However, after the Luftwaffe savaged the Royal Navy during the evacuation of Crete, the escort of convoys to Malta, and in other engagements, it became clear that it would be impossible for the navy to operate without land-based air cover. Air forces may not have been a substitute for navies, but naval power was no substitute for airpower. Even the ground campaigns in the Western Desert were focused to a tremendous extent on capturing key airfields that could in turn serve as bases for aircraft supporting further ground advances while interdicting merchant shipping, which provided a lifeline to the warring armies. This complex interplay between air, naval, and ground operations proved crucial, and one could not succeed without the others. Malta, Gibraltar, and the Suez Canal were all crucial parts of this interplay because by holding them, the British maintained a range of important advantages.

British Priorities

It was good fortune that the British understood the Mediterranean and Middle East theater's grand-strategic importance as well as the complexities of asserting and maintaining control over this huge area, which included the Mediterranean. In fact, they were the only ones to view the Mediterranean Basin and its surrounding areas as a "single geo-strategic unit."[11] Churchill insisted that the British would fight to the "last inch and ounce for Egypt."[12] The desert flank was "the peg on which all else hung."[13] He understood that the Mediterranean campaign could not win the war but might well lose it and used this theater as leverage with Franklin Roosevelt for US support. Churchill believed that losing the Suez Canal would be a calamity "second only to a successful invasion and final conquest" of the United Kingdom.[14] The strategy had to be to conquer North Africa first, reopen the Mediterranean, and then attack Italy and knock it out of the war.[15] These moves were never intended to be substitutes for an invasion of the Continent but rather indispensable preliminaries to shore up Britain's strategic position and weaken the Axis while awaiting US entry into the war. Despite British tendencies to look for new opportunities in the Mediterranean after summer 1943, and the major disagreements these views created with the Americans, they remained focused on the defeat of the Reich by land, air, and sea, to include the invasion of northwestern Europe.

For a brief period after the Italian conquest of Abyssinia, British leaders feared that Italian land-based airpower there and in Libya would make the empire's position in the Mediterranean untenable. Two things changed this view. First, a group of naval officers known as the "Mediterraneanists" argued that the Middle East was a must-hold theater from a grand-strategic perspective and a launching pad for any attack on Italy, should Mussolini side with Hitler. Second, the discovery of major oil reserves in Saudi Arabia in 1938 drove home with even greater force the importance of holding open and developing the sources of this crucial commodity in Iraq, Iran, and now that kingdom. Closure of the Mediterranean to shipping, combined with any serious merchant vessel (MV) shortage, would force the British to rely heavily on US oil. Consequently, they felt compelled to keep British-controlled oil in their hands. Sending oil 14,000 miles around the Cape of Good Hope to the United Kingdom was unpalatable enough; losing the oil sources altogether was unthinkable.[16]

The Italian conquest of Abyssinia caused the British to take actions that paid major dividends after the war began. It refocused their attention on the vital importance of the Suez Canal as the "hinge" in the empire's commerce. The growth of air routes to India, Singapore, and Australia also depended on a secure Middle East, and this development had the added benefit of creating a far-flung network of air bases that proved its worth in the coming contest.[17]

Based on the growing Axis threat, the commanders in chief (C in Cs) of the Middle East theater received modest reinforcements and had time to think about key issues they would face during the war, including basing requirements, logistics, operational planning, and interservice cooperation. They prepared for a campaign they knew they could not lose. Given the theater's huge size and the need to keep aircraft serviceability rates high, air units had to be highly mobile. This required a basing infrastructure, supplies, salvage and repair facilities, weather forecasting capability, and radar. These requirements constituted a huge logistical challenge that the British met by developing, in effect, a parallel Metropolitan Air Force in Egypt.[18] As Humphrey Wynn noted, "uniquely among the [Royal Air Force (RAF)] commands, the Middle East had created a complete Air Force."[19] The distance between London and Cairo was too great to allow for any other solution. The ensuing focus on logistics, intelligence, and command and control (C2) paid enormous dividends. This, Wynn continued, "resulted largely from two factors: the supply of British and American aircraft shipped to African ports; and the support given to the Desert Air Force by the supply, maintenance and repair organization, which was radically overhauled and reorganized during 1941."[20] The RAF's "whole air force" concept conferred important advantages in a theater of operations where complex geographical and logistical realities made land-based airpower vitally important.[21]

Italian Policy Aims

The Italians, too, saw airpower, along with their naval and ground forces, as a key means for achieving their objectives. However, Mussolini sought to achieve them at little cost because he knew Italy could not wage a long war. He believed the Mediterranean was rightfully Italian but also a prison within which the British and French hemmed in Italy. His focus during the 1930s, crowned by the conquest of Abyssinia, was to break out of this "prison" to secure the key geographical points that would ensure commercial outlets and military access beyond the Middle Sea and allow Italy to obtain new colonies. At a minimum, this "New Rome" would include French and Spanish Morocco, Algeria, Tunisia, Gibraltar, the Suez Canal, the Red Sea, the Gulf of Aden, and the Balkans. Only by seizing these areas, the duce reasoned, could Italy become truly self-sufficient economically and able to fend off the threats posed by Britain, France, and, more remotely, the United States. How Mussolini intended to acquire this impressive list of prizes will never be entirely clear, but he sought to employ a combination of diplomacy, perfidy, well-timed alliances, and opportunism. Mussolini planned to wage a short war of expansion on the cheap. He was confident that he could make Italy a de facto Great Power.[22]

Mussolini was encouraged in this belief by his sense that Britain was a declining power without the will to defend its empire. His Fascist legions were tougher. This view shaped discussions with his generals and admirals, who were alarmed that Mussolini wanted a war with the British. His unwillingness to listen, combined with an unwavering belief that Italy must escape its "prison," drove him on. The war in Spain, the conquest of Ethiopia, and the increasingly strong ties to Hitler were simply his first steps in this self-appointed mission to build the "New Rome" on the bones of the British and French Empires. His hollow boast of "eight million bayonets," fanciful notion of the Italian Navy's ability to sink the British Fleet at Alexandria in thirty minutes, cavalier attitude toward Italian raw-material shortages, and the crucial importance of Gibraltar and the Suez Canal in the passage of those raw materials to Italy simply made the entire enterprise more unrealistic.[23]

Germany's Dilemmas

While the British and Italians focused on the Middle Sea and imperial possessions further afield, Hitler concentrated largely on the anticipated conquest of Russia. However, the timing of this attack was not fixed, in part because he wished to take advantage of military successes in France and the

Low Countries to establish German control over Morocco, Gibraltar, and the Canary and Azores Islands. This was part of a long-term strategy to keep the United States away from Germany's newly won holdings in Europe and, should things continue according to plan, Africa. It would also preclude British activity in the Mediterranean, especially if (as Hitler planned) the Germans and Italians also sealed off the eastern Mediterranean.[24] However, relative to the USSR, the Mediterranean was an objective of secondary importance, even if Hitler wished to sew up "loose ends" there prior to his Russian venture. His naval staff, under Grand-Admiral Erich Raeder, sought to win approval for a coherent and achievable military strategy that would facilitate the conquest and control of the entire Mediterranean Basin, with everything this implied for further grand- and military-strategic possibilities. Raeder's vision came from a pragmatic view that a heavy blow to the British Empire in the Mediterranean would leave it moribund. However, as the head of the Reich's most Wilhelmine service, he was also interested in naval control of the Mediterranean and eastern Atlantic, for starters, as the means for regaining Germany's "place in the sun," including colonies. Although there is abundant evidence that Hitler viewed the Mediterranean and Northwest Africa first and foremost as barriers to US and British penetration of Europe and other German holdings, he by no means viewed them as strategic dead ends, and there were moments in the war, especially after Italy's defeats in North Africa, when he considered the merits of a major Mediterranean effort.[25]

Most historians have concluded that the Mediterranean could never have been more than a strategic dead end in the Führer's view. This may help to explain why there are relatively few German-language secondary sources dealing with the Mediterranean theater, but it does not excuse scholars for writing off the Mediterranean because Hitler did not favor heavy German involvement there. The British did engage heavily there, for reasons already set forth, and this alone should elicit enough interest among historians to ask why the Germans did not develop a coherent strategy of their own.[26] Prospects for victory in the Mediterranean, though "politically complex, yet were destined to favor German arms, whose action in the Mediterranean at the right time and place could have changed the course of the war."[27] This was the one place early in the war where the British would come out and fight and thus where Hitler could further weaken or defeat them. In fact, when Hitler wrote to Mussolini on 5 February 1941 regarding the recent deployment of German ground and air forces to Libya, he emphasized that "the arrival of a German unit makes sense only if *by its strength and by its composition* it is really capable of bringing about a turn of fate."[28] This was hardly the kind of language associated with a defensive posture. Libya had to be held, but Hitler's ambitions were greater than this even with the limited force he sent to the Mediterranean.

German historians put forth three arguments about Hitler's putative lack of focus on the Mediterranean. First, they held that Great Britain was not beatable as long as the empire had US support, which meant that the real center of gravity was the United States. Second, German and Italian conflicts of interest were too great to allow for a coherent grand strategy and therefore a coordinated military strategy. Finally, the Mediterranean theater was never a substitute for Operation Barbarossa.[29] All three of these views, although sound, do not take the final step in an analysis of this strategic problem. A defeat of Great Britain was possible, even if "defeat" meant just the loss of its colonies and natural resources or even reliable access to them. However, this would occur only with a decisive British loss in the Mediterranean that allowed the Axis to capture Gibraltar, Malta, the Suez Canal, the Middle East with its oil supplies, and other key assets while simultaneously effecting a linkup with the Japanese after the British position in India collapsed for lack of supplies. A war with the United States was, in Hitler's view, inevitable, so defeating the United Kingdom was simply a prerequisite when viewed in that light—and what better place to engineer that defeat than in the Mediterranean and Middle East, for which Churchill had vowed to fight "to the last inch and ounce"? Had the Germans appreciated the absolute centrality of the Mediterranean to the British grand-strategic position, they could very well have won there *without committing a large number of additional assets. Better use of the ones they had could well have done the trick.* Axis conflicts of interest were inevitable, but even a poorly coordinated effort, when combined with effective German diplomatic and military action, might well have carried the Axis to victory. If the Mediterranean was never a substitute for Operation Barbarossa, it should have been an important complement.[30]

Further, the Axis struggled to obtain the bare minimum fuel requirements. How a country such as Germany, which was so short on petroleum supplies that its leaders resorted to building expensive, inefficient, and vulnerable hydrogenation plants for the conversion of coal to gasoline, could overlook the critical importance of the huge oil supplies in the Middle East (already 15.6 million barrels per year as opposed to Ploesti's 5.8 million) is difficult to comprehend.[31] In point of fact the Germans did not overlook Middle East oil, although they focused first on Romania's Ploesti oil fields and the massive Soviet oil supplies in the Caucasus Mountains.[32] However, it did not have to be an either-or proposition. Given the scope of Hitler's ambitions, even Ploesti, the Caucasus, and the hydrogenation plants together would not have sufficed had the Germans been able to pursue their larger global aims.

They would have needed Middle East oil too, both to fuel their military efforts and to hamstring those of the British Empire. The British knew this. As Michael Howard noted, "Strong forces had to be retained in the Middle East, which was not only a strategic centre for Commonwealth communications, but a source of oil even more significant for the United Kingdom than the Balkans were for Germany."[33] An agile, flexible, and effective German strategic and operational approach would likely have resulted in the capture of these oil assets.

Finally, and perhaps most compelling in terms of the Mediterranean's longer-term importance to Hitler's drive for *Weltmacht* (world power or dominion), were the Führer's own well-documented plans to use the Mediterranean Basin not only as a shield against eventual US involvement in the war but also as a springboard for drives deep into Africa, the Middle East, and Russia intended to result in a partition of Eurasia and Africa with the Italians and Japanese. Operation Barbarossa became Hitler's first step in this effort, but the conquest of the Mediterranean and Middle East were central to his larger ambitions.[34]

Hitler's "lost year," from the fall of France to the invasion of the USSR, thus proved central to the course of the Mediterranean theater and the larger war in terms of its diplomatic failures and their implications for military strategy and operations. After France fell, Hitler tried constantly to woo General Philippe Pétain and Francisco Franco into an alliance. His objectives revolved around control of Gibraltar, Morocco, and the Canary Islands to give Germany a strong position in an eventual clash with the United States. Although the focus was on the western Mediterranean and would have given the Axis a strong position here, they could also, with some effort and imagination, have seized control of the entire Mediterranean Basin. However, Mussolini, Franco, and Pétain all proved unwilling to work with each other, even if they would deal with Hitler unilaterally. Competing claims for territory and resources thus foiled the Führer's plans to capture Gibraltar (Operation Felix), establish military bases in Morocco, and build air bases in the Canaries. By spring 1941, despite increasingly hollow assertions that Great Britain had lost the war, Hitler gave up on any kind of productive multilateral arrangements in the western Mediterranean and was preparing to dispense with the Soviet Union—"Britain's sword on the Continent" and the source of *Lebensraum* and immense natural resources.[35]

German-Italian Cooperation: The "Negative" Alliance

Even efforts to coordinate with Mussolini proved fruitless because the alliance was less about cooperation than it was about maximizing each country's advantage—often at the other's expense. The German official history emphasizes that "one discovers more mistrust than trust" and that talks suggested a "need for political protection from each other rather than for joint military action."[36] The alliance was a "decidedly unproductive partnership."[37] It is in this light that we must view the "Pact of Steel" of 22 May 1939 and all subsequent agreements and military actions. Italy's decision in favor of nonbelligerence on 1 September is further evidence that the alliance was militarily skin deep. However, Hitler did not discount the importance of the duce's friendship and his military's ability to play an important, if limited, role at the right time. He told his senior officers that Germany's fortunes rested to a substantial degree on those of Italy.[38]

Policy issues were not the only impediments to an effective alliance. Economic problems also loomed large. The Germans, already short of oil and other commodities, knew they would have to supply these to Italy after the latter declared war. The Germans would have difficulty supplying enough oil and coal to keep even Italy's limited industry working at full production. The seizure of petroleum, oil, and lubricants (POL) assets thus preoccupied Hitler and Mussolini, but the duce's commanders lacked the initiative to seize the Middle East oil fields, and the Führer lacked the immediate interest, leaving this problem to the "Italian sphere of influence" and "parallel war" while he went after Russian oil. The Middle East oil fields, which should have been a key—if not *the* key—military objective for Axis forces thus remained relatively safe until the opportunity to capture them had passed. Without adequate fuel and coal supplies, the Italians could not engage in unrestricted warfare, particularly because 84 percent of their raw materials came by sea, with the majority coming through Gibraltar or the Suez Canal. This contributed to Axis defeat in the Mediterranean.[39]

Military cooperation was no better. The economic constraints on Italy precluded parallel war beyond minor and short actions. Mussolini dismissed German offers to help block the Suez Canal, take Malta, and send air and ground formations precisely because he remained wedded to the concept of a parallel war in which he set the conditions for peace talks without German assistance. Among the Wehrmacht leadership, only Raeder and his staff pushed consistently for close naval and combined-arms cooperation with the Italians, without any enthusiasm from the latter. Raeder's requests to send U-boats and torpedo-bombers to the Mediterranean without delay came to nothing. The missed opportunities here, if only in the sharing of torpedo, mine, and other ship-attack technologies for submarines and aircraft, dealt a crippling blow during the logistical effort (battle for the sea lines of communication, or SLOCs) in the central Mediterranean.[40]

Consequently, the Nazi-Fascist pact became a truncated alliance with uneven military cooperation and no collective grand or military strategy. This outcome was not foreordained but rather a matter of making choices. The divergence in strategic priorities did not interfere fundamentally with anything but the *timing* for Operation Barbarossa. Raeder had Hitler enthused briefly about a Mediterranean strategy, but this did not last. Hitler soon came to refer to the Middle Sea as "Mussolini's sphere of interest" and focused his efforts on Russia as the two dictators agreed—until the duce had a catastrophe on his hands—to wage parallel war. Even after the Germans arrived in the Mediterranean, there was little coordination. Instead of viewing the Mediterranean as the path to dismembering and defeating the British Empire, Hitler tended to view it first as a bulwark for keeping the Allies off of the Continent—hence his decision to attack the Balkans and Crete. Consequently, rather than taking decisive action despite his desire to control the Mediterranean in due course, he sent a weak force, ill prepared to engage in maritime operations and with shortcomings in terms of the proper employment of air assets.[41]

The Italians had a similarly detached and uncooperative approach. Just before he declared war, Mussolini said, "This time I'll declare war, but I won't wage it."[42] The strategic posture would be largely defensive. When Marshal Pietro Badoglio, supreme chief of the Italian General Staff, informed Major General Enno von Rintelen, the German military attaché in Rome, on 5 June about the intent to declare war on Great Britain effective 11 June, he cautioned that the Italians would be of little military assistance. Also, they would not invade Malta, despite its crucial logistical importance, given the "exceedingly complex" nature of the operation—and despite the fact that Malta had virtually no defenses at that point.[43] As the Axis failed to develop any real strategy or system for combined-arms operations, the British prepared to defend their holdings and attack Italian ones.

British Responses to the Axis Threat

The British understood Italian weaknesses and German predispositions relatively well by 1939, even if they had misread them badly at previous times. Increasingly detailed intelligence underscored how precarious Italy's position was militarily and economically. British strategy was thus to "close the ring" around Italy and defeat it before turning to Germany.[44] Howard said of the Mediterranean, "As a centre of gravity for British forces it was second only to the United Kingdom itself," in large part because it was an ideal conduit for the movement of military and economic resources—one the British would protect at all costs.[45] Churchill grasped the dangers but also the possibilities in the Mediterranean, and he capitalized on successes and failures there as leverage with President Roosevelt to obtain military aid and elicit interest in "closing the ring."[46] British planners understood that airpower would play an important role.

Howard underscored this when he noted that "in the Second World War, the yet harsher arguments of land-based air-power, which made a mockery out of the 'amphibious flexibility' of the British landings in Norway and Greece,"[47] would come to prevail in a fundamental way. He further observed that amphibious operations, which the British had conducted for centuries in their forays onto the Continent, had now, as a result of land-based airpower, become potentially disastrous unless the attackers had air superiority. Airpower had changed the character of combined-arms operations.

The overriding problem the British faced in planning for war in the Mediterranean and Middle East theater (which included the Western Desert and central Mediterranean) was the fact that they had more than 3 million square miles to protect. Further, they understood that it was not a question of whether Hitler would become involved, but rather when, how, and to what effect.[48]

Malta, for instance, became a linchpin in Allied strategy but was within thirty minutes of air attack from Italian airfields. Tunneling, hardening, redundancy, and resupply were critical to ensuring Malta remained active in stopping Axis supplies from reaching Libya. Radar was in place by 1938, but the short distance from Sicily meant that other air-raid warning actions became doubly important. Infrastructure upgrades and new construction, including the airfield at Luqa, accelerated in July 1937. Without these, the island could not have endured the hardships to come.[49]

As the Germans stunned the world with their early victories and the employment of air assets in this *Bewegungskrieg* (mobile warfare), the British learned from their disasters in France, kept fighter command at the greatest possible strength, and employed radar as part of a new C2 structure based on the rapid movement of air-raid data between operations centers to defeat the Luftwaffe. These events set the stage for further British thinking about airpower employment, with important results for combined-arms operations in the Mediterranean.

British Military Preparations

The defense of British interests in the theater would require first-rate interservice cooperation. When Air Marshal William Mitchell arrived as the air officer commanding in chief (AOC in C) Royal Air Force Middle East (RAFME), in March 1939, he saw little of this. He began by reorganizing RAFME and receiving Air Ministry approval to expand it in time of war to include all air assets in Egypt, Sudan, Kenya, Palestine, Transjordan, Iraq,

Aden, and Malta. Centralized control—a now widely employed but then still much-debated principle—was thus in place when Italy declared war.[50]

The services were beginning to work seriously toward interservice cooperation, and the disaster in France spurred the Middle East C in Cs to develop effective combined-arms capabilities. The RAF would play its role here. Its contingent in Egypt increased to three squadrons and received the first of many mobile radar stations providing early warning and target-vectoring. Eventually, these were organized much like the "Chain Home" stations in the United Kingdom with the associated C2 capabilities, including landlines, radios, and group/sector operations centers. The Germans never developed anything on the same scale. Fliegerführer Afrika did have a *Gefechstand* (operations room) and *Bodenstelle* (fighter control) but lacked radar until spring 1942. Reports came from aircrews, ground observers, and army units. Rather than developing a combined operations room, the Italians copied the German version and invited German liaison officers to join them, perpetuating seams in early warning and C2.[51]

Despite their challenges, the British knew the Italians were even more poorly prepared. Italy's small industrial base, reliance on Germany for natural resources, huge military expenses and government deficits, ineffective military reorganization, weapons and equipment shortages, and officer and instructor insufficiencies were all clear to the British. The Italian Air Force (IAF) had reached its peak effectiveness in 1936. Since then, it had fallen behind technologically, its reserves of pilots and materials were low, and it was not fit to fight a major war.[52]

Wing Commander (later Air Commodore) Raymond Collishaw, Air Vice Marshal Arthur Coningham's predecessor as operational commander of RAF assets in the Western Desert, noted just before Italy's declaration of war that a major IAF exercise a week earlier had involved 407 bombers and escorting fighters, which took off from southern Italy and Sicily, stopped at Benghazi and then Tripoli, and flew back to Italy. As impressive as this feat appeared, upon careful inspection it revealed a central flaw in IAF organization. It was divided into two halves—expeditionary and territorial—with the former being further divided into two air corps. The army and navy also had their own dedicated air support, leaving the "independent" air force so splintered that it could not mass, maneuver, or attack on more than a limited scale. General Ritter von Pohl, the air attaché to Italy, made similar remarks from Rome, telling the Oberkommando der Luftwaffe (OKL) that the dispersal of IAF assets would prove disastrous in a war with the British. Only immediate air superiority in Egypt would facilitate victory there—a point he emphasized repeatedly to his superiors in Berlin.[53]

Nonetheless, RAF officers took the IAF seriously and worked hard to increase RAF reserves of supplies and spare parts to facilitate two weeks of intensive effort and ten weeks of sustained effort against it. Nonetheless, the RAF was not yet ready for a desert air war. Airfields were too few, flying training was inadequate, and supply shipments were usually late. Fortunately, local workshops in the Delta staffed with skilled Egyptian tradesmen were soon doing everything from manufacturing refueling ensembles for Blenheim bombers to overhauling propellers. Still, entire categories of specialized vehicles were missing, from salvage trucks and flatbeds to wireless-transmission vehicles. None of the existing ones had been "tropicalized," nor had arriving aircraft.[54] Consequently, air filters and other components failed regularly, spurring tropicalization efforts, driving demands for more supplies from home, and increasing the overhaul work by skilled Egyptian workers. It took more than a year for this effort to mature.[55]

Despite the fragmentation of IAF assets, RAF officers were not about to let the 203 aircraft in Libya get the upper hand. They thought the IAF would operate in strength, but within the constraints of limited logistical and base infrastructures. In case of a war with France, the Italians would be unable to send additional aircraft from Italy. Similarly, their commitment to defend IEA while threatening British SLOCs on the Red Sea would preclude reinforcements from that quarter. So, although the Italians had a numerical advantage, the RAF would, in the C in Cs' view, do well.[56]

However, the C in Cs cautioned that if IAF planes made heavy raids on Egypt, these would be difficult to stop.[57] Therefore, they planned to raid Italian airfields immediately to gain air superiority and thus the larger initiative. This approach proved vital to Allied fortunes in the Mediterranean. Given the unreliability of reinforcements from the United Kingdom, the C in Cs decided to prioritize and attack selected objectives rather than disperse their limited assets in an effort to address every requirement.[58] Because the best way to blunt IAF attacks was preemptive and continuing raids on airfields, RAFME made this its top priority. Mitchell formed a mobile air stores park (ASP), supply and transport column, railhead-handling unit, and repair and salvage unit (RSU) and built landing grounds in forward areas. Fuel and oil supplies were plentiful given the refineries in Iran and Iraq.[59]

To maximize operational effectiveness, RAF and army headquarters would be "in close proximity." A senior air staff officer (SASO) was detailed to serve as a liaison to the army staff. "It must be appreciated," the guiding document said, "that this connecting link between the G.O.C. [general officer commanding] Mobile Division and the Advanced Wing Commander is not an ideal organisation, and the Officer Commanding the Wing must take every opportunity of establishing personal contact with the G.O.C. Mobile Division."[60] Collocation of headquarters became the norm by the time Operation Compass began in December 1940.

With war imminent, the British put an air-reinforcement scheme into motion on 23 August 1939. This movement of aircraft, completed on 28 August, gave the RAF ninety bombers and seventy-five fighters in Egypt. Mitchell planned for an air offensive against IAF air bases, aircraft, fuel, supplies, and maintenance facilities the minute war commenced. However, Italian neutrality delayed the start of operations.[61]

By March 1940, the services were improving their "administration," the process of providing forces with all required items for operations and sustainment. A high degree of ground mobility proved essential in what became a series of military operations designed, in large part, to capture airfields in the Western Desert and operate from them quickly to facilitate the army's further advance. Airfields were also vital for attacking shipping on the central Mediterranean SLOCs and what became a major subsidiary SLOC along the North African coast. The army provided logistical services, but the RAF had its own maintenance organization, including supply, repair, aircraft salvage, and motor transport. The truism that land-based aircraft require a huge infrastructural investment and much lead time to get into position before they can begin combat operations was especially true in the desert. The British grasped this before the war (in part because of their colonial policing operations) and began mastering it. Despite arguments about which service was more or less cooperative, the army and RAF both proved remarkably so.[62]

In addition to its logistical accomplishments regarding a mobile and resilient RAF, General Headquarters Middle East (GHQME) began tackling its most basic problem: the requirement for virtually everything to come long distances by ship. Officers visited South and East Africa, India, Burma, Malaya, and Australia to obtain supplies. This policy of seeking assets throughout the empire gave the British a logistical advantage even after the Luftwaffe made the Mediterranean too dangerous to traverse and underscored the mutual reliance between theaters of war.[63]

As these logistical improvements accelerated, aircraft maintenance and salvage functions followed. Initially, each squadron was responsible for its own maintenance, which led to duplication of effort among ground crews. Only major repair work went to the main depot at Aboukir, but it had not yet received required upgrades and was vulnerable to air and sea attack. The British built a second depot and established an advanced RSU in the Western Desert.[64]

The requirement to hold Malta while hedging bets with a new navy base at Alexandria, 800 miles from Italy, brought additional RAF requirements as Mitchell sought to protect the fleet in conjunction with Fleet Air Arm (FAA) assets. The C in Cs' most pressing operational concern was the inadequate number of aircraft required to meet the IAF raids they expected. They feared that RAFME could not protect both land and sea operations.[65]

When Air Marshal Arthur Longmore succeeded Mitchell on 13 May 1940, he had just less than a month of peace remaining in which to prepare his forces. There were 308 aircraft in the entire theater. The IAF had well over 400 in Egypt and 170 in IEA, although their organizational seams and logistical woes would soon become apparent. Longmore asked for Hurricanes and Blenheim IVs to even the odds. His decision to conserve resources until it was clear when, how many, and what kinds of reinforcements would arrive did not mean he was unwilling to employ his forces. It did, however, mean the RAF would engage in major operations only when reconnaissance and intelligence made it clear the IAF was preparing to do so. Then, Longmore would order a preemptive attack. Otherwise, he would await the right opportunity. Meanwhile, the RAF continued its preparations to gain air superiority. Flexibility became Longmore's watchword. "Throughout the whole period," he said, "on all fronts the policy has been to maintain an active offensive, any defensive measures being reduced to the barest minimum."[66]

Italian Military Preparations

If British preparations appeared hectic, Italy's were nearly nonexistent. Comando Supremo did not begin preparing seriously for war until April 1940, and even these efforts were haphazard.[67] Italy's problems stemmed from the fact that its military was not in the same league as those of Germany and Britain.[68] Despite having the largest navy in the Mediterranean, the Italians had no aircraft carriers and thus relied on land-based aircraft for naval escort, but there was almost no coordination because ships and IAF planes lacked radio contact. Unlike the British, who developed excellent aerial reconnaissance and intelligence capabilities, the Italians never did. This prompted a remark that "poorly conceived and executed aerial reconnaissance influenced almost all missions of the Italian battle fleet . . . and formed the Regia Marina's Achilles' heel."[69] The IAF was not well trained for maritime operations. Pilots rarely had instrument ratings. It was, in effect, a visual-flight-rules, daytime-only air force. By 1940, the regime, senior officers, and air industry had made major misjudgments regarding IAF preparedness.[70] The greatest was the government's allocation of a paltry 13.4 percent of the military budget to the IAF—half of the navy's share and one-quarter of the army's. The military was woefully unprepared even for the short war Mussolini envisioned. Ironically, the duce viewed airpower as a key means to break out of Italy's "prison," but the IAF could not deliver because it was also deficient at the tactical and technical levels.[71]

Italian bombers featured poor bombsights, no radio contact between pilot and bombardier, and few navigational aids. Fighters, though maneuverable,

Air Chief Marshal Sir Arthur Longmore, AOC in C Middle East, 2 April 1940–3 May 1941. Longmore performed with great effectiveness, turning RAFME into a highly combat-effective command and working hard with his army and navy counterparts to improve combined-arms operations. He fell into disfavor with Winston Churchill and was relieved of his command in May 1941. (Imperial War Museum Image No. CM-515)

lacked self-sealing fuel tanks and had radio receivers but no transmitters. IAF leaders had also ignored the problem of attacking ships, which one might consider important for a country surrounded on three sides by water. Obsolete aircraft, the shortage of effective torpedo bombers and crews (there was one squadron of five SM.79 torpedo-bombers in experimental status at this point), most pilots' preferences for biplanes rather than monoplanes, poor tactics, and the lack of high-octane fuel added to this list of problems. Italian aircraft were caught on the ground constantly because of the absence of radar and poor coordination between spotters, fighters, and antiaircraft artillery (AAA). Proliferation of aircraft types, insufficient production based on small shops and handicraft labor, poor maintenance and logistics, and the lack of a salvage function further undermined IAF capabilities. The duce's calls for production of 12,885 aircraft and 22,543 engines in 1939 were somewhat above the 1,750 and 4,191 actually delivered.[72]

There was a major disconnect between the duce's strategic vision for his air force, which was Douhetian in nature, and its capabilities. The rivalry between the probomber "Douhettiani" and the intellectual followers of Amadeo Mecuzzi, who had favored fighters, fighter-bombers, and dive-bombers in view of Italy's limited industrial capacity, produced a combination of production-policy paralysis and randomness. So did bad decision making, including Undersecretary of Air General Giuseppe Valle's decision to stop research on air-delivered torpedoes because it created friction with the navy. Even bringing German aircraft types into the IAF proved largely ineffective given the overarching structural and leadership problems. General von Pohl said the Italians needed capable ship-attack aircraft and munitions, including the 2,200-pound bomb Stukas used with great effect. General Francesco Pricolo, the C in C of the IAF, asked repeatedly for Stukas with heavy bombs as the means for giving Italy a potent ship-attack force. Von Pohl pushed hard for aircraft transfers, resulting in a tentative offer in July 1940. The duce's sensitivities and related political issues shot down this initiative. Von Pohl's view that the IAF could not function effectively without better aircraft, munitions, and communications capabilities proved prescient. So did his insistence on an immediate IAF bid for air superiority after fighting began in Africa. He said simply that no military could win in the desert without air superiority.[73] The Italian view that airpower would be a great equalizer for the smaller Great Powers turned out to be utterly wrong given the expense and complexities involved in producing a capable air arm.[74] The duce's airmen fought hard, but this could not overcome the larger disadvantages under which they labored or the RAF's increasing capabilities.

In the "triphibious warfare" that characterized war in the Mediterranean, the Italian Army was as disadvantaged as the IAF. Only 26 percent of divisions were fully equipped. The rest suffered from serious transport and manpower shortages. More alarming yet was their deployment, with 66 percent in Italy, 13 percent in Libya, 15 percent in IEA, and the last 6 percent in Albania and the Aegean. This was an ineffective disposition of troops and betrayed a defensive stance in line with Mussolini's plans for a short, low-risk, and high-return war. Strength tallies indicated that in a one-year war, the army would have severe ammunition shortages in addition to a 50 percent shortage of vehicles within units whose allotments were already too small for mobile warfare.[75]

The Italian Navy's role in the duce's war was vitally important. Although it engaged Royal Navy contingents with courage and some degree of effectiveness, aggressive British leadership and superior capabilities, such as radar combined with AAA cruisers, sonar, and aerial reconnaissance, gave the Royal Navy key advantages. The inadequate range of IAF escort aircraft and their lack of radio communications added to the vulnerabilities caused by the lack of radar. The loss of one-third of the Italian Merchant Marine at the start of the war, resulting from the Italian government's failure to notify ship captains of impending hostilities, and from a rapid Allied response, was a logistical catastrophe. Combined with the IAF's lack of enthusiasm and ability to help escort convoys, the stage was set for severe supply problems should the war in North Africa intensify and persist.[76]

Despite the major disadvantages under which they labored, Italian sailors, airmen, and soldiers fought hard and often acquitted themselves much more effectively than most histories indicate. However, given their country's inadequate industrial base, lack of raw materials, technological limitations, generally poor senior military leadership, and other disadvantages, the Italian military could not fight with any real degree of effectiveness or for any length of time. The British were the primary beneficiaries of this reality.

The British Advantage Grows

Neither the British nor the Italians yet understood fully what a large influence land-based airpower would have on operations. Nonetheless, Longmore dealt effectively with challenges the RAF faced. He found RAFME's plans and capabilities problematic. There were no modern aircraft and too few spare parts. However, as he knew from intelligence reports, Italian maintenance was inferior. The IAF had inadequate fuel stocks and, in the case of IEA, no way to replenish them after war began. Its serviceability rate was 70 percent in Libya but 30 percent in IEA. By contrast, RAFME's was 80 percent. Longmore believed the IAF would crumble in the face of a determined air offensive. He placed all but one of his bomber squadrons forward in the Western Desert with Collishaw.[77, 78]

The situation was also troublesome in Malta, where none of the promised four fighter squadrons and AAA had arrived. Admiral Andrew Cunningham, who relied heavily on Malta as a naval base and a key point from which to launch reconnaissance and strike missions, was deeply concerned. Although four Sea Gladiators and a radar station installed in March 1939 gave Malta a rudimentary air-defense capability, Longmore and Cunningham committed to an immediate defense of Malta beginning with air-reconnaissance assets. However, the Royal Navy suffered from a weakness that soon became painfully clear: vulnerability to land-based airpower, particularly the Luftwaffe's. The latter's expertise in ship attacks would nearly sweep the navy from the central Mediterranean by summer 1941, leaving the RAF with the lion's share of the effort to control these areas.[79]

In addition to addressing the Malta problem, Longmore had to develop an air-ground coordination scheme where none existed. His first steps represented a major move toward the mature air-ground system in place by summer 1942. After much effort and a number of errors, the British learned to employ airpower in conjunction with ground and sea power to a greater degree than the Axis despite the Luftwaffe's head start and some impressive moments during the struggle for the Mediterranean. John Terraine said of these combined-arms efforts, "When critical land operations are in progress, Army co-operation is not simply a specialized activity of part of an air force. It is the function of *the entire force,* with all of its available strength."[80]

During the interwar period, the RAF had discussed cooperation efforts with the British Army. Airmen believed that large-scale raids over forward areas were too costly and that the major payoff would come by isolating the battle area from enemy reinforcements and supplies. The army disagreed and called for specialized aircraft under the ground commanders' control. Longmore and then Arthur Tedder corrected earlier missteps in France in 1940, and by May 1942, the RAF had a mature air-ground cooperation policy that rejected specialized assets in favor of speed of response, flexibility of application, and concentration of force.[81] Longmore's efforts, and the army's increasing buy in, first became apparent during Operation Compass.

The three-dimensional chessboard was set. Numerically, the Italians had the advantage, but the British C in Cs were not overly concerned. Quietly, professionally, relentlessly, they prepared to execute combined-arms campaigns in IEA and the Western Desert.

PART II

Contest

2

War Comes to the Mediterranean
June–November 1940

The desert flank was the peg on which all else hung.

—Winston Churchill

I will declare war but not wage it.

—Benito Mussolini

England, unlike in 1914, will not allow herself to blunder into a war lasting for years.

—Adolf Hitler

Italy's Grand-Strategic Calculus

AS THE GERMAN VICTORY IN FRANCE shocked the world, Mussolini determined that the time was right for Italy to enter the war. He felt compelled to gamble, before Hitler seized all the spoils, on a series of minor and independent offensives and shows of force against what remained of the French military in the Metropole, paired with an offensive to seize British Somaliland and Djibouti to ensure Italian control of the Red Sea. In Libya, the duce intended to take a defensive posture and let events play themselves out at the peace talks, at which point he would lay claim to his conquests in East Africa and make demands on French and British possessions around the Mediterranean. He also planned to assert claims against Yugoslavia and Greece.[1] Mussolini never intended to engage in a protracted war until at least 1943, if at all, and his military chiefs reminded him constantly that becoming heavily engaged before that would be folly.[2] His efforts were not only a product of Fascist ideas but also of a clear continuity in the imperialist impulse for colonies—an objective military leaders and the royal family supported.[3] The duce's foreign policy rested on clear goals. Nonetheless, it would not be a free ride given that the British had different and more offensive-minded ideas about holding the Mediterranean and defeating the Italians. This became unpleasantly clear to Mussolini in the coming months as defeat piled upon defeat, and his strategy of a short and inexpensive parallel war failed. Italy became, in effect, a vassal state to Germany, which had its own distinct if complex objectives in the Mediterranean.

German Strategy

Hitler's strategy during this period, if vacillating and ultimately unsuccessful, had basic direction. He did not underrate the Mediterranean, as is clear in his plans for a diversionary strategy there but even more so in his attitude following the British rejection on 22 July 1940 of his so-called final appeal for peace. Hitler tried to bring Italy, Spain, and Vichy France together with Germany to secure the Mediterranean Basin under the Reich's lead.[4] This effort to shore up his Mediterranean flank was intended primarily to safeguard the attack on Russia and keep Britain and (later) the United States at bay. Nonetheless, it could have had decisive results in the Mediterranean had a few things gone differently—and they might well have done so with more effective diplomacy and less vacillation. We must reject the official historians' view that German strategy had to be an either-or proposition.[5] With the forces Hitler deployed to the Mediterranean by summer 1941, an effective chain of command, and the barest level of cooperation with the Italians, the Führer could have won there and, by extension, bettered the odds for Operation Barbarossa. Development of a Mediterranean strategy that would give the Axis Gibraltar, Suez, the Red Sea, and inevitably Malta after the Mediterranean was isolated appealed in particular to Grand Admiral Erich Raeder. He saw it as an opportunity to renew the Wilhelmine push for colonies and Germany's status as a world power. Hitler's ideas were of even broader scope, looking toward German air and naval bases in Morocco, the Canary Islands, and the Azores Islands. These objectives, and the follow-on advantages that would likely accrue in Turkey and the Middle East regarding enhanced access to oil and other resources, failed, but not for lack of effort. The German official historians assert:

> Generally speaking, the second half of 1940 could be summed up as the equivalent of a lost battle. Although this did not, in a literal sense, take place, the time lost in failing to achieve a decision, and the result of the unfavourable strategic balance (seen in absolute terms), meant that this superficially so successful year [1940] concluded with the realization that a long war, and hence virtually a hopeless one from Hitler's point of view, had become inevitable. Added to this was the fact that the Reich had proved more vulnerable through its Italian ally than had been expected, that localized conflicts were no longer able to bring about a final decision—even though Hitler and his generals in the summer of 1940 refused to believe this, any more than they did in the following winter—and that the political reshaping of Europe failed to materialize in the face of British resistance.[6]

Hitler's basic impulse regarding the Mediterranean theater—that it was Italy's sphere of influence—may have made sense in a theoretical construct. However, it quickly became clear that any strategic gains would rely on German arms. With the British refusal to capitulate, Hitler turned to Francisco Franco and then Philippe Pétain in an effort to bring them into the war. Franco knew his country was too weak after the civil war, and too dependent on British and US imports of food and other commodities, to risk an alliance. The Germans could not come close to matching these economic inputs. Hitler's efforts to build an anti-British "Continental bloc" facilitating the cooperative delivery of sufficient raw materials to the Spanish, Vichy French, and Turkish also came to grief largely because of the huge collective British and US economic engine.[7]

As the Führer's talks with Franco and Pétain on 23 and 25 October failed, Field Marshal Walther von Brauchitsch, commander in chief (C in C) of the German Army, continued to push for deployment of a panzer corps comprising two divisions to help Marshal Rodolfo Graziani's forces in Libya conquer Egypt, an initiative he began advocating in July 1940.[8] General Alfred Jodl, Oberkommando der Wehrmacht (OKW) chief of staff, supported this and emphasized that the Axis must "not operate for objectives but for victory."[9] This thinking came directly from General Ritter von Pohl and Vice Admiral Eberhard Weichold, the naval attaché to Italy. Both emphasized repeatedly that the Mediterranean would be a key theater of war and that the Germans had to play a major role given Italian military shortcomings. They viewed this as the best means for defeating the British Empire because the British would fight for their Mediterranean possessions. These insights prompted Jodl to submit Brauchitsch's request for the deployment of the panzer corps, Luftwaffe ship-attack and mining assets, air-transport units, and troop-transport ships to Hitler. On 4 October, the Führer offered Mussolini a panzer division (later a corps) and Luftwaffe assets to help take the Nile Delta and Suez Canal. Despite Pietro Badoglio's pleas to approve this, the duce demurred.[10]

Raeder's overtures ultimately failed equally to produce any proactive operations in the Mediterranean and thus any opportunity to bring British forces to battle and defeat them. Hitler favored Raeder's recommendations for airfields in Morocco and Algeria; the dispatch of the army corps to help take Egypt and points beyond; air action to suppress Malta until it fell for lack of supplies; air raids on the Royal Navy and merchant vessels (MVs) to close the Mediterranean to the British; and the capture of Gibraltar, the Suez Canal, and the Red Sea. With the Mediterranean thus firmly under Axis control, Raeder suggested, the Germans would have defeated the British regardless of whether they kept fighting. Hitler saw this as a vital adjunct to the invasion of the Soviet Union but not a substitute. He was confident that the Mediterranean and Middle East would fall in due course—either before or after the conquest of Russia. Nonetheless, Hitler would have preferred to seal off the Mediterranean first and then exploit it after the victory in Russia or, as required, use the Middle East as a base for a second front in Russia.[11]

Raeder was also unconvinced that the Mediterranean should be Italy's exclusive sphere of influence.[12] Earlier than any other German senior officer, Raeder viewed the war as global in extent, with appropriate actions, in the appropriate locations, at the appropriate times, providing the keys to victory. He emphasized that "Germany must on no account allow herself to be pushed into the background in the solution of these [Mediterranean] problems."[13] Raeder asked Hitler to bring Vichy into the war against Britain in order to seal the western Mediterranean and facilitate follow-on operations to take the Suez Canal and the entire Middle East with Vichy assistance from Syria. Finally, he hammered home the reality that a war against Britain was by definition a naval and air conflict even more than a contest on land and that success in the former arenas would ensure victory in the latter. The Führer listened for a time, Hermann Göring and the Luftwaffe engaged in almost no planning for this and in fact clashed constantly with Raeder over control of Luftwaffe antishipping assets throughout the war, and army officers who supported deployment of a panzer corps to Egypt in fall 1940 were stymied first by Mussolini's refusal and then by Hitler as he forbade any such move in the wake of Italy's invasion of Greece. Raeder's clear plans and persistent recommendations, set forth in his 6 September 1940 memo and in his presentation of its contents to Hitler on 26 September, came to nothing.[14]

As Hitler struggled unsuccessfully to find a means to defeat Britain, and as Franco and Pétain remained neutral, Mussolini was about to have the war he declared but not the one he wanted. Despite a few early and minor successes in Italian East Africa (IEA), the Italians did not advance in Egypt. This riled Mussolini, who had changed his mind about offensive action there and ordered Italo Balbo, the governor of Libya, to attack. Balbo was preparing to do so with some energy when Italian antiaircraft artillery (AAA) shot down his aircraft over Tripoli on 28 June. Marshal Graziani replaced him and immediately sent Mussolini and Badoglio a long list of reasons why he could not attack. The biggest was the lack of water required to sustain his large but mostly unmotorized forces. Rather than organizing his army into a mobile component to advance rapidly and engage the British Seventh Armored Division (the only such formation on the British side at this point, and still badly understrength), or engaging in coordinated air-ground efforts along these lines, he moved the army en masse, with most of the soldiers marching and the available vehicles spread among the entire force. This was unimaginative and dangerous.[15]

Combat Operations Begin

At 4:45 p.m. on 10 June 1940, Italian diplomats informed the British and French governments that they would be at war effective one minute after midnight on 11 June 1940. British commanders struck by air, sea, and land early on 11 June. Inexplicably, Italian commanders had not received word. Air raids did grievous damage to the Italian Air Force (IAF). Bombing of Italian airfields in IEA was particularly successful. More than 800 tons of irreplaceable aviation fuel went up in flames at Massawa alone. Arthur Longmore knew he had to keep the IAF on the defensive, and that the Royal Air Force's (RAF's) advantage would fade if he did not receive reinforcements. Only nine planes arrived before the German conquest of France forced a shift to much more circuitous delivery routes. The Battle of Britain also created a severe shortage of replacement aircraft and skilled airmen.[16]

Fortunately, Italian raids were small and ineffective. There were no air attacks on Alexandria or other high-value targets. British soldiers and airmen used the desert to their advantage, operating far from the coastal road and building multiple landing grounds. The Italians stuck to roads and outposts and used existing airfields, providing excellent targets.[17] The IAF was equally ineffective in the war at sea. During the Battle of Calabria on 9 July, following a successful Italian convoy run to Benghazi, IAF aircraft failed to find the British Fleet after having bombed it ineffectually the day before and then bombing their own ships after the naval action.[18] The IAF failed to materialize during the naval Battle of Capo Spado on 19 July in either an aerial reconnaissance or strike role.[19] Nor did airmen find or track Andrew Cunningham's fleet before the raid on Taranto. Although the Regia Marina acquitted itself creditably during its eleven actions at sea with the Royal Navy during 1940, any effective IAF reconnaissance and ship-attack efforts would likely have produced much more favorable results. The same was true of subsequent actions, including the Italian defeat at the Battle of Matapan from 28 to 29 March 1941.[20]

In a review of military strategy three months into the war with Italy, and despite their successes in the air to date, the C in Cs emphasized:

> The Italian (possibly supplemented by German) air forces are likely to constitute the greatest threat not only to Egypt itself but also the Naval base at Alexandria and in certain circumstances to the military forces, and, therefore, *their neutralization is to be regarded in principle as of primary importance.* On

the other hand, direct support for the land and naval forces may from time to time and for limited periods have prior claim on our air efforts.[21]

Senior airmen stuck to this approach doggedly, maintaining air superiority with hardly an exception until they gained air supremacy in summer 1943. The RAF also worked increasingly closely with the army and navy to fulfill the C in Cs' guidance. Air superiority was the enabler; effective combined-arms operations were the ultimate focus. Air and naval efforts in particular had given them the initiative, and they would hold it.[22]

Despite the fact that a major British offensive could not occur for some months, it soon became equally clear that the Italian Army had no interest in a major offensive against Egypt. Raymond Collishaw's No. 202 Group (later the Western Desert Air Force, or WDAF) continued attacking IAF and Italian Army assets while protecting ground troops. RAF bombers raided airfields, ports, and troop concentrations along with fuel, ammunition, and food stocks. Longmore gave the IAF no opportunity to refit.[23] However, the aircraft-reinforcement issue—a nonstop problem for Longmore and later Arthur Tedder—became clear in October, when the first planes from the Takoradi air route, including Hurricanes and Blenheim IVs, started arriving. Even these proved barely adequate after the Italians attacked Greece and the War Cabinet ordered aircraft diversions and transfers to assist the Greeks.[24]

Longmore was also beginning to understand the pivotal importance of Malta: "MALTA has continued to present one of the most interesting problems in the Middle East. Prior to the war the view was officially held that if Italy entered the war there would be little prospect of preventing the Italian Air Force from making MALTA untenable for the operation of aircraft and probably also as any sort of Naval Base except for light forces on occasions."[25] By 1941, a squadron of Hurricanes was in place. Malta's value as a reconnaissance and ship-attack base soon became clear. Air Commodore F. H. M. Maynard was air officer commanding (AOC) Malta, followed later by Air Vice-Marshal Hugh Lloyd and then Air Vice-Marshal Keith Park. Maynard's efforts to build an air-defense capability proved timely and effective.

The IAF raided Malta for the first time on 11 June. During the remainder of the month, there were thirty-six raids. Bombing accuracy was generally poor, but the Italians sank a submarine in port and destroyed the major floating dry dock. This represented a serious loss to naval repair facilities. However, IAF raids had the ironic effect of hardening the resolve of the garrison and populace for the much more severe Luftwaffe raids that began seven months later, improving repair and aircraft-sheltering procedures and speeding reinforcements. The Italians missed an opportunity to capture the island in the war's opening weeks, albeit one that ran a serious risk of Royal Navy interference.[26]

As the Italians continued their desultory efforts, France surrendered, changing the strategic situation. The loss of French naval assets was painful because British warships now had to cover the entire Mediterranean. German plans to attack either the United Kingdom or the Balkans proved equally worrisome to the C in C of the Middle East theater, General Archibald Wavell, because both placed huge strains on scarce resources, the first by constricting reinforcements and the second by forcing existing assets to cover multiple fronts. The loss of French airfields forced an immediate workaround to get aircraft to Egypt. With transit of the Mediterranean becoming too dangerous for anything but long-range planes flying at night, the RAF developed the Takoradi air route across Africa. It was a grueling 3,700-mile transit but the only option other than sailing around the Cape of Good Hope. The British also announced that they would not allow Syria or Lebanon to fall to any power hostile to British interests. Given that the Vichy French reinforced their air forces in North Africa after the surrender to Germany and had major naval forces there, they were in a position to cause serious problems, especially in view of Vichy's pro-Nazi slant. The attack on Mers el Kebir and the destruction of much of the French fleet followed.[27]

With remaining French naval assets at Toulon inactive, the C in Cs knew that they next had to neutralize the Italian Navy to maintain the British position. They received assistance with this from the Italians themselves with the lack of coordination between the navy and the IAF, which created huge problems. Italian land-based aircraft often shadowed British ships without any means to communicate the findings with their own navy, and bombing accuracy remained abysmal. The Royal Navy came to view Italian raids as a minor threat. British sailors would thus be shocked by the Luftwaffe's ship-attack capabilities.[28]

Italian East Africa: Securing the Eastern Flank

Italy's unpreparedness to face a long war was especially evident in IEA. It was isolated, and the supply situation soon became disastrous as a result of concerted RAF attacks. The duke of Aosta and his troops were on their own.[29] The IAF contingent of 260 planes was substantial, but aviation fuel, spare parts, AAA, ammunition, and tires were all in short supply.[30] The RAF's destruction of aircraft and supplies, and its isolation of the colony from replacement aircraft, began immediately on 11 June, placing the IAF in critical condition within a month.[31]

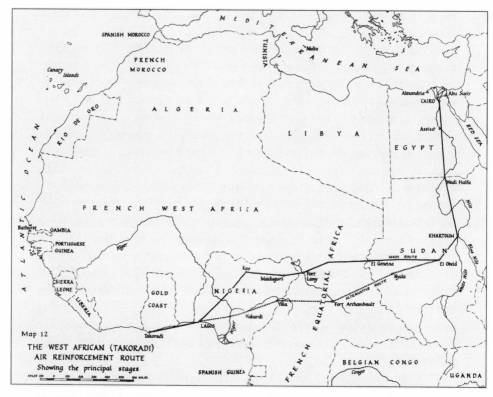

Map 12

THE WEST AFRICAN (TAKORADI)
AIR REINFORCEMENT ROUTE
Showing the principal stages

The Takoradi route. This grueling, 3,700-mile transit for aircraft reinforcement to the Middle East proved one of the mainstays, along with US deliveries to Port Suez and other locations of aircraft to help RAFME maintain air superiority for virtually the entire campaign. From August 1940 through June 1943, the British assembled more than 4,500 Blenheims, Hurricanes, and Spitfires at Takoradi and ferried them to the Middle East. Between January 1942 and the end of the operation in October 1944, another 2,200 US Baltimores, Dakotas, and Hudsons arrived from the United States and were ferried in similar fashion. Most of the US machines went to the Middle East, but some went as far as India. (Originally published as Map 12 in Ian Playfair et al., *The Mediterranean and Middle East*, vol. 1, *The Early Successes against Italy (to May 1941)* [London: Her Majesty's Stationery Office, 1954], 195, now public domain)

Nearly 70 percent of Italian troops in IEA were "Bande"—Ethiopian soldiers under Italian pay and not trained for conventional operations. They were, however, outstanding irregulars and also mostly still loyal to their deposed emperor, Haile Selassie. Wavell emphasized that the key to a rapid victory lay in fomenting rebellion by bringing the emperor back, which he

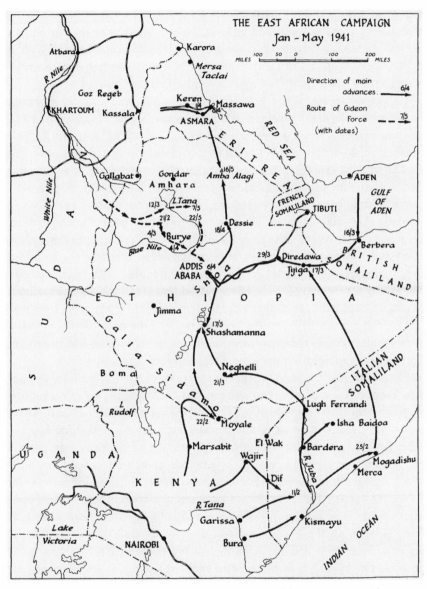

The campaign in Italian East Africa (IEA), December 1940–May 1941. This well-executed combined-arms campaign secured the Red Sea sea lines of communications (SLOCs) and thus opened the way for US lend-lease deliveries because the region was no longer a war zone under the terms of the Neutrality Act. US aircraft and other materials were not long in arriving. (Originally published as Map 25 in Ian Playfair et al., *The Mediterranean and Middle East,* vol. 1, *The Early Successes against Italy (to May 1941)* [London: Her Majesty's Stationery Office, 1954], 391, now public domain)

did when Commonwealth troops invaded. IEA was too big to allow for a decisive victory without adding irregular warfare to the mix. It worked well along with a two-front offensive under Lieutenant-General William Platt from the Sudan and Lieutenant-General Alan Cunningham in Kenya—nearly 1,100 miles apart.[32]

The importance of IEA lay in its location along crucial sea lines of communication (SLOCs). With the Mediterranean becoming an increasingly perilous supply route, the empire's last remaining one for reaching Egypt—the Gulf of Aden and Red Sea—had to stay open. Concerted air raids began on IAF and Italian Navy bases.[33] Fortunately cryptographers at Bletchley Park had already broken the Italians' high-grade ciphers, and code breakers in the Middle East had unraveled the IAF cipher. Commanders thus knew virtually every Italian move. The campaign was the "perfect example of the cryptographers' war . . . the C-in-Cs in Cairo were able to read the enemy's plans and appreciation in his own words as soon as he issued them; indeed, they sometimes received the decrypts while the Italian W/T operators were still asking for the signals to be checked and repeated."[34] RAF units worked in cooperation with ground and naval assets, including antisubmarine warfare (ASW) aircraft flying patrols to keep track of the naval flotilla in Italian Somaliland, fighters for the protection of key cities such as Khartoum and Port Sudan, and light-bombers for direct support of troops.[35]

The RAF's effort to gain air superiority, destroy the IAF's ability to operate, secure passage of MVs through the Red Sea, and provide direct support to ground units, was highly effective. Destroying port facilities also proved important because at least one Japanese merchant ship unloaded fuel and other supplies for the Italians. Air cooperation with the Ethiopian patriots also paid dividends by increasing their morale and speeding recruitment.[36]

Air Commodore Leonard Slatter, an excellent but little-known airman who later headed No. 201 Naval Cooperation Group in Alexandria, commanded No. 203 Group in the Sudan and collocated his headquarters in Port Sudan with the army's. Longmore insisted from the outset that his senior airmen be collocated with their army counterparts. Slatter's group destroyed every IAF unit in its path, silenced dozens of Italian guns, and provided direct support to attacking ground units.[37]

In Aden, on the Horn of Africa, the initial task was to ensure that the eight Italian submarines, seven destroyers, and one hundred or so bombers did not block passage of the Red Sea. A combination of intelligence intercepts, air reconnaissance, Royal Navy attacks, and methyl chloride poisoning aboard IAF submarines after they sortied destroyed or neutralized the entire Italian Navy contingent. From there, aircrews under the command of AOC Aden, Air Vice-Marshal G. R. M. Reid, raided airfields to destroy aircraft, fuel, and munitions. Consequently, IAF raids sank just one MV and damaged another. There were no attacks after 3 November. RAF bombers moved on to Italian trains and the supplies they carried.[38]

The five squadrons in Kenya, the second major axis of ground advance, did equally well. The preponderance of Commonwealth (in this case South African) units was the norm during the IEA campaign. Their performance was superb from IEA to Egypt and Italy. Objectives included airfields, troop concentrations, and the Ports of Kismayu and Mogadishu. Longmore controlled South African Air Force (SAAF) squadrons by agreement with General (later Field Marshal) Jan Smuts, prime minister of South Africa and a member of the imperial War Cabinet.[39]

Two key things were already clear regarding how the RAF and army were learning to wage combined-arms warfare. First, they collocated their headquarters and established joint staffs responsible for everything from intelligence to plans and operations. Second, the focus on gaining and maintaining air superiority was evident from the outset. It was the vital precursor to successful combined-arms operations. These innovations played important roles in later efforts.

During the ground advances into IEA, the RAF gained air superiority, destroyed equipment and supplies, protected friendly convoys and installations, and gave direct support to advancing troops, who reciprocated by capturing air bases so the RAF could move forward with the army. The capture of Mogadishu, capital of Italian Somaliland, in February 1941, yielded 350,000 gallons of motor fuel, 80,000 gallons of aviation fuel, and a handbook with every airfield and landing strip in IEA. The ensuing attack on Addis Ababa included heavy raids on airfields that destroyed more than thirty aircraft and freed up air assets for direct support of ground forces. The city fell on 6 April. In an eight-week campaign, the British had advanced 1,700 road miles and destroyed a much larger force at the cost of 501 casualties and eight aircraft. The IAF was down to thirty-nine serviceable planes.[40]

In Eritrea, the army and RAF harried retreating columns as they fled to Asmara and Massawa. The destruction of Italian naval assets allowed President Franklin Roosevelt to lift shipping restrictions on the Red Sea and Gulf of Aden effective 11 April 1941. Because the area was no longer a war zone according to the terms of the Neutrality Act, US ships began using it to deliver materiel under the Lend-Lease Act, just signed into law on 11 March.[41]

By 31 March, IAF activity virtually ceased. The duke of Aosta surrendered the last major Italian troop formation on 17 May 1941. The RAF destroyed nearly 300 Italian aircraft during the campaign. Although there was nothing foreordained about this decisive victory, proper employment of the air-ground team, combined with Italian logistical and other weaknesses, brought success. With the region in British hands and the Red Sea open to US MVs, Commonwealth in IEA forces moved to Egypt, where the C in Cs were in dire need of them.[42]

Developments in Egypt

As the campaign in IEA unfolded, the C in Cs planned for coordinated operations in Egypt. They expected hostilities to spread across much of the Middle East and thus require a well-orchestrated effort with the control of Egypt central to any larger success. The most fundamental problem was logistics. As the British developed the infrastructure required to support what amounted to a parallel RAF, a major fleet, and an army group, Wavell sought help from the Eastern Group Central Provision Office in New Delhi, which coordinated the shipment of war materiel from various colonies and dominions to the Middle East. This brought to maturity earlier successes in centralizing logistical activities. Winston Churchill also approved a ministerial committee comprised of the secretaries of state for war, India, and the colonies that reported to him and the minister of defense on issues relating to the Mediterranean and Middle East theater. Wavell was clear about the RAF's desperate need for larger numbers of new aircraft models. Air reinforcements soon began arriving.[43] There was, however, no such thing as "rapid reinforcement." The time required to move aircraft from the United Kingdom to Egypt was a month via Takoradi and ten to twelve weeks around the cape.[44]

Despite these challenges, Longmore had advantages. Royal Air Force Middle East (RAFME) was already becoming a capable force with a well-balanced complement of aircraft. The ability to build such a force and keep it in being, despite heavy losses to come in Greece, Crete, and the Western Desert, had much to do with deliveries of British and US aircraft and the support given RAFME by its supply, maintenance, and repair organization. This improved under Longmore and went through a period of radical overhaul after Tedder took command and Air Vice-Marshal Graham Dawson joined him as chief maintenance and supply officer in June 1941. The operational environment demanded excellent logistics and maintenance. Sand, dust, rocks, heat, and eventually the Luftwaffe hampered aircraft serviceability.[45]

The keys to successful RAF operations were rapid movement and resupply, which involved speedy construction of landing grounds to which squadrons could "leapfrog" forward or back as fortunes on the ground dictated. The enablers here were repair-and-salvage units (RSUs), which supplied new aircraft, repaired damaged ones, and rebuilt battle-worn and salvaged ones. Aircraft-maintenance units (AMUs) also increased dramatically in number and capabilities after 1940. They delivered aircraft, engines, fuel, ammunition, and motor transport and conducted overhauls and repairs of airframes and engines.[46]

Italian Actions

The IAF had similar problems but did not fix them. Intelligence and reconnaissance shortfalls eroded operational effectiveness, as the IAF's haphazard target selection illustrated. Enno von Rintelen, who kept Berlin informed about Marshal Graziani's preparations in Libya, was not sanguine. His correspondence prompted German leaders to offer the duce military support (the armored corps discussed earlier along with Luftwaffe units), which Mussolini refused. Of note, von Rintelen emphasized the vital importance of gaining air superiority to keep the SLOCs safe and thus facilitate attacks on high-value targets in Egypt—one of the few Germans who put this equation in the right order. Meanwhile, Hitler and OKW were happy to see how things developed. They did not have long to wait.[47]

Mussolini ordered Graziani to advance into Egypt on 9 September. The Fifth Squadra (a composite air division similar to a German *Luftflotte*) under General Felip Porro had enough mobility and support infrastructure to keep pace with the advance. It included about 300 serviceable aircraft. The IAF's mission was to attack airfields, supply points, and command posts. After the advance accelerated, aircrews were to attack troop formations and vehicles.[48] Heavy air action began on 13 September. The RAF attacked airfields and ground formations, while Italian bombers reciprocated but with little effect. The IAF made no concerted effort to gain air superiority, functioning mostly in small formations as flying artillery platforms for ground commanders. Conversely, the RAF gained air superiority and then devoted sixty sorties a day in round-the-clock attacks on Graziani's supply lines. Navy assets shelled Italian units and mined ports. The British went after thin-skinned vehicles to create a logistical crisis. A dense network of RAF all-weather airfields allowed aircraft to maximize sorties. Meanwhile, Longmore continued to develop a deep appreciation of how crucial Malta was to the entire British position.[49]

Malta: Key to the Central Mediterranean

Although IAF raids on Malta were largely ineffectual, the British knew what might happen should the Luftwaffe appear. Malta was the hinge point, offensively and defensively, in the central Mediterranean, sitting astride the SLOCs to Libya. Only land-based aircraft could carry out sustained ship attacks. Accordingly, the British Chiefs of Staff (CoS) agreed on 9 October to bring the Hurricane and Martin Maryland flights on Malta up to squadron strength (twelve aircraft per squadron, later increased to sixteen). Wellington bombers on Malta began raiding Benghazi. The weight of effort put into preserving Malta as a platform from which to strike the enemy was massive given that the British expected a German invasion attempt on the United Kingdom in spring 1941.[50] Longmore gave Malta reconnaissance, offensive, and defensive capabilities. By the time German aircraft began raiding it in January 1941, the RAF garrison had forty planes. This was enough provided steady reinforcements arrived. Malta was never out of the fight from 1940 to 1943 and engaged in offensive operations even during German raids. Early investment in air defenses paid major dividends.[51]

Although the Italians had decided not to invade Malta, Raeder was of a very different mind, asking Hitler on 18 March 1941 to take the island with an airborne assault before it became too heavily defended. He cautioned, "In British hands this base represents a strong threat to our troop transports to Africa and later for the supply transports. . . . In the opinion of the Air Force, it appears possible to capture Malta by airborne troops; the Navy is in favour of this as soon as possible."[52] Hitler pointed to the island's difficult terrain, asked for more careful studies, and instead chose Crete, but the wisdom of Raeder's entreaties soon became clear.

The Italians Invade Greece

As the British continued strengthening their forces, the Italians complicated matters for everyone by invading Greece on 28 October. It was a military disaster, but the British faced a dilemma. It was vital to come to Greece's aid to meet treaty obligations and to reassure Turkey and Yugoslavia of the empire's commitment to their sovereignty. However, there were not enough air and ground assets to guarantee the security of key locations throughout the theater even without adding Greece to the list. Only naval assistance was available at first. Getting significant bomber and fighter units in place would require three to four months.[53]

At first glance, the contest in the air was one-sided. The Greek Air Force had 160 French and Polish aircraft with few spare parts, whereas the IAF had more than twice as many planes and could attack from Albania, the Dodecanese, and Italy. However, the British responded fast. Heavy air reconnaissance and the improvement of airfields on Crete facilitated RAF engagement. There were four squadrons in Greece on 6 November. The AOC, Air Commodore J. H. D'Albiac, focused on ports and SLOCs to disrupt Italian logistics, but the rapid decline of the Greek Air Force, and the IAF contingent of nearly 200 aircraft in Albania, made this dangerous. Until

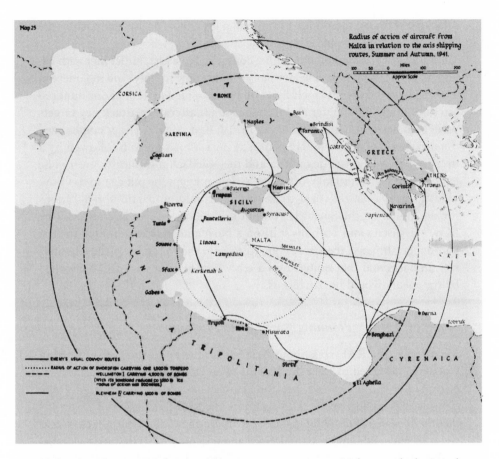

Malta aircraft ranges and major Axis convoy routes, 1941. Malta was the logistical hinge upon which all else ultimately turned in the Mediterranean campaign. The Axis failure even to try to capture this island was one of the major blunders of the war. (Originally published as Map 25 in Ian Playfair et al., *The Mediterranean and Middle East*, vol. 2, *The Germans Come to the Help of Their Ally (1941)* [London: Her Majesty's Stationery Office, 1956], 279, now public domain)

fighter escorts arrived, Wellingtons made night raids with meager results. To add to D'Albiac's woes, there were no all-weather airfields beyond Athens and no prospect of building any until spring.[54]

Andrew Cunningham's raid on the Italian Fleet at Taranto on 11 November with twenty-one Swordfish torpedo-bombers induced the Italians to move all their ships to safer ports on the West Coast of Italy, opening up further opportunities for air and naval convoy attacks in the Adriatic. The first sank four MVs totaling 16,938 tons. This ability to go after MVs sailing to Albania, and the greater latitude the Taranto raid gave the British in attacking shipping bound for Libya, had a long-term positive impact on the logistical and operational balance.[55]

Furious with Mussolini but feeling compelled to assist, Hitler met with Ciano on 18 November, saying Germany would come to Italy's aid in Greece in March 1941. This would include a bomber *Geschwader* (about one hundred aircraft) along with fighter and reconnaissance aircraft to attack key targets while also destroying the bulk of the British fleet. Finally, Hitler emphasized the need for immediate Italian action against Mersa Matruh followed by a decisive thrust against Alexandria and the Nile Delta. Ironically, even as he exhorted Mussolini to take offensive action in Egypt, he put a stop to OKW plans to send a panzer corps to assist. Hitler did not mention preparations for the attack on the Soviet Union but said he needed his aircraft back by May.[56] In discussions with his staff on 4 November, Hitler admitted that he had underestimated the time required to gain the initiative in the Mediterranean Basin and that doing so in a sea where SLOCs were still controlled by the enemy would be dangerous.[57]

Planning for Operation Compass

Meanwhile, the RAF was about to begin raids, in advance of Operation Compass, designed to eject the Italians from Egypt and exploit further opportunities. The arrival of thirty-four Hurricanes across the Takoradi route convinced Wavell to proceed. In addition, the RAF had deployed 1,200 airmen to provide rapid maintenance and repair and a quick turn back to forward airfields. Longmore also received a new deputy AOC in C, Air Marshal Arthur Tedder, who arrived via Takoradi on 2 December, just in time for Operation Compass. Tedder would prove to be one of the most outstanding senior officers of the war.[58]

Longmore had asked the vice chief of Air Staff (VCAS), Air Chief Marshal Wilfrid Freeman, to send Tedder and requested that he travel via the Takoradi route to get a feel for the requirements associated with it. He also gave Tedder responsibility for overseeing air operations in the Western Desert and Egypt. Longmore viewed his own position as one of coordinating policy and military-strategic issues with the other C in Cs, the Air Ministry, and senior airmen in other theaters while ensuring RAF administration ran smoothly.[59] The visit to Takoradi heartened Tedder as he saw how capable Group Captain Henry Thorold and his team were. Despite the rigors of getting aircraft to Takoradi as the U-boat menace increased, the 3,700-mile journey across Africa, and the time involved in constructing aircraft prior to the journey and refurbishing them after arriving in Egypt, this route soon became highly efficient.[60]

Air Chief Marshal Lord Arthur Tedder, AOC in C Middle East, 3 May 1941–14 January 1944. Tedder was one of the great Allied leaders of the war and brought Arthur Longmore's initiatives to maturity while introducing many of his own. He spent five months as Longmore's deputy and replaced him after Longmore was recalled to London in May 1941. (Imperial War Museum Image No. CBM-1176)

The RAF maintained air superiority in advance of Operation Compass. However, reserves were in short supply because Longmore now had five squadrons in Greece and nine in IEA. The growing threat of major operations in the Balkans was also a significant planning factor. So was the fact that Wavell had 90,000 men against the Italians' 250,000.[61] The trick was to sweep the Italians out of North Africa before Hitler intervened, but this presented a dilemma because any major success would bring in the Germans. Nonetheless, the C in Cs continued implementing the basic command-and-control (C2) network, air-ground-naval cooperation, logistical system, deception activities, and intelligence team that eventually underpinned success.[62]

The air plan included a deception scheme over three months in which small raids habituated the Italians to a "standard" pattern. Just before the ground attack, major raids on airfields would render the IAF combat-ineffective. Fighters would keep reconnaissance aircraft at a distance, blinding Italian commanders. Longmore moved three bomber squadrons and one of fighters from IEA. Two new Hurricane squadrons also moved forward. Total air assets on 10 December were 48 fighters and 116 bombers. In addition, a vital, if nascent, army-air component was set up under General Officer Commanding (GOC) Western Desert Force Richard O'Connor, which collocated O'Connor and Collishaw and their staffs at Western Desert Force headquarters. This liaison function grew in size and capability. O'Connor also controlled an army cooperation wing with two squadrons of fighters and a flight of reconnaissance aircraft. These innovations facilitated air-ground coordination and were the first application of the principle of the tactical air force (TAF), which became the norm in the RAF and US Army Air Forces (USAAF).[63]

While achieving air superiority, the RAF inflicted severe losses on supply columns, delaying the Italian advance and allowing Commonwealth forces to dictate the timing and tempo of operations. By the time Italian soldiers arrived at Sidi Barrani, they were shaken, tired, increasingly malnourished, and wondering where the IAF was. The airmen's response was to send most remaining fighters to establish standing patrols over the troops, weakening their airfield patrols and giving the RAF major tactical advantages. Although they still had numerical superiority, the Italians had no idea how to mass their planes to control the air.[64]

The Italians tried to stockpile supplies for the planned advance on Mersa Matruh, but RAF raids destroyed so many trucks that the quartermaster corps was incapable of supplying existing formations, not to mention stockpiling. Bombers raided Tobruk frequently, essentially closing the port and forcing the Italians to move troops, equipment, and supplies from Benghazi, which wore out their vehicles. During Operation Compass, advancing troops found nineteen ships sunk at Tobruk, three in Bardia, four in Gazala, one in Derna, and five in Benghazi. Airfield attacks left 1,100 aircraft destroyed or stranded when Commonwealth troops overran them. Raids also destroyed more than 1,000 trucks. The RAF helped to set the conditions for success in Operation Compass by neutralizing the IAF and creating a supply crisis for the Italian Army. It also forced Comando Supremo to delay the advance on Mersa Matruh until 12 December—two days after the start of Operation Compass.[65]

The Battle for Aircraft

One of the most important "battles" the RAF won was the unceasing struggle to maintain an adequate number of aircraft in the face of constant diversions to Greece and then the Far East, combat losses, and wear and tear. In 1938, William Mitchell called for more aircraft and personnel. Although little happened immediately, the Air Staff was planning for a long war.[66] If aircraft and pilot production surged too quickly and then fell off, said the chief of Air Staff (CAS), Air Chief Marshal Cyril Newall, "We should merely be eating the seed corn upon which we must depend for the crop that will sustain our resistance and build up our offensive capacity in the months and perhaps years to come."[67]

During his assignment as deputy director of plans in the Air Ministry, Air Commodore (later Air Chief Marshal) John "Jack" Slessor told his bosses:

> The immediate essential requirement is not to dispatch additional squadrons from the United Kingdom but to re-equip the existing squadrons with first class aircraft. . . . The collapse of France and our consequent loss of refueling facilities in French territory have created a most serious problem for us. . . . Special measures to meet the shipping difficulties are being examined by the Air Ministry in conjunction with the Admiralty and Ministry of Shipping, and these should be pressed on with the utmost dispatch and afforded a high order of priority.[68]

Given the limits of British production, many RAFME aircraft would have to come from the United States. Charles Portal selected Slessor to help procure these as part of the RAF Delegation (RAFDEL) in Washington, DC. After arriving in the United States, Slessor noted that although British production figures were easy to determine, the US 3,000-a-month scheme complicated matters because the RAF share of these machines was unclear.[69] As Slessor worked this issue, Longmore converted several Hurricanes to work in army cooperation squadrons because the Lysander was extremely vulnerable. Freeman promised Longmore rapid delivery of long-range reconnaissance aircraft in view of the fact that RAFME had a single Blenheim IV for this task. Seven more photographic reconnaissance (PR) aircraft were soon en route.[70]

In an event that threatened to derail US deliveries, Minister of Aircraft Production William Maxwell Aitken (Lord Beaverbrook), refused to provide British production figures, which made it difficult to determine what the United States needed to send. Secretary of the Treasury Henry Morgenthau had already asked for as detailed a statement as possible of anticipated production of planes and number of pilots. Slessor noted with astonishment that Beaverbrook thought the United States would lose control of the figures or leak them to the enemy. Beaverbrook also felt that if the US leaders thought British aircraft production inadequate, they might conclude that the empire would lose and thus withhold support. Slessor said Beaverbrook would listen only to Churchill, telling Portal, "I am sure you will agree in any event that it would put me in a quite impossible position if I were to attempt to undertake discussions with the U.S. Administration on any other basis [than full disclosure of British production figures]. . . . We are in it with the Americans for better or for worse and . . . the only possible basis for dealing with them is one of complete frankness."[71] Churchill told Beaverbrook to deliver the figures.

As Slessor's reputation and accomplishments grew, Churchill told Longmore that he was speeding air reinforcements so they would be in place for Operation Compass. Longmore sent daily reports regarding aircraft arrivals. However, Churchill was "astonished" to find that RAFME evidently already had nearly 1,000 aircraft, 1,000 pilots, and 16,000 other personnel. Tedder combed through aircraft manifests to find those usable for Operation Compass. Churchill ordered Longmore, "Pray report through the Air Ministry any steps you may be able to take to obtain more fighting value from the immense mass of material and men under your command."[72] Predictably, there were no modern aircraft types sitting around. Nearly all were derelict or obsolete, but outmoded Air Ministry books still carried them on RAFME strength tallies. After everyone understood how few capable planes Longmore had, Hurricanes and Blenheim IVs arrived just prior to Operation Compass, playing key roles.[73]

While events unfolded in the United Kingdom and Egypt, Slessor kept at the problem in the United States. The 3,000-a-month plan had proven impractical. Now, the objective was to get 26,000 US aircraft to the British by June 1942—about 1,300 a month. Combined Anglo-American production was adjusted from 3,000 to 2,500 a month.[74] Nonetheless, Morgenthau remained frustrated about the delay in getting US-produced aircraft into combat. He asked for particulars on the P-40's performance. Slessor answered that they were not yet in action. Morgenthau demanded an immediate update and told Slessor the planes needed to get into combat before the US Army Air Corps (USAAC, later USAAF) leadership questioned why they were going to the RAF at all. Slessor worked with Morgenthau to buy Portal some time.[75]

As these discussions unfolded, Wavell and Longmore received orders to send forces to Greece. Longmore ultimately sent nine squadrons but told Portal he was concerned about the impact such a diversion would have on actions in the Western Desert. His position there was already weak. The risks his airmen ran against Italians could not continue after the Germans arrived. He needed more Hurricanes and Wellingtons.[76] In addition, the Air Ministry's optimistic projection of twelve to eighteen months' RAFME supplies on hand would actually be one to four months after Operation Compass. Spare parts were arriving too slowly. RAFME and the Air Ministry agreed to a target of nine to twelve months' stocks for each aircraft type with frequent inventories given the unpredictable and intensive nature of air operations.[77]

Back in the United States, Slessor reported that US Secretary of War Henry Stimson had faced serious questions in Congress about why the British received 300 aircraft in November and the USAAC only 6. This added greater impetus to get US-built aircraft into combat. Enemies of the Lend-Lease Act sought to attack it. Slessor lamented that he was asked constantly why US-built fighters were not yet in action. They would be soon and provided a critical edge. However, this did not happen in time for Operation Compass. The British would make do with their own assets.[78]

3

Triumph and Tragedy
Operation Compass and the German Attack on Greece, November 1940–May 1941

New Grand-Strategic Choices

As THE WAR CONTINUED, and the British refused to capitulate or seek a negotiated peace, both sides engaged in a series of diplomatic and military actions that changed the character and intensity of the war in the Mediterranean along with its importance to each side's fortunes. These developments had the further effect of increasing the role airpower played in the theater. Italian reverses in Italian East Africa (IEA), Greece, and Egypt compelled Adolf Hitler to assist and the British to commit more scarce troops, weapons, and materials to the struggle for the Mediterranean Basin.

British Actions

If the British had certain advantages, including a strong fleet and good commanders, they understood that the Middle East theater required a fundamentally defensive strategic posture, with offensive action wherever lucrative. This would take the form of naval and air action with selected ground efforts, such as the offensive against IEA along with Operation Compass. British leaders hoped to knock Italy out of its African colonies before the Germans intervened. The transfer of part of the home fleet to Gibraltar underscored their determination to hold the Mediterranean. So did the increasing flow of personnel, weapons, and supplies to Egypt. Equally important were the Foreign Office's efforts to keep Spain and Turkey neutral, which succeeded despite many harrowing moments. Initiatives with Spain safeguarded Gibraltar and forestalled overt and large-scale German action from Spanish shores in the Battle of the Atlantic, while the success in Turkey complicated German efforts to control the Middle East and open a second front against Russia through the Caucasus. British diplomats were also extremely concerned about a possible German-Vichy alliance, as were US diplomats, and both engaged in high-level diplomatic efforts to keep Vichy neutral while supporting de Gaulle's Free French movement. Although Mers el Kebir and support for de Gaulle infuriated Philippe Pétain, the marshal was not about to gamble what remained of France on an alliance with Hitler. The most important British effort during this period was success in securing US lend-lease materiel along with other aid and agreements that brought the two countries into closer alignment. Franklin Roosevelt was alive to the dangers the United States would face should Britain suffer a decisive defeat in the Middle East or, worse, on the British Isles.[1]

This concerted effort to secure the Mediterranean Basin while wooing the United States was in line with the British decision to resist Nazi domination, but it also signaled two items of crucial importance for the theater and the war. First, the British were going to strike the enemy wherever they could. The developing Bomber Command effort was the only way to attack Germany, but Italy was more vulnerable. Selective and vigorous action there would give the British desperately needed victories, an ensuing morale boost, and better diplomatic standing. The latter was of the utmost importance with the United States but also with Spain and Turkey. Increasing bonds with the United States remind us that the British were preparing to wage a long global war, which Roosevelt and Winston Churchill believed was inevitable. The British would defend everywhere, attack and win where they could, and await direct US involvement. Meanwhile, they received increasing lend-lease materiel—especially aircraft. Despite Churchill's hatred of Communism, he was even prepared to fight alongside the Russians should the opportunity arise—as it soon would. The character of the war was changing along with the stakes involved. The British Empire would fight not just for its survival but for the destruction of the Axis.[2]

The situation in the Mediterranean brought on by Hitler's unsuccessful diplomatic efforts and Benito Mussolini's calamitous military ones also gave the British time to shore up their defenses. If the Italian attack on Greece and subsequent German offensive actions there to protect Europe's "southern rampart" proved disastrous for the British militarily, they also reconfirmed British determination to hold the Mediterranean and sped the flow of reinforcements. As the hub from which many of the empire's logistical spokes radiated, and through which others passed, Egypt was simply too important to lose. Oil resources were vitally important for rapid fuel deliveries to the Middle East theater, India, and the United Kingdom. The British thus did everything possible to hold and then win in the Mediterranean to set conditions for the defeat of the Axis.

US Actions

The British decision to intervene in Greece further signaled the empire's firm intention to fight on. The US leaders and others took note.[3] Roosevelt's

victory over Wendell Willkie in the 1940 election gave Churchill "inde-
scribable relief" that Britain's greatest supporter would stay in office.[4] Roosevelt's willingness to aid Great Britain while placing Europe and the Atlantic ahead of East Asia and the Pacific in his strategic priorities gave the
British added courage to keep fighting. He put the "Arsenal of Democracy"
to work in support of the British effort.[5]

Roosevelt recognized the importance of the Mediterranean at this junc-
ture in terms of its central geographical position, the place where Commonwealth troops could fight the Axis, and thus a region the British must hold
as a prerequisite for further, offensive operations. The US Joint Chiefs of
Staff's work on Rainbow Plan 5, which called for maximum support to and
cooperation with the British while looking out for US interests, called for
the decisive defeat of Nazi Germany. Aircraft deliveries were a central part
of this growing cooperation, as was the simultaneous expansion of the US
Army Air Corps (USAAC; US Army Air Forces [USAAF] from July 1941).
When Admiral Harold Stark, chief of naval operations, produced his seminal Plan Dog from October to November 1940, and Roosevelt approved
its "Europe-first" thrust, maximum support to the British, and insistence
on full equality in strategic decision making, it became the guiding document for military leaders as they anticipated US involvement in the war. On
16 January 1941, Roosevelt approved Plan Dog formally and emphasized
the importance of immediate and ongoing staff talks with Churchill and
his Chiefs of Staff (CoS). Therefore, "in his extraordinary memorandum
Stark had thus done much more than simply state what would become the
fundamental U.S. strategic decision of the war. He had outlined a balance-
of-power view of American national interests and policies in the present
situation, one that for the first time clearly linked U.S. security to Britain."[6]

Staff discussions, known as the American, British, and Canadian Talks 1
(ABC-1), began on 29 January 1941 in Washington, DC, and continued for
two months. The ensuing "Staff Conference Report of 1941" established
the general military-strategic, resource, and deployment priorities for an Allied military strategy. The plan assumed that if the United States went to
war with Germany, it would likely go to war with both Italy and Japan as
well. It further stated that the security of the Commonwealth, including its
Far East possessions, was of central importance for victory. The plan called
for the earliest possible defeat of Italy, a sustained air offensive to destroy
Axis military production, a rapid buildup of Allied forces for landings in
northwestern Europe, and the strong support of neutrals such as Turkey and
Spain as well as resistance and partisan movements. The Atlantic and European areas were the "decisive theater" and would be the primary focus of
US military efforts, although planners also noted the "great importance" of

the Middle East and Africa as must-win areas for the further, offensive prosecution of the war in Europe. Finally, if Japan entered the war—something almost everyone expected—then military strategy in the Far East would be defensive. Although the ABC-1 agreement was not a formal alliance, it reaffirmed that the United States was preparing for war, that Americans would help to maintain the security of the British Empire, and that their military was modifying the Rainbow Plans to incorporate US military integration, cooperation, and operations with Britain.[7]

The blueprint for Allied cooperation was set. There would be many disagreements as the war continued, especially as US power eclipsed that of the United Kingdom. A general US antipathy toward any more than the bare minimum action in the Mediterranean (the defeat of Italy and control of Europe's southern shore with air and naval power) would prove particularly divisive, but the two sides managed to work together effectively toward their common goal: the unconditional surrender of the Axis powers. The fact that they arrived at common guidance for the conduct of the war—something the United States had never done with another country in advance of an armed conflict—was of fundamental importance and gave the Allies major grand- and military-strategic advantages.

Italian Decline

With Britain's refusal to come to terms, Mussolini's short, low-risk war began to transform into a longer, higher-risk one. The military was utterly unprepared for this, and diplomacy failed to yield the colonial and resource acquisitions Mussolini sought. As German successes continued, and the new German mission to Romania appeared to threaten the duce's aspirations in the Balkans, his last effort at parallel war—the invasion of Greece—followed on 28 October. Confused discussions between Ribbentrop and Ciano in Rome on 19 September, larger miscommunications between the Italians and the Germans during the next month, and Mussolini's growing anger at the German creation of a satellite state in Romania all contributed to the duce's decision. Unfortunately, he gave his military leaders just two weeks to plan the campaign even as his recent demobilization order removed nearly half of the army's manpower, including the most seasoned veterans. Mussolini ordered the attack at the worst time of year given the poor infrastructure and mountainous terrain. These actions speak volumes about the disconnects in the hollow Italian-German alliance. Further, they would highlight the Italian military's inability to achieve its objectives, perpetuating the inexorable shift from ally to vassal. Mussolini's decision to attack Greece also divided limited shipping and military resources at the moment Rodolfo Graziani began his advance into Egypt. Finally, and most disastrously, it convinced Hitler to scrap plans for the deployment of a panzer corps and supporting air units to assist with the attack on Egypt. The Italians had come far, but in the wrong direction, from their leader's assertion about the character of the war: "not with Germany, not for Germany, but for Italy on the side of Germany."[8]

Policy miscalculations and incoherent military strategy were aggravated by lack of vision. It should have been clear that IEA's only hope for survival as an Italian colony depended on a reliable source of supplies. This meant a coordinated offensive from IEA into Sudan and up the Nile River along with a determined advance on the Suez Canal and strict avoidance of the dispersion of effort represented by the attack on Greece. Even with the still-fragile state of affairs in IEA and increasingly successful British efforts to foment an uprising there, this coordinated effort, ideally with Luftwaffe assistance to close the Suez Canal and part of the Red Sea to British shipping, was the only hope for Italy's colonial ambitions. Rather than coordinate their efforts, the Italians and Germans did their own things.[9]

German Dilemmas Continue

Even as Hitler sought alliances with Spain and France and wooed Turkey, his diplomats looked to another source of potential support: Arab nationalism. The Balfour Declaration and rapid increase in Jewish settlers throughout Palestine made many Arabs pro-German given the latter's clear intent to destroy the empire and replace it with something better, or at least anti-Jewish. Arab distrust of Italian aspirations drove them further into the German camp. However, in a striking series of diplomatic and operational failures, the Germans lost this opportunity to foment, support, and capitalize on Arab uprisings. On 5 July 1940—ten months before Iraqi Prime Minister Rashid Ali al Gailani's coup—the Iraqis informed the Germans that Rashid would welcome German support. They also maintained diplomatic relations with Italy in defiance of British demands and welcomed their assistance.[10] However, German Secretary of State Ernst von Weizäcker directed that the Arabs be kept on the hook with promises of British defeat and the "liberation of the Arab world" without providing specifics.[11]

Grand Mufti of Jerusalem Muhammad Amin al-Husayni sent an emissary to Berlin to tell German diplomats that Italy had already promised independence to all Arab colonies, mandates, and dependencies that aided the Axis.[12] On 16 July, he briefed Weizäcker on a "committee for collaboration among Arab countries" that would work with Rashid in Iraq and with the Axis to overthrow the British position in the Middle East. This included 10,000 men and officers, armed by the Vichy military in Syria, who would begin their uprising in coordination with Rashid's seizure of power in Iraq, and the latter's invitation to the Germans to deploy air and ground forces. Further, he asked that the Germans and Italians each fund one-third of the cost of the uprising and promised in return that the Arabs would tie down 30,000–40,000 British troops and prevent the movement of Indian troops to the region. These efforts, he said, would give the Axis the advantages it needed to conquer the Middle East.[13] The Iraqis pressed further, seeking German approval to reestablish the diplomatic ties with the Reich that they had broken on 19 September 1939 in accordance with treaty obligations to Britain. Rashid told the Italian minister in Baghdad that he "emphatically declared his adherence to the Axis powers."[14] On 20 August, the Germans reiterated that Italy would lead the "reorganization" of the Arab world, including the development of independent governments, and beyond that left the matter to Mussolini and Ciano. The Italians remained skeptical, and their attitude influenced the final German decision in December 1940 to forego whatever opportunities an Arab revolt might present. This would cost them dearly the following May, when Rashid's coup and anti-British uprising in Iraq failed in part as a result of the slow and diminutive German response.[15]

As the Arab question festered, so did continuing disagreements with Spain and Vichy. After the conclusion of his visits with Francisco Franco and Philippe Pétain, Hitler issued Directive No. 18 of 12 November 1940, which stated that political and military fortunes in the Mediterranean had not improved and were in fact more dangerous and complex as a result of the Italian disaster in Greece. Hitler's idea of a "Continental bloc" to stymie the British war effort had also gone by the wayside. German actions were thus driven increasingly by Italian misfortunes and the ad hoc support they compelled, by Hitler's diplomatic failures with Spain and France, and most of all by the increasingly rapid and decisive turn to the planned attack on the Soviet Union despite Britain's continuing defiance. German officers thus called for the conquest of Malta and Crete and the mining of the Suez Canal, but not for the deployment of a panzer corps to Libya. The latter recommendation was conditioned by General Ritter von Thoma's assessment during a recent visit to Italian troops in Africa that they were utterly unprepared for military operations. Hitler ordered the army to maintain one panzer division in readiness for deployment to Libya, the navy to convert German ships in Italian ports to troop transports, and the Luftwaffe to plan for concerted attacks on Alexandria to sink warships and the Suez Canal to close the eastern Mediterranean to shipping and thus hamper British logistics.[16]

During this frustrating period, Erich Raeder tried again on 14 November to convince Hitler of the need for an immediate push in the Mediterranean.

He cautioned that even if Great Britain was not currently in a position to dominate the Mediterranean, a German failure to preempt any increase in British strength, especially with US lend-lease materiel now flowing, would court disaster. Further, Raeder reemphasized the highly positive aspects of even a modest German effort in the region, including the high likelihood of sealing off the Mediterranean and Red Seas, seizing immense oil resources, and placing the Middle Sea and its surrounding lands firmly out of Allied reach. Hitler did not agree that a major victory in the Mediterranean would be of "war-deciding importance," as Raeder argued, but agreed that the Germans had to engage. However, this did not amount to anything approaching the scale of operations Raeder advocated. It would involve air action against the Suez Canal and Alexandria, and perhaps Malta, and ground action against Commonwealth troops should the Germans have to assist the Italian Army. Even the timing of possible air action was unclear, and Hitler reminded Raeder that Operation Barbarossa had priority over all other efforts, in effect dashing the grand admiral's hopes for a coherent and proactive military strategy in the Mediterranean and Middle East. Opportunities and victories in the Mediterranean would be more a matter of expedience than of planning.[17]

Even if the Germans viewed opportunities to help organize Arab revolts as being of little value—and events in Iraq four months later proved them wrong—they had little to lose and much to gain. They missed their chance. Hitler had high hopes for the capture of Gibraltar (Operation Felix), but Franco dashed them on 8 December when he decided firmly against an alliance. Hitler's cancellation of the deployment of a panzer corps to North Africa forfeited another opportunity to make tremendous gains in the Middle East. By the time German forces arrived in Libya, six months of opportunity had passed, and the huge logistical and troop demands for Operation Barbarossa were already exerting their influence. Raeder's 20 December appraisal and 27 December briefing to the Führer, in which he pleaded once more to delay Operation Barbarossa until Germany had knocked Britain out of the fight, at least in the Mediterranean and Middle East, came to nothing. He lamented that the "decisive action in the Mediterranean for which we had hoped is therefore no longer possible."[18] The campaign proceeded, with the initiative turning in favor of the British.

Operation Compass Begins

As deception and interdiction efforts continued before Operation Compass, so did reconnaissance. Archibald Wavell and Richard O'Connor knew Italian dispositions, whereas Italian leaders remained blind. The 331 Italian Air

Force (IAF) aircraft in Libya nearly doubled the Royal Air Force's (RAF's) 164, but the Italian machines were scattered as a result of organizational dysfunction and British airfield attacks, whereas the RAF's were concentrated at forward landing grounds. Intensive airfield attacks began on 7 December, damaging 39 Italian planes and placing Italian airmen in a reactive posture.[19] In fact, despite diversions to Greece, the eleven remaining squadrons in Egypt attacked airfields, ports, supply points, and troop concentrations with increasing fury. Night bombing proved particularly valuable because the IAF had no countermeasures available, and RAF aircrews became very skilled at locating and bombing ports given their distinctive visual cues. At this point, only the Hurricane and Blenheim IV were superior to Italian planes, so Arthur Longmore ordered an aggressive and concerted air effort to gain air superiority and then assist the army.[20]

The air offensive profited from densely packed enemy air bases, the lack of Italian initiative, the confusion and demoralization brought on by rapid defeat in the air and an equally serious rout on the ground, and the collapse of what little coordination there was between the IAF and Italian Army. Italian aircraft, fuel, munitions, and spare parts still were not dispersed when RAF raids began—six months after the start of hostilities. Damaging an aircraft was as good as destroying it because the Italian salvage service was ineffective. IAF planes also lacked self-sealing fuel tanks, leading to frequent catastrophic explosions. Within a week, Italian air operations had largely ceased.[21] Raymond Collishaw said,

> The failure of the Italian air force to strike at our aircraft on their aerodromes while the R.A.F. continued their sustained attacks on the Italian aerodromes brought about the destruction of the Italian air force at Cyrenaica. Adequate air stores parks and repair depots did not exist in Cyrenaica and the wear and tear of the air force was not made good. . . . The generals commanding the various parts of the enemy's lines of communication also contributed to the failure of the Italian air force by insisting on having fighter patrols flying over roads to prevent our air force from attacking the M.T. columns moving on the lines of communication. These facts were proved by the examination of the 1,100 damaged Italian aircraft which fell into our hands during the advance into Cyrenaica.[22]

Collishaw said many of the captured aircraft could have been repaired had the IAF possessed even modest salvage-and-repair capabilities.[23]

As soon as troops advanced on 10 December, the problem of moving squadrons forward rapidly became acute given the fast Italian retreat. "Leapfrogging" techniques were not yet mature, although RAF units were

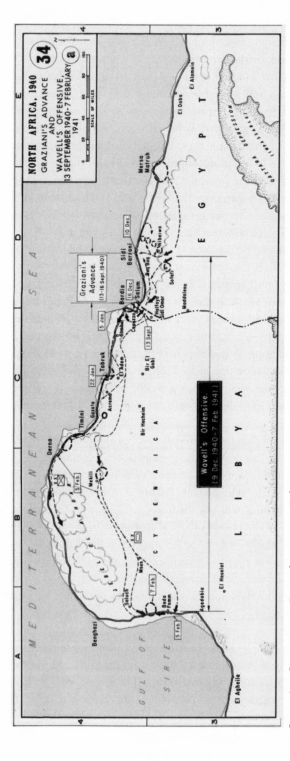

Operation Compass, 10 December 1940–7 February 1941. (Maps Department of the US Military Academy, West Point)

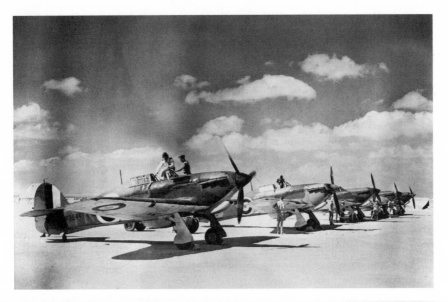

Hawker Hurricane I squadron in the Western Desert. These aircraft, along with the Hurricane II and the venerable Gladiator biplane, clinched air superiority during Operation Compass, allowing the Blenheim bombers to concentrate heavily on airfields, vehicle convoys, and other high-value targets. (Imperial War Museum Image No. ME(RAF) 179A)

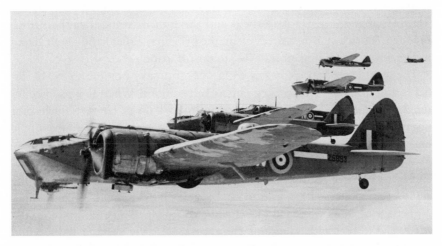

Bristol Blenheim IV bombers over the Western Desert. These versatile aircraft bombed ports, attacked airfields and vehicle convoys, and engaged in frequent ship attacks from low level. Note the chin gun with two .303 machine-guns or a 20 mm gun, which along with its 1,200-pound bomb load did severe damage to targets. (Imperial War Museum Image No. CM 3108)

motorized, allowing for relatively rapid movement. The RAF's logisticians began mastering the establishment of forward landing fields. They also brought an air stores park (ASP), repair-and-salvage unit (RSU), and air explosives and fuel park (AEFP). The AEFP adapted Italian bombs and fuel for use with RAF aircraft. This rapid airfield mobility became institutionalized with the introduction of several units dedicated entirely to this purpose. Accordingly, Royal Air Force Middle East (RAFME), which, like any other large air force, required immense logistical support to be mobile over large distances, became exceptionally agile.[24]

Operation Compass was a combined-arms operation with close cooperation. The army moved tanks into position as the RAF bombed troop concentrations, went after enemy airfields, and maintained air superiority. The navy bombarded enemy positions, causing enormous damage. The successful attack prompted British commanders to continue advancing to the Libyan border and from there to Bardia. The attack on Bardia, from 3 to 4 January, was an effective joint operation, with the RAF providing photographic reconnaissance (PR), bombing of enemy strongpoints, and a bomb "curtain" in front of advancing troops. The Italians lost 45,000 men and 128 tanks captured, along with artillery and motorized vehicles. They had lost more than 20,000 men in previous battles and were now in full flight toward Tobruk.[25]

When Bardia fell, Italian aircrews fled forward airfields only to be jumped as they landed at supposedly "distant" ones the RAF was already raiding. The IAF abandoned Derna airfield and the Italian Army after the British invested Tobruk. It also lost El Adem, the principal airfield and repair depot in Cyrenaica. Surviving aircraft fled to far-off Tripoli.[26]

The assault on Tobruk began with RAF and navy bombardment of airfields and army shelling of key defensive positions. When troops encountered strongpoints, bombers attacked. Tobruk fell on 22 January, the harbor began receiving shipping on 24 January, and the army had a secure supply base from which to continue advancing.[27] Another 20,000 Italians became prisoners. The commanders in chief (C in Cs) followed up each successive victory. After Tobruk, they took Derna. For the RAF, this phase of the campaign represented a logistical challenge as good airfields became rarer. Nonetheless, units occupied airfields at Martuba, El Tmimi, and Gazala. Most of the fuel and munitions came by truck over the Via Balbia, and coordinated operations during the army's accelerating advance continued.[28]

Despite the stresses created by these intensive operations, the C in Cs decided to capture Benghazi and its airfields. This would alleviate growing logistical challenges and facilitate bombing raids on Tripoli, the last Italian port in Libya, and on Sicily and southern Italy. Air operations from Benghazi would also combine with those from Malta to put enemy shipping in a vise. Benghazi fell on 7 February after O'Connor's victory at Beda Fomm, which bagged the rest of the Italian Army in Cyrenaica.[29]

As combat actions continued, RAF armored cars rescued downed pilots with increasing skill. This, combined with the IAF's failure to raid RAF airfields, resulted in low losses for ground and aircrews and translated into a high proportion of experienced crews. They had time to learn their trade against the Italians, and some of the fighter pilots who joined Hurricane units had fought the Germans over Britain.[30]

Italian aircraft were as good as RAF planes (except the Hurricane and Blenheim IV), and there were many more of them. Italian airmen, though relatively undisciplined and inadequately trained, were courageous and capable fliers. Still, the RAF suffered just 302 casualties from 10 June to 31 December 1940, losing 239 aircraft (142 in combat, 27 on the ground, and 70 to other causes). This compared to 501 confirmed and 167 unconfirmed Italian losses, plus the 1,100 captured on the ground. The IAF had sent major aircraft reinforcements to Libya only to see the planes destroyed or captured. The best Italian aircrews died over the desert.[31]

Italian combined-arms failures stemmed from lack of communication and coordination. Dispersal of air assets to home, territorial, air ministry, army, and navy authorities proved disastrous. The lack of radio communication with the army and navy was equally calamitous. On the convoying front, Italian ships often put to sea without coordinating land-based air cover, and Italian aircraft often went after British shipping without coordinating with the navy. Several times, Italian naval commanders turned down promising engagements with British ships because they feared aircraft carriers and the threat their aircraft and radar posed and because they lacked confidence in the IAF's ability to protect their ships from air attack.[32]

While RAF assets succeeded in the Western Desert, Malta-based aircraft flew 94 sorties against Italian ports during November and December 1940. However, a shortage of reconnaissance and strike aircraft meant many missed opportunities. The Italians also sank ten British submarines, nearly half the total on station, between June and December. Consequently, they lost only three ships totaling 15,400 tons. From June 1940 to January 1941, the Italians landed 47,000 troops without loss and almost 350,000 tons of equipment and supplies at a 2.3 percent loss rate. They also flew in nearly 1,000 aircraft. Italian defeats undermined these efforts.[33]

Nonetheless, just as British victory in the desert appeared possible, events in Greece pulled ever more assets away. By 25 February Collishaw's command had sixty-four aircraft. A fighter squadron with orders for Greece stayed put as a most unwelcome new arrival appeared over the battlefield:

the Luftwaffe. Almost immediately after the capture of Benghazi, the first German aircraft began operations that grew rapidly in scope. A sustained effort against Benghazi closed it to British use, exacerbating the logistical problems along an 800-mile supply line.[34]

Even as the Luftwaffe appeared, the RAF continued displaying its qualities as a "learning organization." Luftwaffe raids sped Collishaw's efforts to make RAF units as mobile as their army counterparts. This effort was of mutual benefit because the RAF could not operate for any period of time unless it had a reliable source of army-delivered supplies, robust unit-specific stores, or both. To augment army deliveries, RAFME developed greater logistical capabilities as Operation Compass unfolded. The location of existing railroads and the Via Balbia along the sea meant that large airfields needed to be close by. More remote landing grounds were usable only if highly mobile resupply and radar assets existed. Their increasing availability from late 1940 gave the RAF a vital advantage. Longmore and Collishaw understood the vital interconnections between highly mobile formations, secure supply lines, raids on Axis airfields, and effective combined-arms operations. The RAF's increasing ability to leapfrog, and the mountain of captured Italian fuel, bombs, and vehicles, made continuous operations possible.[35]

To find additional sites for landing grounds, and to ensure maximum air-ground cooperation, three air liaison sections became active at the corps and division levels. The Lysanders were vulnerable, but the Hurricanes replacing them could more than hold their own. With escorts, the Lysanders provided superb reconnaissance and artillery spotting given their slow speeds and long loiter times.[36]

"Informal reconnaissance," which yielded information on enemy forces provided by aircrews after missions, proved important as the joint staffs learned to incorporate intelligence into plans and operations. Intelligence officers debriefed aircrews while communications specialists relayed intelligence to headquarters for action. Artillery reconnaissance also made major strides. However, despite the rapidly growing need for PR and photographic intelligence (PI), there was only one Hurricane equipped with cameras and a small PI section to develop, annotate, and distribute prints. This was problematic given the featureless terrain, which required multiple passes by aircraft to find enemy troop concentrations and give photointerpreters enough prints to produce charts to help find, fix, and engage these forces. It was doubly difficult given the importance of oblique photographs in finding enemy troops. The Hurricane could not take them. Only a handful of Lysanders could. During January 1941, PR aircraft brought back 980 negatives, from which photointerpreters made 15,500 prints—a diminutive effort by later standards.[37]

Air-ground cooperation remained imperfect. Collishaw ordered several attacks on troop concentrations without coordination at Army Corps Headquarters, resulting in an ineffective employment of air assets during mobile phases of the ground battle—a problem the services did not fix until summer 1942. Equally troubling was the difficulty telling friend from foe. Army and RAF liaison officers were just beginning to receive communications gear for vectoring aircraft to target (a solution that awaited the advent of air support controls, or ASCs, during Operation Crusader a year later). For Operation Compass, pilots had to land at forward airfields for briefings en route to target.[38]

Despite these problems, air reconnaissance gave O'Connor the insights he needed to win at Beda Fomm. The entire campaign lasted ten weeks. The IAF made no appreciable impact. It started with 380 aircraft and ended with barely 30 despite massive reinforcement. The lack of direct support for troops proved disastrous. By 15 February 1941, their losses totaled 130,000 prisoners of war (POWs), 380 tanks, and 845 guns.[39]

The Germans Arrive

Just as the British prepared to push into Tripolitania, the situation in Greece drained their strength and stirred Hitler to action. On 20 November, he sent Mussolini a letter suggesting that Axis air forces cooperate in the Mediterranean, Italian troops take the airfield at Mersa Matruh so German bombers could attack Suez, and that the entire "Mediterranean problem" be settled during winter 1940–1941. He predicted that with proper air cooperation, the Mediterranean would become the "tomb of the British Navy" in three to four months. His thinking from there (not shared with the duce) was that Greece would fall in March or April 1941, Russia in summer 1941, and the Nile Delta in fall 1941, after additional assets arrived from the Eastern Front. Hitler's approach was a mix of realism and fantasy about the requirements to succeed in such a huge theater. Raeder and the Oberkommando der Wehrmacht (OKW, the German Armed Forces High Command) had failed to convince him in September 1940, and again in November and December, that an immediate offensive in the Mediterranean was of the utmost strategic importance. They now tried again while the Middle Sea had the Führer's interest.[40]

After their manifold disasters, the Italians finally requested German military assistance. On 27 November, Hitler ordered Fliegerkorps X to move from Norway to Sicily. By 15 January 1941, there were 186 aircraft there. Hitler wanted them back by February. What he thought one air corps could accomplish in a month is difficult to understand. Evidently, he expected quite a bit, since General of Fliers Ferdinand Geisler, Fliegerkorps X commander, had orders to bar the straits between Tunisia and Sicily to British shipping, neutralize Malta, provide air support to the Italians, protect the transport of the Afrika Korps along with reinforcement and resupply convoys to Tripoli, and raid shipping in the Suez Canal. Neither Hitler nor Hermann Göring understood the need for persistence and concentration in an air campaign involving so many tasks, especially one fought in a huge and geographically complex area with a harsh climate and severe logistical challenges. German airmen at the tactical level would fight a determined, courageous, and at times effective air campaign. However, their leadership's failure to employ air assets effectively as part of a combined-arms effort in a "triphibious" theater hamstrung their efforts to gain air superiority and thus to help Lieutenant General (later Field Marshal) Erwin Rommel win.[41]

Hitler was determined to avoid the Axis loss of North Africa. Directive No. 22 of 11 January 1941 said German assistance had become vital "for strategic, political, and psychological reasons."[42] A *Sperrverband* (blocking detachment) would deploy immediately to hold Tripolitania, and Fliegerkorps X would attack the British fleet, disrupt sea lines of communication (SLOCs), and support Italian units.

Operation Sonnenblume (Sunflower) began on 10 February. German troops would be under German command. Fliegerkorps X was directly subordinate to Göring. Rommel arrived in Tripoli on 12 February, the same day Hitler named the German ground contingent in Africa, comprised of the Fifteenth Panzer Division and Fifth Light Division, the Afrika Korps. They were at the front within days. Of 220,000 tons of cargo sent to Libya in February and March 1941, only 20,000 tons failed to arrive as German aircraft began attacks on Malta and British convoys. Despite heavy losses of seasoned aircrews over England, the Luftwaffe was still a formidable instrument, as its raids quickly proved. Nonetheless, it became involved in a new theater of war at a moment when it was at its weakest point since September 1939, and with a third and massive campaign—Operation Barbarossa—on the horizon. German airmen would labor under the exigencies imposed by this three-front air war and the largely ineffective senior leadership directing it.[43]

Operation Sonnenblume relied heavily on airpower to stabilize the situation. Fliegerkorps X had experience in another maritime theater—Norway. Geisler and his chief of staff, Colonel Martin Harlinghausen, had also been naval officers previously and understood air-sea operations. Malta was this unit's most important objective, both to safeguard Axis convoys and to destroy British ones. Despite the centrality of maritime air warfare tasks to success, Hitler refused to send strong antishipping aircraft forces because of

Erwin Rommel on reconnaissance with his chief of staff, Colonel Fritz Bayerlein, 1941. Rommel's arrival in the Western Desert turned the tables dramatically and led to another two years of war in Africa before the Allies prevailed. Had Adolf Hitler and Oberkommando der Luftwaffe (OKW) treated Rommel's efforts as more than just support for the Italians and his victories as serendipitous events rather than major opportunities to exploit, and had the Luftwaffe contingents in the theater performed as effectively as Rommel's troops, the Axis would likely have won a major victory in the Mediterranean and Middle East. However, Rommel's own shortcomings were significant and contributed to the Axis defeat. In particular, his tendency to rush forward to reconnoiter without informing either his own staff or that of Fliegerführer Afrika created serious C2 and combined-arms operational issues. His decision to persuade Hitler to let him continue to El Alamein without waiting for Operation Hercules to secure Malta was a major blunder that effectively guaranteed logistical and thus operational disaster for Panzerarmee Afrika. (Imperial War Museum Image No. HU 5628)

Me. 109E3, late 1940. This fighter and its later F and G variants proved the most capable in the theater until Spitfire Mark 5s arrived, far later than necessary, in April 1942. Had Oberkommando der Luftwaffe (OKL) and its senior officers given Me. 109 pilots clear orders—or any at all—to make air superiority their first priority, the outcome of the campaign would likely have been very different. As it turned out, Me. 109s were tied mostly to bomber escort in support of advancing troops, robbing the Germans of this superb air asset's greatest strengths. (Imperial War Museum image "Bf 109E-3 in Flight 1940," from Wikimedia Commons)

their increasing use in the Battle of the Atlantic and their anticipated requirements in the Baltic and Black Seas during Operation Barbarossa. Inadequate emphasis on such assets, largely as a result of constant arguments between Raeder and Göring about who would control them, also hamstrung the effort. The deployment of a single *Gruppe* (thirty-six aircraft) of He 111 torpedo bombers and one *Staffel* (ten planes) of He 111 mine-laying aircraft proved utterly inadequate to the objectives at hand and contributed to increasing disadvantages during the battle for the SLOCs.[44]

The unit's other commitments included attacking the British Fleet in Alexandria, mining the Suez Canal, and interdicting troop movements. However, leadership failures produced a lack of focus on air superiority. The larger dissipation of effort was equally alarming. Between 13 and 15 February, Luftwaffe units attacked RAF airfields in Libya. Raids on troop and vehicle concentrations in El Agheila and two attacks on Benghazi sought to hamper maneuvers and disrupt port operations. Although each of these raids

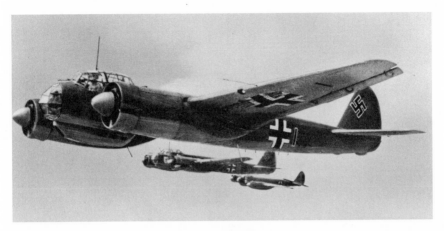

Junkers Ju 88 bomber. This was one of the most capable medium bombers of the war and served in numerous capacities to include bombing, aerial mining, torpedo bombing, and ground attack. More than 100 of these bombers, along with a number of their close cousins, the He 111 bomber, sat partly idle in Crete during Crusader, Gazala, and El Alamein rather than providing maximum assistance to Fliegerführer Afrika. This and the failure to produce enough torpedo-bomber variants of the Ju 88 for the "battle for the SLOCs" (or enough of the even more capable FW 200 long-range reconnaissance and torpedo-bombing platforms) was in large part a result of constant squabbling between Hermann Göring and Erich Raeder. In addition to hampering efforts in the Desert War by restricting the number of ship-attack assets, this deficit of torpedo bombers and long-range reconnaissance aircraft played a role in Germany's defeat in the Battle of the Atlantic. (Imperial War Museum Image No. MH 6115)

produced effects, the Luftwaffe did not gain air superiority, and raids on RAF airfields did not continue with any vigor or regularity.[45]

Mining operations against the Suez Canal, as limited and intermittent as they were, began on 30 January 1941, sinking five merchant vessels (MVs) and closing the canal for nearly three weeks, at first holding up 100,000 tons of shipping and ultimately 1 million. Mines also landed in Sollum, Bardia, Tobruk, Benghazi, and Valetta. Had the Germans concentrated their efforts on heavy mining of the Suez Canal, results would have been substantial. Fliegerkorps X dedicated just 139 sorties—2.5 percent of total effort from 10 January to 31 May 1941—to mining operations. Nonetheless, these missions slowed the arrival of reinforcements just as Rommel began his first offensive and thus played an important role in shifting the initiative.[46]

Fliegerführer Afrika became active on 20 February with General of Fliers Stephan Fröhlich arriving on 1 March. Göring ordered Fröhlich to "direct and commit the elements of the German Air Force employed in the African theater of war—such as flying and antiaircraft units—*in a manner that will*

Junkers Ju 87 Stuka dive-bomber. One of the most feared planes of the early war years, it had limited success in North Africa given the Royal Air Force's aggressive posture and the relatively low effectiveness of its armament against troop formations. It was, however, a ship-killer of the first order and continued to do very well in that role despite the fact that the Germans placed varying and ultimately insufficient emphasis on ship attack and SLOC control. (Imperial War Museum Image No. GER 18)

guarantee maximum support of the Army units employed in that area."[47] Rommel noted in his writings that the German deployment to Libya was designed specifically to give the Luftwaffe adequate space within which to base and operate against the RAF and Benghazi while covering the arrival of Rommel's forces. However, Italian-German staff talks in December and subsequent Führer directives confirmed that its primary purpose was to control the SLOCs while destroying British warships and denying the empire use of its key SLOCs—especially the Suez Canal. From the outset, conflicting priorities and orders pulled Fliegerkorps X and Fliegerführer Afrika in multiple directions. Even though the latter's role was to support Rommel, it relied on Fliegerkorps X to set the larger conditions for victory, and it should have had the flexibility required to engage in all mission types depending on immediate circumstances and requirements. Consequently, these two units failed to provide the full range of mutual support to each other, and to the ground and maritime efforts, necessary to win control of the air, sea, and land in this complex theater. Their force structure was already too small to do so without careful coordination. Subsequent decisions by Hitler and Göring (more regarding these later) changed Fliegerkorps X's mission and thus exacerbated all of these problems.[48]

One of the most serious and persistent problems—and a key indication of the insufficient attention given to air superiority throughout the campaign—was the shortage of single-engine German fighters in Africa. Jagdgeschwader 27 (JG 27) did not arrive until 14 April 1941. Aside from the episodic presence of one *Gruppe* (thirty planes) of JG 53 after December 1941, the hundred or so aircraft of JG 27 carried the entire air-superiority effort. In a theater where the Axis enjoyed overall numerical superiority until July 1942, it had too few of these critically important single-engine fighters in Africa. Despite JG 27's superb pilots and machines, they did not turn the tide in the air when they had opportunities to do so from April to June 1941 and again from January to June 1942 as exhausted RAF formations reduced by diversions first to Greece and then to India were terribly vulnerable. In the majority of cases, German fighter pilots were tied to bomber escort or allowed to engage in *Frei Jagd* ("free hunting"), which took a severe toll on RAF aircraft over time but had limited aggregate effectiveness in the air-superiority arena. Conversely, relentless RAF raids on Axis airfields proved central to gaining air superiority in a different but ultimately much more effective fashion. This failure to control the air proved one of the most important missed German opportunities of the desert war.[49]

Additionally, from the outset, aircraft overhauls occurred far from the front and many on the Continent, a serious logistical shortcoming the Germans never corrected. This reduced readiness rates to around 50 percent, whereas the RAF's, because of the inherent advantage of the Nile Delta workshops and greater innovation, stayed between 70 and 80 percent. Additionally, the modification of Luftwaffe command relationships in the Mediterranean (more on this later) led to major command-and-control (C2) seams in the air and combined-arms efforts. There was also less direct co-operation between Rommel and senior airmen than there was between RAF and British Army officers. While Rommel called constantly for more aircraft to support his troops and protect convoys, he maintained a strained and marginally effective relationship with Fröhlich, but it never evinced the increasing level of British combined-arms acumen.[50]

Meanwhile, however, things went well as veteran soldiers and aircrews went after a tired and ill-supplied adversary at his culminating point. The British could not move past El Agheila because of supply shortages. He 126s reconnoitered them while long-range aircraft on Sicily tracked more distant British positions. Rommel thus knew how dispersed his enemy was. The first dive-bomber attacks on El Agheila occurred on 14 February. After Commonwealth forces retreated, Luftwaffe units could not keep up with the ground forces because Fliegerführer Afrika did not know how to leapfrog and had insufficient vehicles to do so. As a result of Rommel's decision on 31 March to continue his offensive, army units far outpaced Luftwaffe formations, leading to heavy RAF attacks that forced them to disperse.[51]

However, raids on Benghazi closed the port to resupply—another key factor in Rommel's early and major success. Incomprehensibly, intensive air reconnaissance of RAF airfields, which were at that point few in number and highly vulnerable, resulted in only a few sporadic raids by fewer than twenty aircraft. Conversely, remaining RAF assets went after Luftwaffe airfields without pause, setting the pace of air operations and keeping the Germans from winning even greater victories. The constant pounding took its toll on machines and, more importantly, skilled ground and aircrews. This gradual wearing away of German planes and men was one of the crucial results of the different approach to air operations. Along with several other factors, it cost the Axis control of the air.[52]

Rommel's arrival as British assets moved to Greece came at a fortuitous moment, but we must see it within the larger strategic context. The same is true regarding the arrival of Fliegerkorps X and its first air campaign against Malta. These forces were not intended *first and foremost* to conquer the Middle East but rather to keep the Italians fighting. Hitler was more concerned about helping the Italians in Greece than in Libya, especially given his plans to make Greece and Crete the Reich's "southern rampart" against British actions in the Balkans.[53] However, there is no question that he believed German forces would first defend and then conquer the Mediterranean Basin.[54]

By June 1941, German forces controlled southeastern Europe, Crete, and more than half of Cyrenaica, but they found themselves in a costly and see-saw campaign even as the effort in Russia took virtually everything they had. They would fight in the Mediterranean theater on the cheap, and without any grand-strategic vision or consistency, whereas the British threw everything they could into the effort. The RAF contribution was central because the Germans could have won in the Mediterranean had they used their aircraft more effectively—something that will become increasingly clear as the narrative progresses. This was not a failure on the part of German aircrews but rather the product of poor leadership, lack of imagination, ineffective organizational structures and C2, and poor logistics.

Fröhlich, for instance, had orders to give first priority to support of the army. This would have been easier and more effective had he gained air superiority while his fresh formations faced tired and small RAF ones. However, he was an Austrian Army officer who had supported the *Anschluss* and was rewarded with senior Luftwaffe rank. Consequently, although he understood air operations in support of ground forces relatively well, he evidently had no grasp of the crucial importance of air superiority as an enabling action for combined-arms operations. This lack of expertise and intuition, along with Oberkommando der Luftwaffe's (OKL's) restrictive orders, proved costly in the logistics- and airpower-driven Mediterranean theater, where prioritization, concentration, agility, and persistence were indispensable. These shortcomings, along with Fröhlich's de facto subordination to Rommel, made him ill-suited to his job. Logistical problems resulting mostly from inefficient Italian management of shipping assets, but also self-made, dogged the Afrika Korps and Luftwaffe efforts from the start. Preparations for Operation Barbarossa made these worse as OKW commandeered half of the vehicles from Luftwaffe units deploying to Africa. Finally, air commanders seemed to have suppressed any memory of the mauling the Luftwaffe took in the Battle of Britain. To borrow the title for a superb work on that first Luftwaffe defeat, German airmen were once again up against "the most dangerous enemy."[55] The Luftwaffe never showed the flexibility necessary to defeat the RAF in the Mediterranean. As Albert Kesselring noted, it had "become a part" of the army, would "think within its framework," and would "take unconditionally its place in the [army's] struggle."[56] None of these problems was yet apparent, although the Battle of Britain should have been a clarion call for the Luftwaffe to determine the proper uses of airpower *within different contexts against a dangerous foe.*

The intent to conquer the Middle East after Hitler had dealt with Russia was evident, but neither Hitler nor the OKW made any serious plans *before* Operation Barbarossa to campaign there, with their existing force structure, during periods of opportunity. The Germans committed enough forces that had they used them in an effective combined-arms effort, they may well have won. Hitler's willingness to support Rommel's offensive operations over the heads of more senior officers makes it clear that he was not averse to supporting successes in the desert that occurred in parallel with the Russian campaign. However, lack of a clear vision meant that any great victory in the theater would be serendipitous and all subsequent actions largely reactive and ad hoc.[57] Nonetheless, German preparations, as late as 6 April 1941, to deploy an eight-division operational group to Spain and Morocco, another of nine divisions to Egypt, a fourteen-division force to Anatolia, and eighteen divisions to Afghanistan makes it clear that Hitler viewed the Mediterranean as much more than a defensive theater in which the Wehrmacht would simply shore up Italian forces. From closing both ends of the Mediterranean to linking up with the Japanese in the Afghanistan-India region to the acquisition of major holdings in Africa, Hitler's approach was profoundly expansive.[58]

In the short term, though, the RAF had to contend with a serious threat as the Luftwaffe arrived and German airmen went after ports, ships, and army columns. The first air campaign against Malta, despite its episodic nature, shocked British commanders and closed the portal through the central Mediterranean for shipping. It also forced Sunderland flying boats and Wellington bombers to redeploy to Egypt, reducing greatly Malta's reconnaissance and strike assets. However, each time the Luftwaffe ended a major series of attacks, Malta rebounded immediately as Hurricanes and later Spitfires flew in from aircraft carriers, reconnaissance and strike aircraft returned, and naval flotillas once again used it as a base for raiding. Only the island's conquest could put a stop to this dynamic.[59]

In Libya, Rommel and Fröhlich sought to get along, mostly on Rommel's terms. He gave both the Luftwaffe and logistics short shrift. However, he understood the importance of airpower within the context of air-ground operations. He had a dedicated reconnaissance squadron for both of his divisions along with fighters and Stukas attached directly to army units for mobile operations. This effectively "penny packeted" aircraft but made them highly responsive to army tactical requirements because ground commanders had already learned to work with the Luftwaffe in this fashion. Consequently, until the RAF gained permanent air superiority in summer 1942, the Luftwaffe operated with a moderate level of effectiveness during Rommel's major campaigns before El Alamein, even if it never gained a significant or lasting measure of air superiority.[60] Defeating the RAF was the only way the Axis could win in a triphibious theater where logistics were of paramount importance, and the side with control of the air and sea would ultimately triumph—as long as its army did not suffer complete defeat and could reconstitute and resupply to fight another day.

Rommel and Airpower

Despite the challenges he and Fröhlich faced, Rommel was a voracious consumer of airpower—especially reconnaissance—and flew all over the battlefield in his Fieseler Storch to reconnoiter the battlefield. In a discussion with Marshal Graziani on 12 February, Rommel pushed for an immediate offensive against the tired enemy.[61] He also appealed to Fröhlich and Geissler to hold Wavell's army at El Agheila with heavy air attacks and praised them for their success in doing so. After Rommel got the Western Desert Force on the run, reconnaissance aircraft kept him up to date on the enemy's positions and vulnerabilities.[62]

Rommel's grasp of the importance of airpower in the direct-support and reconnaissance roles (if less so in the indirect-support role) appears to have been closely matched by his understanding of its potential to make an important contribution to an Axis victory in North Africa. After his first offensive culminated with the reconquest of most of Cyrenaica, including its crucial airfields, Rommel opined that he could have taken all of Egypt and pushed into the Middle East if Hitler had authorized even modest additional air and ground reinforcements. He was bitter about the Führer's decision to attack Greece, which he claimed occurred without his knowledge and at the cost of diverting most of Fliegerkorps X to Operation Marita when he needed its aircraft desperately to capture Tobruk and continue his advance.[63] So, although Rommel may not have appreciated fully the intricacies and interdependencies involved in combined-arms operations, his writings underscore that he was anything but blind to their importance. Rommel said that a stronger and better-employed air component, combined with the rapid capture of Malta, would have allowed his army to take the Nile Delta and the oil fields beyond. This would also have facilitated a pincer movement against Russia. And although he did not state explicitly that a push further east to meet the Japanese in India would also have been possible, his comments imply strongly that he understood this.[64]

Rommel kept a constant eye on the Russian campaign because he hoped victory there would free up air and ground forces. He believed that the two additional panzer divisions, four or five Italian armored and motorized divisions, and substantial air assets slated for deployment to Africa after Operation Barbarossa concluded would be decisive. Rommel may well have been right, if the suppression or conquest of Malta and the maintenance of the SLOC-dependent logistical system had continued.[65]

As the campaign progressed into 1942 and culminated with the defeat at El Alamein, Rommel came back to these points. OKW, he said,

> still failed to see the importance of the African theatre. They did not realise that with relatively small means, we could have won victories in the Near East which, in their strategic and economic value, would have far surpassed the conquest of the Don Bend [in southern Russia]. Ahead of us lay territories containing an enormous wealth of raw materials; Africa, for example, and the Middle East, which could have freed us from all anxieties about oil. A few more divisions . . . with supplies for them guaranteed, would have sufficed to bring about the complete defeat of the entire British forces in the Near East.[66]

Rommel was a masterful tactical and operational commander who dealt the British Army a series of defeats that portended German victory. The Luftwaffe played a role but proved unable to gain or hold air superiority—something that would likely have put the ground campaign in the bag given Rommel's talents and the British Army's problems adjusting to his style of warfare. Rommel lamented the lack of air assets frequently in his writings. However, he did not appear to understand that the problem was not an absolute shortage of aircraft (the Luftwaffe and IAF had more planes than RAFME in the Mediterranean theater until summer 1942) but rather their lack of concentration and unity of effort resulting from ineffective command relationships and constant micromanaging by OKW and OKL. One example of this division of assets was the reassignment of Fliegerkorps X from Sicily to Greece to support Operation Marita, where it stayed afterward to preside over the more than 50 percent of German aircraft in the Mediterranean that did relatively little to assist Rommel, either with direct support, keeping open the SLOCs, or interdicting Suez. Meanwhile, remaining Fliegerkorps X assets in Sicily, and those of Fliegerführer Afrika, fought tooth and nail against roughly equal numbers of RAF aircraft when the effective incorporation of all of Fliegerkorps X's assets would almost certainly have turned the tide decisively in the Luftwaffe's favor, and thus in Rommel's, from both the logistical and operational perspectives. The direct subordination of Fliegerführer Afrika to Göring and OKL proved costly. Equally problematic, when Field Marshal Kesselring arrived with Luftflotte 2 in December 1941, he did not receive centralized control of air assets. This splintered and largely rudderless Luftwaffe effort robbed Fröhlich and Rommel of the air assets they needed at specific times and places to help turn major victories into decisive ones.[67]

Rommel did have several weaknesses, especially in the airpower, logistics, and C2 arenas. As one scholar has noted, "His disinterest in the dreary science of logistics, his love of action, his tendency to fly off wherever the fighting was hottest—all these may make a good movie, but they are disastrous in an army commander, and they all contributed materially to his failure in the desert."[68] Although Rommel's own writings make clear his growing interest in problems relating to airpower and logistics, which was inversely proportional to how much of each capability he had, there is no indication that he tried to correct them until just before El Alamein. And he never overcame the instinctual reaction to head to the front, often without notifying the Luftwaffe commander or even his subordinate army commanders, which did more harm than good. This frame of mind and the actions springing from it had their roots in German military history and the imperative for decisive defeat of the enemy whenever an opportunity beckoned. If *Bewegungskrieg* (mobile warfare) was a German advantage at the start of the war, it exercised an increasingly pernicious influence on Rommel. One historian has asserted that *"Bewegungskrieg* had evolved over the centuries precisely to short-circuit . . . rational calculation."[69]

It is therefore important to place assertions about Rommel and Luftwaffe commanders, especially Fröhlich and his successors, General of Fliers Hoffmann von Waldau and General of Fliers Hans Seidemann, in perspective. Rommel had seen firsthand the immense value of airpower in his "blitz" across France. He did not undervalue it, nor did he belittle the aircrews that helped his troops win on the battlefield. However, Rommel's decisions in the desert make it clear that his understanding of airpower was truncated. If the Luftwaffe was not overhead, it was not, in Rommel's view, doing its job, part of which was to raise the troops' morale. This is clear in his writings. The concept of flexible employment of airpower that the RAF and British Army ultimately developed was much less evident on the German side. Fröhlich also did Rommel no favors although von Waldau and Seidemann proved more capable if no more free to seek control of the air. Rommel's and OKL's failure to recognize that the RAF was a qualitatively different and more dangerous opponent than any of the air forces the Luftwaffe had yet faced is puzzling in view of the recent hammering over England. The Luftwaffe was a "learning organization" in the war's early campaigns, but it had also faced less capable opponents. In North Africa, the Luftwaffe had an air commander and a high command unable to adjust to the RAF's evolving approach to air warfare, which involved seeking control of the air at every turn. This made it difficult for German aircrews to support their comrades on the ground to the same extent as had been the case in previous campaigns. These problems became evident over time, but for the moment the Luftwaffe's arrival changed the balance in important ways.

The Luftwaffe's First Window of Opportunity

Intensive Luftwaffe raids on Benghazi created major British logistical head-aches. And for one of the few times in the campaign, German airmen also went after airfields, making RAF air operations challenging. The Germans understood that any further Commonwealth advance, and retention of Benghazi's port and airfields, would make an Axis defense of Tripolitania unmanageable. Consequently, this determined first effort by the Luftwaffe, Rommel's 24 March counteroffensive, Commonwealth diversions to Greece, and serious British logistical problems soon restored the Axis situation. Despite the arrival of RAF mobile radar units and a dense network of ground observers, the RAF could not keep the Luftwaffe from closing Benghazi.[70]

To create reciprocal logistical problems, Wellington raids on Tripoli increased, as did Blenheim attacks on Sirte. Both did damage but failed to stop

Ground crews fueling and "bombing up" a Wellington bomber of No. 205 Group. This aircraft served in multiple roles, including heavy attacks on Axis ports, night raids on airfields and troop concentrations in coordination with FAA Albacore flare-dropping aircraft, surveillance and torpedo bombing in the battle for the sea lines of communication, and heavy mining of the Danube River later during the 1944 campaign against Ploesti and the movement of POL products. No. 205 Group was one of the finest air units of the war. (Imperial War Museum Image No. CNA 1272)

Junkers Ju 52 transport plane taking off on the Eastern Front, December 1941, as more than 300 of its cousins worked to resupply German forces in Africa with high-priority materiel. The Tante Ju (Auntie Ju) was a capable aircraft, and the Luftwaffe had by far the best air-transport capability of any major power before 1943. These aircraft delivered immense quantities of troops, fuel, and other supplies to Rommel's forces while evacuating thousands of wounded. The disasters at Stalingrad and Tunis essentially destroyed the Luftwaffe's air-transport fleet. (Imperial War Museum Image No. COL 353)

the German buildup. There were not enough bombers. Compounding this problem, Ju 52 transports flew supplies to Luftwaffe units and Rommel's leading armored units. At this point, with the disasters at Stalingrad and Tunis not yet glimpsed, the Luftwaffe had the best air-transport arm in the world, and it proved its worth in Africa. German units remained relatively well supplied as Rommel engaged in his first offensive. Consequently, what RAF commanders had anticipated as a time for rest and refitting instead became another protracted period of intense air operations. From 1 January to 31 March, RAFME lost 184 aircraft, but only 166 replacements arrived. Diversions to Greece made things especially challenging because most of the Hurricanes and Blenheim IVs had deployed there. The Luftwaffe's airmen were having a significant effect on the course of the battle, but the RAF would not yield the air.[71]

Even after Rommel's attack, the RAF continued heavy raids on airfields. However, air efforts alone could not stop the Axis advance, and by 2 April

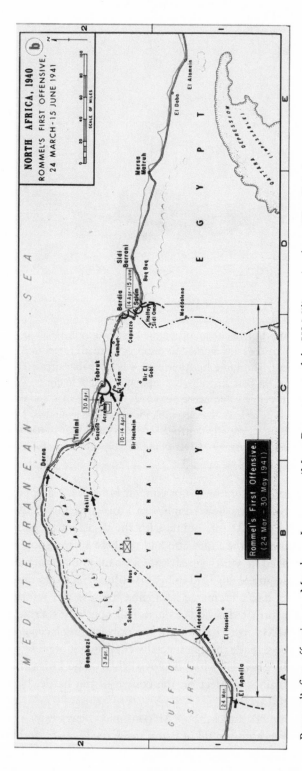

Rommel's first offensive, 24 March–15 June 1941. (Maps Department of the US Military Academy, West Point)

Rommel was about to occupy Benghazi, forcing RAF units to leapfrog back, which they did well. Rommel's continuing offensive broke Commonwealth forces. RAF priorities thus shifted, momentarily, to attacks on enemy convoys and motor transport, direct support of the army, destroying Ju 52 transports, raids on Benghazi to keep supplies from reaching the Afrika Korps, and a fighter defense of Tobruk to protect shipping after Rommel besieged it. This was a tall order for remaining assets, but the airmen made it work.[72]

With Tobruk under siege, Wellingtons could no longer reach Tripoli. The importance of the "battle for airfields" and its major effects on the logistical situation were becoming clear. Tobruk's logistical importance led to the Luftwaffe's almost complete commitment to reducing the bastion, giving Collishaw's depleted forces an opportunity to refit. Incessant Luftwaffe raids suffered greatly because fighter pilots were ordered to fly tight rather than loose escort.[73]

Despite Luftwaffe raids on Malta, the British retained control of the air there as well by reinforcing the island with fighters launched from carriers in the western Mediterranean. Steady infusions of Hurricanes kept the number of operational fighters above forty. And then, suddenly, Fliegerkorps X left Sicily for Greece. The Italians were again charged with neutralizing Malta and attacking convoys. There would be no more Luftwaffe raids until December. Ironically, Fliegerkorps X played a relatively minor role in the attack on Greece and Crete, taking a backseat to Wolfram von Richthofen's Fliegerkorps VIII, so its efforts went largely for naught during this critical period of opportunity in the Western Desert. Only 4.5 percent of the total Fliegerkorps X effort—238 sorties—assisted with the campaign in the Balkans. Most of these were mining missions and thus of potential importance, but they were scattered all over Greece. For this and other reasons, 80 percent of British troops escaped. The real disaster, however, was the permanent move of Fliegerkorps X Headquarters along with most of its aircraft and ground organization to Greece and Crete at the very moment Rommel needed these assets to do the work they were originally sent to perform: safeguard the SLOCs for Axis shipping, close them to British MVs, and provide both indirect and direct support to Rommel's campaigns after the Luftwaffe had the upper hand. The even greater dispersion of effort and C2 seams that followed proved calamitous for Axis fortunes.[74]

Adding to the Luftwaffe's problems, logistical shortfalls soon appeared. Efforts to arrive at an agreement with Vichy leaders for the use of the Bizerte and Gabes rail line failed. Had this succeeded, it would have been much more difficult for the RAF and Royal Navy to interfere with Axis convoys. Instead, attacks became increasingly effective because air reconnaissance was once again able to direct striking forces. Nine Wellingtons returned to Malta temporarily in April, bombing Tripoli and convoys five times until a squadron of Blenheims took their place. From January to May, the Royal Navy sank thirty-one MVs totaling 101,636 tons despite constant raids on Malta—and because the focus on Malta distracted Luftwaffe aircrews from one of their strong suits: ship attack. It sank another 100,000 tons of MVs in other parts of the Mediterranean. Although aircraft sank only two ships, they proved vital in locating MVs for the navy. Malta's air forces were once again as strong in June as they had been when Luftwaffe attacks began in January.[75]

Nonetheless, the Luftwaffe's arrival changed the larger air and maritime calculus. British air reinforcements had to move through Takoradi or around the cape. The first Luftwaffe attacks, against the convoy constituting Operation Excess, severely damaged the carrier HMS *Illustrious* and sank the cruiser HMS *Southampton* on 10 January 1941. After this, convoys sailed to Malta only in dire emergencies.[76]

The Germans also planned to attack the Suez Canal, but the distances were great and the ground-support equipment at Benghazi and on Rhodes inadequate. (The latter could have been upgraded, but there is no record of this having been done.) Raids were episodic and small, yet effective out of all proportion to the effort. A larger mining campaign would have created massive logistical problems. After the fall of Crete, bombers based there attacked the Suez Canal, Port Said, Suez, Alexandria, and airfields. They closed the canal to traffic for up to three weeks at a stretch but raided only intermittently and stopped altogether based on Führer directives (discussed later).[77]

Longmore now faced simultaneous commitments in the Western Desert, Greece, IEA, and in the defense of Alexandria and the Suez Canal. To reduce the air threat to their logistics, the British began moving goods overland from Aqaba to Palestine and then into Egypt with the construction of road and rail links. They also maximized their use of Port Suez, doubled rail capacity between Suez and Ismailia, and ran a pipeline the entire length of the canal so naval fuel could go from Suez to Port Said.[78]

Despite these efforts, the Luftwaffe's virtual closure of the Mediterranean to shipping meant that convoys to Malta and the Aegean, and supply ships moving along the coast to ports in Cyrenaica, required greatly increased protection—more than could be provided by carrier-borne fighters. The carriers' platforms were themselves terribly vulnerable. This required a larger and more modern RAF presence. There was also an urgent need for more defensive measures: fighters, antiaircraft artillery (AAA), and radar; ships fitted with radar; and operations centers. Fortunately, the Führer was already turning his attention to the Balkans as a prelude to his grand prize, Russia. Luftwaffe units gathering in the Balkans, Sicily, and the Western Desert thus had a very short time to accomplish their numerous and often conflicting missions.[79]

Axis Logistics

Meanwhile, logistical and communications difficulties increasingly affected Luftwaffe operations. In December 1940, a Ju 52 *Gruppe* arrived in Italy to resupply forces in Africa. However, logistical planning and convoy scheduling, along with sailing times, rested with the Italian Navy and with the supply and logistics agencies in Rome, which remained at near-peacetime levels of activity. Air escort for convoys was a joint responsibility. This division of labor, like most, led to more friction than effective operations. The German Air Force Liaison Staff in Italy (Italuft) did establish Air Materiel Command Africa, with unloading staffs in Tripoli and later Benghazi. Its mission was to meet Fliegerführer Afrika's logistical requirements—ultimately an impossible task because much Luftwaffe equipment ended up at the bottom of the Mediterranean and much of the rest, especially trucks, on army rather than air force rolls.[80]

Still, Air Materiel Command Africa established fuel dumps at Tripoli and Derna. Air base commands were set up at Castel Benito, Benina, Derna, Tmimi, and Gambut. A field construction agency at Tripoli, with detachments in Benghazi and Derna, built and maintained airfields and billeting. Aircraft mechanics had to clean gasoline, air, and oil filters every two days, and ammunition often had to come back out of magazines for cleaning. The crews endured temperatures of up to 158 degrees Fahrenheit while working to keep aircraft flying.[81]

The Axis faced three major logistical bottlenecks: the passage across the central Mediterranean, the west-east SLOCs along the coast of North Africa, and vehicle movements along the Via Balbia. The distance was 600 miles from Tripoli to Benghazi and 400 more from there to Sollum. Supply convoys were spread across this entire distance, in much like a significant percentage of petroleum or any other long-haul commodity was "in the pipeline" and thus not available to troops. Air-transport aircraft moved high-priority items, but this was insufficient on its own. Rommel complained about Italian failures constantly, from chaotic shipping schedules to the slow movement of supplies.[82]

Efforts to establish reliable Italian-German radio contact for air-to-air and air-to-ground communications failed. Italian military staff organizations resisted close coordination. Their own logistical problems and the inability to replace shipping losses with new production gradually undermined the convoy effort. German dislike of working with the Italians and Göring's failure to make a sufficient effort to develop joint capabilities and assist with Italian aircraft production also limited air-sea capabilities.[83]

Rommel's continuing efforts to seize Tobruk (four times from 9 to 30 April) drained Luftwaffe assets despite a 3 April Führer order to Rommel not to be reckless because most air units were leaving for the Balkans and Operation Barbarossa. One Luftwaffe liaison officer lamented the heavy air and ground losses at Tobruk with no gains to show for them. Me. 109s flying close escort lost their advantages over Hurricanes, making strike packages highly vulnerable. Rommel also noted during his initial efforts to take the port that the Luftwaffe was bringing forward its aircraft and could not provide adequate support. This was once again a result of vehicle shortages created by OKW requisitions for Operation Barbarossa and of Rommel purloining others. The Tobruk air effort marked yet another new requirement in a war in which Fliegerkorps X had already been moved to Greece, and its remaining assets in the central Mediterranean, along with those of Fliegerführer Afrika, were at the limits of their endurance. This marked the culmination of the Luftwaffe's first and initially very successful intervention in the Mediterranean Basin. It underscored the impacts of C2 problems and the movement of ground organizations over long distances and time periods to new operating locations (it took Fliegerkorps X units eight weeks to move from Sicily to Greece during a critical point in Rommel's offensive). Diversions from the original effort to control the central Mediterranean SLOCs, and the failure to gain air superiority when the RAF was at its weakest, hampered Axis efforts. This proved particularly serious regarding the battle for the SLOCs. Luftwaffe overextension and logistical problems were already impinging on Rommel's ability to capitalize on his victories.[84]

Another problem was Rommel's tendency to rely on his own staff's inputs and less on Fröhlich's. After Harlinghausen's departure for the Battle of the Atlantic and throughout the Sollum offensive in March, there was no formally assigned Luftwaffe liaison officer on Rommel's staff. Even after they began arriving, these officers were often ignorant of army intentions until days or hours before (or after) an attack began. Most damaging was the failure to form a joint headquarters. Fröhlich's command post was always at the principal airfield. On occasion, he maintained personal contact with Afrika Korps officers but usually relied on landline and wireless. The latter was highly vulnerable to intercept.[85] Hellmuth Felmy bemoaned the lack of an officer at Afrika Korps headquarters "who could constantly advise and express ideas and *anticipate* the need for aircraft commitments in accordance with the ground situation developments."[86] He continued:

It is true that the flying formations had not enough vehicles and signal equipment, but an advance command post and landing field might have been *improvised* near Gambut or in the Sidi Azeiz area for the duration of the Sollum battle. In that case the Air Brigade Commander Africa would have been able to maintain closer contact with the 15th Panzer Division and might also have been in a position to establish the whereabouts of the 5th Light Division. To pursue the defeated British Forces he would have needed bombing planes with the proper range, which he would have had to request from X Air Corps.[87]

So, Fröhlich could not establish effective C2 internally or with the army and therefore could not maintain regular contact with one German division or even locate the other. His forces were unable to leapfrog as a result of insufficient vehicles. To top it off, he could not get Fliegerkorps X bombers into place for the pursuit phase after Sollum because of seams in the Luftwaffe's command structure. These problems, along with RAF resistance, allowed the British Army to escape for the first of three times. Already, Rommel and his staff were going to ground regularly to avoid RAF aircraft. However, on 20 April Rommel noted that German fighters were finally overhead in sufficient strength to keep most enemy aircraft at bay. The Luftwaffe could protect the Afrika Korps as long as it could keep pace with the advance. This was problematic because Rommel regularly outran Luftwaffe assets.[88]

Events Unfold in the Balkans

As the Germans made their presence felt in the central Mediterranean, troop movements into Romania and agreements with Yugoslavia made clear what they had in store for Greece. The British decision to engage there was risky. Few Hurricanes remained in Africa, and only they stood a chance against Me. 109s.[89] Admiral Andrew Cunningham warned that the naval component would be in extreme danger in the Aegean given Luftwaffe strength. The British C in Cs were pessimistic. However, from a grand-strategic perspective, they agreed with Churchill that it was worth trying even given the high risk of military defeat. The world, and especially the United States, was watching, and US support was vitally important.[90]

General of Fliers Hans Speidel took command of the Luftwaffe contingent within the German mission to Romania. Its role was to guard the Ploesti oil fields and train Romanian Air Force units. A deeper purpose was to give aircrews training in the Balkans before operations against Greece. His force included 50,000 men, a Me. 109 fighter *Gruppe*, and the equivalent of two flak gun divisions. In November 1940, Hungary, Romania, and Slovakia joined the Tripartite Pact, with Bulgaria following on 1 March 1941. In December 1940, seven German divisions moved into the region.[91]

The British deployment to Greece comprised an Australian–New Zealand corps and a tank brigade—58,000 men. The RAF had nine squadrons (200 planes) by April 1941. By then, the RAF wing had flown nearly 300 sorties in support of the Greek Army, contributing to Italian reversals. Wellington night raids on Albanian ports and airfields were frequent and increasingly effective. From January to March 1941, the Gladiator squadron savaged Italian fighters so badly that the IAF rarely appeared again.[92] Despite these successes, J. H. D'Albiac had his hands full. Airfields were his biggest problem because the only all-weather bases were around Athens. Bad weather and the Greek government prevented construction of new ones further north, making large-scale raids impossible. D'Albiac was proud of his airmen but knew they could not hold. The British and Greeks had less than 500 aircraft against the Germans' 1,000 and the IAF's 310. D'Albiac's plan was simply to keep the Luftwaffe away from ground forces when possible.[93]

Luftwaffe tasks were to achieve and maintain air supremacy by crushing the RAF, give direct support to the army by destroying enemy transport and supply services, drop parachute troops on vital locations, and bomb naval forces and ports. Unlike in North Africa, air supremacy was an overriding objective facilitating all further operations, as it had been in earlier campaigns. The interpersonal dynamics between air and ground commanders, and their differing qualities between theaters, played key roles here.[94]

While the Germans prepared for Operation Marita, the British and Italian Navies clashed on 28 and 29 March at Cape Matapan. Italian Admiral Angelo Iachino had no air support other than small reconnaissance aircraft on his ships. Although he received one air-reconnaissance report that the large British Fleet was forty-five miles away and closing, he discounted this, losing two heavy cruisers and several destroyers during the ensuing action. Mussolini's response was to restrict all fleet movements to coastal waters within easy range of land-based aircraft. The British, on the other hand, soon created No. 201 (Naval Cooperation) Group to assist the Royal Navy after disastrous ship losses around Crete, leading to very effective air-sea coordination and attacks on Axis convoys. Fundamental differences in approach to air-naval cooperation and ship-attack priorities persisted.[95]

After the anti-Axis coup in Yugoslavia, the Germans adjusted their troop dispositions to defeat Yugoslavia and Greece in a single campaign, declaring war on 6 April. Luftflotte 4 savaged Yugoslavian targets. For the attack on Greece, Twelfth Army had von Richthofen's Fliegerkorps VIII.[96] As his force moved into position, Directive No. 18 ordered the Luftwaffe to attack all British air bases within striking distance of Ploesti. In addition, Fliegerkorps

X would move its headquarters and most of its air assets and ground organization from Sicily to Greece to support the attack. This changed the unit's primary focus from the central to the eastern Mediterranean, with subsequent negative effects on the logistical situation. Meanwhile, the Luftwaffe seized air superiority, destroying the Yugoslav Air Force. Greek and RAF crews were immediately on the defensive. From there, Fliegerkorps VIII went after major ports and marshaling yards. Then, troops came under heavy attack. RAF efforts to delay the German advance, which had already outflanked the main Greek and British positions in the north, were hazardous. On 14 April, Luftwaffe fighters shot down six Blenheims supporting ground forces. The next day, they destroyed ten Blenheims on the ground.[97]

The timing of the attack on 6 April and the lack of all-weather RAF airfields gave the Luftwaffe major advantages. RAF landline communications were nearly nonexistent, hampering early warning and combined-arms operations. Greek spotters could not keep up with the tempo of air operations. This left RAF fighters blind and unable to protect bombers. D'Albiac ordered all squadrons back to the Athens area, where an intact early-warning system and all-weather bases gave his aircrews time to get airborne. Even this proved inadequate. Blenheims destroyed a large number of German vehicles but could not stop the enemy advance. The Hurricanes intercepted a 100-aircraft raid on Athens-area airfields on 19 April, destroying 22 aircraft for the loss of 5. However, this represented one-third of their strength. The survivors left for Egypt.[98]

Commonwealth and Greek armies were cut off from one another. Faced with manifold disasters, the British decided, with Greek agreement, to withdraw.[99] Luftwaffe reconnaissance missions confirmed this when they discovered ships in Greek ports. D'Albiac used his last Blenheims to delay the enemy's advance. The evacuation began on 17 April as a rearguard force conducted a dogged defense of Thermopylae. By 25 April it was outflanked and retreating, and the RAF's last operational aircraft were lost or withdrawn by 23 April. The main withdrawal began the night of 24 April, going into the next morning. Inexplicably, Fliegerkorps X bombers did not make night attacks, even though more of their crews were night-qualified and Hitler had ordered every precaution against another Dunkirk-like evacuation.[100] Conversely, Fliegerkorps VIII aircraft stayed with the ground forces, flying direct-support missions. Von Richthofen's air corps had its own Ju 52 transport group, which flew a total of 1.2 million miles, transporting 500 tons of bombs; 1,650 tons of petroleum, oil, and lubricants (POL); and spare parts. It provided tactical and operational flexibility. Excellent signals capabilities allowed Fliegerkorps VIII to stay in close contact with ground forces so air intelligence was quickly available to army units.[101]

From 6 April to 1 May, the Luftwaffe destroyed 454 enemy aircraft (340 of them on the ground). From 14 to 26 April, bombers sank 120 MVs totaling 559,000 tons and 5 warships. They damaged another 129 MVs totaling 701,500 tons and 10 warships. Fliegerkorps VIII harried enemy ground units. It outperformed Fliegerführer Afrika as a result of better leadership, clearer and better objectives (air superiority first, followed by direct support), much better mobility, and superior C2 based on an excellent communications network. Of course, unlike Fliegerführer Afrika, Luftwaffe units in Greece had months to prepare for their mission, flew on the Continent rather than overseas, and thus faced fewer logistical and operational challenges—although operating in Greece was no picnic.[102]

To cover the evacuation, the RAF had 43 fighters on Crete. RAF losses in Greece were 209 aircraft, of which 82 lacked spare parts and would otherwise have been capable of evacuation. The RAF destroyed 259 Axis aircraft and claimed another 99 probable. The British did what they could despite severe disadvantages. The debacle continued on Crete.[103]

German Victory and Defeat on Crete

The strategic advantages for Germany of taking Crete were substantial, to include keeping the British fleet out of the Aegean, safe conduct for oil tankers sailing from Ploesti to Italy, a flanking position on British SLOCs, and a block against raids on Ploesti. These benefits were real but also fundamentally defensive rather than part of a larger offensive plan to dominate the Mediterranean and Middle East.[104] With Greece secure, German planes raided Crete heavily, isolating the island. Fliegerkorps VIII's mission was to maintain air superiority, soften up the defenses, and prevent the arrival of reinforcements. Fliegerkorps XI was in charge of the assault itself. To meet the attack, the British had twenty-four operational aircraft. The 28,500 Commonwealth troops were tired and ill equipped.[105] Churchill intended to hold Crete, but the effort was risky given Hitler's fixation on it. When given the option of using his airborne troops to take either Crete or Malta (there were not enough to do both in parallel), Hitler chose Crete despite the fact that all senior officers in OKW favored Malta. The Malta attack thus went on the shelf, for good as it turned out. Churchill's decision to hold Crete thus had unanticipated strategic effects.[106]

The RAF's situation on Crete was bleak. The island was out of range of fighters from Egypt but nearly surrounded by Axis air bases. Air reconnaissance was perilous, so commanders were nearly blind, although they did receive very useful Ultra (high-grade signals intelligence) decrypts. Wavell said on 30 April that the empire would hold Crete.[107] Arthur Tedder realized that

The strategic importance of Crete. Despite the successful but costly German conquest of Crete in May 1941, it became a strategic dead end in three ways. First, major command-and-control seams prevented copious bomber assets stationed there from engaging in a maximum or coordinated effort with Fliegerführer Afrika. Second, Adolf Hitler, shocked by the casualties, never again authorized a large-scale airborne assault, to include the planned Operation Hercules against Malta. Third, the Führer viewed Crete as the "southern rampart" to keep Allied forces away from Ploesti and his Russian flank rather than as an offensive stepping stone to more successes in the Mediterranean and Middle East. Crete could have been an immensely effective offensive base for the employment of land-based airpower in multiple roles. However, the majority of aircraft sat at readiness awaiting Royal Navy sorties and major convoys that rarely materialized. They assisted Erwin Rommel's efforts and the all-important battle for the sea lines of communication only episodically, and their sporadic aerial mining of the Suez Canal ultimately yielded no lasting effects. Crete was only 240 air miles from Tobruk, 340 from Mersa Matruh, 400 from Alexandria, and 600 from the Suez Canal. More effective use of air units on Crete would have helped Axis fortunes. (Maps Department of the US Military Academy, West Point)

sending additional squadrons would do no good. He planned to keep the air contingent at strength by sending replacement planes and also withdrew two Wellington squadrons from Iraq to provide a long-range strike capability against Luftwaffe air bases. It was too little and too late. After the last fighters on Crete were out of action, the Germans began their assault. The loss of the Maleme airfield proved the garrison's undoing.[108]

After Maleme fell, Ju 52s landed three at a time under heavy fire. The two seaborne convoys with 2,500 troops and all the airborne division's heavy and motorized equipment ran afoul of the Royal Navy and were annihilated, but the Royal Navy was itself savaged by the Luftwaffe. The cost was too high to have justified sailing so close to land-based enemy aircraft during daylight. Hurricanes with external tanks flew the 300 miles from Egypt, taking a heavy toll on Ju 52s in the air and on the ground, but this could not turn the tide. Air supremacy gave the Germans a costly victory. The Luftwaffe once again failed to carry out heavy attacks on the evacuation of Commonwealth troops unless they caught them in daylight.[109] By 30 May, most had escaped.[110]

One of the key aspects of the battle, in terms of the role of land-based airpower, was the mauling the Luftwaffe gave the Royal Navy. Bombers sank three cruisers and six destroyers and seriously damaged another seventeen warships. These included the carrier *Formidable,* the battleships *Warspite* and *Barham,* six cruisers, and seven destroyers. Consequently, 59 percent of the British fleet was sunk or out of action for months. Incredibly, the Italian Navy stayed in port despite German pleas for a combined attack on the British Fleet and the protection of the two German flotillas destroyed en route to Crete. The IAF did little more. The RAF lost forty-seven aircraft and nearly as many pilots on Crete, the army 1,800 dead and nearly 12,000 captured, and the navy 1,800 dead. The evacuation saved 18,000 men.[111]

Tedder's view of the disasters in Greece and Crete was that they would have been avoidable, at least in terms of the effects in the Western Desert, had the home government responded earlier. He was bitter about Lord Beaverbrook's policy of maintaining maximum fighter strength in the United Kingdom even after the invasion threat there had lessened.[112] Tedder thought much better of his boss, noting, "Longmore managed, in these dismal circumstances [Rommel's counterattack and the invasion of Greece—he was relieved just before Crete], to remain cheerful. He practiced what I thought an admirable philosophy; if he felt everything possible had been done to meet a situation, it was useless to worry about it, however menacing it might look."[113] This proved fortuitous because, among Longmore's other worries, he had twenty-one serviceable Hurricanes left in Libya on 1 May. Still, he signaled Charles Portal after the loss of Greece that "we will keep going somehow whilst we build up with the reinforcements in sight and promised."[114]

Greece and Crete proved for Tedder that air superiority was the prerequisite for all successful combined-arms operations. He warned Wavell that if the German Army and Luftwaffe operated together in Africa as effectively as they had in Greece, their control of air bases in Cyrenaica would have the same effect as their control of bases in the Aegean. This was, he said, a war for air bases first and foremost, and the side that held the best bases and made proper use of air assets would likely win. In a 30 May message to Portal, he pressed the issue again but also noted that the Germans had employed air-assault assets in Crete with such "criminal prodigality" that their losses had likely kept them from intervening more forcefully in Iraq and Syria (more to follow on these operations).[115]

The Balkans became a major strategic liability for the Germans. Hitler made the region his defensive "southern rampart." There is no evidence that he ever viewed it as a base for offensive operations. The most important effect of the Crete operation, then, is that it played a pivotal role in sparing Malta and perhaps in keeping the Germans from intervening with greater speed and force in Iraq and Syria. Although the Germans rebuilt Fliegerkorps XI, Hitler was so shaken by the losses on Crete that he never again authorized a major airborne operation. Instead, he consigned his elite *Fallschirmjaeger* to ground duty where, ironically, they suffered even greater losses on the defensive in Sicily and Italy than they had on Crete. Wavell believed that the "defense saved in all probability Cyprus, Syria, Iraq, and, perhaps, Tobruk."[116] He might well have added Malta. As one scholar put it, "The beleaguered in the Mediterranean would soon cease to be the British and become the Germans, who had committed themselves to fight in a geographically complex theater with inadequate resources, a poorly adapted force structure, and an undependable ally."[117]

4

The RAF Holds—and Learns
May–August 1941

Allied Grand-Strategic Choices

THE PRIMARY DEVELOPMENTS DURING THIS PERIOD had to do with the evolving Anglo-American relationship and the increase of US lend-lease materiel to the United Kingdom. By August 1941, this had transformed from a trickle to a growing flood and helped the British to hold. General George C. Marshall, chair of the US Joint Chiefs of Staff (JCS), remained committed to Plan Dog but not on a British timetable. US troops would intervene only when ready. Until then, British forces would carry the burden with US material assistance. Marshall became particularly vehement about this after Winston Churchill asked Franklin Roosevelt in May 1941 to bring the United States into the war within a larger British strategic framework. Churchill's subsequent discussions with Harry Hopkins, Roosevelt's principal adviser, in London during July 1941 and the following month with Roosevelt during the Atlantic Conference in Newfoundland, met with US rejection of a number of his proposals. However, Churchill's larger conception of "closing the ring" around the Axis produced general agreement. Naval power and airpower—two growing Anglo-American strengths—would win the Battle of the Atlantic, freeing the sea lines of communication (SLOCs) for a buildup of forces prior to an invasion of the Continent. The Western Desert campaign, in conjunction with proposed US landings in Vichy northwestern Africa, would further set the stage for the defeat of Germany by forcing Italy's surrender. Roosevelt agreed, but his JCS did not, especially regarding a North African invasion, which Marshall opposed doggedly until Roosevelt ordered him bluntly to proceed. US officers accused the British of proposing a flawed military strategy designed to gain political and economic advantages for the empire around Europe's periphery without defeating Germany. The British countered that there was only so much the Allies could do given current realities. An invasion of northwestern Europe in 1942 and perhaps even 1943 was out of the question, so the Mediterranean campaign and Bomber Command's efforts (later augmented by US Army Air Forces [USAAF] heavy bombers) had to do for the present. Although Churchill failed to entice the United States into the war, most of the British delegation's key ideas resonated with Roosevelt. He continued discussing matters quietly with Churchill. The US Victory Program soon followed and set the course for the US contribution to Allied victory, including aircraft- and pilot-production requirements, and heavy-bombing doctrine, set forth in the July 1941 Air War Plans Division Plan 1 (AWPD 1).[1]

This focus on airpower was twofold. First, the United States provided an increasing number of US-manufactured planes, most of which went to the Middle East. Second, the raw materials US merchant vessels (MVs) brought to the United Kingdom allowed the British to increase their aircraft production. In 1941, the British produced 20,100 machines to Germany's 11,030. The British kept too many of their best aircraft, especially Spitfires, at home for longer than necessary. However, US-built aircraft soon reinforced the Middle East theater, and the Royal Air Force (RAF) sent emergency deliveries of UK-produced aircraft to Malta. This heavy mutual focus on airpower paid major dividends in the theater and beyond.

Axis Choices

Germany and Italy were fully committed to their respective "programs" and in Italy's case to something more like survival. The Germans, despite dismissive language about US involvement and military effectiveness, were nonetheless increasingly concerned. General Friedrich von Boetticher, military attaché to the United States, reported that the US war industry was surging and would soon provide the British copious materiel. Although his report remained politically correct, he hinted that time was turning against the Reich. Operation Barbarossa thus had to succeed in the allotted time—one campaigning season—if Germany wanted to avoid the worst effects of a two-front war in which the United States was not only providing aid to Britain and perhaps the USSR but also becoming actively involved.[2]

Adolf Hitler told his senior officers on 9 January 1941 that the defeat of the Soviet Union would compel the British to give in—another misunderstanding of his adversary. Even if Britain fought on, he said, Germany's position would be so strong that there would be no chance of a British victory. As for the United States, Japan would commence hostilities at the right time and keep most of the US efforts focused there—another major underestimation of US productive, logistical, and manpower assets. All major successes in the Mediterranean Basin, including the conquest of Gibraltar and actions against Egypt and the oil fields, would wait until after Operation Barbarossa. Talks with the French to create a sounder defensive "forefront" in the western Mediterranean had failed in December 1940 in part because Hitler would not countenance Vichy efforts to increase colonial holdings.

Dilatory negotiations regarding Vichy provision of arms to Prime Minister Rashid Ali al Gailani's uprising in Iraq, and the equally sluggish response to Iraqi pleas for assistance, robbed Hitler of a very real if risky opportunity to seize much of the Middle East in April and May 1941. Rashid moved too fast, the Germans too slowly, and the French too cautiously. Philippe Pétain and François Darlan continued to insist that the armistice agreement was not an acceptable platform on which to reach further agreements, so the Germans went into Russia without either a western European "forefront" or any gains in the Middle East. By the time the defeat before Moscow occurred, the French were no longer interested in a close association. Hitler gambled everything on his assessment that Russia would collapse after one hard push. Erwin Rommel's recent victory had stabilized the situation in the Mediterranean. Hitler was satisfied with that and confident he could return to the Middle Sea and finish the job there in due course.[3]

The Italian position by now was one of de facto vassalage to Germany. Italian East Africa (IEA) and Italian Somaliland were gone and with them dreams of an expansive "New Rome." The situation was stable in Libya for the moment, but that depended on increasing German involvement. Even Italian "gains" in Yugoslavia and Greece were proving troublesome as Chetniks, Partisans, the Ustashi, and other groups butchered one another and, increasingly, soldiers at Italian and German garrisons. Italy's fate was tied to Germany's. Nonetheless, Benito Mussolini and his subordinates continued avoiding close cooperation or coordination, especially regarding the SLOCs and logistical considerations in Africa. Along with increasing British anti-shipping capabilities, this contributed substantially to the Axis defeat in the Mediterranean.[4]

RAF Innovations

In his postmortem of the campaigns in Greece and Crete, Arthur Tedder concluded that soldiers would have to get used to the idea that airplanes would not be directly overhead all the time because they had interdiction, ship attack, and other missions to complete before the start of any ground campaign. Tedder also resolved to keep his aircraft concentrated and focused on air superiority to maintain an upper hand over the Luftwaffe because "all reports from here [North Africa] and Greece show clearly that the Hun does not like real opposition."[5] He also observed that effective German combined-arms operations in Greece—a product of air supremacy—allowed for a rapid victory. He said the Germans were doing the same thing at Tobruk only without first gaining control of the air—a deadly inversion of priorities.[6]

With Tedder convinced that air superiority was the indispensable first step in any campaign, his units hit Luftwaffe aircraft at Gazala and Derna hard and often. His greatest worry was the shortage of fighters because most had been lost in Greece. Should the Luftwaffe launch a "real blitz" against RAF units, it would be very dangerous. Tedder feared this and made air units highly mobile, dispersing them over wide areas and procuring additional mobile radar sets. Only fighters ready to scramble stayed at airfields. The rest went to dispersal sites. Those needing to strike as deeply as possible into Cyrenaica used forward refueling fields. His advanced RAF headquarters brought together intelligence, logistics, plans, and operations specialists along with army liaison officers. It remained collocated with general officer commanding (GOC) Mobile Division Headquarters. Many of these initiatives were not new. Air operations during Operation Compass had proven their value. What was new was Tedder's growing understanding that the same principles would work against Fliegerführer Afrika. The Luftwaffe he faced was not, fortunately, the same one RAF units had faced against Wolfram von Richthofen, either in quality of leadership, clarity of purpose, freedom of action, or quantity of aircraft. Nonetheless, it would not be easy to defeat.[7]

Tedder's first new innovation after Operation Compass was the creation of an advanced wing comprised of one fighter and four bomber squadrons that gave direct support to troops and cooperated closely with General Headquarters (GHQ) Mobile Division (renamed the Eighth Army on 24 September 1941 under Lieutenant General Alan Cunningham). The GHQ also had an army cooperation component with one Lysander squadron. However, in practice the air officer commanding (AOC) advanced wing exercised command and control (C2) of these planes in close cooperation with the ground commander. This was the vital first step toward truly joint air-ground operations.[8]

British intelligence advantages were also substantial because the Government Code and Cipher School (GCCS) at Bletchley Park had long since broken virtually every Italian code, including those of the armed forces.[9] At the same time, small army units known as air intelligence liaisons (AIL) joined reconnaissance and fighter squadrons. They worked with aircrews to maximize intelligence collection. A relatively sophisticated phone and radio network supplemented these efforts. Each squadron had a direct phone link to the division or corps headquarters with which it worked.[10]

The Long Range Desert Group (LRDG), formed in July 1940, also contributed much air-related intelligence. It had its origins in overland journeys of 6,000–7,000 miles, led by men such as Major R. A. Bagnold (the LRDG's first wartime commander), to map the Great Sand Sea and other areas of

southern Egypt and Libya. Bagnold and Major Guy Prendergast, his succes-
sor, reconnoitered and raided Axis airfields. The intelligence they brought
back, the confusion they sowed, the dissipation of Axis forces they caused
in protective efforts, and the damage they inflicted were substantial. By August 1941, LRDG units were operating in force. Wireless communication
ensured effective C2. The LRDG also rescued hundreds of downed RAF airmen. After Captain David Stirling and his L Detachment Special Air Service
(SAS) began working with the LRDG before Operation Crusader, it was
unparalleled as a reconnaissance and surveillance force and a potent airfield-
raiding asset. As with any special-operations force, the LRDG's purpose was
to conduct tactical missions that created strategic effects.[11]

Interrogation of prisoners of war (POWs) was also valuable and occurred
in three phases: immediate interrogation after capture, detailed interrogations of selected prisoners to obtain specific information, and interrogation
of prisoners with information of strategic importance. In August 1940, the
Combined Services Detailed Intelligence Centre (CSDIC) became active near
Cairo. By the start of Operation Crusader in November 1941, it was obtaining valuable intelligence, sometimes from concealed microphones in senior
POWs' quarters.[12]

Rommel's First Offensive Accelerates

These innovations, although crucial, could not produce victory. Only the
army could do that, and Rommel continued to defeat it. As Rommel's offensive to retake Cyrenaica began, Arthur Longmore sought to disrupt his
flow of supplies—something the British had as yet been unable to do but
that they understood held one of the keys to victory. The main effort took
place against Tripoli. Damaging its port was essential to slowing the arrival
of supplies in Libya. Longmore sent nine Wellingtons back to Malta for this
purpose even as the island was under Luftwaffe attack.

Churchill's directive of 14 April ordered the Mediterranean Fleet to stop
all seaborne traffic between Italy and North Africa. Heavy losses were acceptable. The best available fighters would fly to Malta to protect the navy.
Despite generally good air-naval coordination, Admiral Andrew Cunningham felt the results would not justify the risk. He wanted a battleship at
Malta along with cruisers and destroyers, so they could respond quickly
to enemy convoy activity, but he would not do so until the RAF had two
squadrons of the best fighters in place. Given the Luftwaffe's ongoing raids,
this was impossible, so the flagging effort to destroy shipping persisted.[13]

Meanwhile, Rommel's offensive had destroyed nearly every British tank by 20 April as the tanks impaled themselves on armor and antitank gun ambushes. Churchill responded to the crisis in Cyrenaica by ordering the Tiger Convoy to deliver tanks and aircraft directly across the Mediterranean to Alexandria. Intelligence sources had confirmed that most Luftwaffe aircraft on Sicily would already have moved east for Operation Barbarossa. Axis planes attacked the convoy several times, but the only vessel sunk was claimed by a mine west of Malta on 9 May. Another convoy had sailed from Alexandria to Malta at the same time, which forced remaining Axis aircraft to split their efforts. In addition, both convoys had several antiaircraft artillery (AAA) cruisers that put out immense firepower. These ships and other escorts were fitted with radar and had good radio communications with supporting RAF aircraft. These included a newly arrived Beaufighter squadron that raided Axis airfields in Sicily, attacked ships with great effectiveness, kept reconnaissance aircraft away from convoys, and engaged in direct defense.[14]

Longmore Departs; Tedder Takes Command

Just before the invasion of Crete and the Iraqi coup he had predicted, Longmore returned to Great Britain on Churchill's orders. Tedder became acting air officer commanding in chief (AOC in C) with Air Vice-Marshal Peter Drummond, Longmore's former senior air staff officer (SASO), as his deputy. Longmore left on 3 May, never to return.[15] On 1 May, he told Tedder that Churchill had recalled him for consultations regarding the changed situation in the Middle East. Tedder recalled Longmore's concerns about this development while the Afrika Korps and Luftwaffe were halfway through Cyrenaica, and the Middle East appeared on the brink of an uprising and German intervention.[16] On 30 April, the day before Churchill recalled him, Longmore showed Tedder a copy of a signal in which he asserted that Royal Air Force Middle East (RAFME) was not getting a fair share of aircraft. This pugnacious character was one of Longmore's strengths but contributed to his downfall because his habit of sending direct and acerbic signals found few allies in the War Cabinet. Tedder recalled Wilfrid Freeman's guidance to "tell Longmore that we fully appreciate all his difficulties, but we are becoming tired of moan, moan, moan."[17] Longmore's 30 April message was likely the straw that broke the camel's back—or raised the bulldog's ire. He was downcast and told Tedder that he hated leaving him with such a messy strategic situation. Tedder wrote in his journal, "It is a mess . . . but L. is certainly not to be blamed. If anyone is blameworthy it is the people at home who refused to send adequate forces out here many months ago."[18]

Bristol Beaufighters. These exceptionally capable and lethal long-range fighters were at the top of Tedder's aircraft priority list along with heavy bombers to carry the air effort to more distant Axis ports in Italy. Deliveries of both proved slow despite Tedder's constant pleas and his statement that given the huge theater of operations, "in nine out of ten operational problems at the beginning of my period of command the first and most crucial question was—could we reach it?" By summer 1942, the RAF had sufficient Beaufighters and heavy bombers to do what Tedder intended them to do, from port and ship attacks as far away as Naples to long-range attacks on Axis vehicle convoys heading to the front. (Imperial War Museum Image No. CH 17873)

No work of history has yet given Longmore the credit he deserves for dealing skillfully with a nearly impossible situation. Longmore summarized RAFME's achievements:

1. Malta still held firmly.

2. The Royal Navy had complete ascendancy over the Italian Navy, the fleet base at Alexandria was safe from air attack, and the Suez Canal was fully operational.

3. The Italian military had been completely defeated in Libya, with 180,000 men taken prisoner and another 20,000 killed. This was the army's first victory of the war, and the RAF had played an important role.

4. IEA was in British hands, the Red Sea was safe for shipping, and the United States was delivering lend-lease.

5. The British had done the right thing, morally, in Greece, despite defeat.

He attributed these achievements to effective interservice cooperation, the particularly important role of land-based airpower within the context of the huge theater, and excellent personnel. If he erred in his assessment of Greece's strategic value (Churchill and Alan Brooke later said the mission was a strategic blunder), Hitler's disillusionment with airborne operations against Crete proved vitally important in saving Malta.[19]

In his second dispatch on operations, written in November 1941, Longmore emphasized the close interconnection between occupation of territory along the Mediterranean littoral and the commensurate ability of airpower to interdict and in some cases stop air, sea, and ground movement, or to exact a huge price. This had a more powerful influence on the campaign than anyone had foreseen. Axis reconquest of most of Cyrenaica and its airfields brought home this lesson, as had the near encirclement of Greece and Crete by Axis air bases. The mobility of the desert campaigns had also driven home the importance of having highly motorized RAF units capable of leapfrogging. Longmore had also been sobered by the loss of aircraft on the ground in Greece, especially those the RAF had to abandon, and began improving RAFME supply, salvage, and mobile repair capabilities. After this process began, the RAF lost thirty-one aircraft on the ground during the remainder of the Western Desert campaign—a two-year period—while Axis air forces lost nearly 2,100 in the same way. Finally, Longmore cautioned his readers about the "time-lag factor": the need to plan ahead carefully given the slow reinforcement pipelines.[20]

RAFME had lost 345 aircraft during Longmore's command. In turn, the RAF destroyed 511 Italian Air Force (IAF) machines and another 89 unconfirmed along with 338 German aircraft and 71 unconfirmed. Another 9 unidentified aircraft brought the totals to 858 planes confirmed and 160 probable in addition to the 1,100 overrun on airfields.[21]

The loss of Longmore would have been a serious blow had Tedder not been there. He had been Longmore's de facto operational commander and thus combined that advantage with his deep strategic insight. Tedder told Charles Portal, "Will do my damnedest to justify your confidence."[22] He would not disappoint. Tedder had more than strategic acumen and toughness to bring to the fight. He also had a grudge. The title of his autobiography, *With Prejudice*, gives us some inkling of this. His personal tragedies during the war drove him forward. As a USAAF officer who knew him said later, "Air Marshall [*sic*] Tedder has lost his wife and two sons in the war. His main purpose now is to kill Germans."[23]

Portal placed Tedder at ease, telling him that he should deal with the complex situation as he saw fit. He followed with some thoughts:

1. The Luftwaffe and IAF were "at full stretch" operationally and experiencing fuel shortages as well as severe problems keeping airframes and engines serviceable. However, in spite of these problems, the Germans thus far in the war had managed to achieve good results by concentrating all of their available assets on a definite objective—inevitably air superiority followed by close support of ground forces and longer-range interdiction. [Their failure to do so in North Africa was thus all the more significant for the course of the campaign.]

2. Commonwealth forces had central position in Egypt and thus had a major advantage. However, there did not seem to be a joint-forces plan to capitalize on this.

3. Bold and concerted raids on enemy LOCs, especially shipping, ports, and vehicle convoys, might well make it impossible for Rommel to operate large forces in the desert.

Portal concluded, "I have complete faith in your judgment, determination and ability to lead the Air Force in the Middle East, and you can rely on my not interfering but supporting you in any bold and well-conceived joint operation whatever its result may be."[24]

By background, training, and temperament, Tedder was an excellent AOC in C. A flier in World War I, Tedder was intelligent, energetic, and inquisitive, but also introspective, quiet, and calm. During that war, Tedder authored a book on the German Navy, making him (along with John "Jack" Slessor, who authored *Air Power and Armies*) one of a very few intellectual officers to achieve high rank. Tedder's instructorship at the RAF Staff College from 1929 to 1932 allowed him to share his evolving ideas with his students. From there, he was assigned to the Air Armaments School.[25] Following this, Tedder became director of training at the Air Ministry Armament Branch in 1934. Among his other duties here, Tedder established Civil Flying Schools, adhering to Central Flying School standards to produce reserve pilots. This alerted him, based on peacetime "wastage," to how important aircrew training pipelines would be during another war. In July 1938, he became the director general of research and development for Freeman, whom Tedder credited with leading the successful RAF prewar expansion and its continuation during the war.[26] Tedder saw this job as a defining moment in his development as an officer and in terms of Freeman's mentorship.

Freeman promoted Tedder's abilities and helped to keep Churchill from firing him.[27]

Fortuitously, Tedder also got along very well with Portal, who became a mentor, promoter, and protector. The two men began sharing a "long series of heartening, straightforward, personal messages, which we exchanged at short intervals for the rest of the war."[28] Between Portal, Freeman, and other supporters, Tedder had strong backing.

Tedder understood the grand-strategic stakes and the optimum employment of air assets. Egypt was the single most vital point—the crossroads of the empire. Its loss would be a disaster for the entire imperial position. The safety of Iraq and Iran was equally important given the region's oil, and Egypt once again acted as a block to Axis designs there. The "northern flank," comprised of Iraq and Turkey, was thus equally vital and a constant distraction. Germany's attack on Russia made this vulnerability even clearer. Tedder emphasized that the defense of the Middle East required an "aggressive-defensive" posture and ultimately a fully offensive one. IEA, Syria, Iraq, Iran, and the air-sea battle for SLOCs were or would soon be examples of the former. The effort to defeat Axis forces and drive them from Africa was clearly in the latter category, as was knocking Italy out of the war, opening a second aerial front against the Reich, and exploiting other strategic opportunities.[29]

Taking airfields in Cyrenaica and Tripolitania, Tedder said, would restore control over the central Mediterranean, allow for the relief of Malta and its buildup as a major base for the capture of Sicily, and thus help to force Italy's surrender. To achieve these ends, Tedder had forty-four and one-half squadrons on 3 May 1941—650 combat aircraft and another 400 training and communications planes. There were also aircraft in the pipeline from Takoradi and in repair depots in the Nile Delta. However, the situation was not rosy. The disaster in Greece had taken a severe toll. Seven squadrons from that campaign and one from North Africa were refitting. Replacements were arriving too slowly. The repair organization in Egypt was not as capable as Tedder wished. Many squadrons had obsolete planes and awaited modern types.[30]

The ranges in theater were so great that only Wellingtons could strike distant targets. They had to operate at night to survive. Even more problematic was the absence of the most capable long-range photographic reconnaissance (PR) aircraft, which were still in England along with all the Spitfire fighters. A squadron of Beaufighters represented Tedder's entire long-range fighter force. He lamented that "in nine out of ten operational problems at the beginning of my period of command the first and most crucial question was—could we reach it?"[31] His ancillary problems, given the shortage of

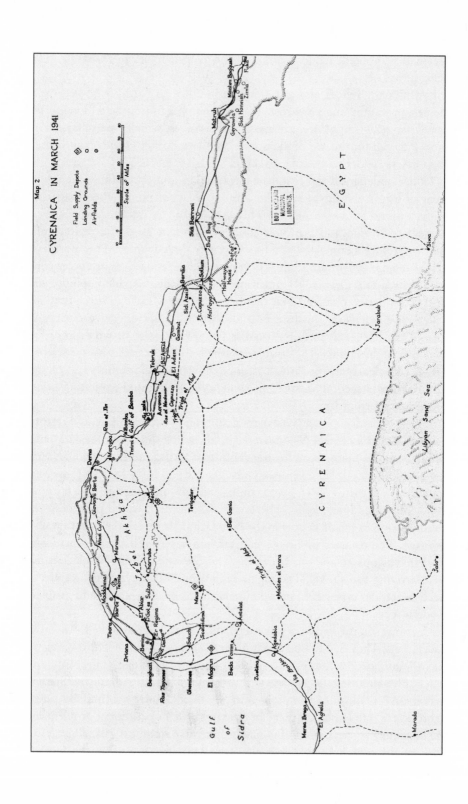

Map 2

CYRENAICA IN MARCH 1941

Field Supply Depots ◈
Landing Grounds ○
Air-Fields ⊙

Scale of Miles
10   0   10   20   30   40   50   60   70   80

reconnaissance aircraft, were: "Can we find it?," "Can we target it?," and "Can we assess the effects and effectiveness of our operations?"

Against Tedder's 650 aircraft (450 operational), the Germans had more than 1,200 in theater during and immediately after the Balkans campaign, and the Italians another 600. Given a 50 percent serviceability rate, this gave the Axis more than 900 planes. But they failed to concentrate and attempt a knockout blow before the withdrawal of aircraft for Operation Barbarossa. Even after these withdrawals, the Germans retained a numerical advantage, central position, and interior lines of operation, whereas every British aircraft had to go through Takoradi, around the cape, or across the perilous Mediterranean route. The Germans also had more than 200 transport aircraft to bring in spare parts and other supplies. Tedder had 32. RAFME's area of responsibility stretched from Malta to Iran and from IEA to Italy. Tedder had to plan for immediate assistance to Turkey should the Germans attack there. Still, there were several bright spots. Malta remained a huge thorn in the side of the Axis. Tedder emphasized Malta's vital role, with its position in range of ports in Sicily and Libya and of the SLOCs linking them. "No investment of fighter aircraft," he said, "would earn a richer dividend than one which ensured the safety of our only base in the Central Mediterranean for carrying out reconnaissance, interrupting the enemy's supplies to Africa, and refueling reinforcement aircraft en route to Egypt."[32] General William Dobbie, governor general of Malta during the first major German air effort, simply said, "By God's help, Malta will not weaken."[33]

On 27 May, Tedder said, "I have been trying for the past three weeks to rub it in to [Archibald] Wavell and [Alan] Cunningham that this war is principally one for air bases."[34] He pushed tirelessly for joint operations to recapture air bases in Cyrenaica so the RAF could make a maximum effort against enemy logistics. Malta became an offensive base once again in May 1941 as Luftwaffe units headed east. In July, as Malta recovered, Axis shipping losses climbed to 16 percent of the total dispatched. The August figure

(*Opposite*) The Cyrenaican hump. Note the heavy concentration of airfields here. Whichever side controlled these airfields tended to have the upper hand in the "battle for the SLOCs" because the RAF had trouble reaching key convoy areas unless it could engage in offensive operations from both Malta and these airfields in Cyrenaica. Every major British ground offensive gave the highest priority to capturing airfields so the RAF could play its full role in the combined-arms effort, whether direct support to the army or ship-attack missions. (Originally published as Map 2 in Ian Playfair et al., *The Mediterranean and Middle East*, vol. 2, *The Germans Come to the Help of Their Ally (1941)* [London: Her Majesty's Stationery Office, 1956], 1, now public domain)

was 33 percent. This increased to 40 percent in September and 63 percent in October. Surviving ships delivered far too few supplies to facilitate offensive operations. Agent reports from Italy tipped off intelligence officers about convoy preparations, allowing them to plan air reconnaissance at the right times and places. Ultra intelligence did the same, and reconnaissance aircraft flew over convoys located by Ultra to keep that source secure. Agent reports also provided information on port operations, making bombing raids more effective. Axis convoys were soon skirting the West Coast of Greece and then dashing for Libya, which complicated intercepts but also lengthened convoy pipelines and sailing times. Consequently, supply tonnages arriving during Operation Crusader were an abysmal 8,510 tons (23 percent of total shipped) in November and 18,360 tons (51 percent) in December.[35]

Regarding Axis air attacks on Allied shipping, Tedder was surprised that as the RAF and Fleet Air Arm (FAA) achieved increasingly impressive results, enemy air forces were not successful in interdicting either Suez or British east-west shipping to resupply Tobruk. Despite geographical advantages and more aircraft, their air units did not engage in a concerted effort against ships at Alexandria or Suez, which German bombers could reach from Crete. Bombers attacked episodically, but dissipation of effort led to failure.[36]

With the battle for the SLOCs in hand by September, Tedder reorganized RAFME to make it capable of controlling the air operations envisioned for Operation Crusader. At his request, the UK Air Ministry Establishments Committee toured his command. The establishment of Air Headquarters Western Desert (from Raymond Collishaw's No. 204 Group), under Air Vice-Marshal Arthur ("Mary") Coningham, was perhaps the single most important change. The development of tactical (Air Headquarters Western Desert—the WDAF), strategic (No. 205 Group), and coastal (No. 201 Naval Cooperation Group) components of RAFME in early October made it more effective. Self-contained mobile wings sped the movement of fighter squadrons. Rapid salvage and repair and mobile field units also evolved. The three operational training units (OTUs) in theater, however, could not train enough pilots, so Tedder relied heavily on aircrews from the United Kingdom. No. 253 Army Co-operation Wing was set up to maximize air-ground coordination. The air-support control (ASC) organization that emerged ultimately drove great improvements in ground-air cooperation.[37]

A central element in this increasing combat effectiveness was leadership. A prominent scholar contrasted leadership styles by employing Rommel and Tedder as archetypes of their two systems. Rommel's bold, impetuous, and exclusionary leadership proved less effective than Tedder's thoughtful,

Italian merchant vessels under attack by Beaufighters. These deadly aircraft were so heavily armed that they often sank ships such as these just with cannon and machine-gun fire aimed at the waterline. The "Beau" carried four 20 mm cannons, six 50-caliber machine guns, bombs, torpedoes, and, beginning in 1943, rockets. It was a superb long-range escort and a feared ship-killer. (Imperial War Museum Image No. C 3902)

methodical, and highly inclusive style. "Commanders," Vincent Orange noted, "should temper boldness with forethought and pay attention to logistics."[38] Orange argued that Tedder, Coningham, Keith Park, and most other senior RAF airmen fit this mold, whereas Hitler, Rommel, and some senior Luftwaffe officers in the Mediterranean theater did not. Senior RAFME commanders and their subordinates also tended to be unorthodox and creative. As one observer noted,

The Desert Air Force, along with its predecessors and successors, tended to produce a host of highly individual, even totally unorthodox, aircrew men; loners who might never have been acceptable to the selection boards of the rigidly disciplined—and often blinkered—pre-1939 mode of RAF life. Yet in the Mediterranean air war such men seemed to find a vocation in the wholly unbureaucratic atmosphere of any operational unit, where the book had little relevance and life revolved around the rock-basis of fighting—and, of course, survival.[39]

Intelligence and Communications

The degree to which commanders appreciated and employed intelligence and communications capabilities also proved vital. Tedder was a voracious consumer of both. Although both sides had substantial capabilities, the British had a more holistic view of intelligence as a vital asset at all levels of war and an ace in the hole with Ultra. Although Ultra was not always important, it was a major force-multiplier, particularly in North Africa where all long-haul communications went by wireless transmission. It ultimately became the most important source of strategic intelligence, helping commanders to understand the pivotal importance of logistics. However, it usually lacked the detail required to plan and execute operations. Intelligence personnel thus relied heavily on other sources, a learning process with frequent setbacks as the Germans changed codes. German Army ciphers were particularly difficult to read before May 1942, although most Luftwaffe traffic, along with some Italian traffic, was wide open, giving the RAF and Royal Navy major advantages. One of the most critical breakthroughs occurred in July 1941 after the British broke the German-Italian C38m machine cipher, which contained information about all convoy sailings. Ironically, the Germans did not trust the Italian convoy cipher, which was in fact unbroken, and convinced their allies to adopt the "superior" C38m. Combined with the departure of most of Fliegerkorps X for Greece and many other units for Russia, this led to a massive increase in shipping losses. Because the GCCS had sent experienced intelligence officers to Cairo in January 1941, they were able to route these and other important decrypts to senior officers with great speed. Their expertise maximized the utility of Ultra decrypts for commanders. Secure wireless channels from the United Kingdom to the Middle East also helped beginning in summer 1941. Most Allied senior commanders became familiar with Ultra while in the Mediterranean and put it to use there and with even greater effect in northwestern Europe and during the heavy-bomber campaigns.[40]

Air reconnaissance was also weak initially but improved rapidly. A shortage of PR planes and a reliance on high-frequency (HF) radios to communicate enemy positions and movements proved problematic. Air Marshal Patrick Dunn noted that although a PR facility was in operation from the start of the war, it took two days to deliver statements (without photographs!) to RAF units. Fortunately, Italian units usually stayed in the same place to be attacked. PR units were not entirely effective until 1942.[41] Effective ground control of aircraft was impossible until very high-frequency (VHF) radio sets arrived in early 1942.[42]

These problems led to a series of crises in battle with the Germans, whose rapid and decisive ground actions often caused a breakdown in C2 as radio sets and lines became congested, and orders, along with air-support requests, failed to get through. They became glaringly clear during Rommel's first offensive. The German Y-Service equivalent (part of the Signals Intelligence Service, or SIS) was superior at this point. Signals and security posed the greatest obstacles to effective C2. The Germans were adept at breaking British (and later US) codes, and their ability to reconstruct orders of battle was superb. They benefitted further from intercepts of Colonel Bonner Fellers's messages originating in Cairo, where he served as military attaché until the United States discovered the leak on 10 July 1942. His reports contained so much operational information that Rommel referred to them as the "good source." In addition, British cipher operations were slow until Operation Crusader, during which messages marked "Most Immediate" could take twelve hours to get to commanders. Poor signals discipline contributed to this problem, as messages flooded in with no prioritization and too few trained operators to process and disseminate them. The resulting delays often made messages irrelevant or even dangerous.[43] Despite these growing pains, the combination of effective intelligence, the LRDG's reconnaissance, and a rapidly improving C2 network became the building blocks of a superb air-ground team. Yet even the best C2 and air capabilities have limited operational effectiveness if the army is outclassed by its adversary. This was the case until El Alamein.[44]

Tedder's and Coningham's flexibility allowed RAF assets to attack the best targets after air superiority was in hand. Until German vehicles had to concentrate at the front to repel ground attacks and became vulnerable to air attack themselves, RAF assets focused on vehicle convoys to weaken enemy logistics. Improvements in communications and intelligence facilitated this increasingly agile employment of airpower and improved combined-arms operations.[45] One of Coningham's wing commanders noted of Tedder's command style, "Except in exceptional circumstances, his instructions to me as OC 258 Wing would be in the form of general directives, often verbal, and that all detailed orders in the squadrons would be my responsibility."[46] This was a major departure from the *Field Service Regulations*, last updated in 1935, which called for meticulous planning and centralized control and execution. The RAF broke from this approach, and the army soon followed.[47]

However, the RAF and later the USAAF suffered initially from C2 and intelligence-dissemination shortfalls. The three services still had separate wireless organizations without an effective liaison function. All had expanded rapidly and overlapped in terms of operations but not in terms of sorting and routing signals. The Middle East Combined Signals Board along with interservice and inter-Allied wireless and line subcommittees soon began fixing this. These augmented the RAF's existing point-to-point system with five comprehensive all-services and Allied-network centers in the United States, United Kingdom, Egypt, the Far East, and India. Each had special cabling and wireless capabilities to handle all traffic, including Y Service intercepts, quickly. The system also proved crucial for tracking aircraft and spare-parts deliveries.[48]

The Y Service profited immensely from these improvements. Its operators required intensive training and were generally not operationally effective until they had apprenticed for a year. Their job was much more complex than that of radio-telephony (RT) interception units, which could often simply deliver their intercepts of enemy radio transmissions to commanders for action. Two Y units and another two RT units stationed at Heliopolis contributed substantially to Operation Compass. Together, they constituted the Combined Bureau Middle East, a functional equivalent of Bletchley Park.[49]

RAF and army Y operators played a major role in all combined-arms operations. They determined enemy order of battle while tracking enemy aircraft outside of radar range, ensuring they did not surprise British aircraft or troops. Their ability to understand Luftwaffe tactical and operational employment was also of vital importance. The relationship between Y and radar units was thus very close as the two tracked enemy air activity and warned aircrews.[50]

Ultra intelligence also became more useful as the war progressed. After initial transcription of Enigma decrypts, relevant ones went forward to special liaison units in theater for briefing to commanders. Bletchley Park read Fliegerführer Afrika's daily signals on aircraft readiness rates and the operational situation. The dissemination and use of this intelligence spurred the formation of a joint headquarters to exploit the advantages thus gained. One RAF officer opined, "Ultra in the Med was really ace. With no land lines, everything had to go over the air and Bletchley got the complete picture. [It was] the one theatre in which it was 100%, if anything can be 100%."[51] Hut 3 at Bletchley Park delivered the complete German air order of battle to RAF commanders, and their army counterparts had Rommel's latest fuel, ammunition, and food returns.

PR and photographic intelligence (PI) were also increasingly valuable and by Operation Crusader were becoming the most important sources of operational intelligence. They located troop and supply concentrations and movements, provided target materials, and were indispensable for damage assessments. Although the Mediterranean theater lacked anything like the Central Interpretation Unit at Medmenham until 1942, the process evolved rapidly. From having no trained PIs in the fall of 1940, commanders had a small but capable PI section in place for Operation Crusader.[52]

In the communications arena, the RAFME chief signal officer (CSO), Group Captain William Mann, found capabilities on Malta barely adequate. Cables had to be buried deep underground to protect them from bombing. The Middle East theater had highest priority for delivery of communications personnel and equipment by this time. Accordingly, the War Office and other RAF CSOs agreed to send 4,000 communications specialists to the theater. A three-man operations research section sent to analyze the signals system in detail also paid dividends in the form of more efficient and denser communications networks.[53]

The theater CSOs' committee also agreed that VHF communications for air-defense and combined-arm purposes must be brought as close as possible to UK standards. Fighters required VHF for tactical communications and reconnaissance, prompting the delivery of seventy-five of the latest sets to Egypt. Hurricane IIs, already wired for VHF, were the first to receive them. More than 400 mobile VHF vehicles soon joined army formations to facilitate communication with VHF-equipped "Hurribombers"—the first dedicated fighter-bomber. Major General William "Ronald" Penney, Mann's army counterpart, worked alongside him to ensure these improvements went into place. All sector headquarters and flying squadrons had these vehicles by summer 1942. Finally, wireless observer units with special vehicles reported on enemy aircraft activity along with the latest air-intercept (AI) radar and ground-controlled-intercept (GCI) radar sites. Three height-finding/direction-finding (HF/DF) radar sets per month also arrived starting in October 1941.[54]

Rommel's First Offensive Culminates

As British air-ground cooperation improved, Rommel's forces pushed forward in the face of heavy air attacks. Blenheims operating in pairs attacked between El Agheila and Benghazi. Single-engine fighters took it from there. Fighter-bombers attacked from the rear, at very low altitude, and usually out of the sun. Collishaw lost only four fighters during this effort. After additional fighters became available in summer 1941, two squadrons made these raids together, whereas two planes had done so during the worst of the aircraft shortage. One squadron attacked while the other provided cover. The Eighth Army had such bad attrition that the RAF, even with its aircraft shortage, played a central role in halting Rommel's advance. His supply convoys could not endure the constant raids and had to disperse.[55] This was the first time Rommel experienced the dual disadvantage of air and logistics shortfalls. He later said, "There is no compensating for the lack of an air force or for shortage of supplies."[56]

The Luftwaffe concentrated its fighters at Gazala and gained air superiority over Tobruk, though doing so yielded no result. Further, it played a purely defensive role in trying to keep the RAF from attacking vehicle convoys. Fighter sweeps between Gazala and the front were predictable and easy to avoid. The Luftwaffe did not have enough fighters to fly sweeps and standing patrols everywhere, or enough ground radar sets, so commanders adopted an air observation system that scrambled fighters only when they received sighting reports. They were thus largely reactive and ineffective. The most useful countermeasure to low-level raids was to station armored vehicles with antiaircraft machine guns at five-mile intervals along convoy routes. These were unlikely to shoot down aircraft, but pilots were at greater risk and less able to make their preferred strafing approaches. However, this also removed key assets from the front. The defensive posture was costly in resources and denied air units any opportunity to defeat the RAF.[57] Conversely, the RAF continued raiding airfields without pause. As one scholar noted, "The most effective support the R.A.F. could render . . . was by increasing its scale of attack against enemy aerodromes; British aircraft concentrated against Derna, Gazala, and Benina, from the second half of April [1941]."[58]

The RAF also denied Axis forces the use of Benghazi, as the Luftwaffe had done to Commonwealth forces. Wellingtons attacked almost nightly and Blenheims during daylight. Most supplies thus came from Tripoli—a 1,000-mile journey. This was difficult because of vehicle shortages after the Italian Army's defeats; the amount of fuel the vehicle convoys used in the process of delivering that commodity along with ammunition, food, and reinforcements; the wear and tear on trucks; and the danger posed by the RAF after the convoys reached El Agheila and Benghazi. Captured Axis airmen said their planes were constantly short of fuel as a result of nonstop attacks on vehicle convoys, ports, transport aircraft airfields, and MVs. The RAF changed attack locations and times constantly to frustrate enemy efforts to provide air cover.[59]

On 18 April, General of Fliers Hoffman von Waldau, Oberkommando der Luftwaffe (OKL) chief of staff, visited Fliegerkorps X and Fliegerführer Afrika to assess the state of affairs. He did not like what he found. His biggest concern was the RAF's clear air superiority over the Western Desert. He called for additional Me. 109s to change this, but none arrived. Further, he emphasized the exhaustion brought on by too many mission types and a very high operations tempo. Supply and maintenance problems were serious as was the shortage of *Baubattalinen* (airfield-construction units), vehicles, and flak guns (Rommel had taken most of the flak and many of the trucks for the army). This proved highly problematic not just for the Luftwaffe but for Rommel, who received word from General Franz Halder, chief of the German General Staff, on 24 April that he would have to fight with what he had and with available air support. Given this impasse, Rommel settled down to a long siege of Tobruk, while the Luftwaffe struggled with its many missions in the central Mediterranean and North Africa. The only good news was an increase of two Ju 52 transport *Gruppen*, but these devoted most of their attention to flying in reinforcements and flying out the wounded.[60]

Operations Brevity and Battleaxe

With Rommel's first offensive over, the Middle East commanders in chief (C in Cs) looked for ways to regain the initiative. Although the operations intended to do so—Brevity and Battleaxe—failed, they reinforced critical lessons. Tedder had incorporated lessons from Operation Compass into RAF operations as he envisioned and built future air capabilities and optimum cooperation with the Eighth Army. He emphasized the importance of technological parity—something the RAF had during Operation Compass only because of the last-minute arrival of Hurricanes and did not have again until El Alamein. Desert warfare was immensely destructive of aircraft and vehicles. This required a large and reliable reinforcement pipeline. Tedder blamed much of the wear and tear on Collishaw's leadership style, which involved "bullocking units about at little or no notice, issuing orders for operations and moves without consideration or warning. Despite repeated pressure Collishaw flatly refused to use the quiet periods to give any of his personnel or equipment any rest. We've had to pay a heavy price in withdrawing squadrons for rest and re-equipment when we ought to have been worrying the Hun."[61] This dissatisfaction with Collishaw's methods was one of several reasons Tedder later replaced him with Coningham.

Tedder also understood the difficulty of sending reinforcements during Operation Compass but could not comprehend why so many P-40 Tomahawks had gone to Greece, Turkey, and China, with 200 held at home "pending development of the situation." Tedder believed the "situation" had "developed" more than enough to justify sending the Tomahawks. He appreciated the importance of keeping Turkey firmly neutral and resisting Japanese aggression in China, but his conviction that victory in the Middle East was crucial to the larger war effort trumped all else.[62]

Tedder's most serious shortage was air transport. He was envious of the

Luftwaffe's Ju 52 fleet and lamented that he did not have four times as many obsolete Bombays as he had.[63] He had been working since June to develop a centralized air-transport capability by combining RAF and British Overseas Airways assets. The "Tedder Plan," though unable to produce anything like Luftwaffe sortie counts in 1941 and 1942, resulted in potent capabilities behind the battlefront. Additionally, the RAF had what the Luftwaffe did not: adequate ground transport. Operation Compass had taught Tedder to seek air and ground transport assets everywhere.

Tedder was satisfied with the RAF's performance during the retreat across Cyrenaica. Airmen had operated without diminution of sortie counts for seven crucial days while moving to new landing grounds each evening. Fighters had concentrated their limited strength against an overly dispersed Luftwaffe and gained air superiority everywhere but over Tobruk. Leapfrogging was working well.[64]

Tedder also detected serious weaknesses in the Luftwaffe's employment of the Stuka without air superiority and also in terms of its basic effectiveness as an attack platform. On 6 May, Churchill asked what kinds of targets Stukas were striking and whether raids were preplanned, close support, or a combination. Freeman told Tedder, "Today I listened to a tirade about how we lost Greece and France through the enemy dive-bomber, and would like you to send me as soon as possible any evidence which you have got to the effect that the dive-bomber is all noise and in actual fact does no marked damage whatsoever. My own opinion is that a dive-bomber does not account for one human being on each sortie. If you can get evidence from the troops to this effect please let me have it as early as possible, but not by signal. Let's have the dope."[65]

Tedder replied that Stukas had traditionally gone after virtually every kind of target, in roughly the following priority: ships, road junctions, transport columns, railways, bridges, AAA protecting airfields, headquarters, artillery, individual tanks or vehicles, and troops. However, in North Africa, they were tied heavily to direct support. Stukas were extremely vulnerable to AAA and fighters but could be very effective when massed against larger targets, such as ships. They often attacked in less than squadron strength, with one or two machines going after a target. This was uneconomical and dangerous, especially without air superiority. Stukas were in close contact with troops and focused almost entirely on supporting them, regardless of the suitability of targets they engaged. Tedder concluded that evolving RAF fighter-bombers and light-bombers should have the ground-attack role. He had no use for dive-bombers.[66]

A number of troops with experience of Stuka attacks in Greece and Libya said they did negligible damage. When Tedder asked troops who had been at El Agheila about this, they said that if the Luftwaffe wished to continue with such an "expensive and harmless amusement," it was welcome to it. They had shot down twenty-seven Stukas in a two-week period and lost three vehicles in return. In the face of air opposition and AAA, the Stuka was of limited value. Conversely, the low-flying Me. 109 was a huge menace, killing and wounding many troops while knocking out numerous vehicles. "We hated them," one soldier said. Tedder pointed out that low-level Hurricane and Blenheim attacks on Axis columns had the same effect and often caused them to retreat. Therefore, Tedder asked Freeman for Hurricane IIs and Beaufighters. As the British pursued fighter-bombers with the utmost energy, the Germans never did. There were only 121 in 1943 and barely half of that during 1941 and 1942.[67]

The gigantic list of Stuka target types underscores Luftwaffe commanders' inability to make optimum use of these vulnerable but potentially deadly planes. They did not understand how to prioritize attacks during air operations and had apparently learned little from Stuka attacks on British ships around Greece and Crete, where they had done enormous damage and forced the Royal Navy back to Alexandria. This latter point is crucial given the central importance of the SLOCs to victory in the Mediterranean Basin. Had the Germans employed their Stukas in coordinated strikes with adequate numbers of other ship-attack assets such as FW 200 and Ju 88 torpedo-bombers, they would have contributed much more to the campaign.

Meanwhile, events in Greece and Libya prompted the army to make another bid to exercise direct control of air units at the corps and army levels. Portal's staff calculated that meeting the army's stated requirements for a specialized ground-support air fleet would require ninety-eight squadrons. Tedder had forty-four and one-half operational in the entire theater. Portal said that such a plan would likely lead to an insufficient focus on air superiority—the enabling factor for all combined-arms operations within the context of the current campaign.[68]

Churchill came to the RAF's rescue in a roundabout way. On 9 June, before Operation Battleaxe, he asked Wavell why he had not mentioned the RAF in his plans and wondered whether his staff was working with Tedder's. He emphasized that destroying Rommel's army was of paramount importance but impossible without close army-RAF cooperation. Churchill told Wavell that his fears about the RAF's inability to gain air superiority were groundless and that he needed to concentrate on joint operations. This ended army efforts to control the majority of RAFME assets and led eventually to clear statements from Churchill that no such thing would happen.[69]

Nonetheless, during Operation Battleaxe, the army found one more opportunity to exercise some measure of control over the Western Desert Air

Force (WDAF), as Tedder's forward echelon was now called. Portal warned Tedder that politicians and generals were ready to make a scapegoat of the RAF, especially if they felt there was any lack of close support. They were afraid the RAF would spend most of its time attacking enemy airfields and LOCs rather than providing protective umbrellas for advancing troops. Portal recommended concerted bomber attacks under heavy fighter escort against enemy troop and supply convoys, which would force German fighters into defensive tactical actions. Tedder agreed.[70]

During Operations Brevity (15–16 May 1941) and Battleaxe (15–17 June 1941), the RAF raided airfields and followed with attacks on ports and vehicle convoys. However, in a sign that the RAF and the army were not yet entirely in step, Lieutenant-General Noel Beresford-Peirse requested that fighters establish defensive patrols over his troops, while medium-bombers remained on call for attacks on enemy columns and vehicles in the battle area. Tedder agreed but quickly saw that employing air assets in this way constrained their flexibility, split them into small packets, and kept the bombers grounded much of the time for lack of escorts. Nonetheless, Hurricanes established local air superiority in conjunction with airfield raids, and the army had air support going into the battle. German air units were split, with Fliegerführer Afrika bearing most of the burden and Fliegerkorps X assets largely absent (although a Me. 110 squadron and two of Ju 87s arrived, but too late to help). Operation Battleaxe started well but failed in the face of excellent German antitank tactics.[71]

Wavell blamed the RAF for the failure of Operation Battleaxe, prompting Portal to ask Tedder whether there was any truth to Wavell's claims that "we never had air superiority" and that the RAF had failed to provide direct support, particularly at Halfaya Pass. Tedder said there was not. Portal needed to engage in damage control with Churchill and the War Cabinet. Tedder obliged by sending a series of detailed signals that proved Wavell's claims groundless.[72]

Tedder noted that the army never asked for close support at Halfaya Pass even though he could have had Blenheims there in ninety minutes. All aircraft radio calls to army units went unanswered, and there was no semblance of a bomb line. Consequently, it was virtually impossible to provide direct support. The RAF thus maintained its focus on controlling the air and attacking troop and supply convoys. Prior to 12 June, raids on airfields, ports, and LOCs accounted for nearly the entire air effort. Fighter attacks on vehicle convoys began on 12 June. There was virtually no enemy air activity. Fighters covered the army's withdrawal on 18 June without incident. Airfield attacks forced Axis planes to fly heavy standing fighter patrols, leaving almost no fighters to escort bombers or carry out offensive operations.

Even the ubiquitous German fighter sweeps were absent. The RAF fighter umbrella, in turn, was far too aircraft-intensive and parceled out into small contingents. When German fighters did appear, RAF pilots were outnumbered and suffered higher losses. Tedder said the tactic made sense for the troop approach but was ineffective over the front. He emphasized, "Increasingly clear that crux of whole problem is communications."[73] Portal sent sixteen air control officers, three of them signal troops and the rest pilots, to fly specially equipped communications-liaison aircraft. This final insight was recognition that the RAF as well as the army had some distance to go in terms of communications, especially regarding direct-support missions. Despite concerted efforts by both services, the problem persisted until summer 1942.

Collishaw provided additional details regarding Operation Battleaxe. A captured German document disclosed that the air-warning system still comprised army lookouts who telephoned all RAF activity they saw to sector stations at Gazala and Gambut, which in turn alerted German fighters. Neither the Germans nor the Italians had radar at forward airfields, which handicapped them in actions against the RAF and made defensive umbrellas vulnerable.[74]

Collishaw called for more long-range fighters now that the Axis controlled airfields in the Cyrenaican hump. Beaufighters soon arrived, and self-sealing external and drop tanks became common as fighters evolved into the fighter-bomber role. Medium-bombers with a heavy complement of machine guns and cannons also began arriving. The Blenheim IV, which could reach Benghazi and also attack supply convoys on land and sea, was a prominent example. The humble twenty-pound fragmentation bomb also proved ideal for attacks on airfields and thin-skinned vehicles. Blenheims carried fifty. They did immense damage to aircraft and inflicted heavy casualties. At sea, the combination of light-case 250-pound bombs and cannons mounted in aircraft proved lethal as the bombs smashed up the ships' controls while cannon fire directed at the waterline often sank smaller MVs on its own.[75]

Finally, aircraft security received emphasis. Aircraft left operational airfields at dusk every evening, landed at "mud-flat landing grounds," and returned at daybreak. Axis aircraft evidently never discovered this ruse because they did not attack a single one of these night dispersal fields during their sporadic airfield raids. Radar, GCI, and observers kept airfields secure during the day. Aircraft dispersal was a minimum of 300 yards between planes. Fake landing grounds with 200 dummy aircraft and real flare paths lighting the "runways" also drew in Axis aircraft. From 19 November 1940 to 1 July 1941, Axis planes destroyed just four RAF aircraft on the ground.[76]

Portal supported Tedder's request to send experienced pilots from the front to instruct at OTUs.[77] Although unpopular with veteran fliers, this produced much better pilots. Neville Duke, the highest-scoring fighter pilot in the Mediterranean theater (twenty-eight kills), disliked OTU assignments but had two of them from the end of Rommel's Sollum offensive in April 1942 until well into El Alamein on 18 November 1942, and again as chief flying instructor of a Tomahawk and then a Spitfire section outside of Cairo from June 1943 to February 1944. "To my horror," Duke said of his second OTU stint, "I am taking over a Tomahawk Flight. Hell—what have I done to deserve this?"[78] The RAF's assignment of such pilots to OTUs at critical junctures (Duke missed Rommel's Gazala offensive, all three of the El Alamein battles, Sicily, Salerno, Baytown, and Anzio) was a painful but proper prioritization of effort. Conversely, the Luftwaffe's tendency to fly its best pilots to death was an equally strong indicator of improper use of scarce human resources.

The Battle for Aircraft Continues

As Tedder struggled to obtain enough aircraft, US MVs began delivering planes to ports along the Red Sea. Nine ships arrived in June 1941, thirty-two in July, and an average of sixteen every month through December. Deliveries increased sharply from there. The first US-made light-bombers reached Egypt in October 1941. Ships delivered fighters to Lagos instead of Takoradi, reducing congestion and allowing for quicker delivery using an alternate air route. The United States sent twenty transport aircraft at the end of October 1941 to move ferry pilots back and forth on the Takoradi route. They built facilities at ports and airfields to manage the flow of weapons, leaving the British free to concentrate on their own deliveries. Runways, roads, railroads, and ports sprang up from West Africa to Iraq. The RAF's front-line strength in the Western Desert doubled between November 1941 and November 1942 largely as a result of these deliveries.[79] Also, RAF salvage crews went after aircraft lost during Takoradi flights or in action, refurbishing more than 800 of the more than 1,000 scattered over 100,000 square miles.[80]

By March 1941, Slessor convinced the United States to send not just more aircraft but also volunteer radio and aircraft mechanics and ferry pilots. To help with the U-boat menace, Roosevelt also offered fifty PBYs (amphibious aircraft) with trained crews.[81] In a crucial development, the United States offered the RAF C-47s. As Henry Self of the British Air Commission noted, "The position in brief is that every transport aircraft we can get we shall take over."[82] The British built far too few transports but realized their great value during Operations Compass and Crusader. Slessor also convinced the

United States to provide single-engine trainers and flying schools for RAF pilots in the southern United States.[83]

In one of the most important meetings of his career on 22 March, Slessor set forth for the United States the complexities of the war in the Mediterranean and Middle East, and the reasons US aircraft were so desperately needed there. He began with the preponderance of obsolete aircraft types in 1940, loss of French air bases for transit, diversions to Greece, and the importance of completing the IEA campaign so US ships could sail the Gulf of Aden and Red Sea. His detailed explanation of the myriad problems facing the RAF resonated with US officials and led to even greater emphasis on aircraft deliveries.[84]

This came just in time. Two weeks earlier, Longmore had calculated that diversions to Greece, equipping the Turkish Air Force, and losses in the Western Desert would leave him with a deficiency of 460 aircraft. He asked Portal urgently for newer types or at least newer models of existing types to arrive as entire new squadrons. He also asked for enough planes to re-equip existing squadrons. Portal agreed to do what he could, including sending Slessor into the fray again to find more US machines.[85]

However, it was not all smooth sailing. In what must have seemed like an April Fool's Day joke to William Donovan, on 1 April Slessor passed him a private note from Portal saying that only the PBYs and Martin Marylands were in action. The Mohawks and Tomahawks were not. Portal explained that there was no way to speed the process given the many modifications required for operations in the desert. The Mohawk alone required about 1,000 modified parts and 700 hours of labor. Despite the US disappointment, they continued helping the British.[86]

However, even if British and US aircraft arrived in the hoped-for numbers, the disaster in Greece made it impossible for RAFME to rebuild its strength to the point where it could engage in offensive action for several months. The few squadrons left were hollowing out. Lower strength meant more intensive operations, more accidents, and disadvantages in air combat. Tedder had just two fully operational fighter squadrons. The rest were badly understrength and awaiting replacements. There was no pilot reserve. Total fighter strength was 84, with 112 Blenheims. Wellingtons were at three-quarters strength. Losses outpaced replacements.[87]

Portal said he would send enough planes to give Tedder 692 aircraft. Projected reinforcements included 224 Hurricane IIs along with 43 Tomahawks to ensure air superiority.[88] Tedder thanked Portal and reminded him that long-range aircraft were vital in his theater given the distances to enemy airfields and logistical assets such as shipping, ports, and vehicle convoys. Low-altitude attack of airfields had proven particularly effective, but at this point Tedder had only Hurricanes with external fuel tanks. He needed Beau-fighters and more bombers.[89]

By July, US representatives were arriving in greater numbers. Samuel D. Irwin of Curtiss-Wright, who toured RAF posts to determine how well his company's engines were performing, was the first. Tedder and Graham Dawson came to rely on him heavily, as they did on many US civilian experts. General Henry "Hap" Arnold, the USAAF chief of staff, sent Brigadier General George Brett to coordinate RAF requirements. Portal asked Tedder to impress upon Brett the importance of establishing depots and providing mechanics for servicing and repairing US aircraft and to assist with airfield construction. Tedder sent his most capable staff officers with Brett on a grand tour. Brett received the message loud and clear that the Middle East must have first call on all lend-lease materiel. Portal commented, "The issue in that area may turn on timely arrival of American aircraft and other assistance."[90]

Brett scheduled frequent industry representative visits and assigned experts permanently to help service lend-lease planes. Douglas Aircraft employees constructed Boston bombers at Port Sudan as Curtiss-Wright specialists assembled Kittyhawks. Similar schemes went into place at Basra and Massawa. The US depot at Gura was a combination of salvage, repair, and maintenance facilities and an aircraft-assembly point. US infrastructure in South Africa focused on getting aircraft at Cape Town ready to fly north across the entire continent to Egypt. Every element of this plan was in place by summer 1942 and often sooner.[91] This proved of great importance for the trial ahead.

The Air Situation in August 1941

With the German intervention, British gains largely evaporated, and a long struggle ensued. Fliegerkorps X and Fliegerführer Afrika had burst onto the scene with incredible energy and high levels of effectiveness. However, the former's move to Greece, the failure to deploy enough Me. 109s to North Africa or use them properly to gain air superiority, logistical and C2 problems, and an astounding dispersion of effort soon evened the odds once again and allowed the RAF to maintain air superiority. Fliegerkorps X's original orders, in agreement with Italian senior officers, were to deploy to Sicily, maintain control of the central Mediterranean SLOCs, and cooperate with Fliegerführer Afrika to help German and Italian forces conquer Egypt and exploit their advantage from there. Multiple changes in orders, along with the other problems already reviewed, placed these original objectives out of reach. Italian officers protested these changed circumstances and warned the Germans about the consequences for logistics. Further, a close look at the composition of the German air contingents also underscores their lack of preparedness either for the sea-control mission or the ground-support effort with which they were charged. Fliegerkorps X and Fliegerführer Afrika looked very much like other similar air units during the opening years of the war. They were highly capable of dealing with air forces such as Yugoslavia's and the small RAF contingent in Greece, but not when it came to defeating what was in effect a parallel and well-led RAF in the Western Desert and central Mediterranean. German airmen had evidently learned little from the Battle of Britain and even their successful campaigns regarding the need to structure air units in accordance with contextual realities rather than accepted norms that had worked in the past for their short, sharp campaigns. This failure to adapt with any speed or deep contextual focus proved a key element in the Luftwaffe's future operational challenges and ultimate defeat.[92]

5

The Back Door Secured and the Front Door Strengthened
April–October 1941

The Grand-Strategic Situation

JUST AS IT SEEMED THINGS COULD NOT become any more complex, a series of crises from April to October 1941, just before Operation Crusader, nearly unhinged the British position. The revolt in Iraq, the subsequent occupation of Syria, the continuing struggle in the Western Desert, and the increasingly vital and costly battle for the sea lines of communication (SLOCs) could all have gone against the empire. However, the British came out of this crucible stronger and with a firm hold on the oil fields, the Suez Canal, and their position in the Western Desert. Conversely, the Axis missed several opportunities to reap major strategic advantages.

Discussions of strategy in previous chapters have set the stage for events addressed here. Perhaps the most consequential development was the Axis failure to assist the coup in Iraq along with the larger Arab nationalist movement, which viewed Germany as a natural ally. This failure resulted directly in the British decision to invade Vichy-controlled Syria in order to eliminate the threat of German action from there. As these events unfolded, Adolf Hitler turned his full attention, and nearly the full might of the Wehrmacht, to Russia. Axis failures in the Mediterranean and Middle East thus reinforced the Führer's view that he had already spent too much time seeking advantage there. Hitler refused Erich Raeder's last plea on 6 June for an offensive in the Mediterranean. After Hitler had dispensed with Russia, the Wehrmacht would move south to bring the entire region under German control.[1]

Iraq Boils Over

With the failure of Operation Battleaxe following the defeats in Greece and Crete, and with Malta having just come through its first aerial siege, the sudden coup in Iraq on 2 April presented a most unwelcome development for the British. However, they dealt with it quickly given the threat to their oil assets. By 1937, no British troops remained in Iraq, but the Royal Air Force

(RAF) maintained bases at Shaibah and Habbaniya as staging points on the air route to India, and for aircrew training. Air Vice-Marshal H. G. Smart, at Habbaniya, commanded RAF forces.[2] By September 1940, Prime Minister Rashid Ali el Gailani of Iraq and the exiled grand mufti of Jerusalem were heavily engaged in pro-Axis agitation. On 31 March, the regent to the Iraqi throne learned of Rashid's plot and fled. Rashid took power and decided to eject the British while offering the Germans intervention and basing rights and requesting Vichy assistance with German operations. However, he overestimated the speed of the German and Vichy responses and their impact.[3] Hitler declared that he wished for a "heroic gesture" to assist Rashid and then promptly refused a Luftwaffe request to send two *Geschwader* (200 planes), choosing instead to send one *Staffel* each of Me. 110s and He 111s (totaling 20 planes) even as he told his commanders that he wanted to see a popular uprising in the Middle East.[4]

At first glance, the odds appeared in Rashid's favor. The RAF had seventy-eight mostly obsolete aircraft at Habbaniya. Six more Gladiators and one Wellington with spare parts for the other aircraft arrived on 7 April. On 30 April, after a ponderous start that gave the British time to prepare defenses and send reinforcements, the Iraqi Army invested Habbaniya. Aircrews attacked enemy positions beginning 2 May. Iraqi troops shelled the airfield and damaged several aircraft. On 4 May, Smart judged that the Iraqis would not attack given the presence of a capable armored-car force and sent his aircraft after the Iraqi Air Force and key Iraqi Army LOCs. Attacks continued on besieging units, which broke and ran when attacked by a relieving force. Archibald Wavell ordered a rapid advance on Baghdad and Basra.[5]

German envoys had tried to delay Rashid's move until after the assault on Crete. They planned, with Vichy cooperation in Syria, to establish an air presence in Iraq and help oust the British. The grand- and military-strategic benefits would have been immense. However, all communications went through the Italian legation, and the British intercepted them. Rashid's premature actions, the Germans' own missteps, and the rapid British response based in part on excellent intelligence, cost the Axis a golden opportunity to seize a critical region and the active help of Arabs throughout the Middle East. It could have been much more than merely a "heroic gesture."[6]

As disaster loomed, Rashid sent urgent appeals. A conference at Oberkommando der Wehrmacht (OKW) on 6 May resulted in a decision to give Iraq all possible support and to accentuate for the Arabs the Nazi efforts against the British in the Middle East. The Vichy government in Syria approved intermediate landings of the Me. 110 and He 111 squadrons, while a small liaison staff to Syria arrived. Because there was no good intelligence on Iraq and no clear understanding of the flying, maintenance, and logistical challenges, the effort was ad hoc. The Me. 110 squadron arrived on 14 May. A day later, six He 111s, overloaded with men, bombs, and spare parts, landed in Damascus. Several aircraft continued on to Palmyra and Mosul on 15 May.[7]

A first reconnaissance mission from Mosul led to a raid on Habbaniya, resulting in one British Gladiator fighter shot down and modest damage to the airfield. However, British aircraft destroyed several German planes on the ground at Palmyra on 14 May and at Mosul on 16 May. Major Alex von Blomberg, the liaison officer to Rashid, was killed by Iraqi antiaircraft artillery (AAA) fire while landing in Baghdad, precluding direct contact between the Luftwaffe contingent and Rashid's government. The Me. 110 squadron moved forward to Kirkuk, but the insufficient ground organization produced between zero and three operational aircraft a day. Nonetheless, the British garrison and Habbaniya endured thirteen small raids or armed reconnaissance missions from 15 May to 29 May. By this time, German aircraft were out of bombs, ammunition, and spare parts. Inexplicably, the thirteen transports assigned to resupply the two squadrons delivered numerous administrative personnel for the anticipated entry into Baghdad but far too few bombs, additives to produce aviation fuel, and spare parts. Colonel Werner Junck, the operational commander, pleaded for adequate supplies as well as seven more tropicalized He 111s and fifteen more tropicalized Me. 110s. OKW denied his requests, leaving him with two operational He 111s and one Me. 110.[8]

While OKW watched, trancelike, as the effort failed, Hitler issued "Führer Directive No. 30, Middle East," on 25 May. He emphasized, "The Arabian Freedom Movement is our natural ally against England in the Middle East. . . . I have decided therefore to encourage developments in the Middle East by supporting Iraq."[9] General of Fliers Hellmuth Felmy had been appointed chief of the Iraq military mission on 21 May. He arrived in Athens on 28 May, after a briefing with Field Marshal Wilhelm Keitel in which the latter told Felmy that the OKW had been completely surprised by the Iraqi coup—hardly an encouraging sendoff. Felmy arrived in Aleppo on 1 June and was promptly withdrawn, along with all other Germans, to Athens on 3 June. The British had threatened to attack Syria if German airmen remained there. Felmy noted tersely, "The German assistance to Iraq had ended quickly and infamously" as a result of Rashid's and OKW's failures.[10]

Hasty preparations; lack of intelligence; rapid RAF, Royal Navy, and British Army responses; and poor maintenance and logistics resulted in failure. With no concrete intelligence, and no longer able to fly combat missions, Junck ordered the remaining aircraft to Rhodes and Aleppo to refit.

However, the installment of a pro-British government in Baghdad on 31 May effectively ended the effort.[11]

The British Occupy Syria

Even before the Iraq campaign, British leaders concluded that Vichy Syria and Lebanon posed too great a threat. Vichy aid to German efforts in Iraq convinced the British to invade. The campaign lasted five weeks, ending on 14 July and proving more difficult than anticipated, but resulting in a complete victory.[12]

The Middle East commanders in chief (C in Cs) remained gravely concerned that Vichy would allow the Germans to establish a major air presence in Syria to attack targets in the Nile Delta along with the Suez Canal and oil supplies. Cyprus would also be in danger, and the Germans could support an advance through Turkey. Arthur Tedder advocated aggressively for action and redoubled his efforts after it became clear on 9 May that German aircraft were transiting through Syrian airfields into Iraq. The German raid on Habbaniya on 14 May prompted the C in Cs to approve RAF raids on Rayak and Damascus. Nine further attacks on Syrian airfields followed during May, destroying several German aircraft on the ground, damaging infrastructure, and destroying the French aviation reserve in Beirut. The British and Free French Armies committed two divisions to the invasion, and the Royal Navy provided two Fleet Air Arm (FAA) squadrons and a cruiser squadron. The RAF contributed eighty-seven planes. Air liaison officers (ALOs) accompanied the ground units. Despite the Vichy Air Force's numerical superiority, it was short on fuel and operational experience.[13]

Tedder was deeply concerned by Wavell's equivocations regarding Syria and said immediate action was necessary to ensure Luftwaffe units did not base there and begin large-scale raids on the Suez Canal, oil supplies from Abadan, and the airfield at Habbaniya in Iraq, which might bring Rashid and his supporters back into the fight with renewed energy. As Vichy forces prepared to disperse remaining aviation fuel stocks, Tedder asked for permission to attack them not just to immobilize the Vichy Air Force but to keep the Luftwaffe from using the fuel in operations from Syria.[14] In response to this impassioned plea, Charles Portal asked Tedder to send absolutely private telegrams so he could engage with Prime Minister Winston Churchill. This marked the beginning of a long and very fruitful "most private and personal" correspondence between the two, and Portal made good on his promise to convince Churchill.[15]

The attack began on 8 June, with the RAF gaining air superiority as troops advanced. Air cover of the naval contingent reduced support to the ground advance, which slowed against Vichy resistance. However, it led to the destruction of numerous German and Vichy aircraft seeking to attack British ships and break the blockade of reinforcements and supplies. Vichy airmen made the cardinal errors of failing to seek air superiority, giving Vichy troops inadequate support, and not providing fighter escort for their bombers. Further, General Jean Bergeret, Vichy secretary of state for air, ordered a defensive effort to minimize losses. His order had the exact opposite effect. Without air support, Vichy troops lost Damascus on 21 June. After that, Vichy fighters concentrated at Aleppo-Nerab airfield, where RAF aircraft destroyed many of them. The Vichy Air Force was soon ineffective. The RAF sank several merchant and troop ships trying to enter the harbor at Beirut, and the Royal Navy did similar work at sea. This blockade and successful army attacks led to Vichy surrender on 12 July. The campaign in Syria safeguarded Cyprus and, more importantly, gave the British a buffer against any future German land campaign through Turkey.[16]

Meanwhile, Iran's position of unfriendly neutrality put oil supplies there at risk. German agents and diplomats had energized pro-Axis Iranians by summer 1941. This prompted a 24 July directive from the War Cabinet to occupy Iran immediately. Key objectives included oil fields in the southwest along with the major refinery at Abadan, oil fields in the northwest, and the Trans-Iranian Railway, running from the Persian Gulf to the Caspian Sea, to facilitate lend-lease deliveries to the Russians. On 25 August, fighters knocked out the small Iranian Air Arm. A few further raids prompted surrender on 28 August. "With the occupation of Iran following on the reassertion of a position in Iraq and the occupation of Syria," Tedder said, "I felt that we had not only succeeded in shutting the back door in the face of the enemy, but had now managed to turn the key and slip the bolts very nicely into position."[17] British forces could now focus on the "front door"—the Western Desert.

British resupply efforts in Egypt were at the top of the list. However, Luftwaffe mining of the Suez Canal in March 1941 caused huge backlogs. At one point, there were more than one hundred ships waiting to unload. This had a cascading effect on operations throughout the theater. Fortunately, the lack of German persistence and the dispersion of air effort meant that the relatively small raids, with long lulls in between, had little aggregate effect. After the raid in March there was not another one until May. From there, attacks began again in July, when mines were scattered all along the Suez Canal and in Suez Bay, which delayed unloading efforts for three weeks. The British countered with the construction of more berths and railways, and they prepared an overland route from Port Sudan as a backup LOC, underscoring deep concerns about German mining. Despite British ingenuity, a concerted Luftwaffe effort would have caused severe problems. However, attacks remained sporadic and soon ceased.[18]

Axis Intelligence Efforts

Having failed to anticipate developments in Iraq and Syria, the Germans overlooked further key developments in part because of their failure to make full use of capable intelligence sources such as the German Signal Intelligence Service (SIS). Before the war, the Germans had ten people a day doing cryptanalytic work—a shadow of the British effort. Code-breaking successes were significant but too late in the Mediterranean. The first serious discussion of additional intelligence assets for the Mediterranean theater occurred on 10 June 1941—nearly six months after the initial German deployment. The SIS broke the RAF's operational code in summer 1942, followed by the US Army Air Forces (USAAF) bomber code in 1943, giving the Germans insights they could no longer use to real advantage because the Luftwaffe had missed its several opportunities to control the air. The bomber code facilitated occasional intercepts, but the area to be covered was so large and fighters spread so thin that most efforts failed. A complete understanding of the most important codes did not materialize until March 1944. The Luftwaffe, army, and navy *Chi-Stelle* (signals organizations) rarely worked together until 1942, when they moved from an exchange of liaison officers in Germany to the establishment of a central liaison office in the Air Ministry—but never within an operational theater.[19]

Consequently, there was very little intelligence available prior to German intervention in the Mediterranean. The *Abwehr* maintained contact with the Italian Intelligence Service, but little of value changed hands. The problem worsened after the Germans intervened because the Italians did not like sharing information. Consequently, German operations relied largely on intelligence collected by German sources. There were too few human intelligence assets, so aerial reconnaissance, radio intercepts, analysis of captured equipment, and prisoner of war (POW) interrogations constituted most intelligence support. Collection efforts focused on Fliegerkorps X's three shifting missions: direct support of ground operations, convoy escort, and neutralizing Malta. Gaining air superiority was not one of the stated missions, even though the SIS could have contributed substantially to such an effort with its capable Y Service equivalent units.[20]

The poor communications infrastructure in the Mediterranean and from there back to Germany made intelligence efforts challenging because *Fliegerkorps* and *Luftflotte* commanders, and even Oberkommando der Luftwaffe

(OKL) senior officers, insisted on staying in direct contact with all missions, sometimes down to the individual aircraft. This overloaded the existing communications channels so severely that the intelligence effort, already overtaxed, could not keep pace with operations. It also gave British signals troops a collection bonanza that their commanders put to good use.[21]

On the positive side, Fliegerkorps X anticipated the demanding conditions its intelligence staff would face and brought a reinforced photographic lab, a reinforced radio intercept company with a decoding section, a naval communications monitoring section, a counterintelligence section, regional experts, and a POW interrogation facility. Photographic reconnaissance (PR) and photographic intelligence (PI) capabilities were very good. These provided excellent coverage of RAF airfields, but airmen rarely attacked them. German PR aircraft discovered part of the Takoradi route and sent additional flights to provide complete coverage of key airfields along its eastern terminus. These were loaded with 300–400 closely packed aircraft. Because these airfields had no radar and almost no AAA, a prompt raid could have destroyed many planes. This would have been timely because these aircraft constituted most of the reinforcements for Operation Crusader. Despite repeated requests from unit commanders for permission to attack, senior officers demurred. One raid consisting of nine Ju 88s eventually went after one airfield. Only three bombers reached it, and they did no damage. Within days of the raid, the airfield was bristling with radar, fighter patrols, and AAA. Radio intelligence had also located hubs on the Takoradi route and kept track of virtually every aircraft arriving in Egypt. The Germans knew where aircraft were going, when they were going to be there in large numbers, and when they would be most vulnerable.[22]

Also, observers near Gibraltar reported on all flights. Had the Germans introduced a capable night-fighter unit beyond the one squadron they had in the entire theater, they could have destroyed many aircraft transiting Malta. This close watch on enemy air units meant that within three to four months, intelligence personnel had a detailed understanding of RAF assets. They often anticipated activities at various airfields. The SIS developed a precise understanding of RAF strength, delivery schedules, and logistical and supply status. Its officers knew where each unit was based and its defensive and warning equipment and reported this to commanders. Still, there were no concerted airfield attacks, serious bids for air superiority, or selective withdrawals to minimize the effects of RAF raids on airfields. As a former SIS officer noted acerbically,

Whether the German operational commands lacked the tactical ability to appreciate the importance of this intelligence, or whether their default in moral courage in the face of dependence on the plans of a single individual [Erwin Rommel] was so abject that they simply disregarded these possibilities, at any rate the German SIS is to be absolved of any responsibility. . . . The most complete signal intelligence situation summary was of no avail if the General Staff failed to adjust its plans to the reality of the situation.[23]

It is important to view this statement in context. The German military saw intelligence personnel as assistants to their operations officers. Consequently, despite successful SIS efforts to assess RAF capabilities, track its movements, and discern its routines, Luftwaffe commanders rarely made use of this intelligence to attack airfields. This microcosm of a much larger German intelligence failure, including the persistent belief that the Enigma cipher was unbreakable, robbed the Axis of what could have been a potent strategic and operational force-multiplier in the Mediterranean, as signals intelligence was for Rommel at the tactical level.[24]

Refining Air-Ground Cooperation

The relative lull in operations during summer 1941, as both sides struggled to build logistical strength, proved a crucial time for British air-ground cooperation and aircraft reinforcements. When Operation Crusader began on 18 November, the RAF had 657 aircraft (554 serviceable). German air strength in North Africa and on Sicily was 244 (121 serviceable), and the Italian Air Force (IAF) contributed another 296 aircraft (192 serviceable). The salient development, however, was an increasingly deep conviction that without effective air-ground cooperation, the British would not defeat Rommel. A series of joint exercises improved air-support efforts, from communications and coordination to calls for air support to optimal uses of aircraft and munitions types in different situations.[25]

Tedder was concerned that the army and RAF lacked clear regulations, training, or skills relating to establishing bomb lines and communicating. After Churchill replaced Wavell with General Claude Auchinleck in July 1941, Tedder and the new army commander found common ground. They immediately formed an interservice committee to improve combined-arms operations.[26] Churchill's involvement in the discussions resulted in clear guidance that bore an uncanny resemblance to Tedder's recommendations for improvements after Operation Battleaxe. "Army Training Instruction No. 6" soon followed. Three RAF mainstays, in order of priority, formed the core of the document: air superiority, isolation of the battlefield from enemy reinforcements and supplies, and attacks against targets on the battlefield. New army air support controls (AASCs) would select and call for attacks on targets in the latter category along with assistance from RAF air support controls (ASCs) collocated with them.[27]

The new air-ground cooperation system that followed was a major improvement but nowhere near mature. Teething problems included a lag time of more than two hours from a request for air support to the arrival of aircraft over target, delays in routing messages through brand-new ASC chains of command, delays in rendezvous between bombers and fighter escorts, and serious problems finding targets. The RAF and army worked out these problems by summer 1942, but providing effective air support during mobile phases of the battle remained problematic. The debacles in France and Greece spurred rapid developments in the Western Desert. However, the entire planning and industrial process of the past five years had been based on the assumption of small-scale air-land cooperation, so making the change was neither swift nor simple.[28]

All three services had to cooperate closely to win. Without air superiority and air support, the army would have trouble advancing. Without an army advance, the RAF could not occupy airfields in the Cyrenaican hump—a requirement for giving RAF units the range to support a further army advance, reach major Axis ports, and engage in joint antishipping missions with the Royal Navy. The first instance of a ground-air liaison team was an air liaison section (ALS) composed of a group captain and a squadron leader, at Western Desert Force headquarters. This heralded many more improvements.[29]

Each corps and armored division received an ASC, which was mobile and comprised of a joint-service staff with advanced wireless communications capability known as a tentacle that linked the ASC to each brigade. An RAF support team known as a forward air support link (FASL) also worked at each brigade headquarters and had two-way radios for talking with aircraft engaged in support missions. Rear air support links (RASLs) completed the picture, connecting advanced airfields and landing grounds with ASC headquarters. The RASLs and air staff at advanced headquarters had radios and other devices to listen in on tactical reconnaissance aircraft communications with the FASLs. Air support gradually became more rapid and lethal, with armed tactical reconnaissance aircraft and brigade commanders using the tentacles to guide airstrikes. Whenever an ASC commander validated a request, his staff told the RASL at a given landing field to get aircraft airborne. Standard codes and phrases sped the process. Aircraft received directions to the target with preplanned coordinates from a reconnaissance aircraft that led them in, by FASL ground guidance, or with a combination. Key ground features defined bomb lines in the fluid environment of desert combat, and colored flares and Verey lights helped pilots distinguish friend from foe. The first two ASCs became operational on 8 October—six weeks before

Operation Crusader. Despite missteps and constant modifications, this air-ground cooperation system marked the start of an effective combined-arms capability that matured in summer 1942 and improved until Victory in Europe Day.[30]

One of the crucial documents to come out of this effort was "Middle East Training Pamphlet No. 3—Direct Air Support," which for the first time set forth clear and comprehensive guidelines for air-ground cooperation. The most important change was the increase in tentacles within each AASC from seven to nine—one for every division and brigade headquarters. AASCs processed requests for air support from reconnaissance aircraft and forward army units through the tentacles. At least one formation of six aircraft in each squadron was at "instant" readiness, and the rest were at two-hour readiness. Bomb lines, ground markings, and vehicle markings were developing.[31]

As command-and-control (C2) infrastructure improved, ground-controlled intercept (GCI) made dramatic advances. New stations were assigned to parent sectors administered by parent groups, which in turn reported directly to Air Headquarters Egypt. All GCI stations were mobile. Range was from ninety miles for high-flying aircraft to twenty-five or fewer for aircraft at or below 5,000 feet. Each also had height-finding equipment to allow for all four elements of information: number of aircraft, range, azimuth, and altitude. They assisted with intercepts by providing air warning through their parent units.[32] Additionally, they received Y Service and other inputs to determine enemy air activity and pass a coordinated picture to supported airfields. After aircraft were airborne, the group controller at the parent group decided which intercepts would proceed.[33]

Meanwhile, Air Commodore Basil Embry, who Portal had sent at Tedder's request to teach aircrews the latest tactics used in the United Kingdom, did exceptional work. Embry, Arthur Coningham's senior air staff officer (SASO), determined that the Germans had sent an elite team of fighter pilots who had bloodied the RAF and also restored morale among IAF aircrews. He helped airmen to counter them. Tedder was impressed with Embry's new fighter tactics and insistent requests for the loan of highly experienced fighter-unit commanders and seasoned fighter pilots from the United Kingdom to help even the odds. Portal sent 105 chosen pilots.[34] Embry's effectiveness became clear after the offensive started. Tedder noted, "Our chaps have for the time being knocked the enemy right out of the air. I had a few seconds to talk . . . with Basil Embry this evening. Said things were very satisfactory, but the Hun won't fly—they can't take it."[35] The Eighth Army commander, General Lewis Anthem Ritchie, said the air situation was marvelous—"like France, only the other way round."[36]

Airfields: A Key to Victory

While the C2 infrastructure matured and tactics improved, the British worked doggedly at the most important component of mobile air warfare—airfields. The construction of airfields from 1934 to 1945 was the largest civil engineering project since the construction of the railways in the 1800s.[37] Airfields and landing grounds, which sprang up by the thousands (about 1,800 airfields and innumerable landing grounds), were part of the vast logistical effort the RAF (with army royal engineer assistance) and then the USAAF put into developing the sinews of airpower.[38]

Building airfields was an immensely expensive, labor-intensive, and specialized process requiring a range of specialists and huge amounts of raw materials for building runways, taxiways, and the roads necessary to reach airfield sites. The supply effort was immensely challenging. Administering such a massive undertaking required huge staffs. One of the greatest efforts was tracking where people and materials were in the pipeline and when they would arrive.[39] Competing for priority allocation of assets was only half the battle. The learning curve, which was steep as warfare became expeditionary and mobile over huge distances, comprised the other half. How well and quickly each air force learned to adjust to and embrace these new realities of combined-arms warfare was a vital determinant of success or failure.

The Airfields Organization (AO), which oversaw the effort, was set up in May 1941. In 1939, there were eight airfields in the Western Desert. Between June 1941 and October 1942, the AO built forty-one more, cleared nine captured airfields, and extended runways to accommodate medium and heavy bombers. From October 1942 to May 1943, it cleared twenty-four captured enemy airfields and built another fourteen. A facility in Cairo tested airfield construction materials to maximize their effectiveness in combination with soil in specific locations.[40]

Building airfields was demanding; clearing captured ones was dangerous. Sappers were killed and injured routinely. At Marble Arch in January 1943, they found 860 mines and nearly 2,000 booby traps around one landing ground. Other problems abounded as the Eighth Army and RAF advanced into Tunisia, with its very different soil types. The rainy season there drove home the need for pierced steel planking (PSP—Sommerveld Track was the RAF equivalent) to build all-weather runways. Given the weather in Italy, airfield engineers knew they would need it there too. Excessively dusty areas were also problematic, increasing aircraft malfunctions and maintenance requirements. Nonetheless, the British often built four landing grounds in three days, in locations designed to minimize environmental impacts such as manmade dust issues and airfield flooding during rainy periods. Each landing ground included dispersal areas totaling 2 million square feet with aircraft aprons at least 200 yards apart.[41]

Airfield construction between 1937 and 1939 also resulted in the completion of four operational airfields on Malta—a pivot on which the campaign turned. When Italian raids began in June 1940, engineers built strong shelters, making planes impervious to anything but a direct hit. Despite a shortage of motorized transport, there were fifteen rollers available to conduct runway repairs and extensions and to build additional taxiways. The lion's share of the effort revolved around the 850 or more farm carts and their Maltese owners engaged daily in the runway-repair effort. A growing network of taxiways combined with this repair force allowed aircraft to move quickly from temporarily unserviceable airfields to operational ones. Even during the worst raids, RAF aircraft kept flying.[42]

The Luftwaffe had its own *Baubatallinen* (airfield construction battalions) and special works companies of 150 to 450 men. Given their experience building large airfields in Germany, these units proved very capable of constructing or improving airfields during mobile campaigns. Before the war, Luftwaffe units moved every six months to new bases in Germany to prepare for a war of movement, and they designed all of their aircraft to operate from grass landing strips. In spring 1941, *Baubatallinen* in Wolfram von Richthofen's Fliegerkorps VIII built new airfields in a week or two, while royal engineer companies took a month or more. Yet the vast majority of this expertise went with the Germans to Russia. The Luftwaffe in the Mediterranean got the leftovers and thus lagged RAF improvements.[43]

To make the Germans' operations particularly challenging, David Stirling and his Special Air Service (SAS) troops raided airfields, often in conjunction with RAF attacks. They also reconnoitered to determine enemy air strength and to let the RAF know if landing grounds were unused and available for occupation. On 24 December, they destroyed twenty-four German aircraft at Sirte, killed a number of ground and aircrews, and destroyed large quantities of spare parts and other supplies. Another SAS party destroyed twenty-seven aircraft at Wadi Tamet on the same day. These losses of aircraft and skilled personnel were difficult to make good. The raids' tempo and effectiveness increased over time. During Operation Crusader, they reduced Axis air strength in advance of the ground offensive. Stirling and his men had orders to destroy as many planes as possible to help the RAF hold air superiority.[44]

Axis Developments

By May 1941, Luftwaffe attacks against Malta declined rapidly as air units moved east. After Operation Barbarossa began, Churchill called for all possible aid to Russia. A key concern was that Operation Barbarossa would include a move through Turkey, which received additional armaments. At this point, when the attack on the Soviet Union appeared likely to succeed, the Germans maintained their view of the Mediterranean as a subsidiary theater. To succeed in their larger aims of dominating the region, they would employ forces moved from Russia after Operation Barbarossa to neutralize Malta, seize Gibraltar, and capture Tobruk. They would follow in spring 1942 with an attack on Egypt from two directions (the "front door" through Libya and the "back door" through Iraq and the Levant states) along with an advance into Iran toward the Persian Gulf. The failure in Russia forced wholesale revisions to this plan. Force dispositions changed, including the episodic and time-intensive Luftwaffe deployments during the long Russian winters and the deployment of U-boats, minesweepers, and motor torpedo boats to the Mediterranean. U-boat and torpedo-bomber transfers drained already scarce assets from the crucial Battle of the Atlantic even as they remained too small to make the difference in the Mediterranean. Additionally, Fliegerkorps X, still headquartered in Greece, was now back to flying escort and strike missions in the central Mediterranean and direct-support sorties for the Tobruk siege. It was, therefore, drifting back to its original roles but with most of its aircraft much further away than necessary and often not committed as the Germans instead kept aircraft ready to raid British Navy sorties that did not materialize. Longer sorties from the Aegean Sea and consequent logistical shortfalls decreased serviceability rates to 41 percent.[45]

While Fliegerkorps X's serviceability rates dropped to new lows, Fliegerführer Afrika's climbed briefly to new highs—71 percent in late June. Two more Ju 52 *Gruppen* had arrived on 25 April. These sixty aircraft played a major role in improving serviceability rates by flying in spare parts. Nonetheless, after the operational pause ended and Tobruk raids began taking their toll, and then Operation Crusader caught Luftwaffe units by surprise, serviceability rates fell once again to around the 50 percent mark and stayed there. Part of the larger problem here was the RAF's constant raids on airfields, which wore down the Luftwaffe's ground organization and killed large numbers of skilled technicians. During the months immediately following Greece and Crete, the RAF lacked the assets to engage in as many of these attacks as Tedder would have liked. This combined with Ju 52 deliveries to produce high serviceability rates for the only time during the campaign. The Germans would have been wise to heed these interrelated factors. Even with high serviceability rates, Stephan Fröhlich's units lacked effective C2, had limited mobility because of vehicle shortages, and suffered from a chronic flak gun shortage as Rommel took most guns to the front (which in turn led to higher losses during RAF airfield raids). Finally, constant calls for another *Gruppe* or more of Me. 109s went unanswered. Despite the generally lower emphasis on air superiority when compared with the RAF's, German airmen knew they needed more fighters if they were ever to achieve their larger combined-arms objectives, including protecting the SLOCs and facilitating direct support of troops. German senior leaders proved unable or unwilling to address these issues to any substantial degree. The result was a paucity of raids against high-value targets such as the railhead at Mersa Matruh, which came under ineffectual attack just three times during August even as Tobruk sapped the larger air-ground effort. Consequently, the Germans were unable to interfere with the movement of British supplies, whereas the opposite was anything but true.[46]

Meanwhile, the RAF raided ports and airfields without letup. During the fifty-one days covered in one report, the RAF raided Benghazi twenty-nine times, Derna twenty times, and airfields twenty-five times. Tripoli suffered twenty-six raids and ports in Sicily thirty-four. These raids produced a steady stream of aircraft and personnel losses. There were also sixteen attacks on convoys resulting in five ships sunk and another seven damaged. In response, the Luftwaffe and IAF attacked Tobruk fifty-five times. Mersa Matruh came under air attack twenty-one times, including the three strikes on the railhead. The Germans did attack Fuqa and other airfields six times, and the Italians fifteen times, but radar warning and poor bombing produced meager returns. Axis aircrews sank seven merchant vessels (MVs) and damaged another twenty-one—the one marginally focused air effort. In the eastern Mediterranean, German planes attacked Alexandria five times and the Suez Canal fourteen times.[47]

Ironically, Axis maritime aircraft produced major results despite the paucity of assets assigned to the theater. Torpedo-bomber and mining units had abysmal 30 percent serviceability rates. Nor had German units grown in number since Fliegerkorps X's arrival. The one brief exception to this rule was the loan of a torpedo-bomber squadron from France from 2 to 17 September. Raids on the Gulf of Suez and northern Red Sea, where German torpedoes could run with the required depth, produced excellent results. The six FW-200 Condor and eleven He 111 torpedo planes joined Fliegerkorps X's operational torpedo and level-bombing assets on 2 and 10 September, sinking two large MVs totaling 14,055 tons and damaging another twelve totaling 66,618 tons. Aerial mining produced its usual damage and long delays, which German PR crews duly photographed. One more raid, on 6 October, sank a 4,000-ton ship and damaged another four totaling 23,000

The port at Tripoli in January 1942. Along with Benghazi, Tripoli was under constant hammering from No. 205 Group Wellingtons and when in range of Blenheim 4 medium bombers, from those as well. RAF crews attacked at night given the greater safety (the Axis had one squadron of night fighters on Sicily at this point and almost no radar). They became adept at following the coastline directly to target. The Italians had no lighters to allow unloading from both sides of the ship, and the Italian-managed convoying and logistical efforts were ineffective. These problems, combined with constant bombing, reduced capacity at Axis ports by 50–75 percent. Leading up to Crusader, and from July 1942 to the victory in Tunisia in May 1943, Allied bombers helped to create severe and ultimately insuperable Axis logistical crises. (US National Archives and Records Administration Identification No. 292579)

tons. For its part, the IAF had activated five new SM.79 torpedo-bomber squadrons by summer 1941. They damaged the battleship HMS *Nelson* and sank an MV of 10,733 tons on 24 September. Despite such successes, torpedo and mining efforts were episodic and dispersed.[48]

In fact, the loaned squadron returned to France, and further torpedo, mining, and strike operations against the Suez Canal and northern Red Sea declined rapidly in number. Ironically, the end of the promising offensive efforts against Suez, the massive increase in Axis shipping losses that began immediately after the move of Fliegerkorps X to Greece, and the reversion of responsibility for convoying to the IAF, prompted Hitler to change Flieger-korps X's mission yet again. Henceforth, it would protect convoys and de-fend Crete and the Aegean rather than flying offensive missions. This de-fensive and reactive role permanently replaced the proactive and offensive one for which it had originally come to the Mediterranean. Every convoy required a minimum of ten sorties for escort, which used up available air-frames quickly in terms of total numbers required, losses, and declining serviceability rates. The only other missions Hitler would allow were con-voying of coastal (west-to-east) shipping along the African coast and occa-sional raids in support of the siege of Tobruk. The inversion of Fliegerkorps X priorities and efforts was complete.[49]

No. 201 Naval Co-operation Group

As the Axis turned Fliegerkorps X to defensive missions, the British worked to exacerbate the Axis logistical crisis and the pernicious effects it was hav-ing on German air priorities. Interactions between the RAF and the Royal Navy were often more important than those between the RAF and army, par-ticularly given the theater's contextual realities. These included the SLOCs feeding Axis forces in Libya and British ones on Malta. Until the disaster at Crete, Admiral Andrew Cunningham was convinced that the Royal Navy could dominate the Mediterranean. German land-based airpower disabused him of this notion. Afterward, navy assets declined gradually and steadily—especially after diversions to Singapore and Ceylon, along with several suc-cessful Axis attacks, left the British without any capital ships in the eastern Mediterranean in December 1941. RAF land-based aircraft thus took the lead in the battle for the SLOCs. Otherwise only submarines and small sur-face contingents could engage enemy shipping with any effect. Land-based planes provided reconnaissance and cover.

Before the official activation of No. 201 Naval Co-operation Group in October 1941, which facilitated effective air-sea action against convoys, Tedder was already focused on this problem because he understood the RAF would carry most of the weight. On 29 May 1941, he asked Wilfrid Free-man for the rapid delivery of a second squadron of Beaufighters. The first had arrived in Egypt, and an additional flight had reached Malta. This was the only fighter with the range to cover navy operations. Otherwise, he said, the ships would "be quite cold meat. In fact as things are at the moment any excursion outside a radius of about 150 miles to the east and north of Alex[andria] is an expensive adventure for the Navy. . . . The air has come into its own with a vengeance in the Mediterranean."[50] Tedder further noted that Admiral Cunningham's complaints about lack of air cover for one of the convoys from Crete were tendentious because the convoy was two hours behind schedule and thus not in position to rendezvous with RAF fighters, which did not receive word of the convoy's delay. This left the convoy in range of bombers during daylight. He further noted that no ship had yet been hit seriously while under RAF fighter escort and that he could provide such escort without having to tie substantial forces to a dedicated and navy-dominated air group. He reiterated this with the case of the AAA cruiser HMS *Coventry*, which Admiral Cunningham sent to Crete without either notifying the RAF or asking for fighter escort. Luftwaffe bombers sank the ship. Cunningham was now calling for at least six dedicated Beaufighter squadrons along with long-range Hurricanes, both of which were in short supply and badly needed for operations in the Western Desert. He wanted these under his operational control—something Tedder and his bosses would not countenance.[51]

Tedder had, interestingly, been a conditional advocate of decentralized air command until faced with the incredibly complex situation brought on by the disasters in Greece and Crete, a major reversal in the Western Desert, a coup and Luftwaffe involvement in Iraq, the campaign against Syria, and a long campaign in Italian East Africa (IEA). "Now that it is necessary to be able to switch rapidly from East to West or vice versa or from West to North to cover the Fleet," he said, "I am afraid some degree of centralized control is unavoidable. During the past three weeks it has again and again been essential and so far as I can see that is likely to be true in the future."[52] Given these contextual realities, he divided his staff into two components: Air Staff Operations, with an operations room and twenty-four-hour watch function to keep him apprised of the latest developments, and Air Staff, Staff Duties to deal with policy, training, and other administrative responsibilities. He and the other service C in Cs agreed to establish a joint-operations center at Ismailia to coordinate air operations among and in support of all three services.

Several things were now clear. First, the Royal Navy could not operate more than 150 miles from Alexandria without air support. Second, victory in Cyrenaica depended on the capture of air bases and, in turn, the ability to strike Axis logistical assets, including shipping, ports, and vehicle convoys. Third, with the Luftwaffe on Crete, Alexandria and other key targets throughout the Nile Delta were at grave risk. Tedder was concerned about Admiral Cunningham's request that his handful of Beaufighters and several other air units be allotted permanently to protection of the fleet, on call at all hours and under Cunningham's operational control. This was in addition to the giant wish list of aircraft Cunningham had already submitted. Tedder asked Portal to intercede with Admiral Dudley Pound, his navy counterpart, adding that it made no sense to have aircraft sitting around when he needed them for other missions, especially during ground offensives.[53]

During these early discussions about No. 201 Naval Co-operation Group, Axis efforts failed to stop several convoys from reinforcing Malta. RAF fighters provided air escort. This development was just one of many that signaled an important and substantial maturation of air-sea cooperation efforts. In fact, the RAF and Royal Navy had been working on this problem for well over a year. The process began in June 1941, when Admiral Cunningham asked Tedder for a dedicated air arm, much like Coastal Command, under Tedder's control but with major input from Cunningham and his staff. Cunningham also agreed to place all disembarked FAA aircraft under this unit's control. Tedder agreed in principle and told Cunningham he was already forming No. 201 Group to fill the role. However, he did not favor a coastal command, with the aircraft held firmly for maritime missions. Instead, he proposed that the unit be flexible so it could support air and land actions when things were quiet at sea. Portal wrote Tedder to assure him that detailed agreements regarding No. 201 Group's mission were in hand. The plan was as Tedder hoped, with an additional statement that as many of Tedder's units and aircrews as possible would learn naval cooperation techniques. Tedder approved as long as the arrangement did not interfere with maintaining air superiority and supporting the army. Portal followed with a note that the Air Ministry and Navy Admiralty had agreed formally on the title No. 201 Naval Co-operation Group and the unit's roles. Portal's discussions with Pound clinched the deal. Tedder would control the unit and coordinate closely with Admiral Cunningham through direct communications and an RAF staff stationed at fleet headquarters in Alexandria.[54]

The increasingly close navy-RAF operational and staff ties that followed resulted in much-improved coordination of operations. Better weather, a concentrated air striking force, better reconnaissance, the Luftwaffe's smaller air contingent after April to include the move of Fliegerkorps X to Greece, and closer cooperation with navy assets combined to double sinkings of Axis MVs after May. Aircrew training and reconnaissance improved. A sustained PR and PI effort began to discern clear patterns in the timing and location of convoy movements. Aircraft and submarines thus took an even greater toll on Axis shipping. The combination of intelligence, planning, and air-sea cooperation began paying major dividends.[55]

By the time No. 201 Group became active in late October 1941, airmen and sailors had long since started working together to maximize their combined effectiveness. From June to November, they sank forty-four MVs totaling 222,227 tons—a massive and irreparable loss for the Italian Merchant

Marine. In its first of several documents on this subject, the Headquarters Royal Air Force Middle East (RAFME) Tactics Assessment Office noted that the protection of friendly shipping and destruction of the enemy's was the most vital single mission in the theater and that No. 201 Group was doing it with increasing effectiveness. It gave credit for these successes to the establishment and evolution of a joint planning staff. Aircraft were vital in protecting their own shipping and attacking the enemy's to achieve a logistical advantage. Seaborne supply was the "paramount" factor in the African campaign, but the navy's losses around Greece and Crete, and later diversions to the Indian Ocean, left it unable to win the battle of the SLOCs on its own. To keep the Luftwaffe at bay, coordinated sea-air operations were essential. During the next year, as the logistical struggle intensified, No. 201 Group grew rapidly under Leonard Slatter, who had distinguished himself in the IEA campaign. And its tactics evolved.[56]

One example of this was the process by which two pairs of Maryland bombers attacked ships from very low altitudes to achieve surprise and cause maximum damage with each aircraft's four front machine guns and eight 250-pound bombs, which often "skipped" off the water and into ships' hulls. Attacks out of the sun or low clouds made them very difficult to stop. Marylands, which replaced the slower and vulnerable Blenheims, also had observers who focused entirely on locating the target.[57]

German SIS officers also noted these tactical innovations and said that No. 201 Naval Co-operation Group excelled at all of its missions: attacking ports and shipping, guarding convoys, engaging in long-range reconnaissance, and mining shipping lanes. They found the operations of 38 Bomber Squadron (No. 247 Wing) particularly impressive. Its air-to-surface surveillance (ASV) radar-equipped Wellingtons proved deadly to convoys at night. They made a very low approach after radar contact, executing simultaneous torpedo attacks from both sides of the convoy and using Verey lights to illuminate targets during and afterward to determine results. After the Wellingtons had convoys in this vise, they rarely survived. As No. 201 Group became more lethal, its aircrews worked with submarines to ensure that RAFME and the Mediterranean Fleet would win the next chapter in the battle for logistics leading up to Operation Crusader.[58]

The Logistical Advantage Shifts

During the pause after Rommel's Sollum offensive and failure to take Tobruk, both sides settled into different routines. While Axis air forces sought to bomb Tobruk into submission and keep supplies flowing across the Mediterranean, the RAF developed new air-ground cooperation capabilities with the army, implemented No. 201 Naval Co-operation Group with the navy, protected shipping to and from Tobruk, attacked Axis convoys, and trained for Operation Crusader. The subsequent period became a race to the next offensive—one determined by the flow of supplies. The British won.

Between the Luftwaffe's major redeployments to Russia and the improved air-sea cooperation led by No. 201 Group, control of the central Mediterranean once again passed to the British. Of the forty ships (totaling 178,577 tons) sunk between June and October 1941, 57 percent (twenty-four ships totaling 101,894 tons) fell prey to aircraft, which were now the primary ship killers. By September, Axis convoys were losing nearly 25 percent of their tonnage and supplies.[59]

Despite these losses, Franz Halder ignored Rommel's pleas for Luftwaffe protection of SLOCs. "Safeguarding transports to North Africa," Halder said, "is an Italian affair. In the present situation it would be criminal to allocate German planes for this purpose. OKW has no means of helping."[60] Keitel echoed these thoughts at a meeting with Marshal Ugo Cavallero, chief of Comando Supremo, on 25 August. He also emphasized that a permanent force had to neutralize Malta. Hitler got involved in September, overriding Raeder's objections to sending U-boats to the Mediterranean when the Battle of the Atlantic was at a critical stage. In addition, on 13 September, in a direct override of earlier OKW orders, Hitler directed Fliegerkorps X to devote all of its energies to protecting convoys rather than bombing and mining the Suez Canal. The Italians were still responsible for escort duties between Italy and Tripoli and between Tripoli and Benghazi (two of the most vulnerable points), with the Luftwaffe covering other convoy routes. The Axis thus got the worst of both worlds: continuing Italian convoy protection along the most dangerous routes and a cessation of what had been small but effective mining operations against the Suez Canal.[61]

As the Axis took these steps, the British continued to install ASV radar with a sixty-mile search track on Wellington bombers. Three were ready by October. The bombers guided other aircraft to Axis convoys and often attacked them as well. In addition, eleven FAA flare-dropping Albacores illuminated the convoys. The navy contributed two light cruisers and two destroyers, known as Force K, to this combined-arms effort, which ambushed convoys often located for them by the Wellingtons.[62] In a single action on 9 November, Force K, tipped off by an Ultra intercept and a subsequent Maryland reconnaissance sighting to cover the original source of the intelligence, sank seven MVs totaling nearly 38,000 tons, including two large oil tankers. It followed on 24 November with another Ultra-guided attack that sank two MVs totaling 4,752 tons loaded with aviation fuel. A third attack on 1 December sank another oil tanker of 10,540 tons. These sinkings, during Operation Crusader, created severe fuel shortages for Rommel and Fröhlich.[63]

RAF Maintenance

As the larger logistical balance shifted against the Axis, maintenance practices made major strides within RAFME. After his arrival in Egypt, Air Vice-Marshal Graham Dawson tackled maintenance, salvage, and repair problems. There were four maintenance facilities when he arrived in May 1941. Dawson decentralized them by placing assets in caves and other hardened and dispersed facilities. Repair depots at Khartoum and Port Sudan ensured newly arrived aircraft received a quick check, tune, and delivery to squadrons. Forward repair-and-salvage units (RSUs) received additional transport and manpower to maximize speed of delivery of damaged machines to the new centralized base salvage depot for distribution to and repair at maintenance units. The worst shortages soon eased as more skilled mechanics and equipment arrived from Britain. This drove major improvements in serviceability rates based on the increased reliability and effectiveness of individual aircraft and their weapons, avionics, and other components.[64]

Dawson's methods were unorthodox and harsh but produced excellent results. With Group Captain Henry Thorold at Takoradi and Dawson in Cairo, aircraft receipts and serviceability climbed. Tedder remarked that without these two men, it would have been virtually impossible to increase air strength enough to hold the Germans and Italians in 1941 and defeat them in 1942 and 1943.[65] Dawson's character was ideally suited to this assignment, which involved working around the Air Ministry on occasion to get things done. He disliked bureaucracy. Dawson had vast technical knowledge and inventiveness, making up for shortages with various innovations.[66] There is no question that he was a maverick. Several senior officers loathed him precisely because he was never reticent about going around them. Portal warned Tedder several times to keep Dawson under control, but at the end of the day he knew the man was too crucial to RAFME efforts and did not fire him.

This may have been due to the fact that Dawson was one of Lord Beaverbrook's favorites. Whatever his other faults, Beaverbrook's support for Dawson paid dividends in the Western Desert. He noticed Dawson's talents after the latter moved to the Air Ministry as director of the Department of Repairs and Maintenance. Beaverbrook put him in charge of engine repairs within his Ministry of Aircraft Production. Dawson upped the output of repaired engines dramatically and as a result became an air vice-marshal in summer 1940. On 11 July, Dawson received a Mention in Despatches for his superb engine-repair efforts during the Battle of Britain. In March 1941, as the situation in the Middle East became grave during Rommel's first offensive, Beaverbrook placed Dawson in charge of US aircraft shipments to Takoradi. By 30 June, a steady stream of 200 US-produced aircraft per month arrived.[67]

On 9 May, the War Cabinet discussed the deteriorating situation in the Middle East and decided to send Dawson at the head of a committee of investigation. This group was to report on the state of aircraft maintenance and repair and to recommend improvements. Dawson became Tedder's chief maintenance officer (CMO) upon completion of the inspection. This made him the head of both engineer and equipment services and principal adviser to the air officer commanding in chief (AOC in C) on technical questions. Dawson also controlled all facets of the Takoradi route. One of Dawson's key functions was to recommend to Tedder all actions required to maximize aircraft salvage, repair, and return to operational status. Dawson flew to Cairo via Takoradi so he could consult with Thorold and brief Tedder on his findings. Before he left the United Kingdom, Dawson contacted Beaverbrook to let him know he had already been studying the challenges at Takoradi and was recruiting technical experts to help Thorold address them. He sought out the handful of specialists with knowledge of US aircraft heading toward the Middle East and secured their transfer orders. Dawson departed on 16 May. At Takoradi, he asked the Air Ministry for immediate transfer of three qualified inspection officers to speed up quality-control and dispatch operations. He also asked for more ferry pilots when he found thirty-five aircraft awaiting them. Finally, Dawson emphasized that Takoradi needed more transport aircraft to carry supplies to intermediate airfields along the route. He obtained all of these resources.[68]

Dawson arrived in Cairo on 1 June 1941. Tedder had served with Dawson during the Battle of Britain and knew how well his unorthodox methods worked. Dawson moved into "Air House" with Tedder and conferred with him continuously. He immediately visited repair and munitions units. Afterward, he asked for a new maintenance group reporting directly to Tedder through him. No. 206 Group became active on 1 September but had actually been in operation since 17 June with Tedder's blessing and intercessions with Portal. Dawson also went to work immediately on equipment shortages from aircraft to vehicles.[69]

On 9 July, the Germans launched one of the few raids they made on Dawson's units, and the only effective one. At Abu Sueir, they destroyed twenty-nine aircraft and damaged another forty-nine along with five heavy workbenches and forty Bristol engines. Dawson moved what was left to the more protected base at Heliopolis. Other units moved into cave complexes.

One key outcome of Dawson's search for safer work facilities was his discovery that Egyptians with small workshops could repair and manufacture parts locally—a huge windfall. Dawson's "make do and mend" policy worked wonders in the availability of spare parts and the repair of propellers and other items.[70]

His other innovations included modifications to Hurricanes so that they could get high enough to destroy Ju 86P photoreconnaissance aircraft. The modified fighters worked in pairs, one stripped of everything except the radio and guns and the other stripped of everything except two guns. One guided the other to within visual range of the enemy so it could destroy the plane or damage it badly enough to force a descent to an altitude where the more heavily armed "guide" fighter could finish it. The first victory occurred at 49,000 feet of altitude. Luftwaffe reconnaissance flights over the Nile Delta virtually ceased. This gave the British freedom of action to engage in various elaborate deception efforts.[71]

Tedder championed Dawson's efforts, which collectively exerted a major influence on the air war. Ideas flowed in from Dawson's shops and the front. Improved air filters increased aircraft performance and flight hours between filter changes. By cropping the rotor on fighter engine superchargers, Dawson's technicians squeezed another twenty miles per hour or more out of the engines. After Spitfires arrived, ground personnel clipped their wing tips to improve rate of roll. By 1942, Dawson had nearly 23,000 civilians, mostly Egyptians, working alongside RAF specialists around the clock. RAF engine experts trained Egyptian workers to repair or replace one or two parts within each engine, forming an assembly line on which specialization of skills led to the output of 245 rebuilt engines per month. Propeller output reached 1,000 a month by 1943. Dawson's troops repaired aircraft armament by cutting off the last inch or so of the barrels after they were damaged in crash landings or began to wear out, resulting effectively in new guns.[72]

After Dawson had shops operating at peak efficiency, he turned to aircraft salvage and repair. His RSUs carried off all aircraft too seriously damaged to be repaired at the squadron or nearest depot. Specialized vehicles and skilled airmen scoured the desert for aircraft to transport back for overhaul, or, if they were too badly damaged, retrieve usable components. This often occurred within range of enemy ground forces as the airmen followed soldiers to crash sites. The Luftwaffe never imitated these efforts. RSUs formed a web of small units and a grid system to cover the Western Desert. This allowed the sixteen mobile repair units and eighteen mobile salvage sections to recover virtually every aircraft. Even after the Gazala disaster of spring 1942 and the loss of Tobruk, the British left behind just six aircraft. Air stores parks (ASPs) further improved logistics, with a generous complement of vehicles and spare parts. They could move within two hours of receiving orders and were issuing supplies within thirty minutes of arrival.[73]

Averell Harriman, Roosevelt's special envoy to Churchill, echoed Dawson's concerns about the supply of US aircraft to the Middle East when he passed through Takoradi just after Dawson. Takoradi had a capacity of 175 planes per month but far too few pilots to ferry them to Egypt. Harriman's concerns received immediate attention, and the pilot situation improved. Dawson's relationship with Harriman was blunt but good. When Harriman asked how well US aircraft were performing, Dawson said, "The Tomahawk is no goddam good."[74] A British wing commander echoed Dawson's sentiments when he referred to the Tomahawk, with its poor rate of climb and maneuverability, as "another American streamlined brick."[75] However, in conjunction with other officers, Dawson spearheaded the development of the Kittyhawk—a much more capable plane built on the Tomahawk airframe. Dawson won over Harriman and set the stage for vast improvements in the receipt, modification, and repair of US aircraft, including the arrival of US engineers and mechanics from aircraft factories.[76]

On 1 August 1941, Dawson reported, "There is now higher serviceability. The output from Takoradi has doubled. . . . Repair and maintenance has been organized. . . . The depots are . . . being dispersed and doing more repair work. The squadrons and repairs on site will do more. Engine overhaul is being developed. Equipment is needed from America."[77] The letter must have gotten into the right hands because more US equipment—and Americans—began arriving, and a formal liaison function was soon in place.

Dawson's maintenance organization had thirty-three major facilities, each with subordinate shops, including fifteen in Egypt. The rest were in the Sudan, Transjordan, Abyssinia, and Kenya. In Cairo's central districts, the Tura Caves, Heliopolis, and Helwan, forty-six separate repair offices and shops churned out spare parts and overhauled salvaged aircraft. There were another twenty-nine in other parts of Egypt. During Operation Crusader, RSUs and reinforcements brought RAF strength to 965 serviceable aircraft. By April 1942, during Rommel's counteroffensive, this number had decreased to 675, but there were 700 salvaged and under repair and another 490 expected to be operational within fourteen days. On 31 December 1942, there were 1,276 serviceable aircraft, with 787 more salvaged and under repair and another 395 expected to be fully serviceable in fourteen days. The RAF salvaged and repaired nearly as many aircraft during these campaigns as it received in reinforcements. At the end of the Tunisian campaign, there were about 1,300 serviceable aircraft, with 890 salvaged and under repair and another 390 expected to be serviceable in fourteen days. The RSUs kept the RAF in the fight despite frequent periods of numerical inferiority. The Axis air forces had nothing remotely similar. Tripoli had too few shops and the Luftwaffe too few salvage assets to move enough damaged aircraft.[78]

German Logistics

One of the most serious Axis failings in the Mediterranean was in the logistical arena. General Johann von Ravenstein, Twenty-first Panzer Division commander, who was taken prisoner during Operation Crusader, said, "The desert is the tactician's paradise, but the quartermaster's hell!"[79] The British, with their long experience in the desert, understood this and adjusted well, especially after Dawson arrived. The Germans never did to the same extent, nor did the Italians, despite their own long experience with the desert. Consequently, the combination of constant logistical problems, both self-generated and brought on by increasingly effective British ship and vehicle convoy attacks, combined to undermine Axis prospects for victory. Rommel's operational and tactical prowess notwithstanding, the Axis could not win without sound logistics and effective airpower. Delivering the latter was itself a logistics-intensive effort.

Problems surfaced with the move of Rommel's *Sperrverband* (blocking detachment) to Libya. One of the most serious for the next two years was the IAF's inability to escort convoys, leaving this job to overtasked German aircrews. In April 1941, the Germans asked the IAF leadership to step up. This resulted in an impressive set of instructions to IAF units to attack warships, protect convoys and ports, and give close support to ground forces. The Germans requested that IAF units detailed to the latter task be subordinated to Fliegerführer Afrika because unified command was essential for effective ground-air operations. This did not happen, nor was there any mention of air superiority. Without it, the Axis had little chance of keeping its shipping safe.[80]

Two days later, Fliegerkorps X noted that the IAF's "waves of attacks" on Malta involved one or two aircraft and often did not occur at all. The RAF had air superiority, in part because airfields on Sicily were crowded and aircraft vulnerable. Hoffman von Waldau said the crowding was made worse by many unserviceable aircraft and vehicles (another sign of logistical problems), leaving limited space for takeoffs and landings. Army convoys and supply dumps along the coast were also extremely vulnerable. There was very little flak. Italian airmen assigned to defend these places tried, but their aircraft were inferior and poorly maintained. Airmen continued to rely on visual observers for warnings. Frequent RAF raids from all directions made this a hopeless effort. Logistical problems hit the Luftwaffe particularly hard because the army controlled all ground transport (along with fuel allocations and maintenance facilities) and did not always give the Luftwaffe a proportional share. The Luftwaffe liaison team in the Armed Forces Motor Transport Office consisted of a colonel and two assistants—far too little rank to wrangle trucks out of army generals. German airmen found themselves reliant on Italian trucks for resupply. Von Waldau requested dedicated vehicle columns for the Luftwaffe without success. He also asked that the many Italian fighters "hanging around in southern Italy and Tripolitania" move immediately to forward airfields and assist Fröhlich's forces. Further, he asked that Italuft place greater pressure on the Italian Navy to perform its escort mission. Finally, he argued that for Rommel to maintain his position in Africa, OKL would have to send another fighter squadron, a flak detachment dedicated to protection of the Luftwaffe ground organization, and an aircraft reporting (mobile radar) company. A long-range reconnaissance squadron and other assets would also be required, as would a labor force under German control to get port and airfield facilities working effectively at Benghazi.[81]

The Italians continued to refuse unified command of long-range reconnaissance assets, making convoys even more vulnerable and forcing the Germans to rely more heavily on their already stretched air transport assets. Improving the Luftwaffe ground organization in Cyrenaica was the most critical task. It had far too few vehicles, spare parts, fuel, and munitions. Supply bottlenecks in Africa continued, brought on by Italian inefficiency and concerted air and naval strikes.[82]

Meanwhile, Enno von Rintelen pleaded for retention of Fliegerkorps X assets in Sicily to escort convoys because the IAF was not up to it. He also asked for more flak. On 15 May 1941, von Waldau said of Fröhlich, "The Fliegerführer can give only limited aid. No long-range reconnaissance data is [*sic*] available. . . . Owing to the supply situation it is not possible to concentrate sizeable reserves and launch counter-attacks."[83] Given Italian inefficiencies and RAF raids, shipping capacity far outdistanced port capacity, leaving Fliegerführer Afrika heavily dependent on Ju 52s to deliver high-priority supplies but with utterly inadequate means of receiving bulk supplies and heavy equipment. On 22 May, IAF units once again took over convoy protection and raids on Malta from Fliegerkorps X as it moved to Greece.[84]

On 15 May, Fliegerkorps X signaled OKL that Fliegerführer Afrika's supplies and LOCs were not secure owing to the British garrison at Tobruk. The Luftwaffe had to use its inadequate stock of mostly Italian trucks to run supplies around the fortress, which increased the threat of air attack. It also had too little cable to establish landline communications (another windfall for British intelligence). Consequently, Fliegerführer Afrika could not make a move forward unless additional vehicles and cable became available. Fröhlich's refusal to advance meant longer flight times and less loiter time. Also, because most of the formation's aircraft were Me. 109s and Stukas, whereas those of the Western Desert Air Force (WDAF) included longer-range aircraft, Luftwaffe machines remained much more vulnerable to airfield attacks.[85] Already, logistical difficulties were affecting Luftwaffe mobility, degrading its ability to leapfrog with the ground forces. This was a perennial problem. As Johannes Steinhoff said of the various campaigns in which he participated, "We had learned to subsist and fight with the help of a minimum of ground personnel—mechanics, armourers and signalers—for the main body, the workshops and the headquarters, seldom caught up with the combat component as it moved from place to place. Altogether a fine example of wasted resources!"[86] As the campaign progressed, even the combat echelons could not keep up with the army because of vehicle shortages.

Part of this problem stemmed from the duce's 27 May statement that although shipping space was adequate, unloading facilities could not handle major reinforcements. He neglected to mention, however, that of the next two convoys totaling twelve MVs, the Germans had two. Given that the Germans were the only ones capable of beating the British, it is stunning to think that in this case they received one-sixth of the total shipping assets. Although the average proportion was 5:3, it never reached 50 percent. Von Rintelen added that the rate of resupply for the Afrika Korps was barely enough to keep the troops supplied, much less increase their strength.[87]

The Germans continued working, ultimately without result, to forge an agreement with Vichy to use Tunis. The French had agreed tentatively to allow the use of the Bizerta-Gabes Railway and to ship 400 trucks from France to Tunisia and deliver them to the Libyan border. The Italians wanted in on the deal. On 2 June, the Germans told them that the Vichy government would not approve and asked them to desist so at least the German military would profit. The Italians retaliated by threatening to keep their shipping entirely for themselves by sending a very large reinforcement of Italian troops for an offensive involving 100,000 men and 14,000 vehicles. Keitel dismissed this fanciful plan, noting that aside from its logistical impossibilities, a smaller force of four armored and three motorized divisions (two and one, respectively, to be German) made more sense. He further reminded the Italians that a new shipping route through Bizerta, down the Tunisian Railway and supported at the far end of the line by 1,500 French trucks "loaned" to the Germans, would make a huge difference. General Francesco Pricolo, the IAF C in C, supported the plan as a means for reinforcing Axis air forces, noting that the most serious problem was the lack of vehicles to facilitate their rapid movement with the army. There were, however, no further discussions, and the deal fell through.[88]

On 19 June, General Cavallero told Keitel that when the Luftwaffe supply situation and ground organization permitted, fighter strength had to increase. He also noted that ground and air formations in Africa could not be reinforced fast enough for a fall offensive. Axis forces had to take Tobruk before that could happen, and Cavallero believed that only heavy artillery and major support from Fliegerkorps X aircraft against vessels supplying the garrison would facilitate the port's capture.[89] Fliegerkorps X Ju 88s rarely appeared over Tobruk, often sitting idle on Crete awaiting Royal Navy sorties that did not materialize or hunting British submarines and leaving the overextended Fliegerführer Afrika to do most of the work.[90]

On 12 August, as the logistical situation deteriorated, Hermann Göring called on OKW again to work with the Italians on the supply problem, which was hitting the Luftwaffe particularly hard. He observed that the Naples-to-Tripoli route was the only one in regular use, which made it easy for the British to concentrate air and naval assets there, especially because the route ran close to Malta. He suggested a greater use of the routes from Taranto and Brindisi to Benghazi, Derna, and Bardia because these would force the British to split their forces, allow Fliegerkorps X to provide more frequent convoy protection, and reduce the threat from Malta. He also said that since the capture of Benghazi, it had been underutilized, while the small but useful harbors at Derna and Bardia remained unrepaired. He asked OKW to get the Italians working on these projects. Finally, Göring noted that the railway line to Athens was still broken at Lamia, making it impossible to move and ship supplies from Greece, but did nothing about it. This, too, led to overcrowding at Italian ports and delays in sailings.[91]

Cavallero allowed the Luftwaffe to keep the loaned vehicles. This had much to do with Keitel's alarming statement that "of the German formations, Fliegerführer Afrika is worst off for vehicles. Its supplies can only be brought up with the help of vehicles placed at its disposal by the Italian Command."[92] Given Italian vehicle maintenance practices, and the high likelihood that they did not give the Luftwaffe their best trucks, one can imagine the challenges German airmen faced trying to move anywhere fast or in force. Finally, Keitel brought up the requirement for radar installations at ports—something still not in place after six months.

On 22 September 1941, in an "I told you so" moment, Raeder emphasized once again his earlier warnings not to move Fliegerkorps X to the eastern Mediterranean. The deterioration in the shipping situation began as soon as the move occurred. He asked urgently for the immediate return of

Me. 110s and other aircraft from Crete to Sicily, or for air units from Russia. Raeder warned that failure to provide heavy air escort immediately would mean that "further deterioration in the Mediterranean transport system will be inevitable and may result in major military reverses in North Africa with corresponding effect on the entire situation in the Mediterranean and on Italy's situation as an ally."[93] Raeder then asked OKW to seek Hitler's permission for Fliegerkorps X to protect convoys along the most vulnerable routes and to give orders for the move of additional Luftwaffe assets to Sicily. He also implored Hitler to negotiate with Mussolini for E-boat and minesweeper transfers and berthing space so the Kriegsmarine could send such assets. Raeder must have received cold comfort from OKW's answer later the same day, which allowed for Fliegerkorps X continuing escort only of the highest-priority convoys along the main SLOCs and no new German aircraft. Sinkings continued to increase.[94]

On 5 April, German liaison officers asked once again that Göring intervene to force more effective IAF operations in support of the logistical effort. The Germans did not even know how many aircraft the Italians were producing but understood the major serviceability problems. In another indication of the C2 fiasco among Axis air formations, German liaisons asked if the fighter and bomber units under the direct command of Italian Army C in Cs in Greece, Rhodes, and Africa, along with the C in C of the Italian Navy, might come under the command of the IAF C in C, General Pricolo. In addition to being dispersed all over the Mediterranean Basin and under multiple air force commanders, IAF assets were also very heavily controlled by soldiers and sailors. There was no centralized C2. Nor were there any night-flying or instrument-flying skills, another thing Italuft's German officers set out to correct with a request that the Luftwaffe oversee the development of these skills at an Italian blind-flying and navigation school run by German airmen. They also pushed for free manufacturing licenses so the Italians could produce Me. 210 and Ju 87D aircraft for convoy escort duty.[95]

Despite these efforts, Italian inefficiency and ineffectiveness persisted. In a 13 November briefing to the Führer, Raeder simply said, "As feared by the Naval Staff since July, the situation regarding transports to North Africa has grown progressively worse, and has now reached the critical stage. . . . Today the enemy has complete naval and air supremacy in the area of the German transport routes; he is operating undisturbed in all parts of the Mediterranean."[96] Hitler's response was to build a large fleet of small ships of about 1,000 tons (Siebel ferries) to compensate for the rapid decline of the Italian Merchant Marine. As further events would prove, he was deeply concerned about the situation and soon ordered Albert Kesselring and Luftflotte 2 from Russia to set things right.[97]

German Maintenance

Although the Germans fared better in the maintenance arena than in logistics, they ran into problems here too, especially regarding the Luftwaffe, which took a backseat to the army. Major depots were too few in number, and the small one in Tripoli represented the entire capability in Africa. There were only two engine-repair shops. Engine, propeller, and instrument work was very difficult in desert conditions. With spares in short supply and limited means to repair damaged parts, maintenance personnel sent parts they could not fix back to repair depots on the Continent. The parts made their way there and back by aircraft or ship, and many ended up at the bottom of the Mediterranean. Propellers were particularly scarce because several of Bomber Command's early large-scale raids had damaged or destroyed the factories dedicated to their manufacture. In fact, by 1942, propellers were top priority for salvage crews. This problem was exacerbated by the Luftwaffe's concentration of all spare parts of a specific kind in the same warehouses, creating huge vulnerabilities to bombing raids. As with the entire German war industry, dispersal of spare parts in Africa followed several WDAF direct hits on major stockpiles, but this approached the absurd as large warehouses shipped spare parts to as many as 100 smaller locations, making tracking and shipment to the front nightmarish.[98] Field repair shops, the next step down from depots, could do only partial overhauls and minor repairs. Increasing spare parts shortages led to cannibalization and the abandonment of aircraft lacking a few parts. Ernst Düllberg, who flew Me. 109s in JG 27, noted, "Often we had to use completely undamaged aircraft for parts to make others serviceable."[99] Desert weather made problems worse. As the war progressed and experienced ground crews became rarer as a result of casualties caused by heavy RAF raids and inadequate training pipelines, less-experienced mechanics misdiagnosed problems and used already scarce spare parts on unnecessary repairs, then circling back to the real problem of using more spare parts.[100]

Problems on the home front made matters worse. Whether as a result of disruptions caused by bombing and subsequent dispersal of industries or by inherent flaws in the system, spare parts procurement programs failed to meet operational requirements. About 25 percent of the aircraft industry's total production went to spare parts, and many of the smaller factories producing these were located in the center of urban areas and thus were the first to be damaged or destroyed in Bomber Command raids. Additionally, critical shortages of spare parts in combat units and ordnance depots became even worse as a profusion of new aircraft models went into production, often in tiny numbers. These many types and variants made it challenging for ground crews to install spare parts because they often learned to do so in the field rather than in a schoolhouse. Even after Field Marshal Erhard Milch, air inspector-general of the Luftwaffe, rationalized production based on the most numerous and successful models, spare parts problems persisted as a result of earlier missteps.[101]

Consequently, OKL offices in charge of spare parts deliveries were so overwhelmed that they turned to a civilian concern, the Aviation Requirements Company, to register spare parts at depots, store and administer them, deliver them to field locations, determine required replacement rates, and oversee the relocation of repair depots when necessary. It was an improvement over the previous arrangement but was not in place until 1944. Further, it underscored the disdain in which combat officers held logisticians. Rather than fix internal problems of their own making, they outsourced them to a contractor. To a great extent, this arrangement was the result of German grand strategy—or what passed for it—gone wrong. The OKL process of rushing new aircraft types into combat worked well for the short, sharp campaigns the Wehrmacht preferred, and the spare parts problem had no time to develop. However, after Operation Barbarossa began and the Germans also committed large air forces to the Mediterranean, the spare parts problems implicit in OKL's organizational structure and production priorities struck units in the field with a vengeance, robbing them of spare parts just as the combat became intense and long term.[102]

In a classic display of twenty-twenty hindsight, specialists involved in this process said,

> In summarizing it can be said that it is essential to insist from the outset of aviation activities on the timely supply of adequate supplies of the proper assortments of spare parts as a fundamental requirement for the maintenance of troops at the highest possible degree of capability. This principle did not receive the proper attention in Germany, neither during the buildup of the German Air Force, nor during the war.[103]

They noted further that the problem had developed despite repeated warnings from the Technical Office to OKL not to overlook spare parts in favor of various aircraft development projects. The plethora of new types and models drove unnecessary production of spare parts for each small production run. This led to an overage of spare parts for these types but a perennial shortage for the most important and numerous combat aircraft. The ensuing imbalance represented a major waste of manpower, productive capacity, and operational capabilities.[104]

One problem of particular importance to the Luftwaffe's fortunes in the

Mediterranean was the fact that an inadequate number of *Schleusen* (transit depots) were responsible for all reconstructions and alterations, including those required for aircraft heading to severe climates such as the desert. This included refitting aircraft to perform missions for which they were not originally designed, such as night-fighters and torpedo planes—both crucial but not available in sufficient numbers in the Mediterranean.[105] The requirement to develop many specialized aircraft types based on the huge range of operating areas and the "fire-drill" method of moving entire *Luftflotten* from Russia to the Mediterranean and back again put an enormous strain on the aircraft industry's most skilled workers, from engineers to mechanics. In fact, the amount of work caused by the profusion of types and variants, and then by the many alterations for service in different theaters, reached such proportions that "one could in reality no longer speak of normal serial production, which was replaced rather by continuously changing improvizations [*sic*]. . . . [This] was not due to technical difficulties or faulty planning, but exclusively to the constantly changing dispositions made by the higher command."[106]

Finally, Göring's order that the development-and-repair bureaus, along with the transit depots, were to make changes based on direct inputs from Luftwaffe commanders in the field, backfired because each commander had different views based on his location and mission, and there was no committee to prioritize inputs. This allowed aircraft firms to pick and choose preferred solutions and play the various bureaus against each other, with disastrous impacts on aircraft development and modifications. This was a result of Göring's preference for pilots over technicians as well as his ignorance of technological matters and his distrust of technical personnel. Thousands of modifications to various aircraft types followed, many of which were a matter of an officer's opinion based on experiences in one theater with one kind of mission.[107]

During postwar interrogations, Milch said 80 percent of the spare parts produced were distributed as follows: 50 percent went to repair shops in the Reich, 30 percent to repair shops at the front, and 20 percent to airfields. The OKL director of supply distributed the other 20 percent of overall production as he saw fit. The director used three different levels of supply—ample, just sufficient, and short—to categorize and track every spare part in the Luftwaffe inventory. At the start of the war, every operational squadron had its own air transport assets (one Ju 52 per fighter squadron and two per bomber squadron) with spare parts kits ready for each move. As the war continued, the units lost these aircraft and relied on supply dumps. This meant fewer supplies, slower delivery, greater risk of the destruction of large stockpiles by enemy air action, and inevitable stocking of too many or too few spares in a given area of operations. The system became less flexible and reliable as the war progressed because of the spare parts travails noted earlier. Milch sought a middle path between organic transport aircraft with spare parts kits versus supply dumps, but by the time he did so the air transport fleet had died at Stalingrad and Tunis, and the Luftwaffe was in rapid decline. Supply dumps prevailed at great cost to serviceability rates and operational flexibility.[108]

In a startling revelation, Milch mentioned that the Luftwaffe maintained separate repair workshops for every kind of aircraft in theater—a major duplication of effort. The British trained their mechanics to service multiple aircraft types in the same workshops. Captured German aircrews gave RAF interrogators a clear sense of the problems their various supply issues caused when they said there was a shortage of maintenance personnel and that consequently some aircraft had to go all the way back to Germany for repair. The absence of mobile field workshops at the start of the war, and their belated creation and inherent inefficiencies, exacerbated these problems. In turn, the perennial shortage of motor vehicles and, increasingly, of air transport assets, meant field workshops were constantly short of spares. Finally, salvage capabilities were a shadow of the RAF's, in part because of vehicle shortages and because they received far too little emphasis at the Luftministerium (German Air Ministry). Nor did they receive adequate attention at *Luftgaue* (Luftwaffe administrative commands), which supplied their counterpart Fliegerkorps and maintained their planes and airfields. The huge numbers of Luftwaffe planes captured on their bases and crashed in the desert drove home these points.[109]

These logistical shortcomings resulted from basic organizational weaknesses. Although the *Fliegerkorps* was the Luftwaffe's basic operational formation, it had no organic supply personnel. These worked in the associated *Luftgau*. Not only were these organizations often geographically separated, but even the various elements of the *Luftgaue* were up to 500 miles apart. Aside from creating coordination and delivery problems, this system also proved to be an intelligence bonanza for the Allies because German units had to communicate their supply needs and find out which ships their supplies were on, when they were sailing, and where they were going. Although certain officers within each *Fliegerkorps* took short training courses on supply and maintenance requirements, and acted as liaisons with *Luftgau* personnel, this proved ineffective. The only *Luftgau* officer assigned to the *Fliegerkorps* who had any understanding of the operations tempo and associated supply requirements was the quartermaster, with his small staff. These officers were often overwhelmed by the pressures involved in keeping a *Fliegerkorps* supplied with spare parts, ammunition, fuel, food, and vehicles. Additionally, many officers in the *Luftgau* and attached from there to the *Fliegerkorps* were *Truppensonderdienst* (special duty officers) and civilians, both held in contempt by most combat officers, who thought their work was detail oriented, nonmilitary, and thus undeserving of close attention. Africa's location also created problems because the Luftwaffe had to stockpile large quantities of supplies where and when it could based on shipments across the Mediterranean, whereas the well-oiled RAF system, underpinned by secure supply deliveries via the Suez Canal, could allocate precise amounts of supplies to each unit based on operations tempo and other factors. Large Luftwaffe stockpiles also proved highly vulnerable to RAF raids and SAS attacks, which exacerbated existing shortages.[110] The *Luftgau* system failed because its main offices were far from the front and therefore tended to be unlauded and unappreciated, which in turn made the *Luftgaue* less responsive to operational commands.[111]

The Luftwaffe supply system was inadequate from the outset and became more so as the war progressed primarily because it received so little emphasis. The air force's overly rapid development, and its senior leadership's disinterest in "mundane" issues such as logistics, made it impossible for Milch and other senior officials to fix the problems or even identify them before the war began. Milch had planned on having eight to ten productive years to develop the Luftwaffe's logistical infrastructure. As it was, he had fewer than five semiproductive years.[112] Colonel-General Ernst Udet, director of air armaments, summed up this problem in January 1938, when he told Milch,

> The movement of squadrons must not be hampered by administrative work. Officers will not be dependent on engineers—such a situation would prejudice the whole morale of the Luftwaffe. All campaigns will be short and German aircraft production will be so tremendous that during such periods of operation no major repairs will be necessary. Damaged planes will be repaired and salvaged at home after the campaigns are won.[113]

Aside from highlighting Udet's own myopic view, this passage says a great deal about the rationality of German combat officers, their cavalier attitude toward logistics, and thus the latter's low priority in the Nazi program for the domination of Europe. These views proved disastrous after the short and decisive campaigns ended and the long war of attrition on multiple fronts began. Even the Luftwaffe's effective practices—and there were many—degraded during the course of the war as a result of this overconfidence and the high attrition that followed. Mobile air equipment issuing stations, for instance, were an excellent organizational development designed to deliver spare parts rapidly to the fighting fronts. Trains, vehicles, and transport aircraft (with gliders attached) brought the goods, but as the fighting fronts expanded, particularly in the case of large and overseas theaters such as Russia and Africa, the demands became too great. The death of the air transport fleet at Stalingrad and Tunis robbed the Wehrmacht of the superb air resupply services the Luftwaffe pioneered long before the Allies.[114]

Axis Command and Control

Among the weaknesses in the Axis air forces, their C2 capabilities and almost complete lack of cooperation proved among the most dangerous. This was nowhere clearer than in the relationship between Rommel and Fröhlich. On 4 July, Rommel complained that whereas Afrika Korps headquarters was at Bardia, Fliegerführer Afrika's headquarters and airfields were in Derna—nearly 150 miles away. Gambut had the only advanced landing ground. Consequently, Rommel noted, "owing to this wide separation and the long approach flights which consequently must be made, there is no longer any guarantee of close co-operation, quick support for the Africa Corps' ground operations and secure and close communication between the two headquarters. In addition to the wireless there is a telephone connection to Derna, but the line is impossible. .... Repeated requests to move up his formations were rejected by the Fliegerführer."[115] Fröhlich based these refusals on the difficult logistical and supply situations, the lack of mobility, and inadequate flak. Rommel neglected to mention that he had commandeered many of the Luftwaffe's vehicles and flak assets and taken off into the desert without his Luftwaffe liaison officer on multiple occasions. One has to wonder how the Luftwaffe was supposed to leapfrog forward with the army.[116]

These problems, however, paled in comparison with the tangled Italian-German C2 structure. An 18 July agreement regarding cooperation between the Axis air forces, approved by both high commands on 20 July, called for close cooperation between the air forces but did not compel it. The ad hoc division of labor made uniformity of action difficult and rare. Reconnaissance aircraft were to cooperate but were not under unified command. PI was supposed to be passed immediately to all interested headquarters, but there was no means for ascertaining which ones were interested because the air reconnaissance process remained fractured. Operations tended to be determined by aircraft and aircrew capabilities, which meant the Germans flew the most dangerous and challenging missions, suffering the highest losses. These could not be made good given the recent attack on the Soviet Union and heavy attrition there. Even though operations against Alexandria, Cairo, the Suez Canal, and shipping targets were best carried out by German platforms, Superaereo had lead authority to order operations against these targets. There was no combined staff in any sense of the word—just a collection of liaisons. These were supposed to coordinate operations but rarely did so. Axis air forces fought parallel, not combined, air campaigns.[117]

One small passage from the agreement serves to underscore the dysfunctional nature of the entire arrangement: "Superaereo, via Italuft Liaison staff, is to keep Fliegerkorps X acquainted with the situation in the Western and Central Mediterranean and to inform Supermarina of the situation in the Eastern Mediterranean as shown in the reports submitted by Fliegerkorps X." In other words, the IAF High Command, located in Rome, was to share information with Fliegerkorps X, located in Athens but with operational control over Fliegerführer Afrika in Derna, through the Italuft liaison organization in Rome. Even more astounding, Fliegerkorps X was to send reports regarding Allied sea movements in the eastern Mediterranean to the Italian Navy Supreme Command, located in a different part of Rome than Superaereo, through the same channels. Completely aside from the enormous waste of time, this nonsystem of command and coordination made even the pretense of effective combined operations untenable. A single, unified staff including senior commanders from each country's air, ground, and sea services—one with the authority to make operational decisions and the responsibility for their outcome—would have made an enormous difference in the effectiveness of Axis air and combined-arms operations, especially regarding air-naval coordination.[118]

This dispersed, confusing, and redundant patchwork spawned a tangled communications web in which overworked wireless, radio, and cipher operators had to send far more messages than necessary, opening Axis plans and intentions even further to intercept, compromise, and exploitation. The Germans and Italians were supposed to share virtually everything from their different headquarters around the Mediterranean, including aerial reconnaissance reports; air, ground, and naval activity; convoy operations and losses; logistical and supply issues; and a host of other insights. What was clearly lacking, however, was the ability or will to do anything with this information that would help to develop or evolve some sort of combined system of command and coordination.[119]

These emerging German and Italian weaknesses in the logistics, supply, maintenance, communications, and C2 arenas, along with uneven but generally stronger British improvements in these areas, played an important role in the outcome of Operation Crusader and every subsequent campaign in the Mediterranean.

6

Preparations for and the Conduct of Operation Crusader
September–December 1941

Axis Strategic Choices

OPERATION BARBAROSSA, an event of central importance to the course of the war, occupied virtually all of Adolf Hitler's time, along with that of the Oberkommando der Wehrmacht (OKW) and the service staffs, during fall 1941. Expectations that the Soviet Union would collapse within six weeks after Germany's 22 June 1941 invasion proved in error. Instead, the Germans found themselves immersed in an immensely costly and lengthy two-front war for which they and their allies had not prepared. This had ripple effects not just on Germany's war effort but on those of all the warring powers. German plans to move reinforcements to the Middle East after Russia was finished did not materialize, with the notable exception of the redeployment of Luftflotte 2 and Fliegerkorps II from Russia from December 1941 to April 1942 with Field Marshal Albert Kesselring moving to Italy as *Oberbefehlshaber Süd* (supreme commander south). As the campaign in Russia dragged into the fall, Erwin Rommel and other German officers in the Mediterranean Basin realized that there would be no major ground or air reinforcements for them. The long war of attrition German commanders feared had arrived. German performance in Russia and the Mediterranean theater, with its immense oil resources and points of strategic importance such as the Suez Canal, Red Sea, Gulf of Aden, Gibraltar, and Malta, would determine the course of the war.

As German prospects in Russia worsened, and Italian ones for a rapid victory over the British had long since disappeared, the Axis nonetheless still hoped to win the war of attrition as a result of Japan's entry into the war, superior will, and by defeating the Red Army definitively in 1942 should this not come to pass in 1941. However, in a huge oversight that soon became clear, the Germans and Italians failed to arrive at any kind of coordinated effort with the Japanese for an Indian Ocean strategy after the latter entered the war.

Anglo-American Strategic Decisions

The reverse in the Western Desert after Rommel arrived, followed by the disaster in Greece, dealt the British serious military blows, but their strategy for the Middle East theater remained fundamentally unchanged. They would hold it at all costs. This proved wise but difficult from May 1941 on, as the disaster on Crete followed hard on the heels of the one in Greece, and from there Iraq, Syria, and Iran all posed challenges to British control of the region. Effective planning and operations, along with a heavy expenditure of assets, brought these areas under firm British control and deprived the Axis of a "back door" into the Middle East, at least for the time being. Fears about a German drive into the Middle East were renewed with Operation Barbarossa, and the British devoted a great deal of effort to preparing for such an eventuality.

After Germany invaded Russia, and the Japanese occupied the remainder of French Indochina the following month, Franklin Roosevelt became convinced that the United States would have to enter the conflict. He expected either a Japanese or German provocation—or both—that would drive him to request a declaration of war. US senior leaders agreed that a peripheral strategy under British lead was still the only effective option for causing significant Axis attrition, and they continued in this belief after Hitler attacked the Soviet Union. US lend-lease materiel began flowing to Russia soon after Germany attacked. Despite initial resistance among US senior officers to help Russia, its survival through the worst of the German attack convinced them that it might become a primary force in bringing about Germany's defeat. Roosevelt thus called for immediate and massive aid to Russia beginning in July 1941—a pledge he kept to Stalin and a decision that contributed in major ways to Allied victory. At this point, the process of "closing the ring" was just getting under way, but Roosevelt and Winston Churchill saw that between efforts in the Mediterranean and Middle East, Russia, and eventually northwestern Europe, they could bring this strategy to fruition.[1] George C. Marshall told Roosevelt on 22 September, "Germany cannot be defeated by supply of munitions to friendly powers, and air and naval operations alone. Large ground forces will be required," noting further that the Allies would have to help the Russians when he insisted, "We must come to grips with and annihilate the German military machine."[2]

As Marshall drove home the importance of destroying the Wehrmacht through ground action, Roosevelt had already ordered the US Navy to escort convoys to Britain beginning in April and slapped major sanctions on the Japanese in late July, in effect giving them a choice to back down or secure their own access to oil, steel, rubber, and other strategic commodities. Russia's entry into the war, followed by Japan's, produced a series of decisions and effects within the Middle East that influenced the course of the campaign.[3]

At the Arcadia Conference in Washington, DC, immediately after US entry into the war, the United States and Great Britain agreed to establish the Combined Chiefs of Staff (CCS) organization to plan Allied global military strategy and direct forces and resources accordingly. The US Joint Chiefs of Staff (JCS) was the counterpart to the British Chiefs of Staff (CoS). These two organizations formed the CCS. The officers constituting this group were exceptionally capable, enjoyed the full confidence and support of their political leaders, and very quickly had a clear objective—the unconditional surrender of the Axis—upon which to focus their formidable talents. The "Germany-first" decision reached earlier also held, with Italy as the necessary stepping stone in Europe. Further, the participants agreed (despite strong US Army dissent) on an invasion of North Africa in 1942 (Operation Torch) to drive out the Axis. The Allies also agreed to the development of joint resource boards to pool, track, and distribute assets to the right locations, at the right times, and in the proper quantities. Nothing remotely like the highly effective combined policy and military effort inherent in the CCS and its frequent and productive wartime interactions with Roosevelt and Churchill existed on the Axis side. Although major disagreements soon emerged between the Americans and British regarding longer-term strategy for the Mediterranean, at this point they were in accord. The peripheral strategy remained in place, and the British prepared to support it further with Operation Crusader, which they hoped would lead to Rommel's defeat and perhaps the ejection of the Axis from North Africa.[4]

Air Preparations for Operation Crusader

Claude Auchinleck and Arthur Tedder visited London in July. All participants in their meetings with the Air Ministry, Imperial General Staff, CoS, and War Cabinet agreed that the Western Desert Air Force (WDAF) needed many more aircraft of the most modern types. These would maintain air superiority while supporting ground forces during Operation Crusader to take back Cyrenaica and advance into Tripolitania. All the Middle East theater commanders in chief (C in Cs) supported a major increase in air assets, having come to realize how dependent operations in the Mediterranean were on land-based airpower. Churchill's guidance coming out of this visit is worth quoting at length:

Nevermore must the ground troops expect, as a matter of course, to be protected against the air by aircraft. If this can be done it must only be as a happy makeweight and a piece of good luck. Above all, the idea of keeping standing patrols of aircraft over moving columns should be abandoned. It is unsound to distribute aircraft in this way, and no air superiority will stand any large application of such a mischievous practice. Upon the military Commander-in-Chief in the Middle East announcing that a battle is in prospect, the Air Officer Commanding-in-Chief will give him all possible aid irrespective of other targets, however attractive. Victory in the battle makes amends for all, and creates new favourable situations of a decisive character. As the interests of the two Cs.-in-C. are identical it is not thought any difficulty should arise. The A.O.C.-in-C. would naturally lay aside all routine programmes and concentrate on bombing the rearward services of the enemy in the preparatory period. This he would do not only by night, but by day attacks with fighter protection. In this process he will bring about a trial of strength with enemy fighters, and has the best chance of obtaining local command of the air. What is true of the preparatory period applies with even greater force during the battle.[5]

There were several clear insights here relating to proper airpower employment. Air superiority was now seen as the key enabler for all other operations. The ground commander would set overall objectives, but the air commander must have command and control (C2) of air assets to maximize their efficiency and effectiveness. Interdiction efforts preceding the offensive would make a key difference in the battle by destroying enemy assets and bringing on a decisive battle for air superiority.

One of the challenges relating to regular Royal Air Force (RAF) presence over the battlefield or anywhere, for that matter, had been a constant shortage of pilots. Prior to summer 1941, there were only three operational training units (OTUs) in the Middle East, slowing the training of combat-ready aircrews. Given high losses in France, the Battle of Britain, and Greece, this represented a serious failure to learn. An Air Staff decision, based on Tedder's strong recommendation, to add a fourth OTU proved crucial, especially because captured German documents commented on the evident lack of flying training and combat experience among most RAF aircrews. Even with an all-out effort, however, fighter-pilot strength in the Mediterranean and Middle East was eighty below the aircraft establishment, with no reserves. Nonetheless, Tedder got this problem in hand by November 1941. The shortage of trained airmen, including ground, maintenance, and salvage crews, was severe enough that the CoS sent the 35,000 airmen waiting for passage to the Mediterranean before their army counterparts in order to expand the number of squadrons to sixty-two and one-half. As Graham Dawson brought maintenance and salvage capabilities to maturity and Tedder addressed pilot shortages, Arthur Coningham took command of the WDAF from Raymond Collishaw in July.[6]

Royal Air Force Middle East (RAFME) probably could not have had a better command team than Tedder and Coningham. Unlike the more reserved Tedder, once described as a "pale, diminutive gremlin," Coningham was tall, powerful, athletic, and an aggressive commander. One officer described him as "big, masculine, confident. . . . He had an easy, attractive personality, a ready and colorful flow of talk."[7] However, Tedder liked Coningham's combination of deep thought, ability to listen, track record for decisive and effective action based on what he learned and intuited, and ability to keep his own counsel. As different as Tedder and his new WDAF commander seemed, they shared the essential qualities of outstanding military leaders.

Coningham began by improving tactics for naval overwatch and offensive fighter engagements. Tedder helped by restructuring flying wings, giving them their own maintenance and logistics organizations along with the three flying squadrons. These improvements were largely in place by the start of Operation Crusader. Greater flexibility and mobility followed. One of the key outcomes of Operation Battleaxe was that both services recognized the need to improve air-ground cooperation. Consequently, joint exercises became frequent from July 1941 forward and increased in frequency and sophistication. An interservice committee was set up to study air support for the army. By September, "Middle East Training Pamphlet (Army and Royal Air Force) No. 3—Direct Air Support," discussed in Chapter 5, was in circulation. This was one of Coningham's most vital contributions. His views regarding the Luftwaffe were straightforward: "Hit them hard and keep hitting them; then they won't ever hit back. Never do anything to the enemy unless you yourself have first discovered the antidote."[8]

Tedder allowed Coningham to appoint subordinate commanders, a process that worked well given the latter's exacting standards and willingness to fire those who failed to meet expectations. Coningham also had substantial experience with the army and thus found and built on common ground with Tedder.[9] Until Tedder and then Coningham arrived, the RAF and army tended to squander the "precious intervals between campaigns for reflection and training."[10] New fighter and bomber tactics took too long to migrate from the United Kingdom. As late as October 1941, Tedder found that most of his pilots had a "village cricket" level of performance. The following year, he upgraded this to a "test-match" level of ground-air cooperation. Tedder told Coningham to "get together" with the army commander as his first task upon arriving in the Western Desert. Coningham said this proved to be a decision of "fundamental importance and had a direct bearing on the combined fighting of the two services until the end of the war."[11] These training intervals were particularly vital for RAFME because there were by definition few of them. As an RAF pilot observed during Operation Crusader,

> Armies spend most of their days preparing to fight and living in comparative safety behind the lines. The soldier thinks in terms of the clash of arms, great victories like Blenheim or Waterloo. Air forces meet the enemy on a routine basis and our psychology is different. We abominate pep talks because every day is somewhat similar. Crisis is normal. Instead of stiffening the sinews for the decisive encounter, we "sweat out" a tour. Decorations are awarded not so much for individual acts of courage as for length of time on the job.[12]

Another author agreed that "air operations, unlike battles on land and sea, are a continuous affair. They go on, like the weather, all the time without the splendid ornament of battle-names. Their battle-honours are measured by months rather than by the breathless hours and minutes which write the names of victories into history."[13] Neville Duke said on 31 December 1941, "I face this new year with some trepidation and have had enough of air fighting, out here at any rate, where there is no rest or respite from continuous fighting. Fighting for your very life each time you go up in a machine which is not equal to the enemy's."[14] This trepidation was well founded when one considers that Jagdgeschwader (JG) 27 alone, the most accomplished Luftwaffe fighter unit in Africa, had 1,300 confirmed aerial victories during the campaign—an important reminder that whatever their leadership's failings, German pilots were excellent.[15]

This constant wearing away of assets drove Tedder to improve maintenance and repair. He and Coningham, with Air Commodore Thomas Elmhirst's assistance, also devised a system in which two existing wing headquarters leapfrogged each other to ensure effective C2 of forward air assets. Each squadron also deployed advanced parties. A standard RAF squadron had three. The A (advanced) party was a refueling team that moved to advanced landing fields with maximum speed. The B (rear) party conducted maintenance at the operational landing field. The C party remained at the primary airfield with the squadron's workshops and logistical assets. The army built advanced landing grounds, which enhanced combined-arms operations. As airfields improved, so did RAF mobility and thus operational effectiveness. Tedder directed the creation of local headquarters in each region of the theater; worked with Dawson to develop repair, salvage, logistics, and supply capabilities at or dedicated to each unit; and secured all the assets required for Coningham's advanced headquarters to move rapidly with the Eighth Army.[16]

Tedder in Peril

Fate nearly gave the Axis an edge when Tedder got into hot water with Churchill and almost lost his job. As planning for Operation Crusader progressed, British intelligence in the Middle East assessed that the Axis had more operational aircraft in the Western Desert and central Mediterranean than did the RAF. It also had even more aircraft in the eastern Mediterranean and could therefore reinforce its air contingent.[17]

Tedder's figures, delivered to Charles Portal on 29 September and again on 13 October, created a political firestorm because the prime minister of New Zealand insisted that his countrymen never again fight without air superiority. Tedder argued that even if the Axis had more planes, their serviceability rates were poor and reserves much more limited. Churchill nevertheless insisted that the numbers soothe rather than aggravate his fellow prime minister given that his troops had borne the brunt of air attacks in Greece and especially on Crete.[18]

Portal sent Tedder a list of questions regarding his estimates and what he could do to increase RAF strength. Tedder's answers did not satisfy Churchill. Portal told Tedder that his 13 October strength comparisons were particularly distressing even if Tedder's point about superior RAF serviceability and tactics was true. Air Ministry intelligence calculated serviceable German aircraft in Libya at 100, with only 30 being fighters.[19] Tedder replied that he was confident the WDAF would gain air superiority, saying, "I am ruthlessly stripping other theatres and formations to give Western Desert all we have. I am not satisfied I cannot do more but I will not (repeat) will not promise until I am sure I can effectively keep promise. I will not put dummies in shop window for D day. Our battle is joined now (r) now and my object is to maintain and increase the pressure to attain air superiority before the Army move."[20]

Churchill insisted that a senior RAF officer go to Egypt immediately and resolve the issue. Tedder's good fortune held when Portal selected Wilfrid Freeman—no doubt to protect Tedder while arriving at the figures Churchill required. Portal told Tedder, "Assure you that this unfortunate development in no way affects my absolute confidence in your ability and determination to win."[21]

Tedder promised to work with Freeman so Churchill and the prime minister of New Zealand would be satisfied. Freeman reported directly to Churchill and Portal. This is the closest Tedder ever came to being relieved.

In a signal to Auchinleck, Churchill blustered that Tedder's strength estimates were misleading and had created serious problems with the dominions. Further, he said, Freeman was a better officer than Tedder and offered him to Auchinleck as Tedder's replacement. Auchinleck signaled back his full confidence in Tedder and asked earnestly that he stay on, especially with Operation Crusader about to begin.[22]

Freeman's visit and initial report satisfied Churchill. The prime minister signaled Tedder that all was in order, and the key now was to win the battle and the campaign. Tedder responded with a note of thanks, saying "Freeman's visit most helpful and resolved many differences. The concentration of the squadrons is nearing completion and they are already taking their measure of the enemy. We have but one aim, to help the Army win this battle."[23]

In a final message to square things with Churchill, Freeman said that he and Tedder had conducted a thorough analysis of comparative aircraft strength in Cyrenaica. He had increased strength by stripping all operationally useful aircraft from subordinate commands. Auchinleck agreed that given the comparatives, and the fact that so many of the Axis aircraft were Italian, the RAF would have air superiority. Nonetheless, in a last snub, Churchill remarked that Tedder's initial figures and the final ones sent by Freeman were only as different as the "difference between white and black." This put an end to the matter and allowed Tedder to concentrate on the upcoming battle.[24]

After the furor subsided, Auchinleck wrote Portal to thank him for doing his part to save Tedder's command. He agreed with Portal that sacking Tedder would have been a "dreadful mistake." In his view, Tedder was a superb joint commander and a primary player in the cooperation that was becoming the norm between the two services. Nobody, he concluded, better understood the air and joint situations. Army officers respected Tedder and his team and found common ground with them.[25] And, in a volte-face after the RAF very quickly dominated the air once Operation Crusader commenced, Churchill told Auchinleck, "Say bravo to Tedder and R.A.F. on air mastery."[26]

Air Operations for Crusader Begin

The air component of Operation Crusader began nearly five weeks before ground forces attacked on 18 November. Reconnaissance, air superiority missions to include concerted raids on airfields, attacks on enemy sea lines of communication (SLOCs) and vehicle supply routes, and attacks on German reconnaissance flights all began in force. The new strategical reconnaissance units conducted long-range, high-altitude collection flights with the Spitfire and Mosquito photoreconnaissance (PR) aircraft now arriving from the United Kingdom, the photographic reconnaissance units focused on specific points closer to the battle area, and the survey flights worked with the army to provide special photos for mapping. Sophisticated photointerpretation and analysis capabilities made full use of the much increased number of aerial photographs.[27]

Despite the arrival of the Me. 109F, which was superior to all RAF fighters in Africa, Tedder was confident the RAF would control the air. His three-phase plan involved gaining air superiority, attacking supplies using night-bombers, and assisting the army's advance. Destroying the German fighter force was of paramount importance. Tedder ordered intensive flying training and tactical evolutions.[28] Efforts to force the Luftwaffe into an air superiority battle were initially disappointing because the Germans refused to take the bait, but this allowed RAF aircrews to concentrate on Axis airfields. Heavy rains helped the British, whose airfields had firm soil, at the expense of the Axis, whose airfields were on softer soil. This further hampered German efforts to discern that a major offensive was about to begin because they remained relatively blind with most of their reconnaissance aircraft mired in the mud or shot down.[29] General Fritz Bayerlein, Rommel's chief of staff, said the Commonwealth approach march and deployment went unnoticed.[30] RAF operations from 14 October to 17 November consisted of 3,000 sorties and did significant damage to Axis logistics.[31]

On 17 November, the day before ground operations began, Tedder wired Portal that "squadrons are at full strength, aircraft and crews, with reserve aircraft, and the whole force is on its toes."[32] He wished his forces "good hunting." By this time, RAF aircrews had set the conditions for success on the ground and now roamed over enemy territory at will. They stayed focused on maintaining air superiority, interdicting logistics, conducting reconnaissance, and engaging in direct support of the army.[33]

Meanwhile, Hitler again changed his mind and decided he must make a bid for German primacy in the theater. "Führer Directive No. 38 of 2 December 1941" ordered Luftflotte 2 Headquarters, together with Fliegerkorps II (General of Fliers Bruno Loerzer) to deploy from the Eastern Front to the Mediterranean under the command of Field Marshal Albert Kesselring as the newly appointed *Oberbefehlshaber* Süd (commander in chief south). His task was to establish naval and air superiority between Italy and North Africa. His air forces were to neutralize Malta and provide support to ground forces in Africa. Kesselring was subordinate, putatively, to the duce and had a mixed German-Italian staff. He commanded Fliegerkorps II and X and could give orders to Axis naval forces, at least in theory. In reality, the Italians resisted combined operations, and Rommel undermined

Kesselring's authority routinely with direct appeals to Hitler. Even more troubling, General of Fliers Ferdinand Geisler, commanding Fliegerkorps X and stationed in Greece, was the air officer in command during the opening phases of Operation Crusader, with General Stephan Fröhlich reporting to him. Fröhlich, stationed in Libya, thus took his orders from Geisler in Greece (another intelligence bonanza for British signals intelligence specialists), whereas Luftflotte 2 had to move whole cloth from the Eastern Front after major overhauls to prepare the aircraft for desert operations. Nearly half of the total available aircraft in theater sat partially idle in the eastern Mediterranean while their opposite numbers fought tooth and nail in the central Mediterranean. Ironically, when Fliegerkorps X Ju 88s and He 111s did appear, always without Me. 109 escort given the distances involved, they put a great strain on Fröhlich's remaining Me. 109s because the latter had to fly close escort.[34]

Before relating the RAF's and the Axis air forces' performance in Operation Crusader, it is worth taking stock briefly at this point of the comparative actions and losses from 12 June 1940 to 31 October 1941, on the eve of the offensive. From the start of operations in the Mediterranean and Middle East, the RAF had gone after air superiority and exploited it. Axis air forces had failed to do so with the notable exception of the campaigns in Yugoslavia and Greece. During this period, the RAF made 1,078 airfield attacks compared with 258 for the Axis. This, in turn, facilitated raids on ground forces, shipping, and ports. While the RAF made 1,031 effective raids of four or more aircraft on Axis ground-force concentrations (mostly supply convoys), the Axis managed 377. They carried out many others, but these involved one to three aircraft and did virtually no damage. Dissipation of effort was debilitating. RAF aircraft on Malta and in Greece carried out 569 raids on ports and shipping. The Axis made 117 such raids. Both sides flew relatively few mining missions (the Axis nineteen and the RAF ten), but concerted mining of the Suez Canal would have made a dramatic difference in the logistical balance. Even after the Luftwaffe arrived in January 1941, the failure to focus on air superiority, and the dissipation of effort that followed, meant the RAF controlled the skies. This had nothing to do with numbers—the IAF had many more aircraft in the Mediterranean than did the RAF before Operation Compass, as the Luftwaffe and IAF together did through the Gazala offensive in May 1942. After Luftflotte 2 arrived, it alone had more aircraft than RAFME. Nonetheless, the RAF outperformed its adversaries, often with inferior aircraft. The operations summaries containing these data hold a wealth of other insights about differences in airpower and combined-arms effectiveness.[35]

As part of this combined-arms effort, RAF and army officers worked together closely to ensure everyone understood the contributions of air-power to success on the ground. On 30 September, Auchinleck and Tedder cosigned an air-ground cooperation document for Operation Crusader and sent copies to all army battalion commanders and RAF squadron command-ers. The document was the result of hard-learned experience combined with intensive joint exercises. Its message was clear: combined-arms operations would only succeed to the extent both services cooperated. It emphasized that coordination between tactical reconnaissance (Tac R), army, and RAF units was vital because the best way to attack enemy ground formations was in defiles, traffic jams, or other places where they could not disperse. Word had to pass quickly through the new army air support controls (AASCs) and air support controls (ASCs) with their tentacles and additional wire-less links. Air reconnaissance assets were relatively scarce, so commanders should call for them only when necessary.[36]

Despite the Luftwaffe's failures to focus on air superiority, Tedder credited it with the ability to change. He looked constantly for a Luftwaffe effort to control the air, assessing that

> on the whole he [the Luftwaffe] would be prepared . . . to strike at us before our air strength had increased to such an extent as to ensure our attainment of air superiority over the battle area. He would undoubtedly arrange for the full force of aircraft in GREECE and CRETE to be used against us in the WEST-ERN DESERT by attacks on our communications, supply organisation and bases.[37]

This did not happen, in part because the Luftwaffe failed to do what Tedder contemplated, and in part because the RAF struck the Germans' airfields first, but Tedder was not about to be caught by surprise.

As the British buildup accelerated, Rommel tried to determine their inten-tions with a reconnaissance in force. It failed as a result of a combination of petroleum, oil, and lubricants (POL) shortages and intensive RAF attacks that caused heavy losses. The Germans should have realized that something was amiss. Huge truck convoys totaling more than 1,000 vehicles moved forward under RAF cover. This activity originated at the Mersa Matruh railhead and was under direct Luftwaffe observation. The same was true for reconnaissance of British airfields, which disclosed major increases in aircraft at forward bases. The Luftwaffe did not raid the airfields. The Ger-man characterization of the offensive as having achieved "strategic surprise" makes it clear that Rommel misread this air intelligence.[38]

Fliegerkorps X's strength had decreased during 1941, mostly as a result of transfers to Fliegerführer Afrika. Ironically, its commitments increased, with raids against vessels throughout the Mediterranean and support of the Vichy regime in Syria. Bombers even raided the oil refinery at Haifa and scored several hits. Small groups of bombers attacked the Suez Canal. Raids on British truck convoys caused losses, and missions against Tobruk continued on an almost daily basis. Fliegerkorps X also protected shipping convoys and unloading operations at Tripoli and Benghazi—a task requiring most of its assets. Although these missions produced notable successes, dispersion of effort was a serious drawback. The Germans seemed incapable of focusing on vital tasks that included destroying the RAF, closing the Suez Canal, and protecting convoys.[39]

By 31 July, German naval officers reported that 73 percent of Axis ships departing Italy were not reaching North Africa—an unsustainable loss rate. Finally, this prompted the German High Command to order Fliegerkorps X to concentrate exclusively on convoy protection. Although this order lowered convoy losses, it robbed Luftwaffe fighters and bombers of their offensive characteristics and gave the British a largely free hand in the air. The Allies unloaded an average of 150,000 tons of supplies per month in Egypt during the summer, whereas the Axis unloaded an average of 82,677 in Libya.[40]

The Ground Offensive Begins

The British Army went forward on 18 November 1941, with twenty fighter and thirteen bomber squadrons in direct support. Attacks on airfields and supply convoys intensified after 18 November, and the Hurribomber fighter-bomber made its first appearance. Despite a German counterattack, the Eighth Army held and resumed the offensive with strong air support.[41] "There was," Tedder said,

> this important difference between the two air forces, namely that our squadrons consistently carried the fight to the enemy and gave him no respite, whereas the enemy's own offensive operations either against our troops or targets further to the rear were on a lesser scale and were comparatively ineffective. Our fighters continued, for example, to make low flying attacks on enemy concentrations and airfields despite the hazards from strong and accurate anti-aircraft fire. The dividends that were paid in destroyed and damaged vehicles and aircraft, and in that less tangible factor, moral ascendancy, were very great. In the result, the enemy was given no rest, for his troops were subjected to the great nervous strain of "round-the-clock" bombardment.[42]

The Axis air response was initially ineffectual. Kesselring was visiting Rommel's headquarters at this time and succeeded in obtaining air transport units to fly in desperately needed supplies. From 21 to 23 November,

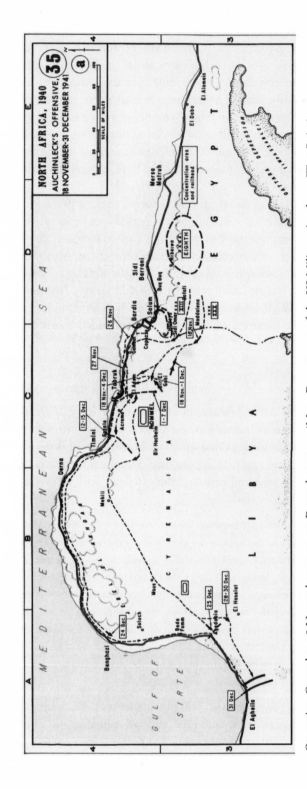

Operation Crusader, 18 November 1941–31 December 1941. (Maps Department of the US Military Academy, West Point)

German air units also became more active, shooting down thirty-six aircraft on the latter two days and doing modest damage to Eighth Army columns. Fliegerkorps X sent reinforcements on 21 November, bringing Frölich's strength to a full *Geschwader* of bombers, dive-bombers, and fighters—about 300 aircraft.[43]

Kesselring recognized Fliegerführer Afrika's problems and established Air Command Libya to remedy them. He assigned an officer to revitalize the Luftwaffe supply system across the Mediterranean and to improve operations in preparation for the arrival of Luftflotte 2. Logistical problems transcended the Luftwaffe. As a result of heavy shipping losses, Axis forces had only two weeks' supplies for full-scale operations. With the RAF destroying many of these, and heavy fighting eating up most of the rest, Rommel decided to abandon the siege of Tobruk and the forces holding the Sollum line, and to defend the Gazala position to the west. By 15 December, Rommel decided to abandon all of Cyrenaica.[44]

On 3 November, Tedder had signaled Portal that hammering the relatively small German Me. 109 contingent had proven the key to air superiority. He emphasized that the Me. 109F was by no means invincible. However, it was a menace. Me. 109F pilots worked in groups of between three and six to ambush larger RAF formations. WDAF losses on 9 December alone were eleven aircraft and perhaps half the pilots with no confirmed German losses. The Me. 109F's superior speed and climb, its 20 mm guns, and the lower experience level of RAF pilots all contributed to the problem.[45] To amplify these points, Coningham said,

> We must stop the people in England thinking that 2nd class aircraft will do us out here. This has become part of the European Front and the German has appreciated it to the extent of sending his best aircraft and units. . . . Air Ministry must respond to the German reinforcement of the African Front with equal reinforcements of the best aircraft and personnel from England. . . . The Kittyhawk alone will not do for this front.[46]

He also concentrated his top fighter pilots in special formations to deal with the German fighters and give green pilots a better chance.

To make matters more challenging, despite Rommel's order to retreat on 7 December, the Luftwaffe was gaining in strength as the campaigning season in Russia ended. The Germans sent first-line aircraft, including more Me. 109Fs. Tedder believed that even with Luftwaffe strength increasing, he could help the army deal Rommel's forces a fatal blow if he could reduce their supplies. His bombers continued going after fuel tankers and other merchant vessels (MVs) along with vehicle supply columns—especially fuel-tanker trucks.[47]

With Rommel on the run and the Eighth Army in pursuit, RAF airfield parties and army royal engineers kept up, repairing the Gazala airfields in two days under artillery fire and bringing in 10,000 gallons of aviation fuel *in advance* of Eighth Army forward units. They repeated this during the move to Mechili, moving 15,000 gallons of fuel to airfields there on every available vehicle, and receiving another 60,000 gallons from two army motor-transport companies. From there, they moved to Msus with another 10,000 gallons of fuel and other supplies. Here, landing parties built a 1,500-foot runway with dispersal points, and four squadrons moved in the next day. This grew to eleven with the addition of a second runway. Coningham said with some pride that agility, improvisation, and risk-taking along with army assistance had underpinned these successes.[48]

On 7 December 1941, the day the war in the Pacific began, Tedder reported on the progress of Operation Crusader, noting that Rommel was ready to fight to the last tank and putting up fierce resistance. His armor, antitank weapons, and tank repair assets were still superior to the Eighth Army's, as was his ability to make rapid attacks. The RAF held the upper hand in the air, but it was a constant struggle. Tedder felt that despite Rommel's tactical advantages, heavy attacks on vehicle convoys, especially those carrying fuel and ammunition, would cause the enemy to break.[49] This is in fact what happened. Tedder had just received word while composing the note expressing these thoughts that Axis forces were retreating under air attack. During this critical period, the army's cooperation squadrons flew more than 150 sorties without a single loss. Air superiority had facilitated the Eighth Army's situational awareness and responsiveness to Rommel's counterattacks.[50]

Rommel's retreat from Tobruk had the paradoxical effect of freeing up Me. 109s for more aggressive actions along the front at the very moment reinforcements were arriving. By 8 December, JG 53 had joined JG 27, bringing the total number of Me. 109s to 120, nearly half of which were the new F4 variant. From this point forward, as diversions and reassignments of RAF aircraft to India began, and as the British offensive neared its culminating point, the balance in the air began to turn from the RAF to the Luftwaffe. However, before the newly arrived fighters could engage the WDAF, their ground organizations had to move into place and improve existing landing grounds or build new ones. Consequently, Rommel's forces suffered a severe mauling from the air as WDAF fighter-bombers attacked them all the way back to El Agheila. RAF assets also provided direct support to troops mopping up the Bardia and Sollum pockets, which fell on 2 and 17 January, respectively, yielding 13,000 prisoners, mostly German, with all of their equipment and supplies.[51]

From 10 to 13 December, Rommel's forces were nearly encircled at Gazala, yet the army did not close the trap, and RAF bombers sat idle for three days because of problems telling friend from foe. Coningham pressed for greater use of fighters in the ground-attack role because they could more easily discern enemy and friendly forces. He knew time was running out for a knockout blow to Rommel's army as diversions of US aircraft deliveries from the Middle East to Russia and from the United States and Great Britain to the Far East weakened the WDAF. This threatened to turn the tables in the air and make reinforcement more difficult than it had been in some time. Seven fighter squadrons diverted to the Far East during the month. Rommel withdrew to El Agheila on 24 December as the British occupied Benghazi and then Sollum. With both sides exhausted, the front stabilized.[52]

Tedder and Coningham knew the WDAF had reached its culminating point. Tedder asked Freeman for help on a number of fronts, from sending the most modern fighters to making yet another plea for heavy-bombers because they would be almost entirely grounded in the United Kingdom during the winter. He emphasized that the torpedo-equipped Wellington was a vital addition to his arsenal and that the air-to-surface surveillance (ASV) model was extraordinarily important given unsustainable Blenheim losses against heavily escorted Axis convoys. He asked for additional deliveries as soon as possible so he could keep Rommel from gaining the logistical upper hand once again. Freeman replied that more ASV Wellingtons and the first six Spitfire Vs would be arriving in the next month and that he had been able to order the move of two PR Mosquitoes to the Middle East—a major force-multiplier given that aircraft's capabilities. However, there would be no heavy-bombers.[53]

Meanwhile, fighter shortages emerged as a result of diversions to the Far East, Axis superiority in numbers as elements of Luftflotte 2 arrived, and the reviving Axis aviation fuel situation as Malta came under attack once again and Axis convoys arrived more frequently at Tripoli. For the first time, Commonwealth forces were suffering high losses to air attacks as the RAF weakened and a reinforced Luftwaffe pitched once again into the battle. Fortunately, the Germans continued to hold back assets in the eastern Mediterranean. Additionally, the transfer of Luftflotte 2 with its ground organization took each subordinate unit nearly eight weeks to complete, meaning that it had its full strength in the Mediterranean for just two months before returning to Russia. Nonetheless, units arrived steadily. Kesselring met with German and Italian airmen several times, emphasizing his orders to regain control of the SLOCs, render Malta ineffective, and prepare to support Operation Hercules, the planned assault on the island.[54]

Luftwaffe reconnaissance units kept Rommel informed about British movements. After his forces retired, and the RAF logistical situation worsened, raids decreased. At this point, Fliegerführer Afrika's liaison officers visited Rommel's headquarters to discuss support for the anticipated Axis counterattack.[55] However, by the time the Luftwaffe began recovering from the thrashing it took during Operation Crusader, Rommel and his staff were once again hiding from RAF planes. His batman remarked that they had already moved several times and were currently sheltering in a small wadi.[56] On 22 December, Rommel followed with a note complaining of ammunition and fuel shortages brought on by the complete lack of air support, which he contrasted with the RAF's control of the air.[57] Nonetheless, the Axis managed to bring up just enough supplies. Auchinleck emphasized that Rommel's army was "greatly helped by the remarkable elasticity of his supply organisation."[58] Whatever else one might say about the structural deficiencies in the Axis supply system—and they were serious—forces in the field used personal initiative to get the basic materials they needed to keep fighting. Improvisation, often in the form of air transport deliveries and the careful timing and spacing of vehicle convoys, kept Panzerarmee Afrika going.

Continuing Innovation and Impending Crisis

Commonwealth interservice liaison improved as Operation Crusader continued. At General Headquarters, the C in Cs' committee dealt with issues of strategic import, while the Inter-Service Intelligence Staff Conference and Inter-Service Operational Staff Conference met daily to exchange information and funnel it to the C in Cs and field headquarters. Air liaison officers (ALOs), who received many of these messages, had to have special training before they received and disseminated them to their army counterparts. Both services placed a premium on clear and timely joint communications.[59]

The AASC process evolved but was not mature until spring 1942. Wings notified Headquarters WDAF of aircraft available for commitment to direct support sorties, which allowed Coningham to assign aircraft rapidly when army requests arrived. Operational experience had made clear the kinds of targets most susceptible to air attack; requests focused on these, including thin-skinned vehicle troop and supply convoys.[60]

During exercises before Operation Crusader, signals from headquarters to flying units took up to twenty minutes to arrive. Often, army formations requesting strike missions did not receive any confirmation that the mission was en route, and if the aircraft found their targets and the units they were supporting, the latter lacked an identification system visible from the air. Pilots were thus often unable to complete their missions. To solve this problem, AASC tentacle units used fifteen-foot white cloth arrows to point toward the target, with bars on the arrow indicating distance to target. Army units painted white Saint George's crosses on a black background on all their vehicles. Later, they replaced these with RAF roundels, which aircrews recognized reflexively. On 29 October 1941, an RAF/army instruction on recognition methods directed that aircraft inbound on direct support missions fire white illuminating flare cartridges. Troops responded with a smoke bomb or canister, a large "T" ground strip, and a "V" sign pointed in the direction of enemy troops. However, the friction involved in such operations, and the common practice both sides adopted of using captured vehicles, made a perfect solution impossible and good ones elusive. Fratricide remained a problem.[61] Efforts to improve air-ground cooperation were still far from ideal. The communications bottleneck remained severe, restricting the flow of orders and air support requests.[62]

In addition to making direct support difficult, unrealistic bomb lines created C2 problems. Army commanders often knew less about the position of their own forces than did the RAF with its advantage of altitude. This resulted in the conservative placement of bomb lines. RAF senior officers and pilots fumed about this but did not understand how chaotic and confusing the situation on the ground was, and how difficult it was for troops in contact with the enemy to provide exact positions, much less take the time under fire to deploy smoke shells, arrows, and other ground markers. The ensuing time lag between calls for air support resulted in a heavier use of fighter-bombers—a major improvement over light-bombers. Most targets attacked from the air were located by fighters flying Tac R missions. Sound joint planning depended on clear processes, and when it came to coordinating bomb lines, there were hardly any until summer 1942. Cooperation during Operation Crusader was still generally on a personal basis, haphazard, or nonexistent. It was not strong until July 1942. The bomb line in 1940 and 1941 was fifty miles from friendly troops, giving thin-skinned vehicles within this area relative immunity until army-air cooperation improved. By the time Operation Crusader culminated, poor weather, decreasing serviceability rates, the lack of sufficient forward airfields, difficulty distinguishing friendly from enemy troops, and insufficient joint communications had reduced the effectiveness of an otherwise sound combined-arms effort.[63]

Nonetheless, the RAF's handiwork became clear early in the offensive. At Derna, Berka, and Benina airfields, the Allies found 172 Axis aircraft abandoned. Many of them were in good shape, lacking one or two key parts. Aircraft losses from airfield raids and a shortage of spare parts proved disastrous for the Luftwaffe and IAF. Tedder visited Derna and Benina from 21 to 22 January. "Derna," he said, an extraordinary sight, littered with aircraft, mostly Hun, in all stages of repair and disrepair! Some, obviously deliberately "demolished," others equally obviously knocked out by our bombing and low shoot-ups. A few in good repair, and behind the remains of a hangar I found a group of our S. Africans hard at work getting an M.E. 109 serviceable. . . . Benina even more of a sight than Derna. Hun aircraft everywhere. A whole string of J.U. 52s all burnt out where a string of bombs had hit them as they were off-loading petrol—beside one of the remains was the remains of a big petrol lorry caught in the act. There is no doubt some of our chaps' bombing had been extremely accurate.[64]

During Operation Crusader, the army captured 458 aircraft in various states of repair. Of these, 250 were German. Losses to RAF raids, David Stirling's Special Air Services (SAS) operations, and other factors brought Axis aircraft losses to nearly 800. Benina had been a particularly valuable conquest given the ground-servicing equipment there. Intelligence officers found advanced wireless radio sets, superb repair equipment, aircraft with new kinds of metal fabrication, new weapons types, and a new bombsight. This was the first time Commonwealth forces had overrun a German airfield, and the RAF sent technical experts to exploit the various treasures. Coningham also remarked that the first Kittyhawk squadron just went into action, destroying five enemy aircraft in combat and doing great damage to Axis vehicles.[65] Despite these successes, logistical difficulties and Luftwaffe reinforcements took their toll. By 16 January, the RAF had ninety-seven fighters and twenty-eight bombers operational in the forward area. Diversions to India further eroded WDAF strength.[66]

Fortunately, logistical difficulties had plagued Axis efforts to launch its own offensive and then to stem Operation Crusader. During November, its forces received only 40,000 of the 120,000 tons of required supplies, and the rate of dispatch and delivery were both falling. Enno von Rintelen said the problem could not be resolved until Luftflotte 2 succeeded in neutralizing Malta—again. This constant dancing around the problem of Malta is all the more perplexing given the exceptionally direct and strong relationship between the island's air activities and convoy losses. The rapid reverse in the war at sea had begun as soon as most of Fliegerkorps X redeployed to Greece.[67]

In addition to detailing the damage RAF assets did to Axis logistics during Operation Crusader, German prisoners of war (POWs), diaries, and other sources made clear the pain and demoralization they caused troops. Captured intelligence summaries emphasized WDAF air superiority all along the front, making air reconnaissance in support of operations dangerous. The RAF and Eighth Army were cooperating very effectively from the German

Me. 109F fighters abandoned at their landing ground. Note the missing propellers in every case. RAF Bomber Command raids destroyed a number of propeller factories in Germany, making this the number one Wehrmacht salvage item in 1942–1943. Shortages were serious even earlier, and many planes captured on the ground throughout the campaign were missing propellers. (Library of Congress, Fsa. 8e00296)

point of view. Repeated air attacks on ports and airfields in Cyrenaica were particularly painful. From 15 November to 5 December, German intelligence officers noted only two Axis air attacks on enemy troops.[68] RAF pilots also took their toll on air transport flights bringing in aviation fuel. By 22 December, Me. 109s were nearly out of 100-octane fuel. German transport pilots managed to deliver enough to get the Me. 109s and other aircraft to Marble Arch. However, the lumbering Ju 52s often fell prey to fighters, even with escorts.[69]

Operation Crusader Culminates

Nonetheless, Luftwaffe reinforcements facilitated serious attacks on Commonwealth troops—a problem also attributable to the long distances RAF fighters had to fly to protect them. Heavily escorted Stukas made repeated attacks on vehicle columns, causing serious losses. British fighters took a heavy toll on the dive-bombers, and soon only Me. 109Fs were regularly in evidence.[70]

The relief of Tobruk on 10 December allowed RAF fighters to move onto airfields and airstrips there. Fighters covered supply shipments to the port. Escort of day-bombers and direct support of ground formations also became less difficult. Fighting now revolved largely around control of airfields to facilitate land-based air operations in support of the battle for the SLOCs and ground operations.[71]

As Operation Crusader wound down, the RAF had been engaged continually in combat since April 1941. It had participated in nine major joint operations (Crete, the Cyrenaica retreat, Brevity, Battleaxe, Crusader, Iraq, Syria, Iran, and the battle for logistics) while struggling constantly to maintain air superiority and training with the army to improve joint operational effectiveness. The RAF gained and maintained air superiority in all but two cases (Crete and the reinforcement of Tobruk, where the Luftwaffe was heavily concentrated). Victories in Italian East Africa (IEA), Iraq, Iran, and Syria were crucial because they allowed Tedder and Auchinleck to concentrate their forces in the Western Desert rather than facing the prospect of simultaneous operations in five places. Both men were pleased with WDAF contributions to the success of Operation Crusader.[72]

Axis Resurgence, Axis Dysfunctions

As Operation Crusader ground to a halt, Luftflotte 2 units arrived on Sicily. This soon gave the Axis major advantages until most of these units returned to Russia in spring 1942. Kesselring arrived in Rome in November 1941 and headed straight to Rommel's command post for an update. Kesselring's mission was to gain air and naval superiority in the area between Italy and Libya to secure the SLOCs and supply convoys. Neutralizing Malta was the key prerequisite for achieving these objectives.[73]

To achieve them, Kesselring had operational command of Luftflotte 2, Fliegerkorps X, and Fliegerführer Afrika. The IAF contributed a number of attack and fighter units along with three new squadrons of torpedo-bombers comprising fifteen SM.79s. Kesselring made immediate plans for the air campaign against Malta and protection of Axis convoys, convincing the duce to release the Italian Navy to escort them—something it had not done as long as aircraft from Malta were able to operate with impunity. On 17 December, as a result of this rare air-sea combined-arms effort, the first major convoy in months reached Tripoli and Benghazi on 17 December without loss. Kesselring expanded airfields in Sicily, brought in good signals equipment, and stocked fuel and bombs.[74]

Field Marshal Albrecht (Albert) Kesselring. After he arrived in the Mediterranean as *Oberbefehlshaber Süd* (commander in chief South) in November 1941, Kesselring made the neutralization and conquest of Malta his highest priority in order to ensure continuing logistical support for Erwin Rommel's army. Despite an effective air campaign against the island and the restoration of Axis logistical security, Kesselring's efforts to gain Adolf Hitler's final go-ahead for Operation Hercules to conquer Malta failed despite the Führer's and Benito Mussolini's initial approval. Ironically, Rommel's great success in the Gazala offensive during May and June 1942 convinced Rommel, the Führer, and the duce that the conquest of the Nile Delta was weeks away and that Malta no longer mattered to the degree that it had. They were proved wrong, and this blunder became one of the most costly of the campaign and the larger war. (Imperial War Museum Image No. HU 51040)

Kesselring had made an impressive start, but he was frank about his unfamiliarity with a theater so complex and alien to his own experience. "I was," he said, "brought up to think along continental lines; I had neither the opportunity nor cause to acquire a detailed knowledge of conditions in the Mediterranean."[75] It was an important statement that contrasted not only Kesselring, but all German senior officers, with their British counterparts, who had been schooled in the empire to consider the complexity and importance of its interconnections. Regardless, Kesselring understood that the conquest of Malta would make or break Axis fortunes in the theater. This went hand in hand with his conviction that the Mediterranean had become a major theater of war after the Germans intervened. He said that major German involvement must result not only in the capture of Malta but also of Gibraltar, the Suez Canal, Aden, and Vichy French territories, along with the recapture of IEA and a move into the Middle East. Kesselring viewed these actions as vital for securing the "southern flank," attacking British shipping in the Indian Ocean, outflanking the Russian front deep into the southern USSR, gaining unequivocal clarity and support from Spain and Turkey, and seizing oil assets. Now that Germany was fully engaged in the Mediterranean, half measures would not suffice.

Kesselring lamented that there was not even a plan on OKW's shelf for a Mediterranean campaign and that the situation with Italy made the entire effort ad hoc. Air command relationships were maddening: "Fliegerfuehrer Afrika was under command [of] Luftflotte 2 (through Fliegerkorps X) and had been ordered to cooperate with Rommel's army headquarters. Fliegerkorps X headquarters was in Athens and later in Crete. Its operational area was Salonika-Athens. Luftflotte 2 itself was subordinate to the C-in-C Luftwaffe and since its Commander was also C-in-C South it was also subordinate to the Duce and O.K.W." Similarly, German Naval Headquarters was subordinate both to Kesselring and Supermarina—another dual chain of command. The same was true of Axis supply services, which were operated by various branches of the German and Italian Armies but all beholden to Italian shipping priorities. General Ritter von Pohl was responsible for Luftwaffe supplies. However, transport through Italy and across the Mediterranean was a Comando Supremo Transport Department task.[76] Wilhelm Keitel tried to modify shipping priorities during talks with Ugo Cavallero on 7 June 1941, when he emphasized that what the Axis needed in theater were smaller numbers of highly mobile and capable units rather than the infantry divisions the Italians kept sending. His insights fell on deaf ears.[77]

There were still enough MVs, but many lacked derricks and thus unloaded slowly. Too few ships were ready to sail because dock services were chaotic. There was no centralized control. German ships sailed with Italian ones, and neither received coordinated air and naval support on a regular basis. Italian naval and civilian maritime offices controlled the dispatch of convoys and were not in a particular hurry. The Germans were used to moving supplies on their *Reichsbahn,* not on Italian-controlled seaborne convoys. Bottlenecks at ports were severe. Shortages of heavy equipment were serious. Bombing exacerbated these problems. As the Axis merchant marine shrank, submarines, gunboats, destroyers, Siebel ferries, and Ju 52s all came into heavier use, but they could not make up for the carrying capacity lost as large MVs became rare. Tankers had been in short supply from the start and were No. 201 Group's primary target. POL dumps in North Africa had the same priority. Fuel became scarce or even unobtainable. As the Italian destroyer force shrank along with the merchant marine, less seaworthy escort vessels took over. Consequently, convoys often sailed without escorts. The RAF and Royal Navy worked closely with their weather and intelligence officers to determine when these convoys were sailing and the routes they would take. Kesselring said the British must have known about virtually every convoy given their effectiveness. The Luftwaffe flew 75–90 percent of all escort sorties, a task for which it had too few aircraft. Finally, Kesselring lamented in a classically Teutonic fashion, "The idea of total mobilization of human and material forces for war was not a characteristic of the Italian people."[78] He argued that the Luftwaffe was wasted in the convoy-escort role, which tied it to a defensive mission rather than allowing it to go after the RAF. This heavy commitment affected all other missions because crew rest decreased and accidents as well as combat losses climbed.[79]

Regardless, Kesselring had a mission to complete. His key challenges in suppressing Malta were its hardened defensive positions; radar sites (which the Luftwaffe never attacked); strong AAA defenses; Spitfires, which perfected ambush tactics for attacking from 32,000 to 27,000 feet; and superb unloading crews for arriving ships. "In Malta," Kesselring said, "the Luftwaffe had met a worthy opponent."[80]

Consequently, he pushed hard for an invasion before or immediately after the return of Luftflotte 2 to Russia. He knew the island's recuperative powers and understood that logistics would determine victory or defeat. Capturing Malta was crucial. He convinced Hitler and Mussolini to approve Operation Hercules.[81] Intensive raids began on 22 December, with 200 aircraft engaged. More than 500 sorties in the first days of January forced the British to consolidate their aircraft at Luqa, the only airfield still operational. On 10 January 1942, German air strength in the central Mediterranean was 523 aircraft (297 operational). This increase allowed a second convoy to reach Tripoli unscathed on 5 January. German U-boats, now numbering thirty-six in and en route to the Mediterranean, also sank the carrier HMS *Ark*

Royal on 13 February and the battleship HMS *Barham*. Earlier, on the night of 18–19 December 1941, three Italian two-man torpedo-swimmer teams snuck into the port of Alexandria, causing heavy damage to the battleships HMS *Queen Elizabeth* and HMS *Valiant* and sinking an oil tanker. With German airpower again present in force and cooperating with Italian naval units under Kesselring's lead, Axis fortunes and Rommel's logistical position improved. By March 1942, less than 5 percent of Axis shipping failed to make the run to North Africa. In April, the figure was 1 percent.[82]

Erich Raeder continued to prod Hitler for concerted efforts in the Mediterranean, noting at a 13 February 1942 briefing that Japan's entry into the war had forced the transfer of Britain's last heavy warships from Alexandria to India. The Axis, not the British, now "ruled the sea and the air in the Central Mediterranean. . . . The Mediterranean situation is definitely favourable at the moment."[83] On 12 March, he followed up with an urgent appeal for the conquest of the Suez Canal, with all the advantages it would confer. "The favourable situation in the Mediterranean," he said, "so pronounced at the present time will probably never occur again."[84] He also noted, significantly, that the Japanese were planning to occupy Ceylon and Madagascar and asked Hitler to intercede with Vichy to allow the latter. However, the British occupation of Madagascar and the Japanese return to the central and southwestern Pacific undid these plans.[85]

By the time the British began running large quantities of supplies into Tobruk, German air strength had increased, and U-boats added to the danger. Together, they sank seven vessels. These attacks, and the extent of the British advance, caused severe logistical problems. Air operations on both sides dropped off as supply shortages, the high operations tempo, losses, and diversions to India took their toll. Aerial reconnaissance became all important as each side tried to determine the other's capabilities and intentions.[86] Nonetheless, the Luftwaffe's efforts turned the logistical balance toward the Axis again, as Rommel mentioned while lauding Kesselring's efforts against Malta.[87] Five days later, Rommel said that mines and the Luftwaffe were hindering enemy pursuit of his army. He had moved it 300 miles to a good defensive line.[88] Luftflotte 2 intervention kept the RAF from doing severe damage to Axis columns.

As the RAF kept pace with the Eighth Army's advance, the shortage of landing grounds and fuel supplies made operations challenging. Tedder pushed his staff to be ready for the scheduled start of Operation Acrobat, designed to drive Axis forces across Tripolitania and, if possible, out of North Africa. However, as supply lines lengthened and fighting units wore down, prospects dimmed. The drain of aircraft to the Far East was becoming severe. "The American consignment," Tedder said, "was vital to me now because I had to send squadrons and aircraft to the Far East and Burma."[89] He warned that the aircraft shortage was having serious operational impacts, as were cumulative strain and exhaustion. The Germans had gained local air superiority and were using their Me. 109Fs to strafe ground forces. For the first time, the RAF could not reliably keep the Luftwaffe away from army units. Coningham said that only major reinforcements of first-rate fighters would turn the tide.

Consequently, Tedder lamented, "the low ebb of the Middle East Air Force at the beginning of 1942 showed no signs of turning for the moment."[90] US aircraft had mechanical problems and often arrived without technical manuals for ground crews. Aircraft inspectors in the United Kingdom often replaced the US manuals with their own versions, which wasted time. "Meanwhile," Tedder said, "we fumbled about, wasting aircraft and equipment."[91] The Takoradi aircraft reassembly unit's habit of removing technical manuals from aircraft and sending them to the Air Ministry for "translation" led to serious delays in aircraft construction and maintenance. In a biting signal to the Air Ministry, Tedder said, "Most grateful you arrange for contractors to pack appropriate handbooks, etc. We can now understand American language, and need not wait Air Ministry interpretation."[92] Packing cases presented additional problems. They had to be unloaded quickly after arriving in port but had no markings on the outside. This further delayed aircraft construction and movement. Portal fixed both problems.[93] Kittyhawks began replacing the Tomahawk and proved nearly as good as the Me. 109F in air combat as well as one of the best fighter-bombers of the war.[94]

Japanese attacks continued forcing the movement of significant assets from the Mediterranean to India. RAFME was unable to replenish losses from Operation Crusader. The ability to control the SLOCs and support the land battle was further degraded. Had the Japanese coordinated subsequent efforts in the Pacific and Indian Oceans with those of Germany and Italy in the Mediterranean and Middle East, the course of the war might have been very different.[95]

Nonetheless, by the time Rommel counterattacked, the WDAF had become a highly mobile instrument. Each squadron's A, B, and C parties were agile. Headquarters WDAF was also split into two parties: Advanced Air and Rear Air Headquarters. The Advanced Air Headquarters operated alongside the Army Battle Headquarters. These advanced headquarters moved forward 200–250 miles before the Rear Air Headquarters followed. Rear Air Headquarters was located close to the fighter and bomber B parties to maximize operational effectiveness. Air stores parks (ASPs) and repair-and-salvage units (RSUs) created a shuttle service for rapid movement of salvaged aircraft. By working twenty-four-hour shifts, salvage crews removed all repairable aircraft before the enemy reached them. RAF units also removed all fuel, munitions, and spare parts. Axis forces never overran any of these formations.[96]

Nonetheless, there were problems. The first was overconfidence in the speed and decisiveness of the ground offensive, which led the RAF to depart from the norm, using a shuttle system with shared rather than unit-specific vehicles. The second was the unexpected intensity of mobile operations, which quickly wore out airplanes, trucks, and other equipment. The third was the still inadequate signals infrastructure, which made communications difficult, sporadic, and often confusing. There was also a seam in the delivery of fuel and explosives. The army was responsible for bringing these materials to within thirty miles of RAF squadrons. From there, the RAF used its own vehicles. The army's delivery schedules were too slow to meet RAF requirements during periods of rapid movement, forcing air units to form ad hoc convoys. RAFME ultimately fixed these problems by acquiring more vehicles. The mechanical transport light repair units (MTLRU) were far too small for their assigned task. A mere seventy-three personnel were responsible for repairing 3,000 vehicles. They had to abandon hundreds of repairable ones after Rommel counterattacked.[97] To correct this and ensure maximum mobility, all vehicle formations practiced packing up and unloading between major campaigns. Ground crews painted numbers on every vehicle to match each one with a specific load.[98]

Operation Crusader was a victory, if a hard fought and incomplete one. The RAF and army were learning to work together but were exhausted by the hard fighting and the long advance. As Rommel observed their difficulties and recouped his losses from the convoys now steaming into Tripoli, he planned a counterattack. Regardless, the RAF's will to continue the fight, once again at long odds with its numbers depleted and in spite of heavy aircrew losses, was as strong as ever. The campaign would continue, losses would mount, and the RAF, with its sister services, would see it through.

7

Rommel Strikes Again
January–February 1942

Grand-Strategic Developments

GERMAN CHOICES

THE GERMAN FOCUS THROUGHOUT this period remained on securing oil resources, with options including an attack through the Turkey-Syria corridor and the Caucasus into the Middle East, outflanking the British position there. However, the defeat in front of Moscow forced Oberkommando der Wehrmacht (OKW) to put these plans on the shelf. Nonetheless, OKW noted that "the oilfields in Iraq and Southern Persia, which have become considerably more important since Japan occupied the British-Dutch oilfields in East Asia, lie behind this line of defence [British positions in Iraq]. . . . It should however, not be forgotten that the disruption of German oil supplies from Rumania may well constitute one of the primary aims of the British High Command in 1942."[1] This concern with Ploesti became a major distraction in the Germans' Mediterranean strategy and drove them to make choices— or reject them—that had a serious cumulative impact on the course of the campaign. The "southern rampart" mentality overrode any possibility of a "springboard" approach to Crete and the Aegean as the gateways to Cyprus and from there, in coordination with other operations, into the Middle East oil fields. Talks with Japan regarding a potential Indian Ocean strategy remained noncommittal and sporadic.

BRITISH STRATEGY

As Erwin Rommel's 21 January–15 February 1942 offensive stopped short of Tobruk (more on this later), the War Cabinet's Joint Planning Staff (JPS) and Joint Intelligence Centre (JIC) reviewed the situation in the Middle East. A rapid resumption of the offensive in Libya was crucial because a junction between Germany and Japan in southern Asia would be a major disaster. The Middle East could no longer be dealt with as a military theater separate from India. They became interdependent as Japanese raids on Ceylon made clear the potential disasters in store should the Japanese adopt an Indian Ocean strategy in conjunction with the Germans. The War Cabinet also recognized the interconnections between the Mediterranean and the Eastern

Front given, among other things, the huge stresses the Luftwaffe endured as it shuttled back and forth. Further, victory in the Mediterranean held the key to Italy's surrender and the opening of a second aerial front against the Reich. Planners also noted that the security of the Middle East now rested largely on continued Russian military resistance and Turkish diplomatic maneuvering. They agreed that the only way to help Russia was to pin down as many German forces as possible with an immediate offensive in Libya while keeping open the lend-lease route through Iran. Unless the offensive occurred soon, Rommel would strike first, and if he were to win and drive the Eighth Army back into Egypt, the Axis could and likely would take Malta. Should Russia collapse, there were not enough troops in the Middle East to hold both Rommel and German armies advancing from the Caucasus. Even major reinforcements would not likely prevent the loss of Egypt and the empire's oil supplies. Planners noted that the Luftwaffe was stretched thin and might not have enough strength in Russia to give adequate support to Axis ground forces. This meant strong air forces were necessary in the Mediterranean to keep the 600–700 German and 500 Italian aircraft stationed there in place.[2]

Building on the War Cabinet's grand-strategic guidance, the Chiefs of Staff (CoS) sent orders to the Middle East commanders in chief (C in Cs) to destroy Rommel's army and drive the Axis from Africa. A key component of this effort would be building Royal Air Force (RAF) strength from forty squadrons to sixty-two (1,010 aircraft) by mid-July, and eighty-eight (1,410 aircraft) by September. Should the Germans win in Russia, they would turn south and try to capture British oil supplies, creating a major, two-front campaign in the Middle East. The C in Cs thus had to defend the Middle East in case of a Russian collapse and Turkish acquiescence to Adolf Hitler's demands. Doing so would be challenging given the shortage of shipping and the need to stem the disasters in the Far East, including the loss of Singapore with its 64,000 men, the threat to India, and the Japanese raids on Ceylon. British strategy for the Middle East in 1942 thus depended upon Russia's survival and on maximum offensive action to take pressure off of the Red Army. The CoS warned that "no matter what situation develops in Middle East, strong air forces there are essential. The only hope of achieving this lies in the reinforcement of our own resources by the American Air Force, especially fighters and light bombers."[3]

US DECISIONS

Despite the largely harmonious discussions at the Arcadia Conference from 22 December through 14 January, and the agreements that followed, the United States found itself heavily engaged in the Pacific even as it shipped lend-lease materiel to Britain and Russia. George C. Marshall and his army colleagues were becoming disillusioned with the peripheral strategy, especially now that it appeared the Red Army might not just survive but even turn the tables during the coming year. Marshall called increasingly for a heavy focus on building US ground and airpower to allow for a major clash with the Germans in northwestern Europe after the Russians had them on the run in the east. Marshall rejected the idea of US ground commitments in Africa and the Mediterranean until Franklin Roosevelt ordered him to proceed with Operation Torch. The US Navy also began backing away from agreements made at ABC-1 and the Arcadia Conference, pushing for a heavier focus on the Pacific theater and a major counterattack there that would involve virtually every remaining US ship—not a positive development for either the Battle of the Atlantic or the planned invasion of North Africa. As these disaffections grew, an as yet relatively obscure figure, Brigadier General Dwight Eisenhower, reminded everyone that only three things were of paramount importance at this point: securing the British Isles and the Atlantic sea lines of communications (SLOCs), holding the Middle East and India to prevent an Axis linkup there, and keeping the Soviet Union in the war. Ultimately, despite often intense disagreement, the Allies stayed focused on all three.[4]

Rommel Prepares His Offensive

As the British sought to reconstitute, Rommel's logistics improved dramatically. Intelligence reports confirmed that the British were disorganized, spread out, and tired. Colonel Siegfried Westphal, Rommel's chief of staff, recommended an immediate attack before British resupply, refitting, and repositioning of units gave them any advantage.[5] Rommel also realized that he had to act while Luftflotte 2 was keeping Malta subdued and had helped Fliegerführer Afrika to gain some measure of air superiority for the first (and, as it turned out, only) time.[6]

Uncharacteristically, the British missed Rommel's intent to begin a counteroffensive. Axis air strength had recovered to a significant degree, making it much more difficult for RAF reconnaissance flights to track enemy movements. Sand storms and rain squalls further blinded these sorties while turning advanced RAF airfields, now consisting of less desirable soil types, into mud wallows. The British managed to fly out four fighter squadrons to better but more distant airfields prior to the German attack, but this limited loiter time over the front to fifteen minutes. There were 515 Axis planes in Tripolitania alone (300 serviceable). The Western Desert Air Force (WDAF) had 445 aircraft remaining (280 serviceable).[7]

Arthur Tedder blamed the inability to begin Operation Acrobat, and Rommel's counterattack, on losing the latest phase of the battle of logistics. The British outstripped theirs, the Axis fell back on theirs, and aircraft on Malta, now under heavy air attack, could not interfere significantly with convoys. However, Tedder believed that despite the unfavorable turn of events, the RAF had done permanent damage to the Axis air forces, while the air-ground team was coming together.[8]

The German counteroffensive began on 21 January. Weak units facing Rommel's forces retreated. He caught and destroyed several at Msus. The counteroffensive made good progress. British units at Agedabia suffered a disastrous defeat from 21 to 25 January. From there, refueled panzer units moved so quickly that they captured thirteen aircraft on the ground and destroyed another ten. Most were already badly damaged, but the repair-and-salvage units' (RSUs') failure to remove them was a disaster in RAF eyes. The Germans ultimately captured twenty-five aircraft. Six were operational. This stung Arthur Coningham and Tedder and drove them to set up the RAF Regiment, which gave airfields dedicated defensive assets and saved many aircraft in the coming months.[9] WDAF raids slowed but could not stop Rommel's advance. Benghazi fell on 29 January. The advance stopped at Tmimi, largely as a result of WDAF raids. Nonetheless, strong German fighter units made these raids costly. Luftwaffe air-ground cooperation was also effective during this series of offensive actions. Bombers inflicted significant damage, and air reconnaissance facilitated the entire effort. After these opening rounds, however, Luftwaffe units had to move to forward airfields, taking them out of the fight for three days. This gave the British a major advantage in the air from 4 to 6 February, at which point Axis aircraft, flying from recaptured airfields at Benghazi, Derna, and Martuba, rejoined the fray.[10]

Coningham left nothing to chance. While all maintenance, heavy units, and air-base parties moved back to bases on the Egyptian side of the border, fighter squadrons stayed far forward to cover the Eighth Army's withdrawal. Coningham had devised a three-stage move to the Egyptian frontier as fortunes on the ground dictated. The plan was flexible, and he could reverse the process from retreat to advance very quickly. A hastily added postscript to a letter he sent Tedder, annotated "12?? Hrs.," said "lots of enemy southwest . . . the whirligig is in full force."[11] The RAF was about to begin the most trying and longest, but ultimately the most successful, chapter of its efforts in the Mediterranean and Middle East.

During the two-week retreat, the WDAF flew more than 2,000 sorties in support of the Eighth Army and kept the Luftwaffe from causing serious damage. Tedder applauded the complementarity of all three services but

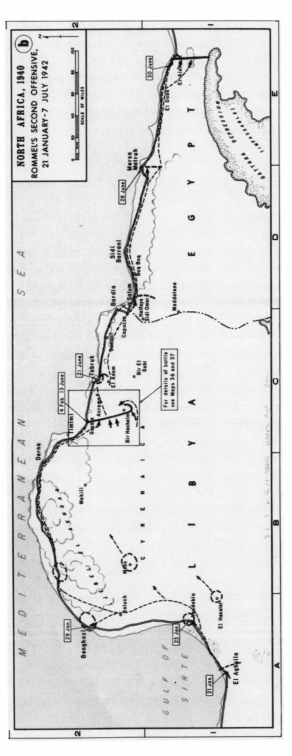

Rommel's second offensive, 21 January–15 February 1942. The effort reached its culminating point near Gazala, between Timimi and Tobruk. This map also depicts Rommel's later offensive, which carried Panzerarmee Africa to El Alamein. (Maps Department of the US Military Academy, West Point)

warned that a major logistical reversal would unhinge the entire Middle East effort. He thus lamented that the Germans once again had the upper hand in the battle for the SLOCs, and that for the first time, the RAF had lost general air superiority as diversions to the Far East and the number of Me. 109Fs increased. "Our forward aerodromes," Tedder sighed at the beginning of March, "lacking good anti-aircraft defences, had been bombed and shot up with impunity by the 109s with heavy losses to ourselves. We could only reply with night-bombing raids."[12]

Axis Logistical Resurgence

As Rommel advanced, so did Axis logistical fortunes. During January, heavily escorted convoys delivered 66,000 tons of supplies with virtually no losses. The first of these, on 5 January, was escorted by four battleships, six cruisers, and twenty-four destroyers and motor torpedo boats. The Luftwaffe provided escort and sent more than 400 sorties against Malta between 30 December and 5 January. At the same time, a British convoy with three merchant vessels (MVs) left Alexandria on 12 January. The Luftwaffe sank two and badly damaged the third. With Benghazi's airfields again in Axis hands, no MVs reached Malta for the rest of the month. The island's most difficult time had begun. Airfields and port facilities were heavily damaged and a destroyer sunk in the harbor. The last thirty-two Hurricanes were falling to Me. 109Fs with increasing frequency, but the arrival of the first Spitfires on 7 March evened the odds.[13]

Of sixty-one MVs that sailed for Malta during 1942, nineteen were lost and ten had to turn back—a reduction of 48 percent in the flow of supplies. However, the British ran in more supplies using small, fast vessels, while aircraft carriers, including the USS *Wasp* (sent to the Mediterranean for two Spitfire deliveries to Malta even though desperately needed in the Pacific), delivered 370 fighters, including increasing numbers of Spitfires. Although Malta was in peril during spring 1942, toughness and good intelligence preserved it as an effective defensive bastion with the capacity to launch occasional ship-attack sorties.[14]

Nonetheless, the Axis ran 67,000 tons of supplies and another 40,000 tons of fuel into Libya in February with a 9 percent loss rate. This dropped to 3 percent in March and 1 percent in April. British efforts to intervene usually resulted in losses from air and U-boat attacks. The British continued having difficulties sending relief to Malta. A convoy set sail from Alexandria on 20 March. Deception operations and RAF and Fleet Air Arm (FAA) attacks on Axis airfields in Cyrenaica gave the convoy breathing space before the Luftwaffe and then the Italian Fleet attacked in force. Disaster struck as dive-bombers sank two MVs. The three remaining MVs were under air attack right into Grand Harbour. RAF fighters surged, flying 475 sorties from 24 to 25 March. Ground crews worked nonstop to repair damage. On 26 March, the two merchant ships that made it to port were bombed in their berths. One, loaded with ammunition, was scuttled to prevent an explosion, and the other yielded little cargo as it flooded. The British also lost two destroyers with another damaged. Of the 26,000 tons of supplies, 7,522 tons—29 percent—were salvaged under constant air attack. An Italian Fleet action had delayed the convoy, allowing the Luftwaffe to deliver crippling attacks. The British evacuated their remaining ships and kept reinforcing their fighter strength. Submarines carrying fuel and kerosene became the key lifeline, but only four arrived during March. By April, Admiral Andrew Cunningham had only four cruisers and fifteen destroyers operational, and although No. 201 Group now had sixteen RAF and FAA squadrons, the loss of airfields in Cyrenaica made it difficult to interfere seriously with Axis convoys. By March, the Luftwaffe had 335 aircraft on Sicily. When it began moving units back to Russia on 28 April, and sent forty dive-bombers and forty-five fighters to Fliegerführer Afrika, effective attacks on Malta stopped. Given that it took eight weeks for Luftflotte 2 to move most of its air and ground-support units from Russia and another eight to move back, more than 700 aircraft were not engaged anywhere for four months. These moves and the debate surrounding them about whether to continue neutralizing Malta from the air or seizing it by air and amphibious assault, in part determined the course of the Mediterranean campaign.[15]

Admiral Cunningham left his post to become the Royal Navy's Combined Chiefs of Staff (CCS) representative in Washington, DC, in April. As he handed over command, there were four cruisers, one small carrier, and eighteen destroyers compared with an Italian strength of four battleships, nine cruisers, and fifty-five destroyers and motor torpedo boats. The Axis had seventy submarines and the British twenty-five. Axis land-based airpower had smothered the Royal Navy and the convoys it escorted whenever land-based aircraft were unable to provide cover. The exact reverse applied every time the tables turned.[16]

Rommel Advances; Malta Holds

Rommel's counteroffensive continued with sufficient supplies. However, whenever German units outran their fighter cover, as occurred again from 4 to 5 February, they sustained heavy losses. As Neville Duke recalled, "It is a terrific thrill to come pelting down out of the sun to let rip at the Huns with the .5s [the Kittyhawk had six 50-caliber machine guns]. . . . You can't help feeling sorry for the Jerry soldier when you ground strafe them. They run, poor pitiful little figures, trying to dodge the spurts of dust racing towards them."[17]

"Own fighter cover," Panzerarmee Afrika's report said, "was not possible since the ground organisation in Martuba could not function before midday on 6 February at the earliest."[18] The Luftwaffe's shortage of motor vehicles once again hampered attacks on the retreating Eighth Army and the protection of Axis troops. In his summary of Operation Crusader and the German counterattacks, Rommel said,

> Co-operation between Panzer Army Africa and Fliegerfuehrer Africa was always good and was strengthened and further improved by the frequent visits of Field Marshal [Albert] Kesselring, who took a particular interest in the constant personal contact with Panzer Army Headquarters. Although the enemy's air superiority was very great at times the Luftwaffe formations always provided excellent support for Panzer Army's operations, except when supply difficulties severely restricted air activity. The absolute superiority of the Germans over the British was also evident in the air. In addition to the notable successes of the German fighters, the indefatigable operations of the reconnaissance aircraft deserve special mention. Working under the most difficult conditions, these aircraft always provided the command with valuable information.[19]

Although this painted a somewhat rosy picture, the Luftwaffe achieved its greatest feats in the Mediterranean during the six months after Kesselring's arrival.

Once Kesselring's airmen had largely suppressed Malta's airfields, they went after dockyards and nonmilitary targets, evidently in an effort to decrease the population's and garrison's morale.[20] As one author noted, "By the end of the month [April] the Germans hardly knew where to drop their bombs. So far as could be judged from the air, every military target had been either destroyed or badly damaged."[21] Bad weather made repairing the heavy damage to airfields a nearly hopeless task.

More than 5,000 men were constantly at work keeping them in condition to allow limited fighter operations. Malta was never out of action, but the RAF's focus was on survival rather than Axis convoys—just what Kesselring intended. By March, only twenty Hurricanes per day were serviceable. British fighters shot down just sixteen German aircraft in January and February because raids were so large and the Me. 109F was superior to the Hurricane II. The arrival of the first thirty-one Spitfires in March caused heavy enemy aircraft losses, with thirty-one downed by fighters and another twenty-eight by antiaircraft artillery (AAA). However, when Tedder visited Malta in early

April, an average of just eight Hurricanes and six Spitfires were ready for action. Even the arrival of forty-seven Spitfires on 20 April, flown from the carrier USS *Wasp*, did not turn things around. Heavy German attacks began within two hours of their arrival, destroying nine and damaging another twenty-eight, leaving ten planes operational.[22]

However, in April, the worst month for Malta, fighters claimed 52 planes and AAA another 101 as the gunners improved. Despite the Luftwaffe's efforts, Malta was not eliminated as an air base. As long as enough supplies arrived, the island held. Even during the height of German raids, more than 400 aircraft transited Malta from Gibraltar to Egypt. "Certainly," Tedder said, "to have let it go would have meant giving complete control of the Central Mediterranean to the enemy. And the consequences of that would have been defeat in the Desert, and the destruction of the whole British position in the Middle East."[23]

Between losses during Operation Crusader, reverses in Malta, and increasing diversions and transfers to India, the aircraft situation was precarious. The Luftwaffe had air superiority. Despite Tedder's urgent appeals for Spitfires, the first squadron did not join WDAF until Rommel's next offensive had started in May. Had it not been for Graham Dawson's efforts, the Royal Air Force Middle East (RAFME) could not have maintained its aircraft strength. Tedder said, "I knew the supply battle was running against us."[24] In April, Axis forces in Africa received well over 160,000 of 170,000 tons of supplies dispatched—the highest proportion of the war.

Continuing Innovation in the Midst of Crisis

To compensate for aircraft shortages, Tedder reorganized fighters into No. 211 Group, which controlled four wings of four squadrons each. In this new organization, each wing's four squadrons fit more effectively onto operational airfields and were grouped in adequate strength to escort bombers without having to coordinate outside the wing. There was also more time for individual squadron training. Tedder also fitted all fighter aircraft to carry bombs in case there was a heavy need for fighter-bombers, which were becoming a potent new weapon. Always focused on air superiority, though, Tedder emphasized, "The primary function of the fighter bomber, however, was to fight and that was never lost sight of: bombs were carried only as an ancillary means of hitting the enemy when conditions were suitable."[25]

By spring 1942, improving signals capabilities, better intelligence personnel, greater air reconnaissance and photoreconnaissance (PR) resources, dedicated fighter-bombers, and better ground-attack tactics were coming together and, in combination with much improved ship-attack procedures, turned the logistical tide again. Even as Rommel advanced, the WDAF helped to extricate the vast majority of the Eighth Army by keeping Axis troops at bay. The supply problems and casualties WDAF raids caused were instrumental in stopping Rommel's advance.[26]

By February 1942, the air tasking system had become agile enough that fighter-bombers just done bombing targets received directions while still in the air to strafe new ones. The arrival of No. 2 army air support control (AASC) and very high-frequency (VHF) radio sets increased these capabilities. On 31 May, a squadron en route to strafe a position in direct support of the army received notification that the position had fallen to British troops. The combination of excellent signals officers, VHF, and a mature command-and-control (C2) system allowed redirection of this unit to another target in fourteen minutes. A new signals plan implemented in early 1942 was behind these successes because all wings now had two operational radio links and one administrative link to Advanced Air and Rear Air Headquarters. With skilled signals personnel, VHF, and landline capabilities in place, the plan quickly proved its worth. With all-terrain signals vehicles in place, tentacles advanced alongside army units and provided terminal attack guidance based on visual acquisition of enemy positions.[27]

Also, in October 1941, the Government Code and Cipher School (GCCS) had opened a Middle East cipher school to improve the utility of Ultra signals intelligence. Construction of the Telecommunications Centre Middle East resulted in fifty radio and fifty teleprinter circuits handling 450,000 cipher groups a day. Its joint staff helped to increase air-ground effectiveness. By the time Rommel attacked Gazala in May 1942, wireless and landline radios linked virtually every Commonwealth unit in the Western Desert down to the battalion level. The separation of operational and administrative links sped the transmission and receipt of messages, reducing turnaround time to ground forces.[28]

Intelligence continued to increase in importance regarding combined operations. Ultra became more valuable as intercepts of German Army Enigma transmissions increased in May 1942. The Royal Army's Y Service became steadily larger, with 1,300 personnel assigned in May 1942 and 2,400 by October 1942. The RAF version grew to 1,000 by the end of 1942. When combined with other sources, Y Service intelligence was exceptionally useful. It accounted for most order-of-battle updates. Prisoner of war (POW) reports played an ever larger role in the intelligence effort, with Axis troops either corroborating what other sources were saying or making clear how WDAF raids were affecting morale and operations.[29]

The combination of POW interrogations, captured documents, air reconnaissance, PR, Y Service intelligence, Ultra, and other sources also gave the

British an understanding of the Axis logistical system's weaknesses. Locations of ammunition, fuel, and supply dumps; routes from these depots to the front; daily travel schedules; and forward delivery points were all well known to WDAF commanders. The wealth of source material, well-trained specialists, and clear dissemination processes made the effort increasingly effective.[30]

Air reconnaissance continued to improve as new techniques for rapid dissemination of information came into use. Squadron intelligence officers and air liaison officers (ALOs) checked information from pilot debriefings against other available sources and sent it immediately to headquarters by radio. Air and ground commanders then apportioned air assets. By April 1942, VHF radio sets allowed reconnaissance pilots to report in real time. Because they flew at low altitudes and thus had a short radio range given the earth's curvature, pilots radioed reports to AASC radio operators, who relayed the information to headquarters for action.[31]

Long Range Desert Group (LRDG) and associated Special Air Services (SAS) operations also provided much useful information on Axis air dispositions and strength while destroying increasing numbers of planes and killing skilled ground and aircrews. David Stirling's men conducted reconnaissance, sabotage, and raids against Luftwaffe bases. Their efforts paid off at Barce from 20 to 21 March, where they destroyed fifteen aircraft. Five days later, they destroyed five more aircraft at Benina. A follow-up raid on Barce on 16 May destroyed one aircraft along with four large workshop trucks and their tools, and a repeat attack a week later on Benina destroyed heavy repair machinery. These raids took pressure off of the airmen at a time when the WDAF was exhausted by Operation Crusader, fighting hard to protect the Eighth Army's withdrawal, repositioning after Rommel's Sollum offensive, and sending nearly half of its aircraft to India.[32]

Concurrently, mature AASCs began operations in March 1942, replacing the earlier Operation Crusader structure. Located at the combined army/air headquarters or occasionally the corps level, they had two elements. The first had two army staff officers and a small staff that controlled a wireless radio network consisting of twelve tentacles. These were assigned to forward brigades and divisions based on need for air support. The second element included an RAF officer with a small staff that controlled eight wireless sets through a forward air support liaison (FASL). In 1942, Coningham added two wireless sets at all RAF units on their airfields. This network distributed air support notifications and intelligence. Changes in bomb-line calculations went hand in hand with these evolutions. Ground units were required to report their positions every two hours at minimum and hourly when on the move. They also radioed in key terrain features.[33]

Reviewing the Air Situation

On 10 March 1942, Tedder had reviewed the air situation for Charles Portal, noting that without Dawson's efforts, the diversion of aircraft to the Far East would have precluded successful operations. With the growth of the fighter force from a low ebb of four understrength squadrons to seven full-strength ones, Coningham was returning to the offensive. Tedder asked again for immediate deliveries of Spitfire Vs to engage Me. 109Fs and a squadron of Mosquitoes for strategic PR. Portal made no promises, and instead ordered more diversions to India.[34]

On 20 March, Stafford Cripps, the lord privy seal, attended a meeting of the Middle East Defence Committee on his way to India. Claude Auchinleck described the overall situation. Cripps then asked for Tedder's views. Tedder said the RAF had lost air superiority as a result of diversions to the Far East totaling 530 aircraft, which came on the heels of five weeks of concerted offensive operations. With replacements arriving and repair-and-salvage efforts ongoing, he expected to regain air superiority within three weeks without further diversions. He reiterated the urgent need for Spitfires. Cripps then asked how the government could assist. Suggestions included heavy-bombers to attack Tripoli, more US-built medium-bombers to raid Benghazi, and a reinforcement of RAFME.[35]

Cripps telegraphed Winston Churchill that night, saying there was no chance of mounting a successful offensive in Libya at this point. Given Rommel's tactical acumen and the superior characteristics of the Me. 109F, the British would need a 1.5-to-1 advantage on the ground and parity in the air, which they had lost. They needed another two months assuming reinforcements arrived on schedule. Cripps said heavy-bombers were absolutely essential for striking Tripoli, and more medium-bombers for raiding Benghazi, because Malta was on the defensive. Spitfires were an absolute must for Malta immediately and for Libya as soon as possible. Cripps concluded with the assertion that there must be no further demands whatsoever upon RAFME for transfers or diversions to the Far East and that it needed to rebuild its strength.[36]

Nonetheless, the CoS called for an offensive by 15 May. To comply, the C in Cs needed more aircraft, tanks, and ground equipment, and Malta had to get back on its feet. Hugh Lloyd and his airmen did extraordinary work but could not stop German raids. When Tedder visited Malta from 12 to 13 April, Lloyd said the Germans' 160 fighters and 250 bombers on Sicily were destroying his fighter force faster than he could restore it. He could only shelter his remaining fighters and use them very sparingly until he received substantial reinforcements, regained air superiority, and could then recall the Wellingtons and Blenheims from Egypt.[37] Churchill's reply was immediate. Malta must be held at all costs, and that meant a major convoy and a British offensive in June. Auchinleck resigned himself to this despite deep misgivings.[38]

However, Tedder was upbeat about the RAF's increasing cooperation with the army. He judged that his air strength for the planned June offensive would be 806 aircraft against a combined Axis total of 1,200, reemphasizing that qualitative factors were more important than numbers. Given the RAF's advantage here, Tedder felt certain that he could support a ground offensive. He kept working with his army counterparts to improve air-ground cooperation. Tedder and Auchinleck moved the air support control (ASC) function from the corps level to the combined headquarters, where air and ground liaisons worked together to relay information from tentacles to airfields, and to let ground units know when to expect aircraft overhead. Also, the services refined the bomb line, agreeing to send ground-force movement and position forecasts two hours in advance and on the hour so RAF units had a good understanding of the army's and the enemy's troop dispositions.[39]

The Battle for Aircraft Intensifies

As Tedder fought to rebuild the WDAF, a US mission to the Middle East had sailed on 10 November 1941 and included vital assets to keep US-made aircraft flying. An instructional school traveled to the Middle East by air on 8 November. It included specialists in radio and engine instruction and repair and propeller, airframe, and hydraulics maintenance. US instructors taught ongoing classes to 100 RAF ground crews. Additionally, one full US Army Air Corps (USAAC) aircrew per aircraft type instructed RAF aircrews at each operational training unit (OTU). A mobile depot shipped on 20 January 1942 with a large quantity of spare parts and engines. Work crews prepared Massawa and Gura to receive US aircraft, spare parts, and personnel. Dawson received a USAAC staff to work with his specialists on the supply, maintenance, and operation of US equipment. Finally, in typical Dawson style, he called for direct liaison with the United States rather than through the Air Ministry. He got his way when Roosevelt appointed William C. Bullitt, former US ambassador in Moscow and Paris, as his personal representative in the Middle East. Bullitt had direct liaison authority with the Middle East C in Cs and reported personally to Roosevelt on all military, diplomatic, and logistical matters.[40]

On 16 December 1941, Tedder signaled Portal that the Germans had just sent additional Me. 109Fs to Libya, bringing their total to about one hundred. The RAF fighter force was working on countertactics for this

machine's hit-and-run attacks, but Tedder warned that he must have Spitfires. However, the first Spitfire flight did not arrive until April 1942—
inexcusable given the numbers in the United Kingdom. Tedder also asked
for the promised air-to-surface surveillance (ASV) radar Wellingtons so they
could guide torpedo-bomber Wellingtons to targets. Portal agreed, and these
were soon sinking large numbers of enemy MVs at much greater ranges than
the aging Swordfish and Albacores.[41]

On 14 January 1942, Coningham asked again for Spitfires to match
the Me. 109s that were strafing Commonwealth forces and causing heavy
losses. The enemy's improved fuel position allowed for a major increase in
air activity, and its superior numbers had the RAF on the ropes. Coningham
told army commanders to expect limited RAF support in the near future
and to disperse their forces. "The vital difference between fighter action in
ENGLAND, MALTA and here," he lamented, "is that there is no water bar-
rier between the enemy and us, and that we are linked to a second service
that can be defeated if our support is withdrawn. . . . We must keep the flag
flying here."[42]

Tedder signaled Portal about the urgency of the situation. He saw three
main factors in the WDAF's predicament. The first was the supply line,
which made moving forward bulk items such as fuel difficult. Second, the
transfer of units to the Far East was painful. Finally, technical troubles with
and delayed deliveries of US-made Bostons and Baltimores were serious because these were replacing the Blenheims. He needed to rebuild his fighter
and light-bomber strength in Cyrenaica without delay lest the Luftwaffe
take and hold control of the air—and the seaborne logistics and ground
operations beneath.[43] Portal assured Tedder that RAFME would receive
Spitfires and said additional ground crews were on the way to fill out the
thirty-one squadrons formed or forming.[44]

On 4 March, Air Marshal Peter Drummond, Tedder's deputy air officer
commanding in chief (DAOC in C) and acting commander when Tedder
caught the flu, cautioned that the German advantage in the air was having a
positive effect on the Italians, who were now flying offensive missions rather
than just standing patrols. He urged Tedder to regain the initiative with
airfield attacks specifically on Luftwaffe aircraft.[45] The Italian Air Force's
(IAF's) resurgence alarmed Tedder. He told Drummond to get a note to Portal. Had the RAF been in the Luftwaffe's position of advantage, Tedder said,
he would have exploited it, but Axis air forces did not focus on maintaining
air superiority.[46]

Drummond informed Portal that fighter squadrons were fighting hard
but understrength, the tactical superiority of the Me. 109F was "out of all
proportion to numbers," and the RAF had to rebuild now. Me. 109s had

attacked RAF airfields a number of times (unusual to begin with), doing serious damage to aircraft on the ground. WDAF airfield attacks now happened only at night to avoid the Me. 109s, with the Albacore-Wellington flare-dropping and bombing combination proving effective. Even when the RAF had large numbers of fighters to go after Me. 109Fs, their superior performance, heavy armament, and backing by Italian fighters gave them unprecedented freedom of action.[47]

However, even if Tedder received backfills for diversions to India immediately, it would take time to regain air superiority. His only steady source of aircraft was from Dawson's operation. The situation was dangerous and would become dire should the Luftwaffe for once engage in sustained and heavy airfield attacks.[48] The salvage organization recovered and repaired every airplane unless it was burnt. However, Dawson emphasized that there were only twenty-eight Kittyhawks en route to the Middle East. Without rapid increases, he warned, no operational Kittyhawks would remain by June even with refurbishment efforts. To receive the promised 60 aircraft per month, 100 would have to be at sea at any given time to make good ship sinkings and delays. On a brighter note, Dawson spearheaded the rapid introduction of the Hurribomber. Because the Hurricane was obsolete as an air superiority fighter but could carry an external tank and heavy armament, it was a successful fighter-bomber and set the Allies on their course toward the widespread adoption of this platform. Tedder forwarded these insights to Wilfrid Freeman, stating that the best way to make up for light-bomber shortages was to use more Hurribombers. Freeman sent thirty-five Hurricanes immediately (they were available in large numbers in the United Kingdom) and followed with regular deliveries.[49]

At the same time, the Kittyhawk situation finally brightened in April. There were 31 en route, another 40 set to embark, and another 40 scheduled to sail before the end of the month. Shipping shortages drove this problem. Estimated deliveries were: April, 140; May, 90; and June, 105. Another 65 with Merlin engines rather than the troublesome Allisons would also arrive in May and June. Nonetheless, the Kittyhawk situation caused Tedder the "gravest concern." He called for the immediate dispatch of 68 by fast freighter and another 68 on the *Ranger*.[50]

In a bit of unwelcome news, all US heavy-bombers would go to the United Kingdom or Pacific unless in exceptional circumstances. Tedder had long pushed for heavies to attack distant enemy ports and airfields. He eventually received a mix of US Army Air Forces (USAAF) B-17s and B-24s after the United States sent combat units to Egypt. They flew concerted port and shipping interdiction missions vital to winning the desert war.[51]

Fortunately, No. 205 Group had already become a superb unit. The

German Signals Intelligence Service (SIS) officers developed what they called a "perfect understanding" of the unit. They watched it improve over time, pioneering the use of pathfinder techniques, mine-laying, ASV-guided anti-shipping attacks, and a host of other missions. It grew from 80 aircraft in 1942 to 120 in 1943 to 170 at the formation of Mediterranean Allied Strategic Air Forces (MASAF) in November 1943. Virtually every night, even in bad weather, 40–60 percent of the group flew missions. The paucity of Axis radar and night-fighters made it almost impossible to intercept these aircraft or to provide tactical warning to their targets.[52]

John "Jack" Slessor helped as Tedder continued his battle for aircraft. In January 1942, the first Henry Arnold–John Towers–Charles Portal conference began in London. The three principal members reviewed progress to date and worked together on a presentation for the CCS recommending the process for coordinating the development and employment of the Allied air forces as well as a long-term agreement regarding lend-lease materiel. Decisions reached at the conference reconfirmed the US determination to assist the British.[53]

The results of this and the next meeting in May would go to the CCS and then to Churchill and Roosevelt for approval. Using a script drafted by Slessor, Portal opened the second meeting with a powerful statement about the interdependence of the two air forces. He was perturbed that the United States had decided tentatively, since the first meeting, to use the aircraft it had promised to give the RAF to instead form and send USAAF units to the Middle East rather than delivering them to veteran RAF units to keep them up to strength. The issue was so crucial that the CoS had asked the Joint Chiefs of Staff (JCS) to send General Henry Arnold and Admiral John Towers to London to receive a firsthand account of the battle for aircraft and in turn to update the British on US air expansion efforts. "I think it is very important," Portal told his US counterparts forcefully,

> that we should not think only in terms of aircraft production, training of crews, ground personnel and so on; we must also think in terms of the practical realities of war. . . . For this expansion and indeed for the maintenance of our existing strength we are relying very largely on deliveries of American aircraft under the Arnold/Portal agreement in 1942, and upon its extension in some form into 1943. . . . So I think you will not be surprised to hear that it was with some dismay that I learned there was a possibility of this agreement going by the board. . . . I am circulating to you a paper which will show you in detail the effect of these proposals on the R.A.F. . . . I am bound to say I am a little puzzled to understand why only five months ago you promised me about 1,000 aircraft a month in 1942, with which to build up the R.A.F.; and now that I

have committed myself to all the work and expense and dislocation of national effort necessary to enable me to use these aircraft, you propose to withhold them on the basis of an American expansion programme which, on your own showing is in bombers about 20% and in fighters about 35% in excess of what you could achieve with the total manufacturing resources of the United States. . . . We have had all too many of our best young men from the United Kingdom and the Dominions killed in the air in 2½ years of war; and frankly, I should be delighted to see American air crews being killed instead of British *provided* they are killed in the right place and *in time*—and that is as I said to begin with, in the first instance, this year. Our own view is that the aircraft allocated to the R.A.F. under the Arnold/Portal agreement would be in action much more quickly against the enemy if they came to us than if they are withheld and formed into American squadrons.[54]

Portal also said that there were twenty-nine RAF squadrons with veteran crews awaiting re-equipment with aircraft that could only be made good by US production. With US-made aircraft, the RAF would expand from 5,500 to between 8,500 and 9,500 operational aircraft by 1 April 1943— Marshall's hoped-for date for the landings in France. By his own admission, Arnold could have no more than 2,649 ready for action—and probably more like 1,300—given the much less developed US logistical and organizational infrastructures. Portal asked, rhetorically, whether the United States would be able to substitute its own squadrons in the Middle East for existing British ones. He further noted that the entire logistical structure in the Middle East had developed to maintain certain kinds of aircraft such as the Kittyhawk. Should the United States stop delivering these types and the British have to replace them with Spitfires, it would create chaos in a logistical enterprise that had developed over the past several years and, more recently, specifically on the basis of maintaining and repairing the promised US-made planes. This would produce a cascading effect in India, where almost all the Hurricanes from the Middle East and their logistical tail had gone. It would, Portal warned, constitute a disaster of major proportions in both theaters.[55]

Portal said the Allies must win in the air by learning to operate together, bringing the maximum and most rapid possible pressure on the Axis air forces, holding and clearing the Middle East, defeating Italy and Germany, and then turning to Japan. The objective must be to bring the maximum number of aircraft into action in the right theaters of operation, with the right timing. He said that more than sixty RAF squadrons now had US-made aircraft. Significantly, he added that they had the organization, bases, bombs, and fuel in place. More than 30 percent of the RAF was now dependent on US aircraft shipments. The RAF would eventually have 100 squadrons equipped with US aircraft, and 9,000 pilots were currently in training to fly these and other US-made machines. The British had withdrawn more than 60,000 workers from industrial jobs and were training them as mechanics to keep the ever larger RAF with its growing complement of US-built planes flying, as had been agreed at the first meeting in January.[56]

On 26 June, Portal told Tedder officially about the updated aircraft agreement with the United States, reached at the second Arnold-Towers-Portal meeting, which would put an end to the formation of RAF heavy-bomber squadrons in the Middle East and instead assign US-made bombers and half of all other aircraft types to USAAF groups that would deploy to the Middle East. The US public was clamoring for more direct combat involvement against the Axis, as was President Roosevelt. Only the two squadrons of Halifaxes already promised would go under RAF operational control. Tedder fought this, saying it was much more effective to give the aircraft to veteran aircrews rather than send them in separate units with inexperienced ground and aircrews while his veteran formations bled out. Portal was confident Tedder could make it work by giving the United States independence on paper while guiding them firmly behind the scenes. Ultimately, the two air forces reached a workable compromise.[57]

Looking back, Arnold told Slessor that the agreements and later modifications had worked out for the best. "We had a good friendly scrap," he said,

> and that kind are [*sic*] always good for both sides. As I explained over and over during your stay here in Washington, my one desire is to have combat airplanes in the hands of combat personnel in the active theaters with first priority the European theater. I will not permit combat airplanes to sit on the ground. They are going to go where they can be used to fight the Axis or to train my units so they can go to the active theaters for the same purposes. The United States Army Air Force[s] and the Royal Air Force I am sure today operate in support of each other or together as a team and it is very gratifying to receive reports from my Air Force Commanders as well as Royal Air Force Commanders that this is the case."[58]

Malta Defiant

As the battle for aircraft continued, attacks on Malta intensified. Axis aircraft flew more than 200 sorties a day on 7 and 8 April. Between 24 March and 12 April, they flew more than 2,000. All the Spitfires were out of action by April, and only about six Hurricanes a day were operational. AAA batteries were heavily attacked. There were on average nine alerts a day. From 12 to 13 April, Walter Monckton (acting minister of state) and Tedder visited Malta. They called for the immediate delivery of more fighters and AAA ammunition, more lighters to unload ships, minesweepers, and a fuel tanker.[59]

Lloyd needed 100 Spitfires a month to cover losses at Malta. Tedder advised strongly that the convoy set to sail be postponed until his fighter position on Malta improved. Even if the ships made it to port, the Luftwaffe would likely destroy them before they could unload.[60] Lloyd had already expressed his concerns to Tedder about the equally serious fighter-pilot losses: "Only fully experienced pilots must come here. It is no place for beginners. Casualties up to now the beginners."[61] Tedder sent this along to Portal and added his hope that "Malta can be treated as special case" regarding reinforcements of experienced pilots.[62]

Tedder was particularly concerned about the severe damage Malta had suffered. It was difficult to launch sorties without a maximum repair effort. Tedder and Lloyd decided to hit back at the Luftwaffe fighter force on Sicily, which was based on three airfields and not well dispersed. Accordingly, ten Wellingtons with double crews went to Malta and made two or three raids nightly. Still, this was a palliative solution. Despite the garrison's efforts, Malta needed more fighters to continue as a defensive base and then to rebuild and resume strikes on Axis shipping.[63]

Tedder pleaded for a strong force of heavy-bombers to raid Luftwaffe airfields. The CoS demurred. Allied shipping losses on all fronts, which reached their peak in June 1942, were disastrous, requiring greater numbers of heavy-bombers to engage in the Battle of the Atlantic. Meanwhile, Malta was nearly out of food, water, fuel, munitions, and spare parts. A continued and strong Axis air assault, followed by an invasion, would very likely have overcome the tired and emaciated defenders. Then, bombing abruptly dropped off after 28 April, as most Luftflotte 2 units departed. This was the closest the Axis ever came to neutralizing Malta and paving the way for Operation Hercules. The almost complete lack of Axis coordination delayed the assault, which was supposed to occur on the heels of the heaviest air attacks in April, when the Axis flew more than 9,500 sorties. Malta still functioned as a defensive base for the now tiny fighter force and for larger aircraft shuttling from Gibraltar to Egypt.[64] A sobering signal noted, "Until further supplies of 100 octane [aviation fuel] reach this island all available 100 octane must be used for defense."[65]

And still the British held. The USS *Wasp* and HMS *Eagle* delivered sixty-four Spitfires to Malta on 9 May. Lloyd had improved procedures for their receipt, which made each one ready for combat within ten minutes of arrival. By the time Axis aircraft attacked, half the new Spitfires were in the air, many with experienced Malta-based pilots. They flew seventy-four sorties.

German bombing was largely ineffectual. A fast minesweeper delivered 300 tons of ammunition and other supplies to keep the new arrivals flying. HMS *Eagle* flew in another seventeen Spitfires on 18 May. The RAF had local air superiority for the first time in five months.[66]

In contrast to the RAF resurgence on Malta, Fliegerkorps II serviceable strength dropped from 154 planes in April to 91 by May, down from 335 in March. The Germans lost 40 aircraft over Malta in May as opposed to 31 RAF machines. The tables had turned with the departure of most Luftwaffe units.[67]

In April, with Axis air superiority over Malta and the SLOCs still complete, Axis convoys delivered 150,000 tons of supplies to Libya with less than 1 percent losses. In May, the totals were 86,000 tons with a 7 percent loss rate. The results of weakening Axis air strength were already evident. Still, the Axis got most cargos through. Efforts to substitute naval vessels for strike aircraft to interdict these convoys without careful sea-air coordination were routinely disastrous. Of four destroyers sent from Alexandria on 10 May to attack a convoy, Ju 88s sank three.[68]

The Axis air campaign against Malta was costly but also effective, and it might well have been decisive had Operation Hercules gone forward. Axis shipping moved freely from January to early May but was once again under increasingly heavy attack after that, as RAF and FAA bombers and torpedo-bombers returned to Malta. Even during the height of the raids, enough bombers had remained on Malta to sink six large MVs and severely damage or sink another eight.[69]

Axis leaders now faced a major decision. The original plan for Operation Hercules had been to conquer Malta immediately after major raids ended and Luftflotte 2 units were still available to cover the assault before returning to Russia.[70] Kesselring and Erich Raeder convinced Hitler that the Axis must conquer Malta, but Hitler was doubtful of Italian capabilities and engrossed in the coming 1942 offensive in Russia. On 18 April, Luftflotte 2 had 742 planes with 467 operational. It could have controlled the central Mediterranean during an assault on Malta, making British relief efforts extraordinarily costly. However, it soon left for Russia. Then, the disaster at Stalingrad and redeployment of aircraft to Tunisia and to the Reich to counter Operation Torch and the Combined Bomber Offensive (CBO) precluded a large-scale return of aircraft to the central Mediterranean. The window of opportunity to capture Malta closed.[71]

However, this outcome was not clear in May 1942. The British expected an invasion, and Kesselring planned to give them one. After staff discussions with Ugo Cavallero and other Axis senior leaders in March and April, and the development of draft plans, on 31 May Kesselring sent Cavallero

Spitfires and F4F Wildcats aboard the USS *Wasp,* April 1942. Despite the urgent need for *Wasp* in the Pacific, President Franklin Roosevelt authorized two Spitfire deliveries to Malta on 20 April and 9 May 1942. Constant infusions of Hurricanes and then Spitfires helped to keep Malta effective as both an offensive and defensive base even during periods of heavy air attack. British carriers such as HMS *Eagle* also flew in hundreds of Hurricanes and Spitfires from 1940 to 1943. (U.S. Navy National Museum of Naval Aviation Photo No. 1996.253.7386.029)

a comprehensive plan for Operation Hercules. It proceeded from the assumption that Rommel would defeat the Eighth Army and bring major combat operations to a temporary hiatus at the Egyptian border as Hercules proceeded. This would free the maximum number of aircraft to support the operation. Fliegerkorps II and X, and most of Fliegerführer Afrika, would concentrate in Sicily. Fliegerkorps XI would move the German paratrooper division to its jumping-off points two days before the operation. Between seven and nine transport *Gruppen* (378–484 aircraft) would deliver them to Malta.[72] Additionally, two Italian airborne divisions, two corps of Italian infantry, six independent Italian battalions, two tank battalions, armored car and self-propelled artillery units, motorcyclists, and some German tanks would participate. After Malta was in Axis hands and supplies and reinforcements were flowing freely across the Mediterranean, Rommel would resume his offensive.[73]

Intensive aerial reconnaissance, including horizontal and oblique photos of proposed beachheads, revealed strongpoints, the best paratrooper and glider landing areas, and myriad other details. After the Germans owned the skies, final destruction of AAA sites would proceed. Coastal guns and emplacements would then come under attack with deception at the center of the effort to keep the British from knowing where the airborne and seaborne landings would occur. Malta was to be out of action as an air base by X-1 (D-1).[74]

As paratroopers and airborne infantry lifted off, Axis aircraft would make nonstop attacks on AAA and other defenses in the landing areas while locating and attacking any Royal Navy relieving forces. The senior signal officer would oversee jamming of radar and verbal communications. Preparatory air operations would commence on 28 June and conclude on 17 July. Direct support operations would commence on 18 July with the invasion.[75]

The Italians agreed that it could be done. Their plan was even more detailed than Kesselring's and showed an uncommon seriousness of purpose. It was realistic in terms of the opportunity and the risks. Hitler had agreed to an Italian operational lead and promised full German support, to include the airborne division and light tanks. The IAF planned to throw 578 aircraft into the effort.[76]

Regardless, serious problems remained. Given the shortage of airfields on Sicily and their congestion with Luftwaffe aircraft, the Italians could offer only thirty sorties a day. The Axis did not build more landing grounds despite German expertise. The Italians also noted that the Luftwaffe's efforts had not completely neutralized Malta's air garrison. In fact, air strength nearly doubled as Spitfires arrived. They said the neutralization of Malta was partial and temporary, that continued heavy pressure was necessary before any invasion attempt, and that there could not be any diminution of air strength. Finally, the Italians noted the high cost in aircraft and crews required to keep a full press going against Malta. Even as the intensity of Luftflotte 2 efforts increased, its strength ebbed as too few replacement aircraft and crews arrived. The unit had 939 aircraft (438 operational) on 3 January and 681 (404 operational) on 10 May.[77]

Hitler and Benito Mussolini agreed that given the Luftwaffe's inability to make a maximum effort against Malta and North Africa simultaneously, Rommel should capture the rest of Cyrenaica in late May or early June, and the airborne assault on Malta should occur in mid-July or mid-August. The attack on Egypt would follow after Axis troops had all the equipment, supplies, and replacements they needed from the now secure SLOCs.[78]

Logistically, the Axis had the upper hand but gave it away. Its leaders could have held it with a moderate and steady employment of bombers stationed in Greece and Crete. Nonetheless, Luftflotte 2's brief all-out involvement had allowed Rommel to refit his forces, which were still understrength but confident of victory, and he did not intend to waste the opportunity. The Luftwaffe was as strong as it had yet been. By this time, however, Allied aircrews were becoming excellent, and their air forces were also more balanced with all aircraft types, including four-engine bombers and a far larger air reconnaissance contingent. Also, by 20 May, six squadrons of Spitfires (about one hundred aircraft) had joined the Malta air garrison. The island was resuming its importance as an offensive base for convoy raids.[79]

Consequently, Kesselring requested a rapid assault and visited Rommel on 21 June after his successful Gazala offensive (more on this later) to remind him of the agreement: Malta first while Panzerarmee Afrika rested and refitted, then Egypt after the logistical situation was secure. Rommel, however, was already planning his advance into Egypt with the Führer's full support.[80] On 1 May, Hitler and Mussolini had agreed at Berchtesgaden that the attack on Malta would precede the advance into Egypt. As one attendee noted, "The whole business [the Mediterranean theater] is now assuming importance after having been regarded hitherto as a subsidiary matter in which victories were looked on as gifts from Heaven, but in which nobody bothered to do anything seriously for the 'Italian theatre of war.'"[81] However, subsequent events, Rommel's requests, and their own lukewarm feelings prompted Hitler and Mussolini to cancel Operation Hercules. This was one of the pivotal blunders of the war. Whether the attack would have succeeded is open to debate, but the necessity of taking Malta to guarantee the Axis flow of supplies is not. Looked at from any angle, this was an operation worth the risks, crucial to the outcome of the Mediterranean campaign, and important to the course of the larger conflict.[82]

8

Axis High Tide
February–July 1942

Axis Strategic Decisions

IN MANY WAYS, 1942 PROVED a pivotal year in the conduct and course of what had now become a global and total war of attrition, waged across multiple theaters with important interconnections. A series of poor Axis strategic decisions, reached without close coordination, interacted with generally sound Allied strategic choices to produce a series of decisive defeats from which the Axis never recovered. The decision not to invade Malta, the collective turning away from an Indian Ocean strategy, and critical operational errors on every front came together to end any hope of Axis victory in the Mediterranean and Middle East—one that had been a very real possibility even with the relatively limited ground assets (but much greater air assets) committed to the theater.

A German-Japanese linkup in the Middle East was a distinct possibility in spring and summer 1942, while the Japanese contemplated an Indian Ocean strategy focused on conquering the Raj, linking up with German forces in the Middle East, and in the process denying the use of the Indian Ocean to Allied shipping. The latter eventuality would in and of itself have spelled disaster for the entire Allied Middle East position. Despite the lost opportunities in Iraq and Syria, the Germans still had an opportunity to achieve their goals, without major reinforcements, if they acted quickly against Malta and destroyed or at least defeated the Eighth Army, following this with an advance into the Middle East.[1] A major objective within Adolf Hitler's Directive No. 41 of 5 April 1942, which set forth the specifics for Germany's 1942 offensive in Russia, was a linkup of German forces driving south from the Caucasus Mountains and north from Egypt to seize oil assets and undermine the British position in the Middle East. His subsequent Directive No. 45 of 23 July 1942 further emphasized the importance of this effort. The failure to achieve any of these objectives, and the Japanese turn back to the central Pacific, resulting in the debacle at Midway, destroyed any opportunity for a larger victory in the Middle East.[2]

Allied Strategic Choices

Although they could not yet realize it, the Allies were about to reach their most serious strategic crisis in the Mediterranean and Middle East, and perhaps of the war, during spring and summer 1942. With US officers trying to backtrack on agreements reached during ABC-1 and Arcadia and their support for a peripheral strategy fading, Erwin Rommel's counterattack after Operation Crusader, and his reconquest of much of Cyrenaica gave the United States further cause for concern. Rommel's major victory at Gazala in May and June 1942 had an even greater impact, although Franklin Roosevelt remained firm in his support of the British. He understood the risks to the empire and Russia should Axis troops overrun the oil fields and cut the lend-lease route through Persia, over which one-third of all aid to Russia flowed. Despite the rapid German advance in Russia during summer 1942, Roosevelt and Winston Churchill remained convinced that the Red Army would survive and eventually gain the initiative. However, without the lend-lease materiel just beginning to flow through Iran in large quantities, the outcome in Russia would be more uncertain.[3]

The importance of lend-lease materiel, not only to Russia's survival but to the Red Army's transformation into an instrument more lethal than the Wehrmacht by 1943–1944, has received far too little attention, but it was central to Red Army successes. Despite Soviet assertions to the contrary during the Cold War, we now know that Josef Stalin himself understood the vital role of lend-lease materiel in Russia's survival, saying several times that without it the Red Army could not have won given the huge losses of industrial plants and equipment, agricultural areas, and mineral-bearing regions in 1941–1942. Marshal Georgi Zhukov said that without lend-lease materiel the Soviet Union "could not have continued the war." The United States delivered more than 500,000 vehicles, 421,000 radios and field phones, 1 million miles of telephone wire, 14,200 aircraft, and immense amounts of food. The military hardware allowed the Red Army to evolve rapidly and beat the Wehrmacht at its own game, and the food kept Russian workers and soldiers strong enough to continue the fight. If the Red Army was going to turn the tables on its enemy, lend-lease materiel through Persia would be of critical importance along with the deliveries to Vladivostok and the much smaller and less reliable ones to Murmansk.[4] Roosevelt and Churchill knew they had to hold the delivery corridor through Persia at all costs.

Staff talks in July 1942, though contentious, kept the effort on course. US Army officers pushed hard for landings in northwestern Europe without any in North Africa whereas US Navy officers advocated a major counteroffensive in the Pacific. Based on the inability to support either effort given resource shortages, the need to keep the British fighting in North Africa, and the critical situation regarding the Battle of the Atlantic, Roosevelt and Churchill demurred. They remained dedicated to winning in the Mediterranean and adhering to the other major agreements made thus far. Roosevelt committed once again to landings in North Africa later in the year. Any offensive action in the Pacific would be based on developments there (primarily Midway, as it turned out) and of second rank to the European theater.[5] Well before these talks occurred, Commonwealth forces were about to begin their most trying period of the desert war, one that nearly cost the British the Middle East.

The Royal Air Force (RAF) Struggles to Rebuild

British loss of airfields in Cyrenaica had given Axis supply routes relative immunity from airstrikes. One of Claude Auchinleck's key objectives was to retake them. However, it was another logistical race to the next offensive, and Rommel won. Auchinleck was not optimistic about his ability to beat Rommel to the punch or defeat him with the available forces. Members of the Defence Committee were unhappy with his assessment based on their view that it was biased in favor of the enemy. Churchill was furious about Auchinleck's contention that he was too weak in tanks and aircraft to launch a major offensive before May.[6]

The RAF reconstituted its strength feverishly as the Air Ministry's June 1941 target of sixty-two and one-half squadrons was just now nearly met— in March 1942. Of note, three heavy-bomber squadrons were in place by May 1942. Given the centrality of air bases and most land-based planes' inability to reach the central Mediterranean without control of airfields in Cyrenaica, the addition of these heavies gave Arthur Tedder his first truly potent long-range striking force. The Wellingtons had done superb work, but there were not enough, and they had a shorter range and smaller payload.[7]

Charles Portal continued to insist that only Tedder's centralized and flexible organization could produce the air superiority in which naval and land forces could operate with success.[8] His views found continuing favor with Churchill, who considered the army's demands for a de facto separate air arm in error. Field Marshal Alan Brooke complained, "We make no headway at all, and are exactly where we were before. The situation is hopeless and I see no solution besides the provision of an Army Air Arm."[9] Nonetheless, the services made headway as discussions continued among the Chiefs of Staff (CoS). Portal and Brooke agreed to build up Army Co-operation

Command to twenty bomber and fighter squadrons of modern types by 1 September. However, they disagreed on command relationships and the numbers of squadrons to assign to army units.[10]

Portal also rejected a proposal to create an RAF army air support group— a composite organization under direct army control, saying the RAF already had too many aircraft dispersed in supporting roles when it needed greater centralized control for the anticipated combined-arms operations in North Africa and beyond. The "[John] Slessor Paper," delivered to the CoS on 21 July, set forth Portal's organizational proposals. Flexibility of action was paramount, and part and parcel of the principle of centralized control.[11]

On 14 September, P. J. Grigg, secretary of state for war, rejected Brooke's plan. Admonishing Brooke for bypassing him, Grigg said, "I would have expressed the view that the Slessor plan was the right one in the present circumstances."[12] Churchill intervened at Grigg's request and chaired a special meeting of the CoS on 5 October, giving the Slessor plan his support. On 14 November the secretaries of state for war and air submitted a joint statement regarding air support to ground operations, in which a unified air command, under a single air officer, would give the army support whenever and wherever required, regardless of other targets. The Western Desert Air Force (WDAF) system was the model both services would follow and build upon in subsequent campaigns. This ended the issues and uncertainties that had long surrounded this problem. Effective air-ground cooperation from summer 1942 to Victory in Europe Day, with very few exceptions, discussed later, proved the wisdom of this decision.[13]

Three things were necessary for perfecting air-ground cooperation: a willingness to cooperate, sound principles and procedures, and reliable communications.[14] The services had already embraced all three. Operation Crusader had been a crucial part of the learning process.[15] Its biggest contribution was to prove to RAF and army officers that effective cooperation was not only workable but a major force-multiplier. By summer 1942, ideas, tactics, techniques, procedures, and equipment merged to produce an excellent combined-arms team. Flexible on attack or defense, the WDAF trained to fight as part of this team but also to gain air superiority in all land and sea engagements to enable the success of subsequent combined-arms efforts.[16]

Tedder viewed every operation as combined. He insisted that air, land, and naval commanders and planners work closely at all echelons.[17] The WDAF's approach proved superior to the German concept as it had evolved by the middle years of the war. In Russia and Africa, German ground commanders tended to exercise increasing power over associated air units, insisting on direct support, often at the expense of air superiority.[18] These emerging weaknesses, however, did not render the Luftwaffe ineffective. The RAF did not have anything like a cakewalk from 1941 to 1943. It had to outthink and outfight a capable opponent whose leadership simply made more mistakes.

Planning the Gazala Offensive

As planning for Operation Hercules continued, Rommel prepared to launch a major offensive, Operation Theseus. With the logistical situation as good as it would ever be, Rommel brought supplies and troops into place. On 30 April, he briefed senior officers on the offensive and said it would begin in early June, after Malta had fallen. If the capture of Malta took longer than expected, Panzerarmee Afrika might attack anyway based on the likelihood of success.[19]

Rommel proposed to destroy British forces in front of Tobruk, take the port, consolidate his logistical situation, and advance from there. In a meeting on 28 April, Albert Kesselring reviewed with Rommel the outcome of earlier talks with Hitler. Malta had to be taken, by the end of May if possible, because otherwise the Allies would win the logistical struggle. Therefore, Panzerarmee Afrika's offensive should begin after the capture of Malta to ensure a steady stream of supplies. Kesselring also noted that despite rapid Allied reinforcement in Egypt, Axis reinforcements were pouring into North Africa at unprecedented rates. From 12 to 14 May, nine ships arrived at Tripoli unscathed with 19,875 tons of fuel and other supplies. Air transports brought in 6,000 men in April.[20]

At a 6 May meeting of senior officers, Kesselring agreed to reinforce Fliegerführer Afrika with 90 planes. A flak *Abteilung* would also come from Italy, Ju 88s and Me. 110s of Fliegerkorps X would support Rommel's offensive from Crete, and more Ju 88s would deploy to Derna. The Italian Air Force (IAF) would contribute another 150 or so planes. Kesselring also noted that, because of delays in preparation, Operation Hercules had to be postponed, meaning that Rommel's offensive would begin before the island's capture.[21]

Luftwaffe involvement proceeded along standard lines. On 16 May, Kesselring and Rommel agreed that it would concentrate on close support to Panzerarmee Afrika and then against Tobruk. At the same time, "waves of attacks" would hold down the western sector of the enemy's fortified front. This latter effort would in practice focus on Bir Hacheim and cause the Luftwaffe heavy attrition. There was no mention of gaining air superiority.[22]

Ultra signals intelligence pinpointed the timing of Rommel's 26 May offensive at Gazala. The Axis had 928 planes in Cyrenaica, Crete, and the Dodecanese with a serviceability rate of 50 percent. It had many more available to call forward from Greece, Sicily, and southern Italy.[23] Tedder had 849 aircraft with a 60 percent serviceability rate—low because of non-stop air operations and insufficient replacements. Malta had 189 aircraft. There were fewer reserves than in Operation Crusader. He was painfully aware that Axis air strengths and supplies were at their peak, and his were not. Still, the aircrews were ready for a fight. They would not yield control of the air without a scrap, and, to their good fortune, Axis air forces stayed focused on ground support. Arthur Coningham commanded the air striking force from Headquarters Western Desert, collocated with Eighth Army Headquarters. Air defense of the Nile Delta fell to Air Vice-Marshal Keith Park, who helped Leonard Slatter, commanding No. 201 Naval Co-operation Group, protect shipping in coastal waters, conduct reconnaissance, and attack enemy vessels.[24]

Tedder finally had one flight of six Spitfires with two in reserve. Graham Dawson's team sent repaired fighters at prodigious rates but could not keep squadrons up to strength after intensive operations began. Tedder had four light-bomber squadrons. His four Wellington squadrons and their Albacore flare-dropping companions were doing exceptionally well. The same was true for long-range reconnaissance with the remaining Marylands and the handful of photoreconnaissance (PR) Spitfires, but air control and tactical reconnaissance (Tac R) aircraft were suffering heavy losses, requiring a shift of fighter assets in that direction. The antishipping effort continued to go poorly because of the extreme range to convoy routes. Air-to-surface surveillance (ASV) radar and torpedo-bomber Wellingtons were progressing well in tests and had the range to do the job but were not yet in action. Airfield attacks continued.[25]

The Gazala Offensive Begins

Operation Theseus began on 26 May and ended four weeks later with the fall of Tobruk. Coningham used fighters heavily in the ground-attack role after gaining air superiority. Enemy aircraft made a substantial appearance in the El Adem and Bir Hacheim areas and drew RAF fighters there in large numbers. In one of their last great showings, Stukas engaged in mass attacks to support Rommel's advance in late May and early June. Piecemeal Eighth Army counterattacks were ineffective and gave the Germans advantages at key moments. British armor losses were high as commanders continued sending tanks to impale themselves on antitank guns. The RAF, already outnumbered, could not keep the army's position from collapsing, but it played a pivotal role in saving it from destruction so it could fight again at El Alamein.[26]

The course of the battle was initially unclear, in part because Rommel failed to notify Fliegerführer Afrika of the offensive's opening date. The

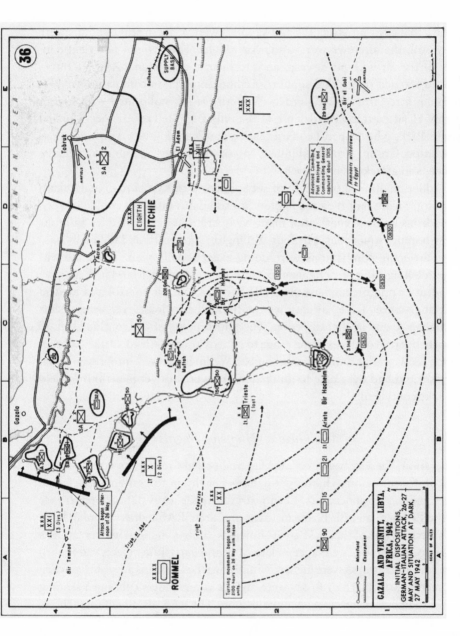

Rommel's second offensive (Gazala and Pursuit), 26 May–7 July 1942. (Maps Department of the US Military Academy, West Point) For the pursuit phase of the offensive, see map entitled "Rommel's second offensive" on p. 181.

Luftwaffe response was thus slow and hampered by RAF airfield attacks. Belated Luftwaffe raids on RAF airfields lasted three days and were ineffective. Tedder lamented the shortage of replacement aircraft, especially Kittyhawks. He said that with close support having played its part for the moment and the Luftwaffe gaining air superiority by default rather than conscious effort, "We must now, for a spell, pay more attention to the enemy air and try and conserve our painfully limited resources against the moment when we can again intervene with decisive effect."[27]

Part of Tedder's challenge had to do with the arrival of a new Fliegerführer Afrika, General of Flyers Hoffman von Waldau, who understood the principles of air operations. Von Waldau also spoke Italian and worked well with General Francesco Pricolo, head of the IAF. On 27 May the Luftwaffe flew missions north and northeast of Acroma based on initial inputs from Rommel. However, Luftwaffe headquarters lost contact with Rommel as he headed off with his staff and left the airmen in the dark.[28]

For two days, von Waldau's requests for information on the ground forces' situation and plans went unanswered. Lieutenant General Ludwig Crüwell, who tried to make contact with Rommel using a Storch aircraft, was shot down and captured, an unnecessary loss of a superb armor commander. Rommel's staff had not included von Waldau's in the preparation of the operations order. Ground commanders failed to consider how much time they needed to brief air commanders on the situation and to disseminate and transmit orders to dispersed airfields. Operations moved so rapidly that requests for air support usually arrived too late. Von Waldau's headquarters reestablished radio contact on 30 May. Air reconnaissance revealed two major concentrations of British forces, against which Fliegerführer Afrika's units flew 326 sorties on 31 May and 1 June. Fliegerkorps X flew night attacks, and the IAF attacked ground units. Rommel acknowledged the Luftwaffe's role in delaying the Eighth Army's offensive activity, which gave him the initiative. German airmen concentrated on Commonwealth forces threatening Rommel's supply lines and on armor. Then, Rommel ordered intensive air support for the assault on Bir Hacheim.[29]

Kesselring was critical of Rommel's tendency to take a small command staff and leave his subordinate and supporting units in the dark. This made air reconnaissance and the use of bombers unnecessarily dangerous. After Kesselring took Crüwell's place as acting commander of all combat forces in Rommel's absence, he understood how difficult it was for units to operate with Rommel out of contact. Kesselring further noted that heavy dive-bomber attacks on Bir Hacheim were of limited success because Rommel did not coordinate them with the infantry assault. "The Royal Air Force" he lamented, "was very active from the outset and tried to achieve air superiority."[30]

Fighters engaged German aircraft in the air and on the ground, and bombers attacked airfields. Coningham placed his forward units at the front, allowing aircrews to fly multiple sorties each day. Maintenance personnel worked eighteen hours and slept four per day.[31]

Rommel persisted with his effort to capture Bir Hacheim and outflank the Allied position from the south. Coningham released his fighters to high-altitude duty to maximize success against air and ground targets. Their operations were so effective that the commander of the Free French garrison signaled "Bravo! Merci pour la R.A.F.," to which Coningham replied, "Bravo! Merci pour le sport."[32] During this battle, friction between the Luftwaffe and army was palpable as a result of aircraft losses in failed operations between 3 and 10 June. Von Waldau blamed Rommel for calling in strikes when there were no ground units ready to take advantage of their effects. Kesselring intervened in an effort to convince Rommel that when he ordered attacks and called for Luftwaffe support, the combination of air-ground power had to be properly coordinated and sufficient to take the objective.[33]

Failure to send in enough ground troops to capitalize on airstrikes led to a ripple effect in which German aircraft, intended to assist with a planned assault on the Gazala line, were instead rerouted to Bir Hacheim, forcing a postponement of the attack further north and leading to even greater aircraft losses. Von Waldau visited Rommel's headquarters the same day to express his objections and to try once again to improve air-ground cooperation. His efforts bore no fruit.[34]

Raids on Bir Hacheim, repeated far too often because of the inadequate ground assault force, led to the premature attrition of the Luftwaffe and to RAF superiority at decisive moments. This continued on 9 June, when another two heavy Luftwaffe raids produced no lasting effects. Rommel moved forward another battalion and supporting artillery. Still, he underestimated. Von Waldau, now completely fed up, informed Rommel that Bir Hacheim had already absorbed 1,030 sorties—all of which, in the case of the fighters stuck in close escort, could have been much more effectively employed to gain air superiority (something von Waldau sought to obtain). On 10 June, three more major raids did serious damage. Finally, behind this attack, ground units penetrated the defenses and forced the garrison to retreat. Von Waldau shifted his depleted units to El Adem and Acroma to help with ground actions there.[35]

Meanwhile, the Special Air Service (SAS) destroyed German aircraft—an assist the RAF needed at this point of crisis. A 12 to 13 June raid on Barce resulted in eleven aircraft destroyed and fifteen Germans killed, while a simultaneous attack on Benina burned three hangars and the numerous aircraft they held. The raiders also destroyed large fuel dumps as a matter of course to exacerbate perennial Luftwaffe shortages. A day later, a team destroyed twenty aircraft at Martuba and Derna, and another burned eleven planes and killed seventeen ground personnel. The payoff continued with three more aircraft and a number of fuel trucks destroyed on 8 July. A raid on Maten Bagush the night of 7–8 July wrecked thirty-seven aircraft. On 11 July, an attack at Fuqa destroyed between twenty-three and twenty-eight machines. Another at Mersa Matruh on 26 July destroyed thirty-six and yet another at Sidi Haneish thirty more. These attacks occurred when the Luftwaffe had already suffered serious attrition and was too far away to help Rommel as a result of transportation woes. A major raid on Barce the night of 13–14 September continued the pattern, torching twenty-four air-craft along with a fuel tanker-truck and a large quantity of forty-four-gallon fuel drums, and damaging another twelve planes. By the time the tide turned at El Alamein, David Stirling and his men had destroyed 350 German air-craft from November 1941 to October 1942 and killed many skilled ground personnel. The destruction of fuel stocks, fuel trucks, and heavy and light tools was also important. Finally, these raids yielded an immense haul of intelligence put to use in subsequent air raids against fields known to hold aircraft.[36]

By 31 May, despite concerted efforts to·delay the Axis advance and reduce its fighting strength, RAF squadrons were down to seven or eight aircraft—half-strength. In operations from 27 to 31 May, the British lost 50 of their 250 serviceable aircraft. Coningham sent urgent requests for replacements. Kittyhawks and other fighters had to move to higher altitudes to reduce losses, but the first six Spitfire fighters and first four PR models had ar-rived and began making their presence felt in the air superiority battle, with improved photographic intelligence (PI) of enemy targets and effectiveness assessments of raids.[37]

As the Luftwaffe bled out over Bir Hacheim, the Eighth Army did so fur-ther north as the last combat-capable Stuka units made an all-out effort in support of German troops. British attacks were piecemeal, with poor use of armor and absent or conflicting command and control (C2). As the army's position crumbled, WDAF pilots maintained localized air superiority and inflicted heavy Axis troop and vehicle losses. The disaster in the Cauldron on 6 June was further proof that air support could not engage effectively without a clear bomb line and in poor weather. Kesselring and von Waldau's all-out efforts to concentrate remaining air assets over the advancing troops proved pivotal in the speed and magnitude of the breakthrough.[38]

RAF planes continued attacking troops around El Adem on 15 June. The diversion of German aircraft to attack the Vigorous Convoy during this period gave the WDAF a temporary advantage and allowed the Eighth Army to retreat with relatively few losses from air attack. The Twenty-first Panzer Division reported gloomily on the "continual attacks at quarter-hour intervals by bombers and low flying aircraft" and called repeatedly for fighter cover.[39]

The enemy advance accelerated, and ground crews at the Gambut airfields could hear their aircraft attacking Axis forces. In an indication that he was beginning to understand how important the capture of airfields was to success in combined-arms warfare, Rommel specifically sent the Afrika Korps after Gambut. "Primarily," he said, "this advance was directed at the R.A.F., who, in their short flight time from neighbouring bases, were being unpleasantly attentive. We intended to clear them off their airfield near Gambut and keep them out of the way during our assault on Tobruk."[40] Coningham ordered an evacuation on 17 June. Airmen stayed in action until the last possible moment. Ground crews generated 450 sorties a day—3 for every available aircraft—as they protected the Eighth Army's withdrawal.[41]

The rapid Axis advance forced Coningham to move his aircraft again on 19 June to Sidi Barrani. This placed all fighters except Kittyhawks out of range of Tobruk. Meanwhile, Axis air forces prepared to support Rommel's attack there. Kesselring gave maximum Luftwaffe support to the 20 June assault. He knew that most British aircraft were out of range, and he was anxious to move on to Operation Hercules. He was willing to give Rommel his victory at Tobruk if he could have his own on Malta.

Rommel and von Waldau met on 19 and 20 June to coordinate their operations. Dive-bomber raids would begin at 5:20 a.m. Fighters and dive-bombers would support ground attacks based on radio communications and dive-bomber grid squares coordinated by the Luftwaffe liaison element to Panzerarmee Afrika. Stukas and Italian fighters would then attack targets previously identified within the inner defense circles, and Ju 88 bombers would go after key crossroads within the fortress and any airplanes still at the airfield. Italian bombers would suppress antiaircraft artillery (AAA) batteries. Artillery would fire smoke shells to mark the key areas of ground penetration and pinpoint areas of resistance, so Stukas and IAF fighters could focus on targets directly in front of advancing troops. Ground forces would use blue smoke generators and grenades to make their positions clear to aircrews, and reconnaissance flights would begin two hours after the start of the attack to give Rommel and his staff constant updates. It was the best-coordinated Axis air-ground operation of the African campaign.[42]

The assault began on schedule on 20 June with every available aircraft attacking the southeastern sector of the defenses. The port fell on 21 June after one day of fierce fighting. Axis forces captured 45,000 troops, more than 1,000 tanks and 400 guns, huge numbers of vehicles, and large quantities of fuel. Excellent air reconnaissance allowed the Germans to pinpoint bunkers and other fortified positions, which dive-bombers attacked effectively, assisting the advance. Hitler promoted Rommel to field marshal and awarded von Waldau the Knight's Cross for the Luftwaffe's key role. The road to Cairo appeared tantalizingly open. Luftwaffe aircrews flew 588 sorties and IAF crews another 177. It was the last time Axis air forces would play a major role in the Western Desert.[43]

By now, the Eighth Army was in full retreat toward Mersa Matruh, where the commanders in chief (C in Cs) sought to establish a line. They soon abandoned it for El Alamein, with its two secure flanks: the Mediterranean and the Qattara depression. Whereas the British chose wisely, Axis leaders made several errors of strategic significance. The first was to allow the escape of the Eighth Army without either a strong air or ground pursuit. The second was Rommel's miscalculation that he could overrun El Alamein. Had he paid closer attention to his intelligence officers, he would have understood that the British had been working on El Alamein's defenses for weeks and had strong reinforcements there waiting to integrate the Eighth Army's retreating elements. Finally, and most importantly, Rommel convinced Hitler and Benito Mussolini to abandon Operation Hercules and thus allowed Malta to continue its rebound as an offensive platform against Axis convoys. Additionally, in a move that spelled trouble for the Axis, the United States delivered an immense quantity of weapons and munitions to Suez after Tobruk fell.[44]

All RAF units worked at maximum tempo and on one-hour notice to move to new airfields. Landing grounds in Egypt allowed for a steady withdrawal just in advance of enemy troops. Seeing RAF bombs burst on advancing Axis units was the sign for retreat to the next line of airfields. As flying units moved, repair-and-salvage units (RSUs) removed everything of value. It was, Coningham said, a "point of honour that there [is] nothing left for enemy. . . . [I am] content that whole machinery is working very smoothly."[45] The WDAF lost no aircraft or vehicles at abandoned airfields.

The enemy air effort against the retreating Eighth Army and RAF had been nearly nonexistent because the Luftwaffe was employing its remaining fighter units to protect German forces and landing grounds. On 20 June, Tedder commented that protection of retreating troops had been outstanding despite instances in which the Luftwaffe had local air superiority. For three days the coast road was packed with retreating vehicles, and enemy landing grounds were only forty miles away. The Eighth Army suffered minor losses. The RAF had fought the entire campaign to this point outnumbered. From 26 May to 17 June, twelve fighter squadrons with an average strength of ten to twelve aircraft flew 4,882 sorties. Serviceability rates averaged 75 percent.[46] Still, Tedder was down to forty-two operational squadrons with no spares and no prospect of major reinforcements. Yet although Axis air forces held air superiority until July, with between fifty-seven and sixty squadrons available at various times, Kesselring saw the writing on the wall as he watched the British resurgence on Malta and in the Western Desert. Additionally, Fliegerkorps II and Fliegerführer Afrika were exhausted, and only eighty-seven Fliegerkorps X aircraft from Greece were engaged in the fight. Kesselring believed Malta would have been easy to capture.[47]

On 21 June, Ugo Cavallero gave Mussolini a personal note for Hitler pleading for approval of Operation Hercules.[48] During a conference at Gambut with Rommel, Cavallero, and General Ettore Bastico (commander of Italian forces in Libya) on 26 June, Kesselring told Rommel he was wrong about the lack of remaining opposition. Most importantly, the Luftwaffe was exhausted. Nonetheless, nearly all parties sided with Rommel (most importantly Hitler and Mussolini, as Rommel himself already knew). Rommel said he would be in Cairo in ten days. "The abandonment of this project," Kesselring said of the decision not to invade Malta, "was the first death blow to the whole undertaking in North Africa."[49] Rommel's culminating point in the desert was 2 July, although few realized it. The Luftwaffe failed to keep up and thus to further weaken the Eighth Army. As Major Eduard Neumann, commander of JG 27 put it, "The general planning seems to envisage a forward push by the Army only. With our inadequate ground organisation we just can't keep up."[50] Lieutenant General Bernard Freyberg, commander of the New Zealand Division, told Tedder, "Thank God you didn't let the Huns Stuka us, because we were an appalling target."[51]

Even with the Bir Hacheim combined-arms problems fresh in everyone's minds, Rommel's headquarters once again kept the Luftwaffe in the dark about its next moves until 13 June. Von Waldau received no insights regarding the reasons for the many disjointed air attacks Rommel's staff required, all of which involved heavy dive-bomber attacks with heavy fighter escort, giving the RAF major tactical advantages. The shortage of fighters also made reconnaissance missions dangerous and often blinded Rommel. A frustrated von Waldau lamented the fact that army requests for air support were not based on any sound conception of combined-arms operations. To make matters worse, two large convoys were heading for Malta, forcing Fliegerkorps X and Fliegerführer Afrika to shift their attention to these, which the German and Italian High Commands said must be destroyed.[52]

Keeping Malta Afloat

The two convoys—Harpoon from Gibraltar and Vigorous from Alexandria—were to resupply Malta, which had reached a low ebb in terms of air capabilities and food supplies from April to May and was critically short of aviation fuel. They were heavily dependent on fighter cover, so another fifty-nine Spitfires flew to Malta from HMS *Eagle,* bringing the total to ninety-five.[53]

Axis raids on 15 June sank one merchant vessel (MV) out of six in the Harpoon Convoy and damaged the cruiser HMS *Liverpool.* By the time the convoy reached Malta, it had lost four MVs and two destroyers. Of eleven MVs in the Vigorous Convoy, two were sunk (along with two destroyers) and nine turned back. The Italian fleet blocked the convoy while aircraft and submarines did the damage. Only two of seventeen MVs reached Malta—a major disappointment and an impediment to resuming sustained antishipping raids.[54] Tedder blamed the disaster on inadequate communications. Escorting fighters received no updates from convoys as they changed course, and they received no information about the scale of enemy attacks.[55]

Another convoy operation, Pedestal, was put under way because Malta would run out of aviation fuel in late August. Air Vice-Marshal Keith Park, who had replaced Hugh Lloyd as air officer commanding (AOC) Malta, had 145 Spitfires and 109 other planes to protect the convoy steaming from Gibraltar. Park managed to keep an Italian surface force from attacking, but the Luftwaffe pressed home deadly raids comprising 272 sorties and sank nine of the fourteen MVs between 11 and 13 July. The five surviving vessels carried a great deal of aviation fuel, so Malta once again became an offensive platform. Between strikes from Malta, those by No. 201 Naval Cooperation Group, and by heavy-bombers, Axis convoys once again suffered prohibitive losses. From late July until Bernard Law Montgomery's attack on 23 October, Axis troops were reduced to living largely off of their dwindling haul of captured supplies. Meanwhile, copious US assistance arrived.[56]

Rommel's Offensive Continues

Despite his troops' exhaustion, Rommel pushed them forward in order to destroy the Eighth Army and conquer Egypt. He believed that captured vehicles and supplies could carry his army to the Nile Delta before Malta once again became a threat to Axis convoys. It was a fatal error in judgment. Rommel underestimated British recuperative powers and downplayed the Luftwaffe's exhaustion.[57] Tedder said, "If only our friend Rommel would run true to form and come bullocking on regardless, there might be a chance of knocking him right out."[58] This is essentially what happened. As Axis forces massed in front of Mersa Matruh, the RAF threw everything at their 5,000 vehicles. Tedder sent forward the only fighter squadron guarding the Nile Delta, stripped training units of many aircraft, and ordered twenty Spitfires from Malta to Egypt.[59] The Albacore-Blenheim team went after Axis vehicle convoys all night, and Bostons took over with hourly escorted day raids. Fighters flew up to seven sorties per day, with Battle of Britain veterans saying they had never seen such intensive operations.[60]

In the next phase of the campaign, fighters kept Luftwaffe bombers almost entirely away from the army, destroying twenty-three Ju 87s and breaking up seven attempted raids. Fighters and light-bombers then turned to unprotected vehicle convoys and destroyed 1,050 trucks. By now, PR delivered excellent prints of such attacks. The guesswork in determining effectiveness, though never absent, was much reduced.[61] Tedder said the Spitfires were an "enormous asset" based on their ability to provide high-altitude cover for bombers and engage the ME. 109F on even terms. German pilots became much more cautious after their nemesis arrived. Ground and aircrews were in good spirits. Tedder was certain that continuing heavy raids on sea lines of communication (SLOCs) would eventually prove decisive after the Eighth Army was able to capitalize on the advantages thus accrued.[62]

Coningham continued reporting on Luftwaffe operations, wondering why German night raids on landing grounds were so small and sporadic and why fighter patrols continued over the battle area and as far east as RAF landing grounds—but without any daylight efforts to attack the airfields. Even heavily escorted Stuka raids were ineffective. Part of the reason for this was the RAF's mobile radar and air defense system, which directed many successful intercepts of bombing raids, making the occasional airfield attacks ineffective. German fighter sweeps were no substitute for a concerted air superiority effort, and the requirement for heavy fighter escort of Stukas was a gross misuse of assets, particularly because the Germans had also discovered the value of fighter-bombers but did not use them with any frequency.[63]

Conversely, the Wellington-Albacore team proved highly effective in its continuing night raids on airfields. In another bleeding of Luftwaffe assets, a 17 June wing-sized raid on Gazala destroyed nearly thirty Me. 109s, reducing fighter activity at a critical point. After the fall of Tobruk, Me. 109s appeared only in small numbers. Axis troops soon outran them in any case. It took ten days, until 30 June, for enough Luftwaffe fighter units to catch up even to make a minor appearance over the front.[64]

Tedder remained sanguine. Salvage units replaced half of aircraft losses with refurbished planes. During June, Dawson sent Tedder nearly 250 aircraft. Tedder told Portal that his squadrons had all been in good strength for the past six weeks as replacement aircrews and aircraft arrived. The wing and squadron organizations were mature, he said, with agile logistics capabilities. Aggressive action against Axis airfields forced such extensive aircraft dispersals that enemy strike packages had difficulty concentrating.[65]

Axis Tragedy

Rommel's troops secured mountains of provisions at Tobruk. However, this small port was 375 miles from the front at El Alamein, 800 from Benghazi, and 1,300 from Tripoli, making the logistical situation perilous should the haul of captured supplies run out. The Axis used between one-third and one-half of their total fuel deliveries just to run the vehicle convoys taking supplies to the front. Vehicles wore out quickly or fell prey to fighter-bombers, undermining Rommel's mobility. During June and July, the RAF gave everything it had to support the army and help stop Rommel's efforts to break the El Alamein line.[66] The War Office said, "There can be no doubt but that the RAF saved the Eighth Army."[67]

The Afrika Korps War Diary for June detailed the damage caused by WDAF raids. As the Fifteenth Panzer Division moved forward on 25 June, nearly continuous day and night raids caused considerable losses. The Luftwaffe was absent. Raids continued throughout the night of 26–27 June, hours after Kesselring met with Rommel and other senior officers, when he had underscored the Luftwaffe's inability to support a major advance, warned that his units were no longer a match for the RAF, and told Rommel that his timetable for taking Cairo (ten days) was far too optimistic.[68] On 27 June, as the RAF withdrew to new airfields, there was a merciful three-day hiatus in the bombing.[69]

These exceptionally heavy raids were a rude surprise to Panzerarmee Afrika, coming out of a major triumph and sensing victory. In reality, Rommel was nowhere near victory. As he prepared to leap at the Nile Delta, the Italians were upset that Malta had become so active again so quickly and asked when Kesselring would be able to send the Luftwaffe back to neutralize it. The answer, as it turned out, was "Never," with the exception of two small and unsuccessful efforts in July and October.[70]

The German advance, spearheaded by the Afrika Korps' remaining fifty tanks, ran into a wall of RAF opposition, with fighter-bombers inflicting heavy losses. All 560 operational planes went into this effort. They could not halt the German advance or prevent the fall of Mersa Matruh, but they punished Axis columns and allowed retreating Eighth Army units to escape and refit. Nonetheless Rommel intended to take El Alamein by assault and open the way to Alexandria, Cairo, the Suez Canal, and the oil fields.[71]

As these developments unfolded, Kesselring made one last effort to get permission to take Malta. In early June, he sent General of Paratroopers Kurt Student to brief Hitler. The Führer listened and then said that even though Student's paratroopers could no doubt form a bridgehead, "I guarantee you the following. When this attack gets started, the Gibraltar naval squadron will immediately take off and the British naval units from Alexandria will also approach. You should then see what the Italians would do. As soon as the first radio messages are intercepted, the ships will return to the Sicilian ports, both the warships and the transport vessels. And then you will sit all alone on the island with your parachutes."[72] The demoralized Student called Kesselring, who in turn told Hermann Göring that after the air attacks of April and May, Malta could be seized with relatively low losses, but that the price would be much higher as postponements continued.[73] Hitler and Mussolini sided with Rommel. The duce flew to Africa for the triumphal entry into Cairo. Hitler forbade any further discussion of the issue.[74]

After the last Malta debate ended, Rommel's army suffered a series of disasters that cost the Axis North Africa. Rommel knew the British were not as weak as he had made Hitler think they were. A detailed map with accompanying photos produced by Fliegerführer Afrika's Intelligence Division in June 1942 illustrated that El Alamein would be a very hard nut to crack, especially without a reliable flow of supplies and strong air support. Rommel had neither. The Eighth Army had escaped. The RAF was operating from major air bases less than 100 miles away from its major depots. Rommel ignored these disadvantages in the air and misjudged the damage airstrikes would do to his troops. Finally, there was no outflanking the position. Every attack would be into the teeth of well-prepared ground defenses and air attacks.[75]

Rommel's gamble was surprising given his constant concerns about supply problems and an equally clear understanding of British logistical strengths. Aside from the fact that the British brought in supplies unmolested—something Rommel bemoaned frequently—he also recognized the inherent superiority of the British system. "Even more important [than the lack of bombing of the Suez]," he said,

> was the fact that on the British side there were men with great influence and considerable foresight, who were doing all they could to organise the supply service in the most efficient manner possible. In this respect, our adversary benefited from a number of factors:
>
> 1. North Africa was the principal theatre of war for the British Empire.
>
> 2. The British government regarded the fighting in Libya as of decisive influence in the war.

3. The British had in the Mediterranean a powerful and first-class Navy and Air Force of their own, while we had to deal with the unreliable Italian naval staff.

4. The entire British Eighth Army, down to the last unit, was fully motorized.[76]

Rommel believed that the poor long-term logistical situation required him to gamble on an offensive. He had beaten the British several times already and just come within a whisker of annihilating the Eighth Army. His troops sensed victory, he had captured immense amounts of supplies, and he felt that it was now or never. However, supply problems, made worse immediately by air attacks, meant Rommel fought at long odds.

Worn-down Luftwaffe units moved onto the Gambut airfields to support the advance on the Nile Delta. Reconnaissance missions made it clear that the Eighth Army would stand at El Alamein. German fighters remained tethered to advancing ground forces. They shot down a number of aircraft but could not stop the continuous raids on troops, or, given their restrictive orders, go after the RAF at its own bases. Their ability to leapfrog remained limited, and the Italians found themselves in the awkward position of giving the Luftwaffe another 500 trucks. Luftwaffe units finally moved onto the airfields around Fuqa on 30 June but were no longer numerically or operationally capable of supporting Rommel's advance.[77]

Meanwhile, von Waldau was flying about in his Storch looking for Rommel. They still maintained separate headquarters. Von Waldau was caught off guard by the decision to advance immediately into Egypt, which Rommel did not discuss with him. The result was an inability to refit, re-equip, resupply, or repair damaged aircraft. The Luftwaffe pushed forward as fast as possible with what it had left.[78]

The withdrawal to El Alamein allowed the WDAF to engage enemy troops from forward airfields and hamper their logistics. As Dawson's shops turned out refurbished aircraft, serviceability rates reached new highs. Also, the US Army Air Forces (USAAF) Halverson Detachment of B-24 Liberator heavy-bombers was assigned to Egypt and joined by a light-bomber squadron of twenty-seven A-29 Hudsons, a fighter group of eighty Kittyhawks, a medium-bomber group of fifty-seven B-25 Mitchells, and a heavy-bomber group of thirty-five B-24s.[79]

Axis forces advanced to the coast road twenty-eight miles west of Mersa Matruh. Panzerarmee Afrika's diary indicated hourly attacks on the Ninetieth Division, repeated attacks on the Italian Twentieth Corps, the halting of the Littorio Division, and difficulty moving fuel forward. Raids that night caused additional casualties. This became a ceaseless ordeal. The WDAF flew 615 sorties on 26 June—2.5 sorties for every available strike aircraft—and repeated this often. Rommel's headquarters reported significant losses, including many fuel trucks. After the Germans reached the El Alamein line, Tedder threw in everything he had to help the Eighth Army keep them from breaking through. On 27 June, Rommel remarked, "The enemy is fighting back desperately with his air force"—a clear indication that the RAF was integral to slowing his advance.[80] Rommel did not wait for the Luftwaffe to reconstitute but rather sent his troops forward on 1 July with the intention of breaking through to the Nile Delta. For the next month, frequent engagements wore out both sides, but the British held. Airpower played an important role in these successes, along with steady improvements in Eighth Army operational acumen and air-ground coordination. It also helped that the RAF had 780 operational aircraft as it fell back on its major depots, whereas Fliegerführer Afrika had 126—a 6.2:1 ratio.[81]

As control of the Mediterranean SLOCs once again passed, this time permanently, to the British, they sank one-third of Axis tonnage bound for Africa from July to November. MVs had to land supplies further from the front, making the already inadequate vehicle pool cover huge distances. These worn-down convoys endured frequent air attacks as they approached the front. By the time Montgomery's attack began and the breakout succeeded, thousands of vehicles were abandoned because of mechanical problems or lack of fuel.[82]

However, the Middle Sea was not yet an Allied *Mare Nostrum*. The Pedestal Convoy, set to run from Gibraltar to Malta, was one of the most critical of the war. The Royal Navy dedicated virtually all of its remaining assets in the western Mediterranean to escort it: 3 carriers with 72 fighters and 28 Albacores, 2 battleships, 3 cruisers with heavy AAA and ASV, and 12 destroyers. There were also 80 serviceable fighters on Malta at the end of July. Additional aircraft flew in from Egypt and the United Kingdom, bringing fighter strength to 100 Spitfires and 36 Beaufighters. The convoy, comprising 14 large MVs, passed through the Straits of Gibraltar on 10 August. The Axis, with excellent aerial reconnaissance and signals intelligence, assembled 600 aircraft to oppose the convoy along with 21 submarines and several newly laid minefields. A U-boat sank the carrier HMS *Eagle* on 11 August, but all 37 of HMS *Furious'* Spitfires reached Malta. Both AAA cruisers were also torpedoed, with one sunk and the other forced back to Gibraltar. The oil tanker USS *Ohio* was hit by the same submarine. This opened up the convoy to heavy air attack as fighter direction and AAA were severely degraded. Although only 5 of 14 MVs reached Malta (including the *Ohio* with her vital fuel oil), the supplies they delivered kept Malta fighting. Pedestal was the last Malta convoy to encounter serious opposition. Despite

Bristol Blenheim bombers destroy a German fuel-tanker truck delivering gas
to Rommel's army. Increasingly heavy raids on "thin-skinned" supply vehicles,
with fuel-tanker trucks specifically targeted to exacerbate constant German fuel
shortages, hampered every one of Rommel's offensives and helped to bring them
to their culminating points. This was one of many combined-arms advantages
conferred by the RAF's unwavering focus on air superiority. (Imperial War
Museum Image No. CM 1500)

horrendous losses, the Malta convoys had given the island enough weapons
and supplies to act as a platform for attack and defense. It also proved a
huge rallying point for Allied morale and a disastrous impediment to Axis
logistics and operations.[83]

Despite the long distances that aircraft outside of Malta had to fly to attack Axis convoys, sinkings increased again. Planes sank 60,588 of 137,814

The SS *Ohio* of the Pedestal Convoy arrives in Grand Harbour, Malta, awash at the gunwales and kept afloat by the two Royal Navy destroyers alongside. Pedestal was a vitally important convoy to keep Malta sufficiently stocked with aviation fuel so its reconnaissance and strike aircraft could continue ship-attack missions before El Alamein. Combined with WDAF and No. 201 Naval Cooperation Group efforts, they did grave damage to Axis convoys. The *Ohio*'s large cargo of fuel proved critical to the continuation of this all-out effort against the SLOCs. (Imperial War Museum Image No. GM 1505)

total tons of merchant shipping from June through September 1942 and shared 11,063 tons with the navy—48 percent of total sinkings. The percentage increased over time. Axis convoys lost 23 percent of general cargo and 17 percent of fuel in June, 6 percent combined in July, 25 percent and 41 percent, respectively, in August, and 20 percent combined in September. The key arbiter of convoy success or disaster in all cases was the relative strength of land-based, offensive airpower combined with naval power and effective reconnaissance and intelligence operations.[84]

Leading up to this logistical crisis, however, the Axis advance made good progress on 1 July despite losses from RAF raids. Rommel's staff contrasted the day-and-night raids with "very slight" Luftwaffe activity. On 2 July, night-bombers savaged supply columns, preventing an improvement in the already serious supply situation. Air attacks continued throughout the day. The promised night fighters never arrived, so night raids remained heavy. Commonwealth artillery was becoming more numerous and lethal every day, and there were no German strike aircraft to go after it because the Stuka was now too vulnerable and the Ju 88 too inaccurate. Nor were there enough fighters left to engage in concerted fighter-bomber operations even if the Germans had pursued that option. On the only high note for the month, the Luftwaffe appeared in strength on 30 June during an Eighth Army attack and kept RAF aircraft away from German units, which counterattacked after three Italian divisions had been destroyed, demoralized by attrition, or fled.[85]

Three raids struck Panzerarmee Afrika Headquarters on 16 July. On 19 July, Rommel ordered that all remaining tanks be surrounded with sandbags and stone walls. On 21 July, the staff recorded that continuous heavy raids exceeded anything they had experienced. Bombing severed phone lines, making contact with subordinate units difficult. Raids continued without respite and soon shifted to artillery batteries, reducing their effectiveness even as that of the Luftwaffe reached an all-time low. Axis forces would have very little air and artillery support in what was becoming a series of set-piece battles.[86]

On 1 August, Rommel asked von Waldau for protection of coastal shipping to Tobruk and Bardia, but the latter said Kesselring would have to provide it because every aircraft in Africa was fully committed to supporting the army. Fliegerführer Afrika had lost 271 aircraft since 26 May, 191 of which were complete write-offs, and the low rate of replacements kept his force hovering around 125. The next day, in a now commonplace event, the RAF sank four vessels at Bardia, prompting Rommel to report that such sinkings were undermining his army's supply situation. Simultaneous RAF efforts against vehicle convoys, and the even more deadly No. 201 Group and Malta-based ship attacks, were undermining Axis logistics at every point. Rommel asked Oberkommando der Wehrmacht (OKW) for more AAA and fighters to protect coastal shipping. Kesselring flew forward as much fuel and ammunition as possible. He also placed 1,500 tons of fuel from Luftwaffe reserves at Rommel's disposal, giving his armies four consumption units.[87] However, because mobile warfare in the desert required a minimum of thirty consumption units of fuel to give Panzerarmee Afrika maximum mobility, and combat units consumed one consumption unit every day during combat operations, it is apparent how tenuous the fuel situation had become.[88]

By the time Rommel's army reached the El Alamein position, the tide had already turned. He recorded that the July fighting, "the main feature of which was continuous and round-the-clock bombing by the R.A.F.," had been costly.[89] Fighting with far too few supplies, aircraft, tanks, and heavy guns, Axis forces were headed for disaster.

9

The Tide Turns
July–October 1942

Strategic Developments

ERWIN ROMMEL'S VICTORY AT GAZALA and advance to the El Alamein line appeared to promise a major Axis drive into the Middle East. However, the failure to take Malta or even make the attempt very quickly changed the situation. In contrast, the campaign in Russia appeared to be going well, and the Germans still hoped that a major victory there would open the Caucasus route to their troops, giving them the oil resources there and an avenue of advance toward those in the Middle East. They would also be able to cut the lend-lease route to Russia. Even with the Japanese no longer in a position to contemplate a coordinated Indian Ocean strategy as most of their Fast Carrier Striking Group lay at the bottom of the Pacific, the Germans still hoped to seize the region's natural resources.

The disaster at Gazala and the fall of Tobruk led to major US deliveries of tanks and aircraft, along with other supplies, to the British. Allied staff talks in London in July 1942, and Franklin Roosevelt's insistence on an invasion of North Africa (Operation Torch), quieted dissent among US military chiefs briefly, but the debate over that operation and the Joint Chiefs of Staff's (JCS's) desire for a Japan-first strategy continued until Roosevelt weighed in again, ordering the JCS to proceed with Operation Torch while holding to the Germany-first strategy. However, the JCS continued to diverge from Roosevelt's and Winston Churchill's views about the importance of and grand-strategic prospects within the Mediterranean, and the disagreements with the British Chiefs of Staff (CoS) became more strident.[1]

For the moment, however, the British were content to push forward based on the agreements reached at ABC-1, Arcadia, and the London staff talks. They entertained the JCS position regarding a cross-channel invasion in 1942 but remained highly pessimistic given the lack of US troops and the disastrous shipping situation. George C. Marshall and his colleagues had argued that just the concentration of forces in England would force the Germans to divert numerous divisions from Russia for the defense of the west. The British countered with the sound observation that shipping shortages, US diversions to the Pacific theater, and the need for Commonwealth forces in the Mediterranean to keep the Germans from conquering the Middle East made such a plan impossible. The US leaders saw peripheral strategies as wasteful of resources and operationally ineffective. The British retort was that the Mediterranean was hardly a peripheral theater from their point of view and was in fact central to the survival of the empire and its ability to continue waging war. Finally, they pointed out that a major cross-channel invasion was impossible without US divisions, heavy equipment, and landing craft. The British were bleeding the Luftwaffe heavily and sinking immense numbers of ships and supplies in the Mediterranean. They were insistent on seeing the desert war to a successful conclusion and knocking Italy out of the war.[2]

The Battle for Aircraft Won

As the Combined Chiefs of Staff (CCS) argued and Rommel found himself cornered, Arthur Tedder won the "battle for aircraft." However, when Rommel had begun his offensive on 26 May, reinforcements were not keeping up with losses. Arthur Coningham was concerned about severe Kittyhawk shortages. His squadrons had an average of 7.5 serviceable aircraft. The enemy had 420 aircraft, and Coningham could not deal with them unless he had enough fighters.[3]

Air Vice-Marshal Douglas Evill, chief of the Royal Air Force delegation to the United States (RAFDEL), told Tedder that, as of 1 June, there were 129 Kittyhawks bound for the Middle East. Another 70 awaited shipping. Nearly 640 more were scheduled for arrival from June through August. Deliveries were increasing, and the newest arrivals had Merlin engines. These deliveries rarely matched estimates but improved over time. As the battle heated up and turned against the Eighth Army, Tedder asked Evill to explore every means of delivery.[4]

In a welcome development, Marshall said Major General Lewis Brereton and his ten or so heavy bombers in India would move to Cairo without delay. The US Army Air Forces (USAAF) plan was to send heavy and medium bombardment groups along with a reinforced light-bomber squadron and a fighter group to Egypt along with a full maintenance tail. These would join the bombers Brereton brought to form Middle East Air Forces (MEAF), which he activated on 28 June. USAAF assets soon played a substantial role in the campaign. The heavies did superb work against Axis ports, convoys, and airfields.[5] Tedder told Brereton and his staff with his usual directness, "We're only here because we both believe certain things are worth fighting for, so let's get on with it."[6] Brereton recalled that his troops were all "sold on him [Tedder], no doubt about that. He didn't talk down to a man. There was never anything in his attitude to indicate that he thinks himself superior to anyone else or to make us feel we'd been in the war for only two months. After that, I always had to let him know when I was going out for an inspection and as like[ly] as not he would come too. He got around to every American unit damn near as often as I did."[7]

By 21 June, USAAF B-24s were bombing Benghazi. They struck thirty-four times between 21 and 30 June, damaging ships and destroying supplies. They freed up Western Desert Air Force (WDAF) assets, including the deadly Wellington-Albacore team, to interdict Rommel's supplies. Brereton's surviving B-17s joined Halverson's B-24s in early July, forming the First Provisional Bomb Group. During the month, 120 B-24 missions and 45 B-17 missions struck ports and convoys. Another four B-24 squadrons arrived, tripling heavy-bomber strength and leading to a series of punishing raids on Tobruk in August, reducing its offloading capacity from 2,000 to 600 tons per day. P-40s and B-25s arrived in late July and went into action, flying individual sorties first with RAF units and eventually operating as USAAF formations, contributing substantially to the battles of Alam Halfa and El Alamein.[8]

On 22 August, Brereton sent Henry Arnold three copies of "Direct Air Support in the Libyan Desert," which covered USAAF operations from 26 May to 22 August. It noted that a key element of the integration of USAAF units into the RAF structure was the assignment of two RAF air liaison officers (ALOs) to each USAAF squadron. They delivered information from the front so fast that squadron intelligence officers found themselves rushing to keep pace as they gave mission briefings.[9] When El Alamein began, the USAAF contingent represented only 10 percent of total air strength but 35 percent of light-bombers and 70 percent of heavies. It packed a punch out of proportion to its numbers—something Tedder acknowledged repeatedly. On 4 November, after USAAF heavies sank two large tankers and a very large merchant vessel (MV) in the opening days of the final El Alamein battle, Hoffman von Waldau noted in his diary that "the fate of the Mediterranean hung more or less on these three ships."[10]

As more USAAF units arrived, maintenance units became active in Egypt and, after Operation Torch began, in Northwest Africa. USAAF depot groups were initially poorly trained and had no equipment, so RAF maintenance units kept servicing US aircraft as they taught their counterparts. Graham Dawson put a stop to the US habit of cannibalizing even moderately damaged aircraft for spare parts rather than allowing the repair-and-salvage units (RSUs) to deliver them for proper repairs. He also developed common engine, propeller, and airframe units to service all Allied aircraft.[11]

Meanwhile, the Kittyhawk situation finally improved. There were 166 expected to arrive for the RAF in December, and another 270 for the USAAF. Although USAAF Kittyhawk pilots were generally inexperienced, they worked in well with their RAF counterparts, first as individuals, then at the squadron level, and finally as entire squadrons within RAF wings. "They are now full-fledged for war," Tedder said, "and have proved themselves fine fighters and loyal comrades."[12] Tedder also rejoiced that US heavy-bombers' destruction of ammunition ships at Tobruk and Benghazi rendered the ports largely unusable.

By 1943, with Allied armies advancing in two directions across North Africa, Tedder finally won his battle for aircraft. Skirmishes would continue, but Arnold had relented on the plan to send only entire USAAF units and had instead agreed to keep sending Royal Air Force Middle East (RAFME) the planes and spare parts necessary to keep veteran squadrons flying. The Allies also reached a formal agreement on the allocation of US aircraft to the RAF.[13]

The Air-Sea Effort Succeeds

As the battle for aircraft succeeded, so did the battle for the sea lines of communication (SLOCs). The agreement establishing No. 201 Naval Co-operation Group, and the tactical innovations airmen and sailors developed together, came just in time. With the war against Japan under way, the Royal Navy moved capital ships to the Far East. Operations around Crete had already gravely weakened the Mediterranean Fleet. The departure of further ships made the RAF responsible for the region's security against a major Italian naval sortie. In December 1941, RAFME and the Fleet Air Arm (FAA) had one squadron each of Swordfish, Albacore, and Beaufort torpedo-bombers. With no reinforcements en route, that meant four light- or medium-bomber squadrons would have to make up the difference. The Swordfish and Albacores did best on Malta given their short range, and the Beaufort and bomber squadrons were based at Alexandria or on Cyprus. To ensure the Italians did not slip out of port unnoticed, the RAF required three long-range reconnaissance squadrons plus a flying boat squadron. This reliance on land-based airpower in light of navy losses invited enemy attacks on air bases (once again not forthcoming), requiring an additional nine short-range and two long-range fighter squadrons for air-base protection and strike escort. Effective control of the central Mediterranean would therefore require something like twenty-three additional squadrons plus control of Cyrenaican airfields. Although a handful of additional squadrons arrived—especially torpedo-bombers and long-range fighters—the RAF did not receive nearly all the items on its wish list. Tedder and Andrew Cunningham would have to make do with less.[14]

On Christmas Day 1941, Wilfrid Freeman told Tedder that the Air Ministry's proposed increases to RAF aircraft capable of maritime strike missions were required to allow the RAF to take over the maritime security of the central and eastern Mediterranean as the navy's capital ships departed. Two long-range torpedo-bomber squadrons soon arrived, followed by additional units. Tedder had operational control over these assets through No. 201 Group.[15]

Proper employment of aircraft in the antishipping and naval escort roles became more critical than ever as the logistical advantage shifted once again to the Axis by 1942. Convoy protection received a great deal of attention given the air assets required to keep the garrison at Tobruk supplied. The huge requirement for fighters to fly defensive patrols was wearing down the force as transfers and diversions to the Far East weakened it. To cover all bases, Leonard Slatter needed at least three long-range fighter squadrons to conduct this effort and take pressure off of Coningham's assets. Ideally, these would be Beaufighters. He also needed more photoreconnaissance (PR) aircraft. Another radar station capable of height finding/direction finding (HF/DF) was also necessary to augment the one in operation. Most of all, though, Slatter needed Malta as a base for antishipping operations. The arrival of Luftflotte 2 in Sicily closed off this option. Coordination between No. 201 Group, Malta, and the WDAF worked well until Rommel's counteroffensive the following month drove the RAF out of its airfields in western Cyrenaica. From then until the arrival of longer-range aircraft equipped with air-to-surface surveillance (ASV) radar and effective torpedoes in spring 1942, Axis convoys routinely made it to Tripoli and Benghazi, and on the west-to-east route along the coast of North Africa.[16]

In an effort to speed the group's operational effectiveness, Tedder sent Charles Portal a detailed note regarding control of the Mediterranean Sea and the assets required to hold it. "With the virtual disappearance of British naval activity in the Mediterranean, other than by submarines," he said, "responsibility falls on the Air Force both for:

(i) protecting our shipping against attack by enemy surface vessels and submarines;

(ii) denying freedom of movement to enemy shipping.

Both of these duties demand:

(i) efficient reconnaissance;

(ii) rapid action by striking forces of adequate strength."[17]

He noted that, regardless of the position in Cyrenaica and the current difficulties on Malta, No. 201 Group needed torpedo-bombers capable of
ranging the entire central Mediterranean. This meant specially equipped
heavy-bombers. ASV and good radio communications were essential to keep
from wasting the striking force's efforts. The training of ASV leader crews
was already helping, but No. 201 Group still lacked the facilities, equipment, and experts to install ASV in strike aircraft. They relied on whatever
arrived from the United Kingdom and had limited means for conducting
maintenance and repairs. Tedder and Portal corrected this. Special training
was also required for ASV crews, which began before ASV-equipped aircraft
arrived in large numbers. The Operational Research Section worked on the
proper employment of technologies and tactics and made detailed recommendations. Finally, No. 201 Group soon received sensitive altimeters, long-
delay-action flares for use in night attacks, and a highly experienced wing
commander from RAF Coastal Command with ASV operational experience
to train crews.[18]

Slatter noted that the only way for his aircraft to reach convoy routes was
to deploy to the most forward RAF landing grounds. This was difficult given
the need to move large numbers of specialized ground crews, munitions, and
spare parts. The key problem was that reconnaissance aircraft could give
only a few days' warning regarding convoy sailings, which meant a hurried
move forward. This was risky because the landing grounds were within easy
range of Me. 109s, requiring a large fighter escort. Slatter recommended
keeping one advanced landing ground ready for ship strike assets, arguing
that it was worth the risk of air attacks to allow his planes to reach convoys.
A small contingent of ground personnel fueled, armed, and maintained aircraft. The effort was highly successful.[19]

Two weeks later, Slatter followed up with his first detailed note regarding the importance of making a maximum effort to sink MVs before they
made it to port, or at the very least sink the west-to-east shipping along the
coast. As No. 201 Group became increasingly formidable with ASV and
more capable torpedoes, Slatter had the force he needed with the notable
exception of ASV-equipped heavy-bombers, which were, in his view, the
final and most crucial element of an effective antishipping air force.[20] As
Rommel's offensive at Gazala routed the Eighth Army and brought Panzerarmee Afrika closer to the Nile Delta, Slatter pushed Tedder for an all-out
attack on enemy shipping now that the battle had reached a crucial stage.
"Two years of desert warfare," he said, "have demonstrated that the only
effective way of achieving this objective [denying supplies to Axis armies]
is to sink the ships carrying supplies between Italy and North Africa, *whilst
they are at sea.*"[21] After large MVs reached port, the RAF faced a profusion

A Bristol Beaufort torpedo-bomber of No. 201 Naval Cooperation Group prepares for a mission from a forward base in the Western Desert. Arthur Tedder ran a calculated risk by allowing Leonard Slatter's aircraft to arm and fuel at forward airbases within easy range of German fighters so they could reach Axis convoy routes until the reconquest of the Cyrenaican airfields allowed them access to bases there. The Luftwaffe never detected these actions. In coordination with continuing strikes from Malta and Royal Navy surface-ship and submarine attacks, ship-attack aircraft created a massive logistical crisis for Axis forces from June–December 1941 and an insuperable one from summer 1942 until the end of the war in North Africa. (Imperial War Museum Image No. ME(RAF) 5506)

of smaller west-to-east sea convoys as well as land convoys, requiring a very heavy commitment of reconnaissance and strike aircraft. Attacks on vehicle convoys, although often effective, were extremely inefficient in terms of air assets committed and damage done. For these reasons, and with the aerial siege of Malta over, Slatter concluded that Malta was the key location for No. 201 Group's best and longest-range antishipping assets, with the rest going after west-to-east coastal traffic. Tedder agreed, and from this point forward Axis shipping losses skyrocketed. The battle for logistics shifted radically and permanently in the Allies' favor. From virtually no losses in April, sinkings became catastrophic by October. From 1 August to 5 September, the Axis lost sixty-one MVs totaling 243,737 tons with another eighteen totaling 73,767 tons so badly damaged that they were out of the fight for many months. This marked the virtual end of Italy's contingent of large MVs, forcing a shift to smaller vessels such as Siebel ferries and an even heavier reliance on air transports. By 20 August, five of the Luftwaffe's eighteen transport groups—28 percent—were supporting logistical efforts in the Mediterranean. Enno von Rintelen told Oberkommando der Wehrmacht (OKW) that existing assets could no longer ensure the adequate delivery of supplies and materiel to Rommel's forces.[22]

As RAF and Royal Navy assets hunted down the remaining large MVs, Hudson torpedo-bomber planes with ASV sank numerous small vessels moving supplies along the coast, exacerbating the logistical crisis. The destruction of several major convoys in October, including large oil tankers, was disastrous for the Axis, as were heavy raids on Tripoli, Benghazi, and Tobruk. A German prisoner of war (POW) claimed that when RAF aircraft sank the most crucial of these convoys, Rommel, Albert Kesselring, Hans Seidemann (von Waldau's replacement as Fliegerführer Afrika), and General Wilhelm Ritter von Thoma witnessed the entire action. The convoy consisted of a large tanker, two freighters, and four destroyers. Tedder signaled the men of No. 201 Group, "My sincere congratulations on the destruction of the enemy convoy yesterday right under the noses of his shore-based defenses. It was a magnificent example of sheer courage, tenacity and determination to kill."[23]

No. 201 Group and RAF Malta made the fullest use of their now numerous ship attack aircraft. In addition, the navy received reinforcements and operated in joint strike efforts again after the RAF controlled the central Mediterranean. The combination of air and sea attacks proved deadly. From its humble position in October 1941 with four squadrons, No. 201 Group had grown by January 1943 to eighteen squadrons with ASV, advanced torpedoes, and superb tactics, all supported by outstanding air reconnaissance. This force undermined Axis logistics and thus produced major advantages for the Eighth Army and then for Allied forces in Libya and later Tunisia by starving Axis troops of supplies, causing heavy losses of equipment and personnel, and blinding the enemy by sweeping maritime air reconnaissance assets from the skies. Conversely, by November 1942, convoys began unloading at Malta without loss, and smaller vessels moved supplies west along the North African coast without fear of air attack and with little risk from submarines, which No. 201 Group and the navy were hunting down.[24]

On 30 June, Kesselring assigned Fliegerkorps II, with 146 aircraft, entirely to convoy escort as shipping losses skyrocketed. On 24 August, the effort became officially and entirely defensive in nature with an order to protect convoys and defend Sicily from growing RAF raids. Despite aircrews' courage and dedication, they had too few aircraft covering too many convoys and miles of ocean. The effort failed to reduce MV losses or RAF raids on Sicily and southern Italy. Even an encore effort to reroute convoys to areas mostly out of range of ship attack assets failed as the number, combat radius, and capabilities of No. 201 Group assets increased along with the Royal Navy presence in theater.[25]

The Headquarters RAFME Tactics Assessment Office concluded its major report on antishipping operations with a note that from the start of the war to 30 June 1943, the Axis had lost 1,659 merchant ships—897 in all areas outside the Mediterranean and 762 within it—46 percent of the total. Of the 5.7 million gross tons sunk, 2.4 million—42 percent of the total—went to the bottom of the Mediterranean. This theater of operations thus became a graveyard for Axis merchant shipping as well as its air forces—a sobering reminder that "subsidiary" theaters can become anything but that without the proper level of grand-strategic insight and focus.[26]

Commonwealth Resurgence

Throughout the battle for aircraft and the SLOCs, Tedder, who was promoted to air chief marshal on 3 July 1942, remained determined to maintain air superiority and precipitate an enemy logistical crisis. During the first clash along the El Alamein position on 1 July, the RAF inflicted heavy losses on Axis troops, who had as yet only a handful of fighters and Stukas within range. On 1 and 2 July, raids occurred every hour, around the clock. Ground crews kept serviceability rates at 67 percent the first week of the campaign, 75 percent in the second week, and 84.8 percent by the fifth week. Part of this achievement is attributable to the fact that the RAF was falling back on its major base infrastructure. However, the lion's share of the credit belonged to the ground crews and Dawson. As serviceability increased, so did the number of aircraft. Single-engine fighter squadrons went from an average strength of 15.3 aircraft at the start of Rommel's offensive to 16.9 on 6 July. By 5 July, relentless RAF raids and improving army defensive operations stopped Rommel's advance and forced him to disperse his forces to reduce losses to air raids.[27] The first Spitfires were performing even better than the Hurricane in desert conditions. US-built Boston and Baltimore light-bomber squadrons were at full tilt.[28]

Axis POWs were stunned by the RAF's effectiveness, emphasizing the chaos and panic (Flucht-Psychose, or "flight psychosis") that raids caused in addition to the heavy loss of vehicles and men. An artillery battery started June with sixty vehicles and ended it with twenty. All but two of those hit were unsalvageable. Bombing was bad enough. Strafing was unnerving. It was almost always completely unexpected, highly accurate, and the cause of heavy losses.[29]

Rommel tried again to break through between 13 and 16 July. By 17 July it was clear that air attacks and strong Eighth Army defensive positions had undone the attack. Luftwaffe air-transport assets continued flying in troops to replace heavy losses.[30] Rommel also noted that Luftwaffe bombers had given his troops a momentary but significant opportunity to break through on 14 July, when massed raids on key enemy positions opened holes in the line. However, a slow advance and RAF raids ended this development.[31]

Claude Auchinleck counterattacked the night of 21–22 July with heavy air support, pushing Axis troops back. Aircraft struck vehicle convoys and supply dumps, causing heavy losses, but by 23 July the lines were back where they had been in part because of German air attacks. The Luftwaffe threw in everything it had left.[32] A British follow-up attack from 26 to 27 July failed. The situation at the end of July was thus a stalemate in which the RAF owned the air and Rommel's logistical position deteriorated.[33]

The United States now had complete units comprising its own aircraft and airmen. Based on early USAAF operations, especially with heavies, Tedder said that "the Americans were already fitting in very well." In another signal just before El Alamein, Tedder told Portal that they had "already shown up well in combat. . . . They are learning from us, and we are learning from them—I was glad to hear this from both sides."[34]

Churchill and Alan Brooke arrived in Egypt on 3 August for an inspection and to make changes in the army's leadership. Churchill relieved Auchinleck and named General Harold Alexander overall commander, and General Bernard Law Montgomery took command of the Eighth Army. Montgomery had an electric effect, and morale began to improve along with operational acumen. Coningham was impressed with Montgomery and happy to learn that he agreed entirely with the importance of gaining air superiority first, of unified control of air assets, and of a strong joint command-and-control (C2) system. One of Montgomery's first acts was to keep his primary and forward headquarters collocated with those of WDAF.[35]

Cooperation between the services improved even faster. Air Chief Marshal Frederick Rosier (then a group captain) remembered a briefing to his unit by Montgomery, indicative of his close working relationship with Coningham, in which he said,

I have brought you together to tell you that I have made a plan—and when I say I've made a plan it's not quite right because I've made a plan in conjunction with the Air Force. Every plan has to have an intention—mine is to go to Tripoli, and it's the intention of the Air Force too to go to Tripoli. In fact we're all going to Tripoli together.[36]

On 7 September, Tedder signaled Portal that he had "returned Saturday from visit to Western Desert. General feeling is that threat to Egypt has been scotched."[37] He continued, "Difference between this land battle and previous ones is that in this one soldiers have refused to play enemy game and send tanks against guns. Enemy has been forced to send his tanks against our guns."[38] Liaison with the army was much improved. Portal replied, "We are deeply impressed by the remarkable effort put out by your squadrons and delighted by their success especially the splendid work done against the German troops and the Axis shipping. . . . Delighted to hear of your good relations with the Army. Best wishes to you all."[39]

As planning for El Alamein progressed, Tedder hoped to deliver the Luftwaffe such a heavy blow at the outset that it would become combat-ineffective. He was heartened to see the army developing operations rooms in which they gathered, analyzed, and acted on information from the field. Air-ground cooperation was maturing.[40] Montgomery and Coningham made it clear that the air and ground forces were a team. Coningham said, "The soldier commands the land forces, and airman the air forces; both commanders work together and operate their respective forces in accordance with a combined Army/Air plan, the whole operations being directed by the Army commander."[41] In the same talks to RAF and army officers, Montgomery instructed his forces, "The commander of an Army in the field should have an air headquarters with him which will have direct control and command of such squadrons as may be allotted for operations in support of his Army. Such air resources will be in support of his Army and not under his command."[42]

Montgomery enhanced his C2 capabilities by creating satellite headquarters staffed by soldiers and airmen. The main headquarters handled most administrative work, whereas the advanced headquarters, referred to as tactical headquarters (Tac HQ), responded to immediate requirements. The increasing use of radios to deliver orders increased the operations tempo and proved well worth the risks associated with German interception of certain messages.[43]

Montgomery was also an enthusiastic consumer of Ultra signals intelligence. The breaking of the German air liaison officers' Scorpion key in July

1942 revealed Panzerarmee Afrika's intentions, order of battle, unit positions, supply situation, and ground-air efforts. The Y Service was also useful here. Ten mobile field units, including two attached to No. 211 Group, gave commanders near real-time intelligence. All RAF air reconnaissance units (and army cooperation units) merged into No. 285 Air Reconnaissance Wing, a vital step in centralizing the PR and photographic intelligence (PI) processes. The PR Detachment, Strategic PR Flight, USAAF's Survey Squadron, the Middle East Interpretation Unit, and the Army Air Photo Interpretation Unit all came under No. 285 Wing's operational control.[44]

In March 1942, the commanders in chief (C in Cs) oversaw publication of "Middle East Training Pamphlet (Army and R.A.F.) No. 3A—Direct Air Support." It contained important changes in air-ground cooperation activities. First, "indirect" and "direct" support replaced the older and unclear "close" support as standard terms. Indirect support included "air action against the approaches to a theatre of war such as sea communications, and against ports, land communications, base installations, etc., within a land theatre of war. It is in fact strategic air support directed against any target which has an effect, though not an immediate effect, on the battle between ground forces."[45] Direct support was focused on the battlefield and further subdivided into prearranged and impromptu support. Both types were designed to give ground troops assistance when no artillery, mortars, or other means were at hand. Aircraft could not deliver attacks over time like artillery but could be highly effective when properly employed. Air superiority was the indispensable prerequisite to all other operations.[46]

Key targets included troop and vehicle concentrations, headquarters, artillery positions, supplies (particularly fuel, ammunition, and rations), and crossing points over obstacles. Sustained effort against a given target was crucial. Air support controls (ASCs) were the means by which the RAF did this. Each had a small staff and wireless organization consisting of nine tentacles, nine more wireless sets for listening to tactical reconnaissance reports, and three more sets at the control headquarters (usually located with an army division) for direct communication with the tentacles. The RAF contributed two liaison officers (wing commanders) to the division headquarters and a wireless organization comprising two forward air support liaisons (FASLs) for the liaison officers to direct support aircraft where needed, another wireless set for controlling tactical reconnaissance assets, two more for communication between the ASC headquarters and the combined army-air headquarters, and eight more at combined army-air headquarters for immediate communications with the FASLs. Tentacles were allotted to brigades. This system sped the process by which tactical and special reconnaissance aircraft, as well as brigade commanders, requested air support.[47]

Army air support control (AASC) and ASC commanders emphasized that there were still too few RAF liaison officers at the division and brigade levels, leading to inadequate use of the communication infrastructure within the ASCs. They further noted the inadequate use of FASLs, summarizing the problem with an assertion that in the nonstop retreat of the past few months, the ASC processes detailed in "Middle East Training Pamphlet No. 3a" had not yet had time to prove themselves. The report's author, Wing Commander J. R. Wilson, was one of Tedder's best signals officers. The trick, he said, was to determine how best to employ these units with input from operational commanders, especially when it came to closing seams in the bomb line. Army and RAF liaison officers became more numerous and increasingly capable from this point on. Chosen initially based on their general knowledge of the other service's operations, candidates attended ALO courses, which included four weeks of air-liaison training at operational training units (OTUs), ten days of photointerpretation training, and a period of understudy with an experienced liaison at an operational unit.[48]

As the air-ground coordination system developed, so did the employment of its favorite aircraft—the fighter-bomber. A 25 August report, based on studies compiled from Operation Crusader forward, said it was superior to the escorted bomber because it could carry out five missions, if necessary, on the same sortie: armed reconnaissance, indirect support, prearranged direct support, impromptu direct support, and air superiority. The German method of throwing colored smoke grenades to inform Stukas of their own forward positions in tactical engagements was soon copied by Commonwealth troops. Colored smoke grenades and artillery smoke rounds became the norm in the North African, Italian, and northwestern Europe campaigns.[49]

Reports from friendly aircraft, and inputs to them from the AASCs and the FASL tentacles, routinely put aircraft over the target in thirty minutes as opposed to three hours during Operation Crusader. Attacks became increasingly effective as troops employed code letters, ground panels, illumination shells, and colored smoke projectiles to pinpoint friendly and enemy positions. Night-bombers and their Albacore pathfinders used night landmarks to find their targets. Enemy losses increased as friendly fire incidents decreased.[50]

By summer 1942, Wellington-Albacore night missions were highly effective, with the latter dropping magnesium flares from 5,000 or 6,000 feet to illuminate the target. Colored flares marked high-value targets. Parachute flares acted as beacons at the target for successive waves of bombers. This night component of round-the-clock air action left German units bloodied, exhausted, and demoralized.[51]

Rommel's Last Roll of the Dice

Rommel's logistical problems worsened. On 1 August, several barges were sunk trying to deliver supplies to Tobruk. In response, remaining Luftwaffe units provided continuous defense of coastal shipping. Also, around 30 percent of Axis vehicles were in for repairs at any given time, and 85 percent of the total fleet comprised captured trucks with few spare parts. Rommel said attacks on shipping would stop supply deliveries altogether unless he received immediate air and flak reinforcements.[52]

Italian units still received more reinforcements than German ones. An entire Italian division arrived in July. Shipping priorities continued out of kilter as nearly 1,000 German trucks awaited transport. On 4 August, Rommel appealed to Wilhelm Keitel for better convoy protection, prioritization, and coordination, but neither he nor other senior officers had any real influence. There was no centralized department for directing and carrying out coordinated convoy escort. Neither Luftwaffe nor IAF and Italian Navy units were privy to the full shipping schedule, making reliable air cover impossible.[53]

Rommel persisted in the belief that he could win a major battle. However, the Germans could not make good their losses. From 1 to 20 August, the Italians received 15,000 tons of supplies and the Germans 8,500—the standard 5:3 ratio. Heavy raids stretched unloading times for ships to eight days or more, making them extremely vulnerable to damage or sinking. In comparison, the British were unloading 400,000 tons a month in Suez.[54]

On 25 August, after excluding von Waldau and his staff from the planning for his upcoming offensive at Alam Halfa ridge, Rommel briefed Fliegerführer Afrika one day in advance of the attack. He directed von Waldau to engage in five key missions, every one of which required more planes than were available. The first was continuous fighter patrols along the front to keep RAF reconnaissance at bay. To this he added nuisance attacks on coastal roads. Continuous fighter cover over German mobile units and fighter-bomber attacks on British Army forces facing those units also received priority. Protection of convoys and port facilities rounded out the list. The final directive was to harass retreating enemy units after the army broke through.[55]

As Rommel planned the offensive, he sent constant messages to the C in C of the German Naval Headquarters in Italy, Vice Admiral Eberhard Weichold, warning that Panzerarmee Afrika had fuel for only four to five days of combat operations. On 23 August, he warned again that fuel and antitank ammunition were critically short. Without immediate resupply, he could not attack. Weichold replied that heavy shipping losses had further delayed the delivery schedule. At the same time, air reconnaissance proved that the amount of shipping in the Suez Canal had increased by 150,000 tons during the past month—clear evidence that Rommel's last chance to launch a successful offensive was slipping away. Still, he persisted, but the lack of supplies meant his effort, which began 31 August, was limited in scope and very unlikely to make any gains.[56]

On 17 August, Ultra warned of an Axis attack scheduled to begin in nine days. Coningham loosed the WDAF on 23 August, well before Rommel's troops moved forward. Concerted night raids kept the Germans tired. After the offensive began on 30 August, the RAF attacked troop convoys and concentrations. "Heavy air raids commenced," the Afrika Korps war diary said, "directed mainly against 21st Panzer Division."[57] Between reconnaissance and intelligence sources, the British knew exactly when the Germans were coming.

Rommel launched his offensive at Alam Halfa with grim determination, but he must have known the odds were extraordinarily long. Hans von Luck noted laconically that "our own Messerschmitt fighters stood on the airfields with no fuel" as RAF aircraft savaged Axis forces. Bombers managed one night of ineffective attacks. From there, the few Me. 109s that got airborne were tied to close escort of ineffective and costly Stuka raids, with a handful engaging in the equally ineffective but still standard *Frei Jagd* ("free-hunting") tactic. Even with better tactics, the Luftwaffe could not by this time overcome the major reversal in terms of numbers, with 902 operational WDAF aircraft facing 171 serviceable Luftwaffe planes—a 5.3:1 ratio.[58] Rommel had to call off the attack almost immediately and go into a defensive posture. "Before long," he recalled, "relay bombing attacks by the R.A.F. began on the area occupied by our attacking force. With parachute flares turning night into day, large formations of aircraft unloosed sticks of H.E. bombs among my troops."[59] Panzerarmee Afrika impaled itself on strong Commonwealth defenses. Rommel made it easier by fighting on the enemy's preferred ground without any advantages given his lack of fuel and ammunition and the need to achieve a rapid breakthrough.[60]

By 2 September, nonstop raids on advancing German columns had caused heavy casualties. These consisted of twenty to thirty bombers with fighter escort and occurred almost hourly. Flak was often unable to engage effectively as a result of the near absence of searchlights and radar. There was one night-fighter *Staffel* of twelve aircraft, far too few to protect Rommel's army and convoys, and one Freya radar unit at El Daba.[61] Rommel remarked, "Between ten and twelve o'clock we were bombed no less than six times by

British aircraft. . . . Swarms of low-flying fighter-bombers were coming back to the attack again and again, and my troops suffered tremendous casualties. Vast numbers of vehicles stood burning in the desert."[62] He continued, "Our badly outnumbered fighters hurled themselves again and again towards the British bomber squadrons, but rarely succeeded in penetrating to their targets, for they were intercepted every time and engaged by the tremendously strong fighter escorts of the 'Party Rally' bomber squadrons."[63]

After Rommel called off the offensive on 3 September, strike aircraft harried the German withdrawal, making ten raids comprising 180 aircraft. Heavy night bombing followed. Vehicles, artillery, and antiaircraft positions suffered heavy damage. The most heavily hit units lost almost all of their vehicles. At Rommel's request, Afrika Korps prepared a report on air attacks between 30 August and 4 September. "The casualties and loss of equipment caused by these attacks," the report concluded, "necessitate an immediate improvement in the day and night [air] defences."[64]

Kesselring ordered Luftwaffe and IAF units to cover the army's withdrawal to its start positions and to make a dawn fighter-bomber attack on British airfields on 4 September along with dusk-to-dawn night raids from 4 to 5 September.[65] Kesselring ordered that every crew capable of night operations participate. Airmen went all out in a last major effort to protect the army.

Fuel shortages, already severe, nearly prevented Axis troops from getting back to their start lines. Even when ships got through to deliver more, vehicles used three consumption units to get one to the front. The Axis had received 4.2 consumption units of fuel and 443 tons of ammunition, but another 5.5 consumption units of fuel and 350 tons of ammunition went down with their ships. So, even the 4.2 consumption units that arrived meant only 1 for the troops by the time supply convoys reached the front. Without immediate improvements in delivery, the army would be out of fuel by 5 September.[66]

Rommel finally came to understand something profound about airpower within a combined-arms effort, and particularly the value of air supremacy, when he lamented,

> We had learnt one important lesson during this operation, a lesson which was to affect all subsequent planning and, in fact, our entire future conduct of the war. This was that the possibilities of ground action, operational and tactical, become very limited if one's adversary commands the air with a powerful air force and can fly mass raids by heavy bomber formations unconcerned for their own safety.[67]

He further emphasized "the paralyzing effect which air activity on such a scale had on motorized forces; above all, the serious damage which had been caused to our units by area bombing."[68] Later, he recalled that "British air superiority threw to the winds all the tactical rules which . . . had hitherto applied with such success. There was no real answer to the enemy's air superiority, except a powerful air force of our own. In every battle to come, the strength of the Anglo-American air force was to be the deciding factor."[69]

The same day, Rommel told von Rintelen that the army's bread ration had been cut in half, and sickness was prevalent as a result of malnourishment, lack of sleep, and strenuous operations. British artillery was outfiring surviving Axis guns by a ratio of 10:1. Consequently, Rommel ordered a major change in army dispositions to create greater depth and breadth, thereby reducing losses. Italian troops filled in the gaps. This worked well in its stated purpose but placed the army in a distinctly unfavorable position to repel a major offensive.[70]

Food shortages worsened. British torpedo-bombers were becoming so deadly that they made the run from Italy to North Africa extraordinarily dangerous. If they missed, submarines often did the job.[71] As the logistical death spiral accelerated, Rommel spoke bitterly in his memoirs, after having pushed so hard for an immediate advance on the Nile Delta, of the crucial lost opportunity at Malta. He had volunteered to lead Operation Hercules himself, he said after the fact, and would have succeeded with decent air and naval support. "Malta," he lamented, "has the lives of many thousands of German and Italian soldiers on its conscience."[72] Meanwhile, his pleas for additional Luftwaffe assets came to nothing.[73]

Although Rommel's tale of woe makes clear the magnitude of the disaster inflicted on his army, Coningham said at a Middle East air officer commanding (AOC) meeting that a change in enemy air tactics provided "evidence of infusion of new blood at the top." General Seidemann, like von Waldau before him, was a major improvement over Fröhlich. He had been Kesselring's chief of staff and with Wolfram von Richthofen's Fliegerkorps VIII in the Balkan campaign before that. He arrived too late. Nonetheless, Coningham's key lessons for continuing to outfight the small but capable Luftwaffe contingent were as follows:

1. Fighter governs the front.
2. Air Commander must have control of all air forces in forward areas.
3. Bombing by day in battle area involves permanent fighter escort.
4. Within range of shore-based fighters Stuka is dead.
5. Necessity for continuous bombing by day and night.
6. Germans very susceptible to attacks with delay-action bombs on their landing grounds.

7. Use of fast fused bombs at night means that Germans have to check aircraft for damage by blast and splinters before they can fly next day.

8. Need for communication aeroplane which can land and take off from smallest possible space.

9. Fighters need speed and performance below 15,000 feet.

10. Over-specialisation of fighters to be discouraged.

11. Need for larger proportion of fighters to bombers: say 4 to 6 fighters for one bomber.

12. In combined operations Army and air commanders must live together and see each other daily.

13. Need for forward planning on a pessimistic basis.

Summing up, attendees agreed to plan for forward movement after the offensive began with the object of overcoming the difficulties inherent in lengthening lines of communication.[74]

On 5 October, in another "I told you so" moment, Erich Raeder reminded Keitel and Alfred Jodl,

> The enemy has recognised sea supply communications as the weak point of our North Africa operation and is carrying out constant heavy attacks with aircraft from Malta and Egypt, as well as with submarines, causing us daily losses. As regards range, he is commanding the entire sea area, which is even controlled at night by excellent radar. Our defence against submarines is barely sufficient and against air attacks is inadequate. Continuing shipping losses cannot be replaced. Supply loss cannot be retrieved by strengthening defences later. The danger therefore exists of continued weakening of our forces in North Africa and the ultimate loss of the entire North African position with dire military and political consequences. Measures to hold down Malta are therefore, in the opinion of Naval Staff, Operations Division, of the greatest urgency.[75]

The Axis could only wait for the blow to fall.

During the lull from early September to late October, British combined-arms initiatives came to fruition. When the offensive began, the WDAF and the Eighth Army were at full strength with good reserves. RAF and USAAF headquarters were collocated, and aircraft of the same types operated within the British group structure. Tedder continued to praise Montgomery's approach to combined-arms warfare. He said the RAF had to go all out to defeat remaining Luftwaffe assets. This was crucial because Hurricane squadrons still made up much of RAFME's total strength. Until the skies were relatively safe, Hurricanes faced heavy losses. Most tactical reconnaissance (Tac R) aircraft were Hurricanes, so this was also important from a reconnaissance and intelligence perspective. However, in the first of what would be many bitter disagreements with Montgomery, Tedder ended by asserting that the Eighth Army commander's plan would not force the Germans to concentrate to the degree necessary for airpower to play its full role in the battle. Montgomery's methodical style may have allowed the remnants of Rommel's army to escape, but he was not about to risk another reversal while Rommel still had troops and any fight left in him.[76]

Aircrews kept raiding ports and convoys. Attacks on Tobruk were so effective that seaborne convoys diverted to Benghazi, 250 miles to the west, forcing trucks to drive an extra 500 miles roundtrip. USAAF Liberators attacked Benghazi without pause. No. 201 Group and a reviving Malta turned up the pressure even further. In April and May, while the Luftwaffe controlled the central Mediterranean, Axis convoys lost only 2.7 and 5.8 percent, respectively, of their tonnage during the passage to Libya. In June, losses shot up to 35 percent. The Axis did slightly better in July as the Luftwaffe made an unsuccessful attempt to reverse Malta's revival. In August, 35 percent of supplies did not reach Libya. In September the total was 30 percent. In October more than 50 percent, including most of the fuel, went to the bottom. A last and disastrous effort from 11 to 20 October to suppress Malta failed. It also kept more than 600 aircraft on Sicily occupied when the longer-range assets could have shifted to Crete, which was less than 350 miles from El Alamein, and assisted Rommel.[77]

Despite their training efforts, the British took advantage of heavy rains over German airfields at Daba and Fuqa, which were waterlogged by 9 October. Full-scale day and night attacks followed. While the Germans struggled to resume air operations, Coningham hit them again on 19 October, causing further heavy losses.[78] As the RAF set the conditions for victory in the air and in terms of supporting the Eighth Army's advance, Montgomery prepared to strike at El Alamein on the night of 23–24 October. Allied landings in Algeria and Morocco on 8 November, on the heels of Montgomery's victory, placed Axis forces in a vise from which they would not escape.

10

El Alamein and Operation Torch
October–November 1942

Preparing for the El Alamein Offensive

As BERNARD LAW MONTGOMERY put the finishing touches on his offensive (Operation Lightfoot, to be followed by Operation Supercharge during the breakout phase), Arthur Tedder and Arthur Coningham kept pace. The Me. 109F and Me. 109G were still superior but badly outnumbered. Nonetheless, only the seventy-eight Spitfires now in theater (fifty of them in September) and the perennially understrength Kittyhawk units could contest the air. The day-bomber picture improved as B-25 Mitchells joined Bostons and Baltimores, but there were only fifty operational Wellingtons in No. 205 Group as a result of transfers to India and growing Bomber Command requirements.[1]

Montgomery's successful deception scheme relied on allowing the Germans to see only what he wanted them to see. This was never entirely possible, but with the Royal Air Force (RAF) in command of the air, it kept the enemy focused on the deception and away from areas containing the real buildup of forces. To the greatest possible extent, fighter and day-bomber units stood down until 19 October, while night-bombers in Egypt and on Malta kept striking enemy logistics.[2]

On 19 October, the RAF began intensive attacks on airfields and targets of opportunity. From 19 to 23 October, the Western Desert Air Force (WDAF) and No. 205 Group flew 2,469 sorties—about 540 a day. Even this tempo was restrained and designed simply to ensure air superiority for the ground battle. Consequently, the WDAF gave unprecedentedly heavy support to the Eighth Army. These operations also gave reconnaissance aircraft a free hand over enemy lines, allowing for detailed intelligence appreciations. Guns and minefields in critical sectors were now well known.[3]

Coningham's plan for the night of 23–24 October was to illuminate and bomb known gun positions, attack troop and truck concentrations with fighter-bombers, jam armor units' radio transmissions with electronic warfare Wellingtons, and otherwise create maximum friction by laying large smoke screens and dropping dummy parachutists. At daybreak on 24 October, light-bombers and fighter-bombers would attack prearranged targets while smoke-laying aircraft stayed on call to meet the army's needs. Fighters would provide air cover, escort bombers and reconnaissance aircraft, attack vehicle convoys, and conduct visual reconnaissance.[4]

The Battle Begins

The artillery bombardment at the start of the ground offensive was a model of effective centralized control, with guns massed against key targets revealed by aerial reconnaissance, signals intelligence, sound ranging, muzzle flash spotting, and other means. The ratio of fire at key points was between 10:1 and 22:1. Wellingtons following Albacore pathfinders contributed, with forty-eight bombers dropping nearly 125 tons of bombs on gun positions. The electronic warfare Wellingtons jammed virtually all radio communications, delaying the Axis response.[5]

On 24 October, the day after Operation Lightfoot began, the Afrika Korps staff said that Hans Seidemann had to turn down its request for fighter-bomber support of ground forces because only twenty fighters were serviceable as a result of airfield attacks.[6] Even these had too little fuel to fly at full intensity.[7] After heavy rains halted the offensive on 24 October, the Eighth Army picked up momentum. RAF transports delivered 100 tons a day of supplies to advancing troops and an RAF aerodrome reconnaissance party and army aerodrome construction party just behind the front. These units would sprint forward to occupy captured airfields and service arriving aircraft. Everything was focused on speed so the retreating enemy would get no rest from air raids. This occurred seven times in two weeks without loss of operations tempo.[8]

The Eighth Army's objective was to rip a hole in the Axis line with a combination of infantry, engineers, artillery, and airpower and exploit the breakthrough with armored units. The British were methodical and still not up to German tactical standards, but the latter's need to fight a defensive battle with limited mobility tilted the balance. General Georg Stumme, who had replaced a convalescing Erwin Rommel, reported that he lacked ammunition for counterbattery fire during the opening nights of the offensive. He lamented the lack of mobility against the relatively methodical attack, which a sharp counterattack could have stopped. Instead, isolated and infantry-heavy fortified garrisons were left to stop attacks wherever they developed.[9]

Rommel returned at once after Stumme died during a strafing attack and launched Twenty-first Panzer Division in a counterattack. RAF strikes, the new six-pounder antitank gun, and armor stopped it, destroying one-third of its tanks. Hourly air attacks along the line caused losses, fatigue, and demoralization. By the time Montgomery launched Operation Supercharge

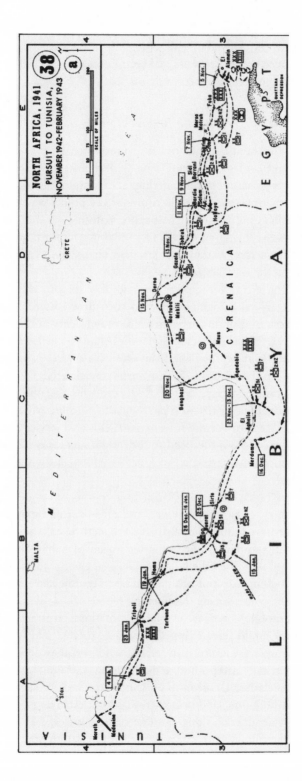

Defeat at and pursuit from El Alamein, November 1942–February 1943. By the time Rommel reached the Mareth Line in Tunisia in February, he had 22,000 troops left and far too little heavy equipment, fuel, and air support. He lost nearly 90,000 troops and thousands of vehicles and guns along the way. (Maps Department of the US Military Academy, West Point)

on 2 November, RAF assets had dropped more than 200 tons of bombs a night on enemy positions and 352 tons in a concentrated area just before the assault. The combination of Eighth Army attacks and incessant air raids proved too much.[10] Rommel commented that "ceaseless bombing attacks hammered down on the German-Italian forces. The Luftwaffe tried all it could to help, but could achieve little or nothing against the tremendous numerical superiority of the enemy."[11] On 1 November, his staff recorded thirty-four air raids on the pivotal defensive position at Hill 28 while fighter-bombers savaged his dwindling supply convoys.[12] The loss of five large fuel tankers in a week left Rommel without any fuel reserves and none on the way. Of 4,244 tons of fuel allocated to Panzerarmee Afrika between 27 October and 1 November, 893 tons arrived.[13]

On 2 November, Rommel wired Oberkommando der Wehrmacht (OKW) that he faced overwhelming enemy superiority. The fuel shortage would not allow a major withdrawal. There was only one road available, and his army would come under unceasing air attack. On 3 November, Rommel asked Adolf Hitler for permission to withdraw. The Führer refused. On 4 November, Rommel asked again. In a letter to his wife dated 2 November, Rommel closed on a sobering note: "Air raid after air raid after air raid!"[14] Burning vehicles choked the battlefield as Rommel decided on 4 November to ignore Hitler's orders and withdraw his units to the last defensive positions.[15]

Rommel Retreats

After Axis forces withdrew, the WDAF flew 206 sorties against retreating vehicles. The Luftwaffe had too few aircraft left to attack concentrations of Commonwealth vehicles on the coastal highway. As the Germans retreated, they stayed far from the major roads and moved in highly dispersed groups to minimize losses to air attacks. Nonetheless, the Afrika Korps war diary noted heavy losses of vehicles. Occupation of forward Axis airfields yielded nine brand-new Me-109Gs, fifty other aircraft in various states of repair, and thirty-nine wrecks.[16]

Tedder had just visited the front and said his airmen were "in magnificent form and out to kill."[17] He pushed for an aggressive pursuit and the destruction of Rommel's army to avoid another stalemate and reverse at El Agheila. This did not happen.[18] WDAF aircrews, who believed they could make hash of Rommel's forces, were shocked that they were not given orders to fly maximum sorties. The sortie rate dropped from 1,219 on 3 November to 741 on 4 November and 416 on 5 November. Montgomery's methodical style of battle had its drawbacks, and one of them was now becoming clear. Tedder was not happy. The only way to bag Rommel, he thought, was to give Axis forces no respite.[19]

Axis air forces were in wretched shape, having lost almost all of their ground equipment and reparable aircraft. By the time reinforcements were en route, Operation Torch forced their diversion to Tunisia. The fifty Ju 52 transports Albert Kesselring dedicated to resupplying Rommel suffered heavy losses as Ultra decrypts facilitated aerial ambushes. Of 250 tons of fuel promised by air, only 60 tons arrived during November. Rommel's efforts to maintain an orderly retreat and effective resistance devolved into a constant effort to secure fuel. Fortunately for him, the difficulty of telling friend from foe on the ground during the confused pursuit, along with worsening weather, limited WDAF effectiveness. However, five B-17s attacked Tobruk, sinking a large merchant vessel (MV) and damaging several other ships. Later the same day, twenty-one Liberators attacked Benghazi, sinking a small MV and setting a large fuel tanker on fire. The lack of fuel continued to cause significant heavy-equipment losses during the retreat.[20]

General Wilhelm Ritter von Thoma, captured on 4 November, said of the RAF, "We were impressed by the great superiority of your forces. Only a few days before the battle, I had a telephone conversation with Kesselring, who asked me for my opinion about the work of the Luftwaffe. I told him I felt that our chaps were so outnumbered by the English that they seemed to be reduced to a purely defensive role. He replied that this opinion agreed with the reports he had received."[21] Von Thoma also suspected that fuel shortages kept the Luftwaffe from making full use of its remaining planes. Further, he said the major and sustained WDAF raids on airfields did severe damage and made a profound impression, and nonstop attacks on ground forces inflicted serious losses, restricted mobility and combat effectiveness, and had a major impact on morale. Air raids put all of von Thoma's artillery out of action and caused severe confusion at his headquarters and throughout Axis positions. When asked why the Luftwaffe often dispersed its efforts during the desert war and failed to concentrate on key targets such as the RAF (to gain air superiority) and the Suez Canal (to interdict Commonwealth supplies and reinforcements), von Thoma replied that Kesselring, rather than Rommel, controlled air policy and was to blame. This was far from true but also irrelevant by now.[22]

The Axis suffered a disaster from which it could not recover given the now immense Allied airpower and logistics advantages. The defeat heralded many more to come and the gradual ending of *Bewegungskrieg* as an option for conducting war. As one scholar has noted, at El Alamein, as at Stalingrad, the Wehrmacht lost not just men, material, and initiative but "something far more precious, something that went to the heart of its identity and to the core of its ability to function as a military force: its mobility."[23]

Despite the enormity of the defeat, after Rommel's forces neared their sources of supply, while the Eighth Army moved away from its own, the opportunity for a deathblow evaporated. Rommel preserved a kernel of his army to slow the British advance for the next six months. Although Coningham's airmen were adept at leapfrogging, doing so with a force now comprising more than 1,200 aircraft was a mammoth logistical undertaking, and Montgomery was not prepared to let the Eighth Army entirely off the leash to finish Rommel in coordination with the WDAF. Nonetheless, Coningham did what he could. Airfield reconnaissance parties travelled with leading army columns, had a wireless radio link to Advanced Air Headquarters, and sent word when newly occupied landing grounds were operational.[24]

The Pursuit Continues

Despite the failure to deliver a knockout blow, air raids continued. However, these never slowed Panzerarmee Afrika enough to satisfy Montgomery that he could or should finish the job. Supplies from Italy were insufficient to keep Rommel's army in any condition to counterattack. Only 12,981 tons of supplies left Italian ports in December, of which 47.4 percent—6,153 tons—made it to Libya. Rommel could no longer rely on captured materiel. Coming on the heels of the 86,000 troops lost from all causes between El Alamein and the Mareth line (where he had 22,000 remaining in January 1943), this was an irreversible disaster.[25]

Coningham kept after remaining fighters and airfields along with shipping and air transports. Beaufighters and Beauforts savaged coastal vessels. Liberators struck MVs with increasing success, and Halifaxes bombed Tobruk and Ju 52 airfields because the old transports were now the primary delivery platform for supplies. Raids continued against supply columns. In a particularly effective one against Derna on 17 November, Kittyhawk pilots destroyed at least thirty-seven aircraft as they caught some preparing to take off and the rest lined up wing tip to wing tip. More than half were Ju 52s and Italian SM79 and SM82 transports delivering fuel.[26] Tedder emphasized that the maximum air effort against logistics targets would continue.[27]

Still, the WDAF's move to newly captured airfields was hampered by a combination of truck-heavy columns, bad roads, bad weather, and mines. The RAF thus made greater use of air transport to move ground parties forward. Forty Hudsons ferried supplies forward as soon as airfields were operational. This was generally quite fast because the chief engineer of the Eighth Army gave airfield clearance and operational refurbishment efforts top priority so the WDAF could continue supporting the Eighth Army's advance.[28]

Royal engineers connected Tobruk to the Nile Delta by railroad on 1 December and used water-efficient US diesel locomotives to move 2,500 tons of supplies per day there. These continued forward on small MVs. These coastal convoys moved forward 800 tons of supplies per day. After the Eighth Army captured Benghazi on 20 November, engineers had it receiving cargoes by 26 November. These logistical achievements gave troops the supplies they needed to keep Rommel on the run. The end of diversions to other theaters also helped. Far from being short of supplies, the Eighth Army and the WDAF replenished their stocks as they advanced.[29]

With Benghazi's airfields again in RAF hands, coordinated air and naval action protected a westward-bound convoy to Malta that sailed on 16 November. The entire convoy made it through, bringing much-needed aviation fuel. Another ten ships arrived unscathed in early December, bringing more fuel and food. Rations increased from 50 to 70 percent, and the siege ended. Tedder made maximum offensive use of the island immediately to keep the logistical pressure on Panzerarmee Afrika as Rommel tried to establish a defensive position at El Agheila on 23 November. Rommel realized he could not hold and withdrew to Buerat on 13 December under constant air attack. Reconnaissance aircraft tracked him. As RAF air transport became widely available to move air units forward, "bounding" extended to a distance of 100 miles from 40 (the limits of road-based movement each day), with entire wings moving forward together.[30] The end in Libya began coming into view.

Strategic Decisions Regarding Operation Torch

With Rommel on the run, the Allies opened a second front in North Africa with the Operation Torch landings to help drive the Axis from the southern shores of the Mediterranean, entice Vichy administrations and their resources to the Allied side, and bring the Allies closer to knocking Italy out of the war. Anglo-American discussions in London in July 1942 and leading up to Operation Torch epitomized the US preference to close with the enemy immediately in France. George C. Marshall advocated a decisive action that, in conjunction with the Red Army's efforts, would put Germany in a vise and lead to its rapid defeat. The more circumspect—and experienced— British preferred to win in the Mediterranean theater first, knock Italy out of the war, secure the sea lines of communication (SLOCs), and safeguard oil supplies. They prevailed but agreed to Lieutenant General Dwight Eisenhower as commander in chief for Operation Torch, commander of the Allied Expeditionary Force, and head of Allied Force Headquarters (AFHQ) as of 6 August 1942.[31]

During the July conference, the British pushed for a Mediterranean strategy while assuring the United States that they would focus all efforts on Japan after Germany was defeated. The United States had no better options. Taking Sicily would open the Mediterranean fully to merchant shipping. If the Allies could pass thirty ships through the Mediterranean every ten days, they could stop shipping around the cape. This would release 225 MVs from that "pipeline"—a huge increase in available shipping. The conquest of Sicily would also divert German strength from Russia, bring Italy closer to collapse, loosen the enemy's hold on the Balkans by causing an Italian surrender and consequent withdrawal of troops, and perhaps bring Turkey into the war on the Allied side. All but the latter proved at least partially successful.[32]

Viewed in light of these efforts, the decision to invade Sicily and then Italy, confirmed later at the Casablanca Conference, was intended not just to knock Italy out of the war and create a second aerial front against the Reich but to tie up major German forces away from the Eastern Front. If the Russians were killing most German troops (two out of three killed perished on the Eastern Front), the Allies would destroy the Luftwaffe, making German troops in the east even more vulnerable to Red Army attacks. The Allies would also launch a massive heavy-bomber interdiction campaign in southeastern and central Europe—the largest of the war—to speed the Russian advance.[33]

On the German side, Kesselring took command of all German forces in theater in September 1942 based on intelligence of a landing somewhere in the Mediterranean. His pleas for one division to help defend Sicily went unheeded, as did his view that agreements with Vichy France to begin logistical preparations in Tunisia were vital to effective resistance there. Oberkommando der Luftwaffe (OKL) sent two torpedo squadrons and made significant improvements to the worn-down and underequipped ground organizations in Sicily. The convoluted air logistics chain of command still ran through Comando Supremo and from there through the Italian Navy, even though most of the aircraft dependent on this arrangement were Luftwaffe Ju 88s.[34]

Kesselring saw the grave risks posed by an Allied landing in North Africa:

1. Total loss of German-Italian Army of Africa.

2. Loss of Tripolitania.

3. Peaceful occupation of French North African colonies and recruitment to Free French forces.

4. An ideal base for staging a landing on Sicily or mainland Italy.

5. Possibility of eliminating Italy as an Axis partner.

6. Commencement of a bombing campaign against southern Germany in summer 1943.

7. Possible further operations in Italy, southern France, or the Balkans.

Available counters included delaying the landings with air and submarine attacks, seizing and holding a bridgehead in Tunis, extending the bridgehead to a position Axis troops could hold, and establishing an effective logistical system. Rommel had 2,000 miles of open terrain across which to retreat, so the Axis could trade time for distance. Priority thus went to making the lodgment in Tunisia rather than reinforcing Rommel. The bridgehead in Tunisia and Rommel's army would require twice the supplies previously needed to keep just Rommel's forces in the fight. The Italians no longer had enough large MVs for this, so small vessels came into ever greater use along with the air transport contingent. Kesselring implored Comando Supremo for more rapid building and repair of MVs. He also set two German construction battalions to work on the Mareth line so Rommel's forces could refit there.[35]

Kesselring was concerned that the Allies would gain air superiority. He believed ground forces would bear up well under air attack, but continuing Allied ship attacks would be a logistical disaster. The threat, he said, "could and had to be obviated. . . . There was a very great danger of wasting the Luftwaffe; their duties were too many and their numbers too weak, especially since the Italians were useless over the sea and unreliable elsewhere."[36] Protection of shipping, air transports, and port facilities was paramount. From there, aircraft would raid Allied shipping and ports and attack air bases and Allied troops.

Allied Air Preparations for Operation Torch

Brigadier General James H. "Jimmy" Doolittle was the senior US Army Air Forces (USAAF) commander, in charge of the Twelfth Air Force, with Air Marshal William Welsh in charge of the RAF Eastern Air Command (EAC). Both were directly responsible to Eisenhower but had separate commands. Eisenhower had preferred a unified air command, but Henry "Hap" Arnold and Major General Carl "Tooey" Spaatz, Eighth Air Force commander, talked him out of it on the grounds that the USAAF needed to prove itself without undue RAF influence—a flawed and costly argument. Similarly, Eisenhower championed an independent air force from the outset, not subordinate to army commanders, and he backed his preference over the long run by refusing demands for permanent aircraft apportionments to army divisions.[37] This took time to achieve, with operational failures along the way. Additionally, almost none of the lessons the WDAF learned in the desert, including unified command, centralized planning and control, and coordination with land and naval forces, seems to have been applied to air operations for Operation Torch. Nonetheless, Eisenhower insisted that all AFHQ actions would proceed "as though all its members belonged to a single nation."[38] He came as close as anyone could in late 1942 given his forces' logistical and operational shortcomings. Yet, as one scholar has noted, "the reluctance of tribes, nations, and armed forces to learn except from their own experience"[39] largely undermined the supreme commander's guidance. Eisenhower could not achieve his three objectives for seizing North Africa—establish lodgments on the coast, exploit inland, and move east to crush Axis forces between his troops and the Eighth Army's—without agile and effective air forces. The air contingents' shortcomings also hindered the ability to carry out Eisenhower's ancillary tasks: protect friendly logistics, deny supplies to the enemy, and prepare for operations against Italy.

The EAC, formed on 1 November 1942 from RAF No. 333 Group, supported the Algiers assault force, while the Western Air Command (WAC—the US Twelfth Air Force) supported the Oran and Morocco landings. The boundary between air commands was a north-south line running through Cape Tenes. Both commands were initially headquartered at Gibraltar. The Allies planned to fly 160 fighters to Oran, 160 to Casablanca, and 90 to Algiers within three days of the landings. The goal for 1 January 1943 was 1,244 aircraft in WAC, including 282 in reserve and 454 in EAC. Long-range reconnaissance aircraft kept track of all French and Spanish ports and air bases. A massive air effort involving Coastal Command, Bomber Command, and the USAAF Eighth Air Force got under way to suppress U-boat activity and safeguard troop convoys coming from the United Kingdom and United States that sailed between 22 and 26 October.[40]

On 22 October, Bomber Command began seven weeks of night raids on targets in the Genoa-Turin-Milan area to tie down Italian aircraft and anti-aircraft artillery (AAA) and bring home to the Italians the increasing might of the Allied war effort. Wellingtons also bombed airfields on Sardinia three times from 8 to 11 November. By 8 November the Allies had more than 350 planes at Gibraltar. Many of these were "reverse lend-lease" USAAF Spitfires.[41] The operation profited from the fact that Hitler expected landings elsewhere and, concerned about possible Vichy reactions, did not take precautionary actions in North Africa and dissuaded the Italians from doing so.[42]

The French had about 500 planes in North Africa. The Germans had 298 on Sicily and Sardinia and the Italians 574. The Luftwaffe also had numerous transports. This gave the Germans a significant initial edge in theater mobility. Had the French Fleet in Toulon joined the Axis, the tables might have turned, but this threat never materialized. For the combined-arms operations envisioned in Operation Torch, air superiority and the rapid capture of good airfields were vital. However, launching the operation at the start of the rainy season created serious problems.[43]

Allied deception kept Kesselring's staff guessing about the location of the landings. False reporting, frequent forays from Gibraltar, and other actions left the Germans in the dark. Air reconnaissance could only ascertain a major buildup of ships, troops, and supplies in England and Gibraltar. However, low activity in eastern Mediterranean ports convinced Kesselring that the landings would occur in the western Mediterranean. He placed the vast majority of reconnaissance assets there after he knew that a large convoy had left England. A U-boat reported another large convoy crossing the Atlantic toward Gibraltar.[44]

Operation Torch Begins

On 8 November, Allied armies landed where nobody, with the partial exception of Kesselring, expected them. The rapid capture of Maison Blanche and Blida, outside of Algiers, gave EAC two all-weather bases. Within a day of the invasion, two squadrons of Hurricanes were operational at Maison Blanche and several Martlets (F4Fs) at Blida. Carrier-based planes and those on Gibraltar neutralized the Vichy Air Force. The race for airfields at Oran included La Senia and Tarafaoni, both of which fell after short but fierce engagements. Twelfth Air Force medium-bombers escorted by Spitfires arrived at the latter base just as Allied troops took it. From 11 to 12 November, German bombers sank two transports that tried to land troops at the Djidjelli airfield, and further attacks on unloading operations at Bongie caused losses and delays in deployment. On 14 November, British troops, including paratroopers, took Bône and its airfield, which the Luftwaffe promptly bombed. Nevertheless, the next day British troops pushed east to within fifty-six miles of Tunis, while paratroopers further south seized the airfield at Youks les Bains.[45]

The Germans Respond

Kesselring reacted quickly. Troops established strongpoints at the mountain passes along Tunisia's western border, backed by a motorized force to repulse breakthroughs. Axis hopes hung entirely on logistics. The odds were long. Supply shipments depended on control of the air. If Kesselring had a broader grasp of the war's current course, he must have understood that

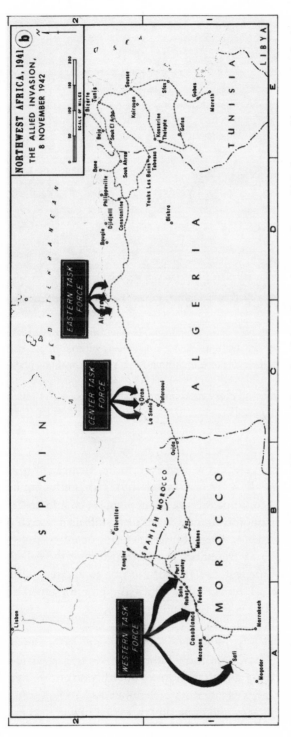

Operation Torch, 8 November 1942. This "second front" in Africa placed Axis forces in a vise from which they could not escape. It also increased their logistical requirements and put excessive pressure on what remained of their shipping and air-transport assets. (Maps Department of the US Military Academy, West Point)

FAA Martlet (Grumman F4F) taking off from HMS *Formidable*, 1942. After RAF carriers could operate once again in the Mediterranean without major risks from U-boats and bombers, their aircraft, including Martlets and Seafires, played crucial roles in securing the beachheads, attacking airfields, and supporting the troops during Operation Torch and every subsequent amphibious assault. (Imperial War Museum Image No. A 11644)

such a thing was unlikely. OKW was even more disengaged from reality, approving the transfer of seven armored divisions to Tunisia but having apparently overlooked the impossibility of supplying such a force.[46]

On 10 November, photoreconnaissance (PR) confirmed 100 German and 20 Italian aircraft in Tunis. Fliegerführer I was in place by 15 November.[47] From 6 November to 25 December, in an effort to halt the Allied advance on Tunis, the Luftwaffe lost 340 aircraft out of a total strength of 877, including 201 aircrews. Interdiction of Allied shipping for Operation Torch was intense and costly. When the landing began, 11 of the 150 Luftwaffe aircraft in place to repel it were torpedo-bombers—just 13.6 percent—for an effort that would rely, ultimately, on how many MVs and transports the Germans sank. Two more torpedo-bomber *Kampfgeschwader* (attack wings), KG 26 and KG 30, both veterans of interdiction efforts against shipping to Murmansk, arrived late on 8 November. KG 26 had transferred to the Mediterranean in May and done superb work but with too few aircraft to produce required levels of effectiveness. German airmen set to their task with a will, although there were too many ships, too well escorted by fighters. Nonetheless, antishipping raids did serious damage from the outset. From 7 to 13 November, they sank 183,000 tons of shipping and damaged another 234,000. Additionally, they sank three cruisers and four destroyers and damaged another twenty-five warships, including three carriers and a battleship. The torpedo planes did most of the work but were too few in number and could not turn back the invasion force. Even Hitler's major transfer of aircraft (by 30 November, Luftflotte 2 had 1,646 aircraft compared with Luftflotte 4's 1,177 planes around Stalingrad) could not change the course of events given the size of the invasion fleet and Allied air forces.[48]

Allied Efforts

The Allied advance toward Tunis was under way, but the Germans had local air superiority, attacking advancing columns. Allied units occupied the airfield at Djedeida, seven miles west of Tunis, on 26 November. A single 88 mm AAA battery stopped the lead armored columns there. Despite this reprieve, the Germans soon felt the first harbinger of defeat when heavy-bombers raided the port and airfield at Bizerte on 29 November, followed by another raid against Tunis on 30 November. Aircrews on a resupplied and reinvigorated Malta intensified their raids on convoys, with night torpedo attacks under flares doing grave damage. Hitler ordered an immediate increase in the number of aircraft escorting convoys, but they could not stop night attacks, and the diversion of effort from North African air operations began slowly to shift the balance. The constriction of Axis air operations and logistics had already begun.[49]

Still, early Allied raids on Tunisian airfields had little effect because the Germans had a *Luftgaustab* (special air duty district staff) that prepared alternate airfields. However, the Germans were surprised by the speed with which the Allies established airfields in Algeria. An aerial photograph revealing the use of pierced steel planking (PSP) for runways cleared up the mystery. Yet a marginally effective and vulnerable Allied air-ground effort gave the Luftwaffe advantages from November 1942 to February 1943. The relatively late maturation of the Tedder-Coningham system may have made Allied planners for Operation Torch skeptical of its utility, although there were many useful insights to be had. Poor command-and-control (C2) arrangements caused severe problems. The USAAF's overreliance on *Field Manual 31-35* was also problematic. It was a reasonable document but allowed ground officers to control air assets, hindering the USAAF's effectiveness early in the campaign.[50]

Despite the USAAF's challenges, it was hardly the basket case some historians have claimed. Both organically and with British support, it grew into a formidable instrument. US airmen had long thought seriously about all forms of airpower employment—a "general air strategy," as Richard Overy called it. Although strategic bombing eventually became the mechanism for pushing USAAF independence, airmen developed a full range of capabilities. The RAF assisted, but the US airmen were fast learners and had already engaged in the intellectual if not the practical exercises necessary to become highly combat effective. The simplest expression of this understanding was Spaatz's remark to Arnold on 7 March 1943 that "in order for the Army to advance . . . the air battle must first be won."[51]

Field Manual 31-35 might have been comprehensive, but until the aviators had authority commensurate with statements in the document, airpower employment and combined-arms operations suffered. For instance, the document emphasized,

> Designation of an aviation unit for support of a subordinate ground unit does not imply subordination of that aviation unit to the supported ground unit, nor does it remove the combat aviation unit from the control of the air support commander. It does permit, however, direct cooperation and association between the supporting aviation unit and the supported ground unit and enables the combat aviation to act with greater promptness in meeting requirements of a rapidly changing situation.[52]

Aside from its tactical focus, this passage uses variations of the word "support" five times—all in reference to direct air support of ground troops. There is no mention in the document of the ways in which the two arms are mutually supporting. The term *air support commander*, if standard for the period, was equally loaded contextually and made it clear that, at the end of the day, what the ground commanders said would go. This is precisely what happened in Northwest Africa until a combination of the steep US learning curve, some pointed advice from Tedder and Coningham, development of communications and logistical infrastructures, effective and concerted raids on enemy airfields, and ultimately the centralized control of air forces under one airman allowed air assets to function with high effectiveness. The most crucial ingredient was "buy in" on the part of US ground commanders.[53]

This problem regarding air-ground operations was clear from the start. USAAF air assets had not participated in the major army prewar exercises, making it difficult for anyone to envision what they might accomplish. Although ideas about combined-arms operations existed in the minds of US airmen long before Operation Torch, it does not obviate the fact that the

USAAF went into combat in northwestern Africa tied excessively to ground units; without an adequate study of the two-year RAF effort in the Western Desert or even of Lewis Brereton's Ninth Air Force and with inadequate logistical, communications, and intelligence support. The aircraft, doctrine, and leaders were there, but they had to go through the same learning curve the RAF and British Army did. The USAAF ultimately became an excellent learning organization.[54] Nonetheless, airmen could not assert themselves until Eisenhower gave them complete operational control of their airplanes. The ultimate issues thus revolved around who would control the airmen and their aircraft and how quickly the USAAF learned to use its strength in a combined-arms context after it had the latitude to do so.[55]

Spaatz emphasized, "The correct use of air power was not really close support, but rather air superiority and interdiction operations, hitting enemy airfields, tank parks, motor pools and troop convoys—in effect interdicting enemy supplies, equipment, and troops *before* they reached the battlefield. If he [the air commander] maintained a constant umbrella over one small portion of the front, then his available force would be dissipated without any lasting effect."[56] Coming from one of the foremost "bomber barons," this reminds us that US senior airmen were pragmatic in the current fight even as they maneuvered for independence in the long term. One of the reasons Eisenhower called for Spaatz to serve in North Africa was his penchant for pragmatism along with his ability to work with army and RAF officers. His tour in the United Kingdom in 1940 had resulted in friendships with RAF counterparts, which paid handsome dividends.[57] Accordingly, Spaatz told Arnold, "The proper coordination of air effort with ground effort depends to a large extent on the personalities of the commanders," and effective command relationships relied on "mutual respect for each other's capabilities and limitations."[58]

This leads to the question of whether *Field Manual 100-20* had its origins in RAF concepts. The speed with which this new doctrine document proliferated in 1943 had much to do with the fact that its basic assertions were based on RAF experience. The RAF had led the way in developing and promulgating these concepts, even as US airmen intuited them before the war and embraced them after it began.[59] In a telling statement, B. Michael Bechtold noted, "American airmen had developed a sound doctrine of their own, but they were unable to convince the Army hierarchy of the applicability of their ideas. . . . This was where the British came in. . . . The American airmen became disciples of the British doctrine not because of its originality, but because they could use its success to convince the Army leaders of its utility."[60] This illuminating passage underscores the crucial point that nobody could ignore the combined-arms successes in the Western Desert. It was the only example for the air forces in Tunisia—and more importantly the ground commanders—to follow. In effect, RAF exploits in combination with the Eighth Army's convinced the US Army to change its corporate view of proper airpower employment.[61]

As the campaign unfolded and the immense importance of heavy-bombers flying interdiction missions became clear, a running fight ensued between Tedder and Eisenhower on the one hand and Arnold and Ira Eaker on the other. Eaker asserted that the diversion of Eighth Air Force planes and crews to North Africa for Operation Torch set the USAAF portion of the combined bomber offensive (CBO) back by at least a year. Even if this were true—and it was not given the Eighth Air Force's inability to strike targets in Germany without prohibitive losses in 1942–1943—it is more important to focus on how the diverted units contributed to major victories in the Mediterranean. The USAAF and Allied leadership did the right thing by diverting some heavy-bombers to North Africa.[62]

Eisenhower merits much of the credit for the air campaign that followed the Operation Torch landings. Despite its flaws, which were not of Eisenhower's making, he insisted that air forces stay focused on air superiority and interdiction while ensuring army commanders did not have private air forces. Even Eaker, who opposed heavy-bomber diversions, was "tremendously impressed with Gen. Eisenhower's keenness for air operations."[63]

As the value of heavy-bombers in the interdiction role became clear, so did that of airborne operations. The initial ones, to capture airfields in Algeria, were not successful but improved considerably and soon became crucial to the rapid Allied advance into Tunisia. On 8 November, airfields at Tafaraoui and La Senia fell to a combination of US airborne and ground action. Three days later, British paratroopers took the airfield at Bône. On 14 November, US paratroopers took the key airfield of Tebessa. Two days later, they took Souk el Arba.[64]

However, even as this innovative and effective employment of army and USAAF assets got under way, by 30 November air units suffered from a lack of replacements and an ineffective logistical infrastructure. Fighter squadrons were at half-strength or less. Brigadier General John ("Joe") Cannon, the Twelfth Air Support Command (ASC) commander, established an air depot in Casablanca and hired every French aircraft mechanic he could find to keep aircraft flying. Doolittle said, "Joe Cannon has done an outstanding job" and that he was getting along "beautifully" with George Patton—an achievement in its own right.[65]

Although employment of heavy-bombers in the air superiority and interdiction roles was not exactly in keeping with senior airmen's views of their ideal use, it proved highly effective, *contextually,* in helping to defeat the Axis in the Mediterranean. Air assets became increasingly sophisticated, nimble, and effective. They focused on high-value targets such as airfields, ports, and MV convoys. During January, the five largest ports in Tunisia (Tunis, Bizerte, Sousse, Sfax, and Gabês) discharged a combined total of 2,200 tons per day of Axis cargo. Their prebombing capacity was 7,500 tons.[66]

In an indication that some lessons from the WDAF were making their way west, the Twelfth Air Force soon had an army air support control (AASC) with five air stores parks (ASPs) to accompany leading ground units in the Central Task Force. These teams transmitted all requests for air support to Twelfth Air Force Headquarters. EAC had an AASC and nine tentacles. Either the US Thirty-fourth Division or the British Seventy-eighth Division controlled Welsh's EAC air assets depending upon which one had active operations ongoing. This problematic system was compounded by the fact that the division "in control" of air assets at the time could ask only for tactical reconnaissance or fighter protection. These were the only available options. As Allied units advanced, the AASC moved to corps headquarters and its tentacles to brigades. The poor landline infrastructure and a shortage of wireless sets made communications with airfields almost impossible as the troops moved east. In most cases, requests went to First Army, from there to AFHQ, and finally to airfields in the form of orders. This system was slow, cumbersome, and largely ineffective for anything but preplanned missions.[67]

As German aircraft poured into all-weather airfields in Tunisia, they were close to their sources of supply with serviceability rates approaching 70 percent. They quickly gained and held air superiority in the Tunis region. By December, the Luftwaffe had 420 aircraft in Tunisia, all of which were concentrated and with excellent Signal Intelligence Service (SIS) support. The Allies had 639 planes and another 230 on Malta, but the former often operated from improvised and fair-weather fields, whereas the latter were at extreme range. As a result of Allied planning and doctrinal shortcomings, poor airfields, and weak C2, the Germans had the advantage in the air until February 1943.[68]

Allied airmen tried to do too much with too few assets, leading to dissipation of effort and higher losses. Ground commanders continued to use fighters in "penny-packet" patrols over Allied troops, and they sometimes refused to release these aircraft to assist neighboring units calling for air support beyond what their own air assets could provide. The clear need for improvement led Eisenhower to name Spaatz overall commander of Allied air forces, with Air Vice-Marshal James Robb as his deputy, but the parallel air efforts and ground commanders' tight grip on airplanes persisted.[69]

By 19 November, the Twelfth Air Force had four fighter groups, a light-bomber squadron, two transport groups, and two B-17 squadrons. However, there were insufficient maintenance personnel and airfield equipment. Consequently, the Twelfth Air Force's contribution was limited with the notable exception of heavy-bomber raids. Ironically, the USAAF also had a truck shortage for leapfrogging, even though the US automobile industry was in the process of giving the Russians, and to a lesser extent the British, all the trucks they needed to become heavily motorized. The fact that there were only four all-weather airfields between Casablanca and the Tunisian border (Port Lyautey, Tafaraoni, Maison Blanche, and Bône) did not help matters. It took a series of reverses and missteps, rapid development of ASPs and repair-and-salvage units (RSUs) along RAF lines, and British-style C2 and combined-arms processes to transform the USAAF from a struggling junior partner to a formidable force.[70]

Axis troops repulsed the British advance on Tunis and maintained local air superiority, but supply and communications problems were serious from the outset. Mobility was also problematic, with far too few vehicles. Still, Luftwaffe and Italian Air Force (IAF) units remained very active during December because they operated from all-weather airfields close to the front. Fliegerkorps II attacked Allied airfields and ports at night and on cloudy days and flew many direct support missions.[71]

Eisenhower knew that taking Tunis would be impossible during the rainy season given supply shortages, enemy air activity, and a rapid increase in Axis troops. The thrust toward the city had occurred with too little attention to logistics. Enemy dive-bombing raids against forward troops were frequent and costly. Conversely, Axis troops had freedom of action conferred by air superiority. Stukas were often engaged in support of ground troops within five to ten minutes of calls for assistance.[72] Eisenhower emphasized German air superiority when he noted,

> Because of enemy domination of the air, travel anywhere in the forward area was an exciting business. Lookouts kept a keen watch of the skies and the appearance of any plane was the signal to dismount and scatter. Occasionally, of course, the plane would turn out to be friendly—but no one could afford to keep pushing ahead on the chance that this would be so. All of us became quite expert in identifying planes, but I never saw anyone so certain of distant identification that he was ready to stake his chances on it. Truck drivers, engineers, artillerymen, and even the infantrymen in the forward areas had constantly to be watchful. Their dislike of the situation was reflected in the constant plaint, "Where is this bloody air force of ours? Why do we see nothing but Heinies?"[73]

With Allied troops stalled, heavy-bombers stepped up raids on ports and airfields in concert with attacks from Malta. The fight for Tunisia was a battle of logistics, and whoever controlled the air controlled the supply routes. Allied torpedo- and medium-bombers began sinking large MVs berthed at ports such as Naples and Palermo. Torpedo-bombers and submarines preyed on ships crossing the Tyrrhenian Sea. Naval flotillas worked with aircraft to make a dangerous situation perilous for Axis shipping. Finally, belts of mines sown by aircraft and ships took their toll. All of this had its basis in control of the air and thus the SLOCs. If they were lucky enough to reach port, even smaller ships of 1,000–5,000 tons took a full day to unload, and larger ones could take three days. In a major oversight, the Italians again failed to provide lighters, which restricted unloading operations to one side of the ships. These factors made MVs and their cargo susceptible to damage or destruction in port. Far too many Axis aircraft had to protect the convoys, and the chaotic conditions in Italy slowed sailings. Small vessels such as tank landing craft and Siebel ferries often got through because they were small targets for aircraft and impossible ones for submarines, whose torpedoes passed below their hulls. Unloading was usually complete in two hours, but these vessels had small carrying capacities, and there were not enough to keep Axis forces supplied. Even worse, Italian shipyards, already in chaos, received exclusive contracts to build all small vessels, ensuring an inadequate supply. The political necessity of holding some part of North Africa to keep Italy in the war, and Hitler's aversion to giving up ground, meant the Axis fought in Tunisia despite these logistical woes.[74]

By 30 November, continuing C2 and logistical problems limited Allied air effectiveness. When Air Commodore George Lawson joined EAC's Army Support Group Command (ASGC), he found communications

> in a chaotic condition, Advanced AFHQ's [Allied Forces Headquarters'] Command Signal Section working but [could] not get communication with any of the forward aerodromes or AHQ. The Signal personnel here convinced it [was] due to the fact that the receiver stations not told of the existence of Command Post or of the frequencies on which they [were] working. . . . [I am] astonished at . . . the lack of knowledge of the operational setup and of the urgency or drive in getting proper communications established.[75]

The fact that units in the field were unaware of the ASGC's existence or the frequencies on which this hub for air support operated is stunning.

The most serious military-strategic and operational failing of the Allies' disjointed air, ground, and intelligence organizations was the inability to concentrate on Axis troop and supply movements from Sicily and southern Italy to Tunisia. By the time commanders fixed this, the Axis had brought in

150,000 men, heavy equipment, and a large stock of supplies. The employment of fighters in protective umbrellas led to severe maintenance and pilot fatigue problems in air forces that had arrived in theater with poor logistical capabilities and too few aircrews. Serviceability was often less than 40 percent as air units foundered in the mud at forward airfields.[76]

On the Axis side, in one of the great missed opportunities of the campaign, the Luftwaffe committed far too few of its exceptionally capable FW 200 and Ju 88 torpedo-bomber aircraft to the ship attack effort. KG 26 demonstrated what a dedicated antishipping effort could accomplish. From their arrival in May 1942 through October 1943, the unit's aircrews flew 2,139 sorties and launched 1,653 aerial torpedoes (including new noise-homing and circling versions), of which 342 (21.1 percent) hit their targets. They made numerous attacks on Allied MVs supporting Operation Torch. The result was 77 MVs totaling 552,000 tons sunk. Another 165 MVs totaling 1,100,000 tons suffered damage. To highlight the effectiveness of torpedo-bombers, in August 1943, regular bombers flew 1,140 sorties, sinking 33,000 tons of shipping and damaging another 506,000. During the same month, 65 torpedo-bomber sorties sank 88,000 tons of shipping and damaged another 106,000. Both KG 26 and KG 30 did well against the convoys supporting Operation Torch. Had Erich Raeder and Hermann Göring not had their many contretemps about who should control these planes, resulting in a severe shortage of torpedo-bombers at the moment they were most needed here and in the Atlantic, Axis fortunes might have been much better. Luftwaffe raiders did major damage but not enough. Italian SM.79 torpedo-bombers also proved very effective but were too few in number (seven squadrons comprising about thirty-five aircraft). Italy's best airmen took to this mission and did justice to the torpedo arm's motto, *Pauci sed semper immites* ("Few, but always aggressive"), yet the motto itself underscored the IAF's failure to place adequate emphasis on this capability.[77] This major failure in the dedicated ship attack effort spanned the entire Mediterranean campaign and was an important reason for the Axis defeat in the battle for the SLOCs and, because they diverted substantial numbers of torpedo-bombers but not enough, the Battle of the Atlantic. They had too few of these assets and spread them too thin, at the wrong times and places, to be decisive.

As Allied forces consolidated their holds over Morocco and Algeria, "fast" (troop) and "slow" (heavy-equipment) supply convoys arrived every two weeks, with four days between them to allow for offloading without congestion at ports. Allied aircraft kept the Luftwaffe from halting these deliveries. EAC strength reached thirty-one squadrons. The RAF also brought an air observation post (spotting) squadron and ground-defense units of the

RAF Regiment to defend air bases. No. 4 Photoreconnaissance Unit (PRU) provided strategic reconnaissance, and units arrived ahead of schedule.[78]

Stalemate

The Allied effort to take Bizerte and Tunis before 1943 had four phases: contact, failure of the offensive, the German counteroffensive, and Allied plans for a second offensive, which they had to postpone until spring 1943 because of bad weather. Luftwaffe bombers struck back, attacking airfields at Bône and Maison Blanche. From 20 to 21 November, they destroyed thirteen Allied aircraft, damaged nine others, and obliterated the ground-processing station for the PRU. Airborne-intercept (AI) and ground-controlled-intercept (GCI) radar, long in use with the WDAF, were absent. EAC aircraft took a beating at forward bases, with only nine aircraft (of sixteen) on average operational within each squadron. Spare engines and parts were scarce. Distance to the front from Bône, the only all-weather airfield in the forward area, was 120 miles. The closest airfield, Teboura, was 60 miles away but not capable of sustained operations in the mud. Aircraft from Youks les Bains were 140 miles away. Conversely, Axis aircraft at El Aouina were 20 miles from the front.[79]

As the Allies bombed ports and airfields, the Axis destroyed eleven Spitfires on the ground at Souk el Arba on 22 November (again, no radar stations or AI-equipped aircraft in place) and another seven at Bône on 28 November. However, the Allies also destroyed a number of German aircraft on the ground. The first flight of AI-equipped Beaufighters went into action on 27 November, and German losses in night attacks soon increased. By December, Allied aircraft were attacking the smaller ports of Sousse, Sfax, and Gabês. Heavies increased their raids on airfields and ports around Tunis. Railroads also came under attack.[80]

A key development during this period of Allied introspection and reorganization was the rationalization of heavy-bomber forces. All B-17 units were centralized on Algerian airfields, whereas all B-24s went to Tedder's and Brereton's control. This reduced logistical problems and increased operational sortie rates. Arnold supported this reorganization and in fact pushed for a maximum use of heavies in the Mediterranean until good weather and adequate numbers of bombers in the Eighth Air Force allowed it to raid Germany.[81] At the same time, Arnold gave Spaatz command of USAAF assets in Northwest Africa. On 15 November, he had asked Eisenhower to make Spaatz his overall air commander. He followed with a note to Spaatz asking, "How much of our heavy bombardment strength should we retain in England during the bad weather months where the weather prevents operations more than once or twice a month and how much should we send down to the Northern Mediterranean where the weather is normally good and we can operate at least two or three times every week with greater effectiveness?"[82] This represented an evolution in Arnold's thinking.

On 25 December, Doolittle, frustrated with the air effort, sent a pointed note to Spaatz, saying, "Let's stop our wishful thinking, abandon our 100% bitched up organization, stop trying to win the Tunisian War in a day, and through forward planning, sound organization and an appreciation of what air power, when properly utilized, can do, put the God Damn thing on ice."[83] Doolittle recommended specific fixes. His plan to gain air superiority and then go after airfields, ports, and convoys without letup mirrored Tedder's combined-arms approach. Only after these actions were in motion, Doolittle continued, should ground forces advance on Tunis. By January, as a result of heavy-bomber raids brought on by Doolittle's cajoling, Tunisian ports were operating at between 20 and 50 percent of capacity. The raids also did immense damage to airfields.[84]

The wisdom of employing heavy-bombers in the Mediterranean theater during the winter months soon became clear. From January through October 1943, the Twelfth Air Force's heavies launched 13,974 sorties—5.8 sorties per aircraft assigned. The Eighth Air Force managed 17,187 sorties—2.8 per aircraft. Regarding effective sorties, the numbers were 11,675 (83 percent effective) and 11,599 (67 percent effective). With far fewer bombers, the Twelfth Air Force completed more effective sorties in a vital interdiction role that proved central to the defeat of the Axis in Tunisia.[85] This success was one of many reasons Eisenhower and Tedder soon resolved to establish a unified air command within AFHQ.

11

Building a New Air Command and Clearing North Africa
November 1942–May 1943

Allied Strategic Choices

LATE IN 1942, WINSTON CHURCHILL and the Chiefs of Staff (CoS) pushed for a continuing effort in the Mediterranean to force Italy's surrender and continue bleeding the Wehrmacht—and especially the Luftwaffe—while the Allies built up forces for a cross-channel invasion in the summer. Franklin Roosevelt agreed and told the Joint Chiefs of Staff (JCS) to increase forces in the Mediterranean and Pacific while beginning to do so in the British Isles. The JCS response was vehemently negative. Its members believed the Mediterranean had lost its value as a major theater of war with the impending defeat of Axis forces in North Africa. Further, they argued that the continued diversion of forces there and to the Pacific would make a cross-channel invasion impossible in 1943. Only General Henry "Hap" Arnold of the US Army Air Forces (USAAF) saw complementarity in a continued Mediterranean effort, in part because of the option it offered of a two-front bombing offensive against the Reich and Ploesti. George C. Marshall was in favor of Europe because he wanted a cross-channel invasion.[1]

However, given that such an invasion was probably not feasible in 1943 even with a full-scale buildup in England, Roosevelt and Churchill were unwilling to stake everything on that option. Nor were they willing to entertain the absence of any major offensive at all in Europe during 1943. The Grand Alliance finally had the initiative, and the Allies had to do their part in keeping it. Josef Stalin was already furious about what he saw as the Allies' insufficient contributions to and sacrifices for victory. Even though the Battle of the Atlantic would soon go in the Allies' favor, the shipping crisis would continue until US Liberty ships and other new merchant vessels (MVs) appeared in sufficient numbers to bring US troops and equipment to England for the cross-channel invasion. The CoS thus countered JCS proposals with an assertion that US wishful thinking about a cross-channel invasion in 1943 would leave the Allies with a small and unsuccessful effort on that front and no major gains elsewhere.[2]

In January 1943, four months before the end in North Africa, Roosevelt, Churchill, the Combined Chiefs of Staff (CCS), and senior theater commanders met at Casablanca to chart the next steps in their strategy for the defeat of the Axis. At this point in the war, the British were still contributing the majority of forces, and the CoS proposals at Casablanca were clearer and more persuasive than those of the JCS. The British dominated the conference and got almost everything they wanted. It was the last time they would do so. Nonetheless, there were points of agreement, including an overriding focus on winning the Battle of the Atlantic and implementing a combined bomber offensive (CBO) against Germany. They also agreed on an invasion of Sicily in July to knock Italy out of the war and a final postponement of what became Operation Overlord until 1944. Finally, they reconfirmed and made public, largely for Stalin's benefit, the policy of the unconditional surrender of the Axis.[3]

Planning for a Combined Air Organization

To seek a solution for the divided air command and uneven performance of Allied air forces in Tunisia, Arthur Tedder visited Dwight Eisenhower in Algiers in late November 1942. He said, "I could not help being deeply disturbed by what I saw and heard" during a tour of operational units.[4] Communications were rudimentary and relied on the outmoded Vichy phone system. Airfields were utterly inadequate and dangerous given the lack of dispersal sites. New construction was nowhere in evidence. Tedder felt compelled to vent regarding a situation he characterized as "dangerous in the extreme."[5] There was no semblance of combined air operations, with William Welsh and James Doolittle running separate air wars. During a private interview with Eisenhower before he departed, Tedder recommended establishing an operational air headquarters in Algiers to facilitate better coordination and the inculcation into Allied Forces Headquarters' (AFHQ's) airmen of a basic understanding of mobile air operations during a ground advance, airfield security, communications, maintenance, and salvage. Charles Portal felt that the only effective solution was to place all air forces in the Mediterranean theater under one airman's command. Tedder agreed, adding that the "relationship between British and American Air Forces must depend on personalities."[6]

Eisenhower knew the headlong rush into Tunisia had led to heavy aircraft wastage and low serviceability. However, he did not want to change the air command structure in the middle of a campaign and asked Tedder to make the existing system as effective as possible. Tedder declined Eisenhower's offer to make him his air adviser because, as he said succinctly, "advice without responsibility would be of no practical value."[7] The CoS agreed.

From the outset, Tedder told Eisenhower to give Axis resupply efforts his air forces' full attention and to maximize effectiveness by centralizing command and control (C2). After he was back in Cairo, Tedder cabled the CoS, who favored his proposal for a single air commander. Eisenhower demurred and made Carl Spaatz his deputy for air operations, in the hope that a system built on coordination between Tedder and Spaatz could achieve the desired results.[8]

The Middle East Defence Committee proposed a combined air command. At the very least, it noted, all Royal Air Force (RAF) assets in theater must come under Tedder's command. Axis air forces operating in Tunisia were flying out of Sicily and southern Italy, and the heavy-bombers that could reach them were in the Western Desert Air Force (WDAF). Tedder lamented, "It was clear enough that the existing air organisation was almost crazy, with two air forces but no effective command."[9] By the end of December, Eisenhower realized that he must have a single air command. Spaatz was the first commander-designate for such a post, but the British planned to counter with Tedder during the Casablanca Conference.[10]

During his first visit with Eisenhower, Tedder told him frankly during their final meeting how his air command should look. Eisenhower said he would do something about it, and Tedder saw that he meant it. Eisenhower asked Tedder to return and advise him again in December. Continuing airpower failures in Tunisia ultimately convinced the United States to go along with Tedder's proposal for a unified air command, but not until Roosevelt and Churchill provided the policy impetus at Casablanca in January 1943, and not fully until the debacle at Kasserine Pass the following month, by which time the new Mediterranean Air Command (MAC) was just in place.[11]

Tedder saw increased understanding among US commanders regarding command relationships for and employment of aircraft. His objective was to bring two years of RAF and British Army organizational innovation and operational experience to the USAAF and US Army. The fight would be converting ground commanders to his views, or at least getting them to go along by weight of evidence. Accordingly, Tedder queried Portal regarding a second visit to Eisenhower: "If you approve, propose return to Cairo tonight taking U.S. General Craig with me with view to educating him on control of Air Forces, which is quite chaotic here, and to improve coordination between here and M.E., especially Malta. At present Americans are practically ignorant . . . and naturally feel they are doing the bulk of the job. Real co-ordination . . . in this theatre is practically non-existent. . . . The whole affair requires grip and drive."[12]

Before Tedder's second visit to Algiers, Portal told him that the primary difficulty appeared to be US reluctance, for political reasons, to place air

assets under RAF command. He suggested that Tedder address this by sub-
stituting "control" or "direction" for "command." Portal insisted on an
even and fair distribution of senior airmen in command and staff positions
and that they must have recourse to someone from the same nationality up
the chain of command.[13]

A major focus of Tedder's second visit was air support to the land battle. Tedder felt the air forces had done "magnificently" despite their severe
handicaps and given more than anyone had a right to expect. The United
States was beginning to understand the difference between strategic and
tactical uses of airpower and between centralized and decentralized control.
However, airfields remained mired and with too few vehicles to deliver basic
supplies, much less pierced steel planking (PSP) for all-weather runways.
Tedder discussed the situation with Eisenhower, pushing again for a unified
command and telling Portal that virtually anything would be better than
the present arrangement.[14] On 12 December, Tedder hosted a conference in
Algiers during which airmen and soldiers agreed that "tactical" air assets
in Algeria should be under the control of an air support command (ASC),
whereas "strategic" assets such as heavy-bombers would be under the direct
control of AFHQ. This was the best Tedder could do, but Eisenhower began
to give way on the question of centralized command.[15]

By January 1943, Eisenhower was gravely concerned about the numbers
of troops, weapons, and supplies reaching Axis forces. He sought to resolve
the air command issue during the Casablanca Conference on 15 January.
British ideas prevailed, and the CCS named Tedder air commander in chief,
Mediterranean. MAC had three subordinate commanders. These included
Spaatz as commander, Allied Air Forces, Northwest Africa (known as
NAAF); Air Chief Marshal Sholto Douglas as commander, Royal Air Force
Middle East (RAFME); and Keith Park as air officer commanding (AOC)
Malta. Spaatz's command was further subdivided into Strategic, Tactical,
and Coastal Commands. General Harold Alexander became Eisenhower's
deputy in command of ground forces, with Arthur Coningham collocated
with him to deliver air support to Allied armies.[16]

Tedder began planning for the formation of MAC beginning 11 January
1943. "With the reins in one pair of hands," Tedder said,

> air power could be concentrated to the confusion of the enemy either in the
> front line of battle, or along his lines of land and sea communications into his
> base areas, ports of supply and marshaling yards. The air forces were neither
> parceled out to naval and land commanders, nor tied down to particular sec-
> tors. If emergency arose, I should have the knowledge that I was able to con-
> centrate my total force to oppose the enemy on the vital issue.[17]

Tedder felt that the MAC-NAAF organizational structure offered the flexibility that was the essence of air power. Churchill and Roosevelt approved the new air command arrangements officially on 26 January. Tedder's headquarters was collocated with Eisenhower's at Algiers.[18] With the policy work done, airmen began building the organization. Tedder insisted that it be completely integrated, with personnel from both countries holding collective parity within the various directorates and with senior airmen in charge of MAC's three subordinate commands each having a deputy from the other country's air service. This "complete union" of the two air forces worked well. Tedder's immediate priority was the creation of an effective logistical structure. Graham Dawson was soon at work. Tedder assumed command of MAC on 15 February 1943. Eisenhower activated it officially on 17 February.[19]

MAC grew out of the operational successes of the RAF in the Western Desert, but just as importantly as a result of Roosevelt's and Churchill's decision to exploit the military situation in the Mediterranean by ejecting the Axis from North Africa, driving Italy out of the war, and making a bid for a rapid advance up Europe's "soft underbelly." The senior leadership agreed that there could be no invasion of France in 1943 but that the Allies must pin German forces away from the Eastern Front. The attack on Sicily and invasion of the Italian peninsula would thus follow. Accordingly, Tedder set the following objectives for MAC:

1. Neutralize remaining enemy air forces in North Africa and Sicily, giving the Allies air supremacy while also forcing the Germans to move additional aircraft from the Eastern Front during the critical summer campaigning season.

2. Give maximum support to ground troops in Tunisia.

3. Engage in strategic bombing of enemy ships, port installations, and air bases to disrupt sea, land, and air communications within North Africa, Sicily, and southern Italy. Enemy naval forces in a position to resist the invasion of Sicily would be found and destroyed.

4. Protect Allied ships, ports, and base areas with reconnaissance, fighter, and ASW [antisubmarine warfare] assets.

5. Establish a central organization for supply, maintenance, and all other MAC logistical requirements.

6. Train and put in place the RAF and USAAF ground personnel required to ensure maximum serviceability rates for aircraft.

Spaatz took command of NAAF, the operational element of MAC and key executor of these six broad air objectives.[20]

To carry out CCS direction from Casablanca regarding Operation

Pointblank and the CBO, Spaatz placed Doolittle in command of the Northwest African Strategic Air Force (NASAF), comprising the Twelfth Air Force's heavy- and medium-bombers. He would oversee raids on high-value targets in Tunisia, Sicily, and western Italy. Coningham took command of the WDAF, Twelfth ASC, and No. 242 Group, which combined to form Northwest African Tactical Air Forces (NATAF). He collocated his headquarters with Alexander's in Constantine. NATAF assets included MAC's fighters (with the exception of escorts assigned to NASAF) and light-bombers. Hugh Lloyd took charge of Northwest African Coastal Air Force (NACAF), with orders to protect friendly shipping and sink the enemy's. He established his headquarters in Algiers alongside Admiral Andrew Cunningham's, with whom he shared an operations room. Lloyd briefed the AFHQ chief of staff meetings on naval targets, and Spaatz had a naval operations officer assigned to him for provision of operations and intelligence pertaining to enemy shipping. Colonel Elliott Roosevelt commanded the Northwest African Photographic Reconnaissance Wing (NAPRW). Transport of airborne troops and supplies fell to the Northwest African Troop Carrier Command and its fleet of US C-47s. A new Northwest African Training Command trained RAF and USAAF aircrews, and the Northwest African Air Service Command handled USAAF supplies. The British established a parallel but heavily integrated supply organization. The arrangement proved effective. RAF Malta, under Park, reported directly to Tedder and was responsible for operations against shipping and airfields in and around Sicily and for air defense of Allied ships within 100 miles of the island. Park also had orders to prepare Malta for the influx of aircraft and airmen prior to the invasion of Sicily. Douglas commanded RAFME, but without authority over the WDAF or RAF Malta. His No. 205 Group bombers coordinated with NASAF, attacking shipping and ports. Douglas was also responsible for air operations in the eastern Mediterranean.[21]

Dawson tore into logistical problems, reaching immediate agreement with the local Air France leadership to use its factory at Maison Blanche for overhauling aircraft and engines. He established a major depot at Algiers, an aircraft assembly point at Casablanca, and a number of air stores parks (ASPs) and repair-and-salvage units (RSUs). His primary salvage unit had eight highly mobile sections that scoured the countryside for damaged aircraft. Before Dawson arrived, there was almost no salvage capability. Hundreds of downed but salvageable aircraft sat until Dawson's crews reached them in March 1943. Dawson also pushed maintenance officers to fix the many aircraft sitting idle at airfields for want of a spare part.[22] Officers on the receiving end of his unannounced visits to depots and other facilities referred to themselves as being "Dawsonized."[23]

During meetings to establish MAC, Tedder insisted that Spaatz take on a senior RAF officer, Air Commodore George Beamish, as his chief administrator, in effect Spaatz's senior air staff officer (SASO). Tedder complimented the RAF chief signal officer's efforts. William Mann created capable signals facilities where none had existed. Tedder ended a signal to Portal with a guardedly optimistic passage: "The whole situation both from the operational and organisation [*sic*] point of view is quite incredibly untidy and it will undoubtedly take a long time to get it tidy. . . . One does feel however, that as far as the air is concerned, we have got the real goodwill of the Americans and we are determined to make a job of it."[24]

At the dinner where Tedder and his senior commanders sealed agreements regarding command arrangements, he spoke briefly about what a combined Anglo-American air effort would mean. "You know," he said,

> we British are very proud of our Air Force. We think it is the very best in the world, and that it saved England and the world—all of us. We have our own way of doing things and I suppose we feel we are justified in keeping these ways. But we also know that you Americans are equally proud of your splendid Air Force, of your magnificent aeroplanes and equipment, and that you feel justified in doing things your way—as well you are. However, it will be the fusion of us, the British, with you, the Americans, that is going to make the very best Air Force in the world. And now, gentlemen, this is the last time I shall ever speak of "us," the British, and "you," the Americans. From now on it is "we" together who will function as Allies, even better than either of us alone.[25]

Allied Challenges Continue

These sentiments notwithstanding, much remained to do before the new air organization was mature. The Luftwaffe continued to inflict serious losses at forward airfields. Coningham's address to Allied senior officers in Tripoli on 16 February regarding the proper use of airpower began to correct the causative elements of this problem. His comments, along with those from a subsequent speech, formed the basis of an Air Ministry pamphlet on the proper employment of airpower. There were four broad principles: (1) The army and air commander must act together in accordance with a combined plan; (2) the fighter governed the front and thus had to be under an airman's centralized control to exploit its flexibility—no more "penny-packet" defensive umbrellas; (3) air superiority must be constant and overwhelming to give ground troops freedom of action and hamper that of the enemy; and (4) the battlefield was to be isolated from resupply and reinforcements to the greatest possible degree by destroying access—roads, bridges, railways, canals, and ports.[26]

Coningham's Tripoli speech was magnified in its effect by Air Marshal Trafford Leigh-Mallory's report to Portal, after a March–April 1943 inspection visit, that Coningham's organization was a superb model to follow in setting up a tactical air force for a cross-channel attack. The Air Ministry revamped Army Cooperation Command into Second Tactical Air Force (TAF; Coningham's was renamed the First TAF). Meanwhile, Brigadier General Laurence Kuter, Coningham's deputy, carried *Field Manual 31-35* back to Washington, DC, with him and used it to indict the previous employment of US air power. Air Corps Tactical School (ACTS) faculty and students weighed in, leading to publication of *Field Manual 100-20* on 21 July 1943. From this point forward, whether or not as a result of *Field Manual 100-20*, joint operations, already improved by the experience in Tunisia, became even stronger.[27]

Allied Victory at Malta

Malta was a key area of emphasis after MAC became active. The sea route from Sicily to Tunisia was short, protected by mines, and difficult to attack except by air. Malta was the perfect point from which to strike but needed additional aviation fuel and other supplies. The four-MV Operation Stonage Convoy, covered by ships and aircraft under control of an RAF–Royal Navy Combined Operations Room at No. 201 Group, had arrived unscathed on 20 November 1942. The group had three general reconnaissance squadrons, three for ASW duties, four torpedo-bomber squadrons, and one Beaufighter squadron to provide protection from air attack until the convoy was within range of Malta. The WDAF also covered much of the route, to a point forty miles west of Benghazi, at which point the Malta-based fighters would take over. US B-24s at Gambut and twelve squadrons of aircraft on Malta (including five of Spitfires) added their weight. Improvements in tactical communications, based on very high-frequency (VHF) radios, yielded unprecedented effectiveness. Another convoy of five MVs and one tanker arrived safely on 5 December. The 56,000 tons of cargo these ten ships brought to Malta ended the siege. Attacks on Axis shipping increased rapidly after this.[28]

From 26 October to 2 November, aircraft sank six MVs and tankers in seven days, driving Axis fuel deliveries below 500 tons. Heavy minefields between Sicily and North Africa made it difficult for surface vessels and submarines to do much damage, so aircraft filled the gap. The RAF sank numerous small vessels at Tobruk and Benghazi in November, and the Fleet Air Arm (FAA) sank a large MV the same month. The USAAF also sank four ships at these ports. Malta intensified its offensive operations as the number of Wellingtons there increased to thirty-four. By January 1943, a squadron

Senior air commanders in Tunisia, 1943. From left to right: Arthur Coningham, Carl Spaatz, Arthur Tedder, and Brigadier General Laurence Kuter. (Imperial War Museum Image No. CNA 408)

of Mosquito 2s, another of Albacores, and one of Beaufighters had arrived. Two new Wellington squadrons arrived in Algeria along with more USAAF heavy-bombers. Six cruisers and an equal number of destroyers reconstituted Force K and Force Q, bringing together a potent air-sea striking force on Malta for the first time since January 1942. These ships quickly sank a convoy of three MVs and two escorting warships on 1 and 2 December. The next night, they finished off the remnants of another convoy already hit by FAA planes. From 20 to 21 December, Wellingtons from Malta guided two destroyers to an enemy MV, which they sank. After this, the Axis either ran night convoys behind heavy minefields or large daylight convoys with heavy air cover.[29] Neither solution worked.

Continuing WDAF Innovations

In Libya, the WDAF became expert at moving assets via air transport, facilitating rapid operations from forward bases. USAAF C-47s delivered fuel, flying in 130,000 gallons for the WDAF in advance of the battle at El

Agheila. The Marble Arch landing ground received another 152,500 gallons in December 1942 and January 1943. Ground crews flew in on Hudsons escorted by the fighters preparing to use the landing strips. Water, spares, and equipment for ASPs also arrived by air. Forward salvage units (FSUs) supported advanced squadrons, stayed in close touch with those coming forward, and had salvaged parts ready when aircraft needed them. These specialized units were highly mobile, allowing ASPs to move with two hours' notice and be ready to issue spare parts and other supplies within thirty minutes of reaching a new site. Repair shops in the Nile Delta turned out nearly 1,600 refurbished planes and 2,400 engines from November 1942 through January 1943, reducing WDAF dependence on new aircraft.[30]

Coningham developed air plans for supporting the attack on the El Agheila position and larger air support requirements for covering the advance to Tripoli. Strike aircraft hounded the Luftwaffe's remaining 85 operational aircraft. From 25 to 26 November and from 12 to 13 December they also sank two large MVs at Tripoli along with many small vessels. The Allies captured another 408 Luftwaffe and 111 Italian Air Force (IAF) planes as they occupied abandoned airfields.[31] Erwin Rommel blamed the continuing lack of air support and his army's fuel shortages on the German occupation of Tunisia, which now required the majority of airlift assets in theater. Rommel and Hans Seidemann received a maximum of 200 tons of fuel per day by airlift, slowing the retreat and allowing the RAF to destroy many vehicles.[32]

Rommel withdrew to Buerat. Bernard Law Montgomery delayed the Eighth Army's advance until units had refitted and the RAF had forward airfields. The WDAF occupied Marble Arch on 18 December just two hours after royal engineers cleared it of mines. A total of sixty-one transport aircraft from the recently formed No. 216 Group, escorted by fighters, delivered 160 tons of cargo and key personnel in seventy-eight sorties. The first fighter-bomber sortie took off at 1:30 p.m. the same day. WDAF aircrews flew sixty sorties against vehicle convoys whose drivers did not expect them to be in range. The Luftwaffe was caught flat-footed, having moved its remaining fighters too far back to provide cover. Radar sites and wireless communications capabilities also went in on 18 December. As the Kittyhawks came forward with external fuel tanks, they took up defensive patrols over Marble Arch. After their fuel was low, they landed, traded fuel tanks for munitions, and joined previously arrived aircraft in convoy raids. This operation became a model for future moves.[33]

Occupying advanced landing grounds quickly was a principal part of the WDAF's growing effectiveness. Aerial reconnaissance of three good sites in addition to Marble Arch resulted in RAF airfield detachments accompanying soldiers there so that air transport squadrons and fighters could arrive immediately after the fields were in friendly hands. The WDAF had three fighter wings, two day-bomber wings, and a reconnaissance wing dedicated entirely to direct support of the Eighth Army. These air and ground components needed enough supplies to reach Tripoli in a single bound. Montgomery and Coningham were determined to take Tripoli and hold it, making leapfrogging more important than ever. The ensuing rapid advance sometimes encompassed two moves in a twenty-four-hour period.[34]

From 12 to 14 January, the WDAF attacked enemy airfields, ground targets, and vehicles around Tripoli. It engaged in direct support from 15 to 19 January as the Eighth Army advanced. Day-bombers switched to night attacks as more Axis units and supply columns moved after dark. On 17 January, photoreconnaissance unit (PRU) Spitfires reported nearly 200 Axis aircraft on the ground at Castel Benito in Tripoli. Bombers attacked that night, followed on 18 and 19 January by two more raids. Fighters and fighter-bombers from new forward airfields joined on 18 January. Air reconnaissance confirmed on 19 January that the handful of surviving enemy aircraft had fled. The Eighth Army occupied Tripoli on 23 January.[35] Advancing troops captured another 114 German and 327 Italian aircraft, bringing the total since El Alamein to nearly 1,000 airplanes.[36]

As WDAF assets moved to Tripoli, they worked with the Royal Navy and air units in Tunisia to attack Axis supply lines. PRU aircraft provided intelligence on ship movements. Results were soon evident. During December, 70 percent of ships made it to Tunisian ports, but sinkings of the last few large MVs resulted in a loss of more than 50 percent of the supplies and vehicles scheduled for delivery (6,913 tons of the 12,784 tons shipped). Aircraft, surface vessels, and submarines took an even more serious toll in January, sinking twenty-four MVs totaling 17,450 tons. They also sank twelve smaller vessels at sea and many more in port. This represented well over half of total sailings.[37]

The Germans struck back, but their bomber force was worn out, their best crews dead, and their ability to strike distant targets much more limited. And yet their airmen kept on with their customary courage in the face of losing odds. The Luftwaffe's tasks were to hinder Allied forces operating in North Africa, maintain control of the Sicilian narrows, and counter any attack on the German sphere of influence in southern Europe. These tasks were so massive as to preclude success. The additional requirement to make "constant attacks on enemy air bases, roads, railways, transports, flak emplacements, concentrations of armour and vehicles, bivouacs, front lines, and troops on the march" further underscores the lack of focus. The Germans continued attacking nearly every target category, sometimes with good results, but the effects and aggregate effectiveness of such dispersion were minimal. By this time, the Luftwaffe no longer posed a major threat to Allied operations.[38]

In retrospect, and within the bounds of what constituted acceptable criticism of the regime, Luftwaffe staff officers made some useful statements about how things could have gone in Tunisia and throughout the Mediterranean campaign. In a telling statement, they said, "All the various types of air combat had a decisive influence on the outcome of the land battles. The final decision was fought out on the ground, but the effect of the air war on supplies and morale had already determined the outcome of the battle."[39] They underscored the overriding importance of air superiority. Additionally, they remarked that Allied convoys made the Luftwaffe's task difficult by sailing just outside of Ju 88 range during daylight hours and heading into Algerian ports after dark. By this time, there were few German bomber crews with night-flying skills, and these came under night-fighter attack. German fighters did receive orders to gain air supremacy over the Tunis bridgehead along with its airfields and port installations, and they did so. Their second task was to escort bombers attacking Allied supply points while defending Axis convoys. Luftwaffe attacks on air bases were relatively rare although sometimes highly effective. As Allied air forces moved to closer bases and then became bogged down by the winter rains, the Axis missed an opportunity to inflict severe losses on aircraft as well as ground and aircrews. *Operativer Luftkrieg* ("operational air warfare") had been corrupted in the Mediterranean theater, with an inverted set of priorities driving the Luftwaffe's operations from beginning to end.[40]

In a key passage, the Luftwaffe historians emphasized that fighters could not be used to their full capacity because so many of them were engaged in convoy escort, airfield defense, and port security operations that "they could play no part in the battle for air supremacy."[41] They continued this criticism, noting, "The fighters could only accomplish their task by taking the offensive."[42] Finally, they said, in contrast to the Luftwaffe, Allied air forces concentrated their strength, went after airfields, gained air superiority, and then focused on Axis supply efforts.[43]

The Allies began winning the battle of the buildup in North Africa. Two fast convoys per month arrived from the United States and another two slower convoys from the United Kingdom. By early January, hundreds of miles of landline allowed the Allies to divide the coastal areas into fighter sectors, each with its own early warning radar. German raids on ships in Algerian and Tunisian ports thus led to heavy losses as radar and night-fighters took their toll. The most potent remaining Axis weapon—U-boats—had little freedom of action because of ASW patrols and almost no air reconnaissance to help them find ships. On 30 January, Admiral Erich Raeder resigned, frustrated, among other things, by Adolf Hitler's failures to grasp the importance of the Mediterranean.[44]

The Ground War Continues

A limited German offensive from 18 to 25 January to optimize ground positions—Operation Eilbote—achieved its objectives but at a high cost imposed largely by aircraft. German reports note that air attacks were frequent and caused heavy vehicle and supply losses. The Luftwaffe went after troops, tanks, artillery, and antiaircraft artillery (AAA) rather than logistical targets, with limited effect. The German Army complained about "inadequate" air support.[45]

Eisenhower made army protection of airfields at Souk el Khemis, Tebessa, and Thelepte the highest priority to ensure ground forces received continuous and reliable air support. Eisenhower understood that questions of "support" worked both ways. His writings throughout the war indicate that he much preferred the concept of coordination to that of support. As a result of Eisenhower's order, following Operation Eilbote and leading up to the Kasserine offensive, the Allies flew minimum sorties as the force rested and refit. The arrival of the first Spitfire IXs in January was welcome given their ability to deal with the FW 190. The US experience was more frustrating as USAAF fighters engaged in heavy ground strafing at the behest of ground commanders but without gaining air superiority first, giving the Germans opportunities to bounce escorting fighters or the strafing aircraft while exacting a toll with flak. Albert Kesselring boasted that German airmen "still reigned supreme," but his bravado rang hollow as Allied heavy-bombers attacked ports and airfields with unprecedented ferocity from mid-January forward.[46]

Kasserine Pass and the Mareth Line

General Walter Warlimont, Oberkommando der Wehrmacht's (OKW's) senior operations officer, visited Tunisia on 8 February to assess the situation. What he found was not comforting. Battalions covered fronts of six to seven miles each. The supply situation was perilous. Requirements in Tunisia were 90,000 tons per month. The senior supply officer was a colonel—far too junior for an effort of this magnitude. The Italians could deliver at most 80,000 tons per month, making allowance for a loss of 20,000 tons en route. The 400 Italian and Siebel ferries were far too few in number to carry this burden. Warlimont said the logistical situation would soon cause the entire position to collapse "like a house of cards."[47]

The only alternative to a rapid collapse, Warlimont continued, was to shorten the lines from 390 miles to 93 miles, giving up much of Tunisia, including several all-weather airfields, and allowing Allied troops to link up and form a single line. Warlimont warned that trying to hold this line and repel an Allied offensive would require 140,000 tons of supplies per month—much more than the Italians could deliver. He ended with a warning: "Only by preparatory planning of our future course can we avoid serious repercussions . . . and only thus will it be possible at the same time to make preparations for the defence of Southern Italy."[48] His veiled request that Axis forces evacuate Tunisia fell on deaf ears.

As the logistical calamity worsened, air transport units comprising 440 planes assisted the faltering effort. During January, they delivered 15,816 troops, 4,742 tons of weapons and ammunition, 162 tons of fuel, and 8.5 tons of food. These numbers were a drop in the bucket given the 90,000-ton monthly requirement. Still, the Italians operated convoys on a fixed schedule that the Allies had long since discerned. A fast convoy sailed every three days, and a slow convoy every four or five. At this point, the loss of one large ship created a logistical crisis. The Italian Merchant Marine was down to a capacity of 84,620 gross tons. Only twenty-six large MVs and tankers remained.[49]

On 6 February, to assist Rommel's pending offensive, the Germans reorganized their air forces into Fliegerkorps Tunisien in order to delineate and better control ground support and bombing operations. The increased operational flexibility made units in Tunisia available for immediate response. This worked well, as did the administrative and logistical improvements brought to bear by the development of Air District Tunis, which had three mutually supporting ground organizations able to move quickly between airfields and provide the required repair and other expertise. No such organization had existed in the Western Desert. Even with the much greater distances and speed of operations, it would have been extremely helpful. Effective airfield dispersal efforts provided maximum protection. Wireless radio communications were the norm in signals units, meaning lower maintenance and manpower requirements, but greater vulnerability to intercepts. Still, the redundancy they introduced made air operations more agile and effective. Freya radar identified Allied raids 160 miles out, and Würzburg radar delivered the altitude, azimuth, and airspeed data for interception.[50] These assets would also have played a key role had they been used in the Western Desert before 1942.

On the air transport front, Germans and Italians escorted their own planes without any coordination. Still, the air forces' supply situation was satisfactory in February 1943. Serviceability rates were between 50 and 60 percent. The improved organization, in combination with a small operational area that allowed for maximum agility, made a significant contribution to the tactical successes in February. The unfortunate news was that the radio monitoring units, in coordination with intelligence assessment, determined that there were 2,769 Allied aircraft in theater, compared with 837 Luftwaffe planes and perhaps another 100 Italian ones. Most Axis sorties supported the February offensives and attacks on ports and convoys, but too few were left to support the Mareth line in the south. German airmen would have been particularly downcast to know that on 1 January 1943, the entire Luftwaffe had 4,911 aircraft—just 772 more than when Germany invaded Poland. Their days of numerical and qualitative advantage in the Mediterranean theater had ended in summer 1942 and never returned.[51]

The Kasserine Offensive

The German attack at Kasserine Pass began on 14 February and initially went well. For the last time, Stukas did serious damage to troops in North Africa. The link between German tactical reconnaissance and air attacks was vital, but increasing Allied air superiority undermined this cooperative effort. The Germans overran Thelepte airfield on 18 February, driving US aircraft back and forcing US ground crews to destroy thirty-four unserviceable planes. Night attacks increased as Allied airpower became more dominant.[52] Rommel was stunned by the relative levels of air and logistical support US troops received. "The Americans," he said, "were fantastically well equipped and we had a lot to learn from them organizationally. One particularly striking feature was the standardisation of their vehicles and spare parts."[53]

Coningham took command of NATAF on 17 February—just as Rommel began his offensive—and made immediate changes. He was stunned to learn that Twelfth ASC light-bombers had not engaged in any actions and fighters were flying defensive umbrellas. Coningham issued immediate orders for all NATAF units to use aircraft offensively. In addition, he steered aircraft away from attacks on armor to strikes on troop and supply convoys. Most of Doolittle's heavy-bombers also came under Coningham's command during this emergency, striking convoys and troop concentrations. These heavy raids, a feud between Rommel and Hans-Jürgen von Arnim in which the latter essentially withheld support, and a deteriorating supply situation forced Rommel to call off the attack. From 23 to 24 February, good weather permitted all-out air attacks on retreating German units.[54] Rommel noted, "The bad weather now ended and from midday [23 February] onwards we were subjected to hammer-blow air attacks by the U.S. air force in the

Feriana-Kasserine area, of a weight and concentration hardly surpassed by those we suffered at Alamein."[55] Despite tactical defeats at the start of the battle, air support helped to turn things around, giving Allied troops time to regroup and counterattack. Tedder was pleased with Coningham's performance but furious about what had come before. He decried the penny packeting of fighters.[56] The remedy, he continued, was proper organization and control, which showed immediate results as Rommel's forces stalled and then retreated.[57]

After Kasserine Pass, as the Allies regained the initiative, Cannon published new objectives for the Twelfth ASC. They were "(1) to defeat the enemy by fighter sweeps and escort for intensive bombardment of airdromes, (2) to provide visual and photo reconnaissance along the II Corps front, and (3) to give direct support to ground forces by attacks on enemy supply columns, vehicle concentrations and enemy armor units."[58] With two new heavy-bomber groups and additional P-38 escorts in place, raids on airfields and ports reached a new level of intensity in April, driving the Axis supply situation to collapse and forcing the Luftwaffe to evacuate virtually all of its aircraft and ground personnel.

After Rommel's offensive failed, von Arnim began his own rather than having helped Rommel as ordered. From 1 to 5 March, First Army, supported by more than 1,000 sorties, stopped von Arnim's troops in their tracks, inflicting heavy casualties. Rommel, meanwhile, flung his already weakened forces against Montgomery near the Mareth line in an effort to disorganize and delay his planned attack there. Ultra signals intelligence had identified Rommel's line of advance and planned points of attack, and Y Service units tracked individual formations, allowing air and army assets to inflict severe losses. The Allies destroyed 52 of Rommel's 145 tanks and damaged virtually all the rest. This tripartite disaster—Kasserine, von Arnim's foiled attack, and Rommel's failure before the Mareth line, marked a rapid Axis decline in Tunisia, as did the increasingly deadly attacks on ports, airfields, and seaborne convoys.[59]

For the major battle that soon followed at the Mareth line, the Allies had approximately 780 aircraft and the Axis 200. However, serviceability differences gave the British a 6-to-1 advantage, along with a 5-to-1 advantage in tanks. Diversionary attacks began the night of 16–17 March, with the main force attacking on 19 March. For a week before the ground assault, the WDAF attacked troop concentrations and gun positions. German aircraft were pinned to the ground on 17 March, forcing bombers from Sicily to provide what support they could. Heavy raids hit Axis positions nine times on 19 and 20 March. Attacks continued on 21 and 22 March, and air units in northern Tunisia kept the Axis from moving aircraft south in support.

The upcoming ground battle, on 26 March, would depend heavily on air support.[60]

Before the battle began, Montgomery told commanders that he wanted a rapid advance and that heavy air support would facilitate an afternoon attack with the sun in the Germans' eyes.[61] The air plan called for an unprecedented concentration of aircraft in space and time to maximize firepower over a narrow front and prevent an effective enemy defense at the breakthrough point. The air phase, which the new WDAF commander, Air Vice-Marshal Harry Broadhurst, planned in great detail with his staff, began with intensive bomber attacks on 24 and 25 March. The objectives were to destroy motor transport and telephone lines, deprive the enemy of sleep, and create maximum disruption and dislocation. Airstrikes continued on 26 March, when three large bomber formations attacked from low level, dropping their bombs in a pattern to maximize the effects already created by earlier raids. Then, as the artillery barrage began and troops advanced, fighter-bombers arrived in relays every fifteen minutes, flying ahead of the spearhead New Zealand Division to drop bombs on known enemy positions and strafe targets of opportunity. Spitfires patrolled overhead while NATAF fighters, fighter-bombers, and day-bombers made all-out supporting attacks on more distant Axis airfields to destroy or ground remaining aircraft.[62]

By this time, RAF forward air controllers were expert at guiding aircraft to targets. They established clear landmarks, perfected target marking with red and blue smoke, and learned to watch for the orange smoke indicating friendly troops' most advanced positions along with smoke rounds fired in particular patterns from artillery to guide their strikes on enemy ground forces. Air attacks did heavy damage, caused communications outages, and created an almost complete inability to move anything toward the front.[63]

Air attacks reached a crescendo along the northwestern portion of the line, where the New Zealand Division carried out a flanking movement. The Germans managed to halt this advance the next day and begin a withdrawal to positions on the Salt Sea, avoiding encirclement. However, they used their last ground reserves, the weather was improving rapidly, Allied aircraft had moved to forward airfields in large numbers, and the supply situation was critical.[64] The last German strongpoint had fallen.

The End in North Africa

With Rommel's army retreating, Coningham ordered a continuous air offensive to prepare for and support the final offensives. The Luftwaffe would be attacked everywhere and destroyed. Numerous small fighter formations

tied closely to radio direction-finding (RDF) and ground-controlled intercept (GCI) capabilities, rather than large fighter sweeps, formed the core of
this effort. Night-bombing received high priority because crews could now
often find, illuminate, and attack vehicle convoys.[65]

By the time the US offensive against Gafsa in southern Tunisia began on
March 16, and Montgomery attacked the Mareth line four days later, the
air forces gained immediate air superiority by attacking enemy air bases every
fifteen minutes during the run-up to the offensive and its opening stages, and
by shooting down fighters that tried to interfere with bombing raids. Coningham directed the WDAF to support the Eighth Army directly, and No. 242
Group and the USAAF Twelfth ASC went after enemy airfields and aircraft.
Hurricane night-bombers strafed enemy positions constantly, and their raids
deprived enemy troops of sleep.[66]

From 30 to 31 March, the German First Army gave the following supply
returns: rations three days with troops, seven days in dumps; petroleum,
oil, and lubricants (POL) 0.6 consumption units with troops (about forty
miles of travel in good terrain); ammunition 0.8 issues with troops, 1.5 issues in dumps. For the Fifth Army, rations were at six days with troops, no
reserves; POL 0.3 consumption units (about twenty miles of travel in good
terrain); and ammunition at 1.2 issues with troops, 1.6 in dumps. Even after
a few small tankers reached Tunisia, the First Army still had only enough
fuel to travel sixty miles. Despite von Arnim's repeated requests for guidance
should adequate supplies fail to reach Axis forces, Alfred Jodl refused to
give orders while the Italians insisted, unrealistically, that they could deliver
25,000 tons of supplies in the next week.[67] However, by 5 March, USAAF
B-25 ship attack aircraft with radar and heavy armament had arrived, making the already perilous crossing for Axis convoys even more so, as these
specially equipped Mitchells joined air-to-surface surveillance (ASV) radar-
equipped Wellingtons, Beauforts, and Beaufighters.[68]

Allied air raids left the Luftwaffe combat ineffective. In addition, air
reconnaissance was so intensive that it became impossible even for small
vessels to escape attack. Yet with the handwriting on the wall, the Axis
continued pouring assets into Tunisia in March, disembarking 30,000 men,
35,380 tons of materiel, 1,861 tons of fuel, and 1,114 motor vehicles. Losses
to raids amounted to 13.5 percent of the materiel dispatched, 26.9 percent
of the fuel, and 27.9 percent of the vehicles. These totals, sustainable earlier,
were now crippling given the limited carrying capacity of remaining vessels. The only other option was to increase supply deliveries with airlifts.
In April, with defeat an absolute certainty, German bombers flew in fuel
as air transports delivered additional supplies.[69] By April, 200–300 Allied
heavy-bombers at a time, with fighter escort, bombed ports and airfields as

far north as Naples. German Freya radar systems could detect these raids with ample warning, allowing planes to disperse, but damage to airfields was heavy, and their serviceability declined.[70]

When the Eighth Army attacked Rommel's position in the salt marsh at Wadi Akarit on 5 and 6 April, precise raids on gun positions facilitated a rapid breakthrough and another major victory. During the final series of Axis retreats, air attacks took an immense toll on trucks, making enemy ground formations increasingly static and ineffective. Fritz Bayerlein, who would feel the effects of Allied air superiority again in northwestern Europe as a division commander, recorded heavy troop and vehicle losses as good weather allowed hundreds of Allied aircraft a day to make devastating low-level gun and bomb passes. Between NASAF and Middle East Command, the Allies had thirty-nine squadrons of heavy-bombers by April. Attacks on airfields in Sicily and Sardinia were constant and often kept Axis planes from getting airborne. There was little air support for the rapidly failing ground forces.[71]

German forces regrouped on the Bon Peninsula. The Luftwaffe was utterly on the defensive, and air transport missions succeeded only with careful planning, defensive measures, and rapid unloading. Engine overhauls and other major repairs were now done in Sicily. There were no significant maintenance capabilities left in Tunisia. Surviving dive-bombers departed in April as remaining fighters escorted air transport and seaborne convoys. A heavy-bomber raid on Grosseto devastated the German torpedo-bomber base and school, destroyed multiple aircraft, and put further operations on hold for several months—another major blow to already insufficient anti-shipping operations.[72]

As Axis forces retreated into a pocket stretching from Bizerte in the north to Enfidaville in the south, round-the-clock air attacks inflicted heavy losses. As the Germans looked desperately to air transport planes for a way out of the logistical conundrum, Y Service units were already tracking routes, flight times, and stopping points for the 250 sorties per day. The aerial massacre that followed, Operation Flax, destroyed 432 aircraft (over 400 transports) at the cost of 35 fighters. Along with the debacle at Stalingrad, this broke the back of the German air transport force for the rest of the war.[73]

Neville Duke recalled the spellbinding if horrific scene during the Palm Sunday Massacre on 18 April, when his squadron flew cover for forty-eight USAAF Kittyhawks attacking a huge formation of Ju 52s:

> It was one of the most amazing sights I have seen in this war; 48 Kittys attacking this enormous gaggle of transports right on the deck. A 52 would start to glow then become a ball of fire and, still a mass of flame, would continue a

The end of a once-great airlift capability. These Ju 52s at El Aouiana Airfield in Tunisia were destroyed by heavy-bomber raids from January to May 1943. Along with the aerial massacre during Operation Flax in April and the Stalingrad disaster before that, bombing raids effectively destroyed what had been the greatest air-transport service in the world. (Imperial War Museum Image No. CNA 710)

> little further across the sea trying to make land and then crash or blow up. This happened a dozen times, in the last light of dusk—these burning aircraft showing up for miles. Some make the coast and crash-land but not many. Poor devils. 73 Squadron went out at night and strafed some on the beach.[74]

After this final supply line dried up, only 20,000 tons arrived before the Axis surrender. Air transport, once a vital Luftwaffe asset, had become an Allied one as the German Fleet disappeared and Allied C-47s and other aircraft continued arriving in droves. By May, hardly any small vessels got through. The Luftwaffe was defeated, and German troops were running out of fuel and ammunition as the final Allied offensives began.[75]

After 6 May, when von Arnim relocated his headquarters for the last time using a single remaining jerry can of fuel, Allied troops and air forces engaged in a running fight with very few preplanned targets. On 6 May, despite launching 1,000 sorties against enemy lines before 9:00 a.m., senior airmen could not get the enemy to fight. One pilot reported, "We have nothing to bomb. The enemy have [*sic*] dragged all their remaining aircraft off the airfields and hidden them under the trees. There are no enemy aircraft in the sky. There is no movement of vehicles along the road. There is no sign of German activity at the front. There is practically nothing we can see to hit, nothing to strafe."[76] The last Axis aircraft fled for Sicily and Pantelleria on 9 May. In Johannes Steinhoff's JG 77, mechanics fled by wedging themselves into the fuselage compartment behind the pilot's seat. A number of German paratroopers, among the best line infantry units once Hitler forbade further large-scale airborne operations after Crete, escaped by tying themselves to the undercarriages of Ju 52 transports. They might have noted the irony here given that they were poised to invade Malta, using more conventional air delivery methods, less than a year earlier. Attacks on the island of Pantelleria intensified during this period. Heavy-bombers continued major raids on key ports and rail facilities on Sicily and the Italian mainland. On 13 May 1943 at 1:12 p.m., the last German radio station in Africa closed down. At 1:16 p.m. the same day, General Alexander cabled Churchill: "Sir, it is my duty to report that the Tunisian campaign is over. All enemy resistance has ceased. We are masters of the North African shores."[77] The three-year campaign in North Africa was finally at an end.

The Reckoning

The period of contest in the air was over. The period of exploitation had already begun. Getting to this point required a superior Allied employment of air assets. The impact of airpower on the battlefield was neither uniform nor linear in its development, but its processes and procedures, in conjunction with the army's and navy's, became increasingly effective. Although the brutal slogging match in Italy, with its ideal defensive terrain and bad winter weather, would make clear the contextual limits of the ability of airpower to help deliver results in the combined-arms arena, it also made clear that even in the most difficult operational environments, air supremacy conferred substantial advantages. It forced enemy troops onto the defensive, robbing them of the ability to engage in mobile warfare, keeping them short of supplies, and creating friction throughout the operational and logistical arenas.

The Axis lost between 238,000 and 250,000 men in Tunisia. Allied records note that 101,784 Germans, 89,442 Italians, and 47,017 unspecified—a total of 238,243—had been captured unwounded. There is no tally of the number of wounded troops taken prisoner, but there were substantial numbers. The Eighteenth Army Group did its own count of unwounded prisoners and arrived at 157,000 Germans and 86,700 Italians—a total of 244,500. From 8 November 1942 to 7 May 1943, the Axis lost 1,696 aircraft in combat and Allied troops captured another 633 on the ground—2,329 aircraft to an Allied loss of 657 planes. The Tunisian disaster was as great as the one at Stalingrad and another hinge point in the war.[78]

Total Axis air losses in the Mediterranean and Middle East campaign were disastrous for a theater in which the Germans had simply planned to assist their Italian allies but could and should have seized the strategic initiative. Aircraft losses from June 1940 to May 1943 were 5,211 destroyed outright, 1,420 probables, and 3,320 damaged. Within each category, 2,494, 918, and 2,410, respectively, were German. The remainder—2,717, 502, and 910, respectively—were Italian. Because the majority of probables and damaged aircraft were captured by advancing Allied armies as a result of perennial spare parts and vehicle shortages and the resulting inability to move damaged airframes, between 70 and 80 percent of the total machines noted above were write-offs. The total lost, then, was on the order of 9,700 aircraft. German fighters and some of the better Italian machines took a severe toll on their RAF counterparts, destroying more than 1,400 in air combat from April 1941 to December 1942. RAF fighters destroyed about 1,200 Axis aircraft in the air but lost virtually no aircraft on the ground. This was the key difference between the two air superiority efforts and, given the severe Axis aircraft and skilled manpower losses on their airfields, underscores the greater effectiveness of the RAF approach. From January 1942 to May 1943, the Luftwaffe committed 40 percent of its new aircraft production to the Mediterranean theater.[79]

The RAF made its share of mistakes, including sending Spitfires far too late, assigning too few of the best pilots from the United Kingdom to Africa, and employing tactical formations that led to high fighter losses (although they were very successful at protecting the bombers that went after German airfields). German fighter pilots wondered why the Spitfires, their nemesis from 1940, were nowhere in evidence until April 1942, and they often commented on the less capable pilots they faced. Nonetheless, RAF pilots relied on group cohesion and were driven by an overriding mission to gain and hold air superiority, especially with concerted raids on Axis airfields. Conversely, the free-hunting tactics of the best German aces led to a highly individualistic approach that tended to give wingmen few opportunities to score kills, reducing effectiveness at the tactical level even as the lack of any larger focus on air superiority placed German fighter units, and in fact all Axis air units, at a major disadvantage.[80]

Regarding Axis shipping losses in the Mediterranean, the Headquarters RAFME Tactics Assessment Office concluded its major report on antishipping operations with a note that from the start of the war to 30 June 1943, the Axis had lost 1,659 merchant ships—897 in all areas outside the Mediterranean and 762 within it—46 percent of the total. Of the 5.7 million gross tons sunk, 2.4 million—42 percent of the total—went to the bottom of the Middle Sea. This theater thus became a graveyard for Axis merchant shipping as well as its air forces—a useful reminder that "subsidiary" theaters can become quite costly without the proper level of grand-strategic insight and focus.[81]

The Italian Navy had fought hard and delivered nearly 100 percent of reinforcements embarked along with 86 percent of materiel to Libya and 71 percent to Tunisia. These figures, although impressive, came at the cost of the Italian Merchant Marine and most of the Italian Navy. Both suffered immense losses at the hands of the RAF and Royal Navy. More importantly, the delivery of reinforcements and supplies to Africa was only the first step in an extraordinarily costly and painful process to move them forward hundreds of miles either with small vessels or road convoys. Reinforcements and supplies also tended to arrive episodically and were often destroyed in their unloading berths. These constant supply, materiel, and manpower crises, epitomized by perennial and severe fuel shortages, make it clear that delivering men and materiel to ports was not enough. Until those human and other resources were integrated into combat operations in an effective manner, they were of no use.[82]

The experience in the Western Desert and Italian East Africa (IEA) had been at first exhilarating, as the British destroyed Italian Army and IAF units. It then became painful as the Afrika Korps and the Luftwaffe became heavily engaged by spring 1941. Ultimately, however, the RAF proved more effective than the Luftwaffe, which was less ably commanded, suffered from serious logistical problems as a result of mismanagement and missed opportunities (especially taking Malta and mining the Suez Canal persistently), and supported the army before it gained air superiority—a fatal inversion of priorities.[83]

After Allied units in Tunisia adopted the template developed in the Western Desert, the tide turned quickly there. With Allied aircraft in complete control of the skies, as they had been in the Western Desert from summer

1942 onward; under strong leadership; and with excellent operational, logistical, intelligence, and air-ground coordination procedures in place, the campaign's outcome was not in doubt.

The new capabilities that emerged in the crucible of the campaigns in North Africa carried forward into Sicily and Italy and from there to France with Major General Pete Quesada's Ninth TAC and Coningham's Second TAF. Quesada's experiences as deputy commander of NACAF gave him firsthand experience of the outstanding air-ground system the Allies had developed and of the many roles and missions of airpower. He would put these insights to excellent use in northwestern Europe.[84] Coningham's appointment as Second TAF commander in support of Montgomery's Twenty-first Army Group for the Normandy campaign, and Tedder's as Eisenhower's deputy supreme commander, highlighted the degree to which Allied soldiers and airmen had come to value the coequal and interdependent roles of ground and airpower.

Several authors have noted that senior USAAF officers in the interwar period understood every tenet of the doctrine Arthur Longmore, Tedder, and Coningham developed in the Western Desert. That is true, but the difference between understanding a doctrinal concept and implementing it is substantial. The US air system (and the RAF's EAC) initially fared poorly in Tunisia because senior officers had not insisted that their forces put doctrinal tenets into practice. They were also hamstrung by *Field Manual 31-35*, which contained many of the WDAF's doctrinal and procedural tenets but nonetheless wedded airpower to the army in a subordinate, inefficient, and ineffective fashion. Those errors ended in Tunisia. At the end of the campaign, Tedder sent a message to all of MAC:

> By magnificent teamwork between nationalities, commands, units, officers and men from Teheran to Takoradi, from Morocco to the Indian Ocean, you have, together with your comrades on land and sea, driven the enemy out of Africa. You have shown the world the unity and strength of air power. A grand job well finished. We face our next job with the knowledge that we have thrashed the enemy, and the determination to thrash him again.[85]

The exploitation phase of the air-ground effort, built on lessons learned in Africa, was about to get into high gear. The full range of air capabilities, available as a result of air supremacy, now went into action against Italy and the Reich.

PART III

Exploitation

12

Operations Husky, Avalanche, and Baytown
May–December 1943

Exploiting Control of the Air

PREVIOUS CHAPTERS FOCUSED ON THE WAYS in which the Allies' use of airpower played a pivotal role in bringing them victory during the long contest for the Mediterranean's southern shores. After that success, the Allies turned to their longer-range goals for the Mediterranean campaign: forcing Italy's surrender, holding as many German forces as possible away from the Eastern Front, and opening a second aerial front against the Reich. The latter included contributing to the combined bomber offensive (CBO) and Operation Pointblank—in the former case through concentration on oil targets and fighter factories, and then through degrading transportation networks in southeastern and central Europe to speed the Red Army's advance.

Operation Husky, the invasion of Sicily, was an important example of the utility of airpower in exploiting air supremacy, as were Operations Avalanche at Salerno and Baytown as the Eighth Army moved across the Straits of Messina onto the Italian toe. These complex combined-arms operations involved all three services and brought the Allies closer to victory in the Mediterranean and the larger war. Operation Baytown resulted in the capture of the major airfield complex at Foggia, from which the Mediterranean Allied Strategic Air Forces (MASAF), including the US Army Air Forces' (USAAF's) Fifteenth Air Force and the Royal Air Force's (RAF's) No. 205 Group, would fly its heavy-bomber campaigns in the second aerial front. By then, the Italian campaign, though its importance continued in the successful effort to tie down substantial numbers of German divisions, had played its major role regarding the value of air supremacy in an exploitation role.

Although Operation Shingle at Anzio and the Cassino offensive were poorly conceived and executed, airpower kept Anzio from becoming an abject disaster. It could not, however, render any meaningful assistance at Cassino, where bad weather, some of Italy's roughest terrain, and a dogged German defense made achieving the operation's objectives impossible, especially because the operation went forward after Operation Shingle's failure even though it was to have gone forward in concert with the break-out from the Anzio beachhead. The air effort in support of the Allied spring 1944 offensive to take Rome and points northward, Operation Strangle, had mixed results but worked well after ground forces compelled the Germans to use up fuel and ammunition during Operation Diadem—the combined-arms offensive following Operation Strangle. When the Allies shifted their attention to Operation Dragoon in southern France, they won a major victory with a mature combined-arms organization. More than 40 percent of all Allied supplies for the campaign in northwestern Europe flowed through Marseilles and Toulon as an effective combined-arms campaign secured those ports undamaged.

Although Operations Shingle at Anzio, Cassino, Strangle-Diadem, and Dragoon form important elements of the larger air-ground-sea effort—in the first two cases highlighting airpower's inability in many cases to obviate the results of poor planning and execution on the ground as well as the challenges of weather and ideal defensive terrain—they did not represent fundamentally different uses of airpower from those developed in North Africa, as will become clear in this narrative.

Conversely, the heavy-bomber campaigns were a key departure from the norm, helping to destroy the Luftwaffe and engaging in highly effective raids on oil and transportation assets to demotorize the Wehrmacht on all fronts. This helped to speed the advance of Grand Alliance armies, in particular the Red Army. The Balkan Air Force's assistance to Josef "Tito" Broz's Partisans also aided the Russian advance and the destruction of Axis formations re-treating from the Balkans. We now turn to these crucial aspects of airpower exploitation in the war's final two years.

The Grand-Strategic Context

After the Allies had the strategic initiative in the Mediterranean, senior leaders met again at the Trident Conference in Washington, DC, in May 1943, where they reconfirmed their decision to capture Sicily. The Quadrant Conference in Quebec in August, after the Allies had taken Sicily, was decidedly more turbulent. By this time, the Joint Chiefs of Staff (JCS) and their planning staff had become substantially more effective in dealing with the British, in part because of major improvements in their staff processes and also because US power was eclipsing that of Great Britain. Arguments between the JCS and British Chiefs of Staff (CoS) became so heated that staff officers were told to leave the room several times so senior officers could speak with absolute candor. At the end of the meetings, the Combined Chiefs of Staff

(CCS) agreed on a compromise strategy that included follow-up amphibious assaults on Italy to secure major airfields for the second aerial offensive against the Reich and Ploesti and to tie down as many German divisions as possible. From there, the focus would shift heavily to the cross-channel landings and the advance into Germany in 1944, along with the two-axis advance on Japan from the central and southwestern Pacific theaters.[1]

Yet differences persisted. Franklin Roosevelt preferred a light commitment in Italy—just enough to remove that country from the Axis and establish airfields for long-range bombers. Winston Churchill sought a somewhat more energetic course but assured the president that he, too, sought only Italy's surrender along with ports and air bases sufficient to hold part of the country. Roosevelt's reasons were always predicated on the need for a cross-channel invasion. So were Churchill's to a point, but he saw opportunities in Yugoslavia and other areas of the "weak underbelly," as he called the area from Italy through the Balkans. When the talking finally ended, the president and prime minister selected Operations Baytown (the Eighth Army's landings on the Italian toe) and Avalanche (an amphibious assault at Salerno).[2] From this point forward, disagreements about the continuing value of the Mediterranean as a major theater of war intensified. Nonetheless, with these agreements in hand, the stage was set first for Operation Husky and then for Operations Baytown and Avalanche.

Axis Responses

Adolf Hitler, alarmed by the invasion of Sicily on 10 July, visited Benito Mussolini in Italy on 19 July, insisting the Italians fight harder and improve their command structure. He refused to send aircraft or tanks to the Italians but instead promised two more German divisions for Calabria if the Italians showed more fighting spirit. He already planned to send these divisions as part of Operation Achse to secure Italy with German troops after its surrender. Mussolini sat through the tongue-lashing sullenly and fell from power on 25 July. Most of the Italian Army surrendered to the Germans, but the Italian Fleet sortied for Allied ports. On 9 September, Do 217 bombers attacked it with radio-controlled bombs, sinking the flagship *Roma* and slightly damaging the *Italia*. All other vessels, including *Italia*, arrived safely at Malta on 10 September. The Italian Air Force (IAF), also to come under Allied control by the terms of the armistice signed on 8 September, had about 3,000 total aircraft, of which one-third were operational. The Allies ended up with 329 of the best aircraft, the Germans grabbed 400, and the Fascist Republican Air Force obtained 225. The Italians sabotaged many more, and the Germans destroyed the rest along with vast quantities of IAF stores. By the time Dwight Eisenhower and Pietro Badoglio signed the formal Italian surrender on 29 September, hopes for a unified Italian military defection had faded.[3]

Mussolini's fall and the immediate German moves to control Italy (Operation Achse) prompted Churchill and Roosevelt to call for a rapid invasion in the Naples area. Planning for Operation Avalanche got into high gear. The most serious problem was a shortage of shipping and landing craft. Combined with aircraft range limitations, this caused serious difficulties.[4]

Planning for Operation Husky

The shift from air superiority to air supremacy allowed the Allies to exploit their advantage in the skies in several crucial ways. The balance of this work studies the most important of these in turn, beginning with air-ground cooperation in Operation Husky. As planning for the operation accelerated, Arthur Tedder had to contend with the 1,560 Axis aircraft on Sicily, of which 810 were German. Fortunately, airfields were packed together in three locales: Gerbini in the east; Comiso and Ponte Olivo in the southeast; and Castelvetrano, Milo, and Palermo in the west. This made dispersal difficult and aircraft vulnerable. Also, fighters at airfields in the first two regions, and those in the third, could not support each other because of range limitations. However, with proper placement of aircraft, the Axis could in theory launch 730 on D-Day.[5]

The Allies had to choose their landing beaches based on the proximity of ports and the range of air cover. This ruled out Messina, at Sicily's northeastern corner. The eastern coast of Sicily south of Syracuse, the southern coast, and the western end of the island were within single-engine-fighter range, although these had to come initially from Malta and Tunisia, meaning only minutes of loiter time. The early capture of airfields and ports was thus crucial. The plan called for seven major groups of airfields to be in Allied hands by D+3 to facilitate the move forward of air units. British airborne troops were scheduled to drop onto three airfields. Bernard Law Montgomery was particularly concerned about the Ponte Olivo group of airfields, stating that the risk from the air to Allied troops would be excessive without their early capture. In a 24 April message to Harold Alexander, he said the air plan should give the Eighth Army excellent support. Inexplicably, he stated in the same signal that he wished to drop the landings at Pozallo and Gela, which meant neither the Ponte Olivio nor the Gela airfields would fall quickly into Allied hands. This contradictory statement is puzzling given Montgomery's concern with having air superiority.[6]

His proposal of an entirely new plan six weeks before Operation Husky was set to commence would concentrate all forces in southeastern Sicily, toss out the landings on southwestern Sicily, and leave thirteen airfields in enemy hands. Tedder said the Allied Air Forces, Northwest Africa (known as NAAF) could not cover them all, and Admiral Andrew Cunningham emphasized that enemy control of these airfields would pose a grave danger to the invasion fleet. Eventually, the senior commanders reached a compromise in which they canceled the Palermo and other western landings in favor of a major one in the southeast, followed by a rapid drive to capture as many airfields as possible to maximize air support and all the ports capable of handling the unloading of supplies.[7]

Commanders agreed that the value of attacking Sicily, and thus Italy, lay not just in knocking it out of the war, but more importantly in forcing the Germans to garrison areas previously under Italian military control, including parts of the Balkans. The Germans would also have to place troops at Brenner Pass, along the Riviera, and on the Spanish and Italian frontiers for fear of Allied landings. This dispersal and pinning of German troops would help the Russians and then the Allies as they made their cross-channel invasion. Montgomery's new plan was not popular, but the other commanders felt that it would still allow them to achieve these larger aims.[8]

Of thirteen divisions or brigades making the initial landings, six had orders to take specific airfields so fighter-bombers could fly in to give direct support. The final plan had five objectives: preparatory naval and air action to ensure sea and air superiority, seaborne and airborne assaults to capture key airfields and the Ports of Syracuse and Licata, establishment of a firm base from which to capture the Ports of Augusta and Catania along with the airfields around Gerbini, the capture of these ports and airfields, and the final reduction and conquest of the island. Ground, sea, and air forces would have to cooperate closely to succeed. This was the largest amphibious operation yet attempted, with 2,590 ships delivering 180,000 troops.[9]

The air plan had three phases: from the fall of Tunisia to D-7 (13 May–3 July), from D-7 to D-Day (3–10 July), and from D-Day to the conquest of the island. The first involved bomber raids stretching from Sicily to northern Italy to weaken Axis air forces, reduce Italian industrial output, and lower Italian morale. RAF Bomber Command would assist by attacking targets in northern Italy. The second focused on neutralizing Axis air forces and interdicting lines of communication (LOCs) to Sicily. Aircrews would try to destroy German fighters on the ground before the landings. Just before D-Day, electronic warfare aircraft would spoof or jam radar stations on Sicily and Sardinia to obscure exact landing sites. After troops were ashore, direct support would increase steadily as Sicilian airfields fell into Allied hands. Protection of shipping was also crucial.[10]

Tedder established a forward fighter control aboard each headquarters ship holding corps and division staffs, which would go ashore with their army counterparts. Fighter-direction ships with very high-frequency (VHF) radios would relay information within each assault area. Three ship-borne ground-controlled-intercept (GCI) radar stations would control night-fighters from sea until they could establish themselves ashore. To round out the command and control (C2), fighter sector operations rooms would be set up in the Gerbini, Comiso, Castelvetrano, and Palermo areas. Mobile radar units with height-finding and direction-finding (HF/DF) capabilities would land immediately, with longer-range radar units following. Reconnaissance would be constant and heavy, air-to-surface surveillance (ASV) radar-equipped aircraft would report constantly on all movements of enemy warships and other vessels, and night-bombing operations would create maximum delays in the movement of enemy troops and supplies. The plan's one great weakness was the vague guidance for antiaircraft artillery (AAA) units, which were to fire by day at all aircraft below 6,000 feet of altitude not recognized as friendly and by night to engage any aircraft not positively identified as friendly, except during periods when Allied night-fighters were known to be overhead. This shoddy arrangement would cause the Allies great grief.[11]

Air planning suffered moderately from a requirement to move directly from the Tunisian campaign to planning for Sicily with less than two months before the start of the operation. There was a period of frenzied but generally effective planning and a huge move of assets to Malta and northernmost Tunisia to put fighters in range of Sicily. The narrow land front on Sicily, along with the short loiter times over the island prior to the capture of air-fields, meant that the entire air armada had to be treated and commanded as a single air force under Tedder's direct command. He had 3,462 aircraft available for the campaign.[12]

German Preparations

The Germans were concerned about the entire Mediterranean Coast from France to the Balkans. Operations Achse and Konstantin, the plans to occupy Italy and Italian-controlled areas in the Balkans, were in place. Albert Kesselring remained *Oberbefehlshaber Süd* (supreme commander south), but command of Luftflotte 2 passed to Field Marshal Wolfram von Richthofen in June. He reported directly to Oberkommando der Luftwaffe (OKL). This created friction and continuing seams in C2 structures. In one good decision, OKL named Colonel Ernst-Günther Baade the German commandant,

Straits of Messina, on 14 July. He would command Luftwaffe flak superbly during the island's evacuation.[13]

Axis air forces could not approach Allied strength levels. By July, there were about 1,750 aircraft (excluding transports and coastal patrol aircraft) in Sardinia, Sicily, and Italy. Of these, 960 were German. There were additional planes under the new *Luftwaffenkommando Südost* (German Air Force Command Southeast), but these were intended to defend southeastern Europe, including Greece, Crete, Romania, and Croatia. Yet another C2 seam developed. The perceived requirement to defend every inch of the Mediterranean's northern shore diluted Axis air strength, even though, as had been the case in the North African campaigns, it meant that hundreds of aircraft would sit largely idle while hundreds more, at a hopeless numerical disadvantage, tried to stop the invasion. Consequently, by July the Germans had just 775 operational aircraft in range of Sicily.[14]

Adolf Galland, head of the Luftwaffe day-fighter arm, had visited Sicily a year earlier, in May 1942, to determine whether and why, as Hermann Göring asserted, Luftwaffe fighter units had lost the major air campaign against Malta. Galland discovered that JG 27 and JG 53 were terribly understrength and that Lieutenant General Bruno Loerzer of Luftflotte 2 was doing nothing about it. When he returned to Berlin, Galland attempted to modify fighter escort tactics for shipping convoys to allow for much looser formations and offensive tactics (another missed learning opportunity from the Battle of Britain), but by this point the fighter force in the Mediterranean was in steady decline. Inexplicably, he did not advocate any kind of sustained raids on Allied airfields as the best means of keeping German bombers safe. Galland returned to Sicily in 1943 to command fighter units in defense of the island. His conviction that they had to operate in large formations to counter heavy-bombers and other raiders was the right solution from a *Luftflotte Reich* perspective, where time and depth combined with a strong radar network gave the Germans time to scramble and form up. However, in Sicily, it proved disastrous as Allied aircraft swarmed crowded Axis airfields, surprising their occupants, knocking out the majority of Galland's fighters on the ground, and causing heavy casualties to ground crews. When Galland briefed Johannes Steinhoff on his tactical approach, the latter said, "That's damned dangerous." Galland felt he had to gamble anyway, although the odds proved far too long.[15]

Efforts to combine German and Italian fighter units under a single command failed again. "Only during the air offensive against Malta in May–June 1942," Galland said, "did joint fighter action materialize and besides the technical difficulties this led to so much friction not only between the higher commands but also between the units themselves that there was a reluctance to repeat the attempt. The fact that the Italians thought more of stunt flying than raiding led moreover to personal differences between the individual fighter units."[16]

German fighters had to cover Sicily, Sardinia, and southern and central Italy. They were rarely able to locate strike formations, which had Y Service and other inputs to change course and avoid contact. Escorted bomber formations created a huge challenge for fighter pilots in terms of basic survival. Göring berated his airmen for their failure to stop the overwhelmingly superior Allied air forces. He ordered Galland to threaten pilots with courts-martial if they had no successes in air combats or returned without shot-up aircraft. To make matters worse, both flak divisions in Tunisia had been lost, to include the large-caliber guns capable of reaching heavy-bombers. German signals units relied on a single landline from Rome to southern Italy and Sicily. Air raids often cut the cable, and even when it was operational, senior-officer communications took precedence, which meant headquarters and units often could not communicate. Radar coverage was excellent, but most of the wireless sets and other equipment required to vector fighters were lost in Tunisia. The development of a night-fighter force was one of the few positive developments.[17]

Göring sought to change the balance in the air by bringing in von Richthofen and Galland. Von Richthofen was a capable combat leader, but he arrived in Italy with a false sense of the situation. Initially, he believed he had only to straighten out the leadership and the airmen, implement a proper operational approach, and turn the tables. As Paul Deichmann recalled, "He arrived believing that, as he himself expressed it, the business would be cleared up in a few weeks and he would be able to go back to his old Luftflotte in the East."[18] However, von Richthofen clearly understood that things were more serious than this. Whatever the merits of Deichmann's statement, von Richthofen's efforts in Sicily and later in Italy, where he served until November 1944, underscore his seriousness of purpose.[19] Von Richthofen's first order was to reduce the number of bomber sorties to a minimum so this force could refit and receive new crews, but the time was too short. New fighter tactics to conceal concentration in the air, direction of travel, and objectives also went into effect, but these relied on too few aircraft and inadequate communications assets.[20]

Efforts to disperse aircraft ran into Italian landowner and government objections on the grounds that such actions would damage private property. The Oberkommando der Wehrmacht (OKW) supported the duce, leaving fighter units tightly packed on their airfields. The political imperative to show a "forward defense" further consigned fighter units to heavy attacks. Rather than determining the maximum combat radius of Allied fighters and establishing bases outside of it, Luftwaffe leaders stationed their units forward and thus condemned the fighter force to rapid defeat. Had they employed the other option, resistance would have been much more effective because Allied fighters would have fought on German terms with little fuel remaining, and Allied bombers would have had to attack airfields without regular fighter escort. In cases when the relatively few fighter units assigned further north engaged bomber formations, they did well.[21]

Air Action: Prelude to Operation Husky

The Northwest African Strategic Air Force (NASAF) continued raiding ports, airfields, and marshaling yards. From 16 May to 6 June, airmen focused on general operations without special attention to Sicily, shifting in the period from 6 to 13 June to raids on Pantelleria, Lampedusa, and some sorties against Sicilian airfields. From 13 June to 3 July, attacks shifted to Axis air forces on the ground. The last phase of NASAF's work prior to the invasion involved systematic attacks on air bases that destroyed 122 German and 105 Italian aircraft on the ground and damaged another 66 and 117, respectively. Raids on ports and marshaling yards often drove Italian laborers to flee, forcing Germans to load and unload. They were equally unwilling to repair airfields, which were under constant attack. B-17s did grave damage to the marshaling yard at Palermo, and B-24s destroyed the Luftwaffe's central telephone exchange at Taormina on 9 July, one day before the landings, creating massive C2 problems.[22] However, there were limits to what airpower could achieve regarding the interdiction of German reinforcements. From June to early August, the Germans shipped nearly 60,000 troops along with 13,700 vehicles and 40,000 tons of supplies to Sicily.[23]

From Malta, Keith Park's Spitfires covered the fleet. Fighter strength there on D-Day was 600. These planes were ready to move onto Sicilian air bases after troops captured them. This staging effort relied on the steady forward movement of squadrons from Malta to Sicily and from Africa to Malta. During the first seven days of Operation Husky, Malta-based aircraft destroyed 151 enemy aircraft with another 28 probable and 74 damaged. Allied losses were 35 fighters. Of the total enemy aircraft destroyed, at least 53 fell to night-fighters—a dramatic increase over previous campaigns. The intensive weeklong attack on airfields was highly effective. Serviceability rates were 95 percent. Universal fitting of interrogation friend or foe (IFF), along with heavy use of VHF radios and the most advanced radar and operations room capabilities, conferred major advantages.[24]

By D-day, only two Axis airfields were fully operational, and half of all aircraft on Sicily had fled to the Italian mainland. Axis strength, C2, and morale were steadily ground down. From 30 May to 11 June, the Allies flew more than 4,000 combat sorties against the heavily fortified island of Pantelleria, which the Italians had dubbed an "unsinkable aircraft carrier." After a massive naval bombardment on the heels of the air effort, Pantelleria's garrison capitulated without a fight on 11 June. Lampedusa and two smaller islands fell on 12 June. Pantelleria soon housed five Allied fighter squadrons and a radar array to provide warning and control functions. Antisubmarine warfare (ASW) aircraft on Pantelleria kept submarines away from the landing force.[25]

Photoreconnaissance (PR) was central to success. PR aircraft photographed major airfields every four hours to determine optimum times for attack and reattack as repairs neared completion, provided detailed poststrike photographs from heavy-bomber raids, and photographed 10,000 square miles of Sicily to give ground troops detailed knowledge of landing beaches and inland terrain. In the course of flying 1,086 sorties, these specialized aircraft provided a wealth of vertical and oblique photographs.[26]

Operation Husky Begins

The landings on Sicily during the night of 9–10 July were something of an anticlimax, with casualties much lower than expected. Convoy routes deceived the Germans, with those from Alexandria "threatening" a landing in Greece before heading to Sicily and others doing the same for southern France. Given their shortage of aerial reconnaissance, the Germans were in suspense until late in the game. A strong force of warships delivered immense firepower on the lightly held Axis positions, and aircraft went after troop concentrations. The power inherent in naval gunfire was immense. Cruisers were roughly equivalent to a medium regiment of artillery and with a 19,000–29,000 yard range; destroyers equaled two field batteries with a range of 16,000–20,000 yards. This formidable concentration of firepower generally did more than airpower and artillery to stop and turn back enemy attacks within range of the guns.[27]

Axis raids on invasion convoys began at 4:30 a.m. but achieved minimal successes. In most cases, fighters drove off the attackers. Heavy attacks on Axis airfields, combined with constant fighter-bomber strikes on LOCs, slowed the German response to the landings. Suppression of Axis air activity forestalled Luftwaffe attacks. Logistical problems resulting from delivery shortfalls and the near absence of any effective ground organization (almost

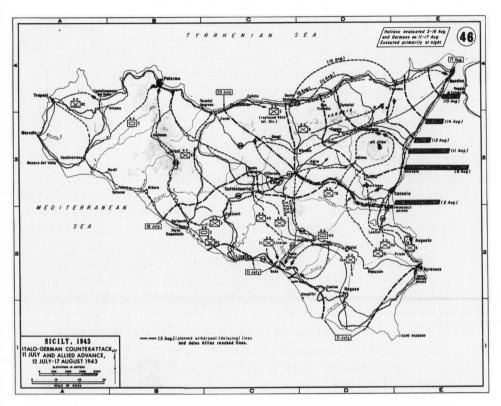

Operation Husky, 10 July 1943. Despite the successful conquest of Sicily, the Allies missed a golden opportunity to cut off 60,000 German troops and another 60,000 Italians, who ultimately escaped. Poor command and control of air assets led to heavy-bomber missions against other targets just as Axis forces crossed the Straits of Messina. The Germans also brought out 14,100 vehicles, 94 guns, 47 tanks, and 23,100 tons of supplies. (Maps Department of the US Military Academy, West Point)

every vehicle, tool, and other asset had been lost in Tunisia) compounded these other crises. Steinhoff could only lament that "air fighting on the eastern front was little more than a harmless game compared with this hell."[28]

The Axis had 481 aircraft in action over Sicily on 11 July but only 161 by 15 July. They switched almost entirely to night-bombing. On 15 July, Kesselring told ground commanders that they must no longer count on air support during daylight hours. The Luftwaffe began moving the rest of its serviceable aircraft, aircrews, and ground crews to southern Italy. Consequently, German troops, now without air support, very quickly learned to maximize their concealment and move at night. The "German glance," as the constant look around the horizon for Allied aircraft was called, would persist and intensify for the rest of the war.[29]

As the initial air effort proceeded, so did the planned airborne landings at key approach points to the beachheads. Unfortunately, this effort showed the same signs of confusion apparent in the small landings in Algeria, but with additional negative results. And yet, despite failings in the airborne landings, they played key roles in securing the beachheads. The first of several paratrooper contretemps, in which troops seized the Ponte Grande Bridge despite appalling losses, started the process. Another followed with the US 505th Parachute Regimental Combat Team, which flew directly into heavy Allied ship-based AAA that had been firing all day at German attackers and was now firing at everything that flew overhead. Once again, the paratroopers took their objectives despite severe losses. Finally, the British First Parachute Brigade at Primosole Bridge also took its objective after suffering heavy losses in the landings. Allied AAA fire downed eleven aircraft, and three others fell to other causes. Navigation errors meant only thirty aircraft placed their troops in the drop zone, nine dropped near them, and forty-eight dropped wide by distances from one-half mile to more than twenty miles. Alexander and Eisenhower agreed to suspend further major airborne operations until the Allies worked out better techniques. Despite these intensive efforts, the Normandy drops would demonstrate once again that even with a heavy emphasis on training, much would go wrong. However, unlike the Germans, the Allies persisted with the large-scale employment of airborne troops, and despite the inevitable imperfections in execution, their landings played key roles from Algeria to Germany.[30]

Meanwhile, as Axis air forces fled, Allied aircraft streamed in. Within five days, fifteen Allied fighter and PR squadrons were operational on Sicily. Arthur Coningham controlled all air operations in direct support of the assault forces. However, Park retained control of air assets in Malta, even though they were supporting the landings, until Air Vice-Marshal Harry Broadhurst, air officer commanding (AOC) Western Desert Air Force (WDAF), established his headquarters ashore to control all tactical air operations. At this point, however, things became even more complex with the arrival of the Twelfth Air Support Command (ASC), which took effective control of all US air assets, leaving Broadhurst with control only of RAF assets. A de facto split in C2 developed that was neither foreseen nor condoned by Tedder, and it took time to fix. Despite these problems, the air effort remained effective. The severe disruption of Axis air activity, combined with the Luftwaffe's requirement to keep aircraft spread thin to guard against any invasions elsewhere, reduced air strength dramatically from the outset.

The Allies lost only two warships, one hospital ship, and seven merchant vessels (MVs).[31]

A series of very effective raids on marshaling yards on 13 and 19 July left a 200-mile rail gap between Rome and Naples that stopped traffic between central and southern Italy for several days. This created serious German logistical difficulties. The Allies now understood the vital importance of marshaling yards, but they continued to spread their attacks too widely and carry them out with too little persistence, choosing instead to move on to new targets after one or two major raids on the first group. Intelligence personnel highlighted these errors in time for the transportation and oil campaigns in 1944 and 1945, but in the meantime the Germans found ways around short-term logistical blockages while they repaired them.[32]

The death throes of the Luftwaffe in the Mediterranean theater were simply the proximate causes of two years of poor employment. Failure to focus on air superiority, lack of joint and combined operations and C2, dissipation of effort, logistical shortcomings, maintenance shortfalls, and inferior employment techniques were the root causes. Conversely, Allied success in the air and as a combined-arms force had its foundations in how senior commanders approached the question of air superiority as an enabler for all further operations, the effective air-ground cooperation that developed among British and then US forces, good intelligence, capable ground and aircrews, efficient industrial planning and practices that turned out new generations of aircraft in minimum time, better C2, better employment of technologies such as radar, and several other factors. As the Allied air effort destroyed the Luftwaffe on Sicily, one of Steinhoff's pilots said, "I've just been reading a manual, usually the last sort of thing I do. LdV 16 is the one and it describes exactly how to set about it. Written by General Wever himself."[33]

While operations progressed, the geography provided a foretaste of what armies and air units would experience in Italy. Rough terrain gave the enemy superb defensive positions, slowed Allied advances and made air attacks difficult. The Germans claimed that Allied airpower, although creating friction, did not seriously hamper defensive operations. Aircraft rarely engaged in direct support given the enemy's elusive nature but instead degraded LOCs. Although major successes were infrequent, there were enough to hamper the Axis supply situation. On 22 July, for instance, thirty-six US Kittyhawks caught a supply convoy on the road and destroyed sixty-five vehicles. During the same period, Allied aircraft twice intercepted Ju 52s flying in fuel and other supplies, destroying all nineteen. Heavy-bombers continued their nonstop attacks on marshaling yards and airfields. Raids on Foggia, a vital rail center controlling north-south and east-west movement from the eastern side of the Italian peninsula, brought all traffic to a stop for several days. They obtained similar results at marshaling yards as far removed from each other as Salerno and Bologna. Although these efforts did not preclude an effective German defense, they created a substantial measure of friction.[34]

As the Allied armies engaged Axis positions, fighter-bombers destroyed gun positions when they found them. Whenever ground forces brought large enemy formations to battle, airpower and artillery inflicted serious losses. The Fifteenth Panzer Grenadier Division, for instance, lost 1,600 men killed in a sharp fight with the US First Infantry Division. Many fell to air and artillery attacks. As Allied airpower grew in strength, German counterattacks, so effective earlier, became more infrequent.[35]

By 1 August, the Germans had decided to evacuate the island. Nearly all escaped during Operation Lehrgang. General Hans-Valentin Hube, the senior officer on the island, directed the effort with skill. The general principles were that troops would exercise the strictest movement and noise discipline; men took precedence over materiel. Colonel Baade had forty heavy and fifty-two light flak guns on the Sicilian side of the Straits of Messina, and eighty-two heavy and sixty light ones on the mainland. He also had a large complement of heavy naval guns. The Germans had 120 boats and ferries of various sizes. Between 1 and 10 August, the Germans moved 12,000 men, 4,500 vehicles, and 5,000 tons of equipment to the mainland. The evacuation then went into high gear and continued until 17 August. Coningham, who had not conferred with Tedder to any extent about an interdiction effort, was unwilling to send aircraft by day, holding to night attacks as the only feasible option. Inexplicably, after Wellington raids proved highly effective at night and forced Baade to move to a daylight evacuation, Coningham released the B-17 units that had been at his disposal for a week to bomb other targets. Simultaneously, he ordered the Wellingtons to raid the Italian side of the straits rather than the heavy German troop concentrations on the Sicilian side. These were major errors not in keeping with Coningham's expertise. His limited experience with heavy-bombers may have played a role, but these lapses in judgment proved costly. From Admiral Cunningham's perspective, naval action to stop the evacuation was bound to fail with heavy losses, so he refused to commit ships to the effort. He remembered his experiences as a junior officer in the Dardanelles campaign.[36]

The Germans Escape

The interdiction effort to impede the German evacuation failed for four reasons: Tedder's and Coningham's failure to plan for it even though the Germans would clearly have to evacuate their army, favorable geography for the defenders, poor Allied interservice coordination, and a paucity of heavy-bombers. One of the most difficult things to fathom is the relatively light effort against the Straits of Messina, the two-mile-wide chokepoint over which Axis assets passed. German flak was thick and deadly at lower altitudes where fighter-bombers and mediums operated but could not have reached the heavies had there been enough of them.[37]

Heavy-bomber raids on Rome, followed by their withdrawal to prepare for Operation Tidalwave—the 1 August low-level attack on Ploesti—made these assets scarcer. The survivors of Operation Tidalwave did not return until 13 August—too late to impede the evacuation. Eisenhower and the CCS had attempted to delay Operation Tidalwave until after completion of the Sicilian campaign. Ira Eaker convinced Henry "Hap" Arnold to oppose this. Even the B-17s assigned to Twelfth Air Force could have hammered the evacuation, but a second raid on Rome, carried out on 13 August with 107 B-17s during the height of the evacuation, and a 180-plane raid on Luftwaffe airfields north of Marseilles, precluded this. No. 205 Group's Wellingtons did severe damage to evacuation efforts from 8 to 13 August, forcing the Germans to return to daytime operations, which, ironically, made them safer given the absence of USAAF heavies and the inability of other aircraft to penetrate the dense flak curtain.[38]

Although the assertion that the failure to do greater damage to German units evacuating Sicily was in large part a result of the increasing emphasis of heavy-bomber operations in support of the CBO and Operation Pointblank is true, we must balance this failure there against what the heavies did accomplish. The raids on Rome and Naples were highly effective in terms of delaying and disorganizing Axis reinforcement and resupply efforts in advance of Operation Baytown. Similarly, Operation Tidalwave, which was poorly executed and resulted in heavy B-24 losses, nonetheless reduced the production capacity of the Ploesti refining complex by nearly one-third, which, as will become clear later, allowed the major offensive against Ploesti in 1944 to cut into production four weeks sooner than would otherwise have been the case. The view that the 1 August Ploesti raid was a waste of heavy-bomber and aircrew assets that did nothing of substance for the larger war effort even as the Germans escaped from Sicily is open to debate. Although it is true that Axis fuel supplies did not suffer any major interruption in late 1943 as a result of Operation Tidalwave, this overlooks the raid's major impact on the massive and rapid decrease of German fuel supplies from summer 1944 to Victory in Europe Day. The view that Operation Tidalwave, although feasible, was poorly timed and executed is also correct as far as it goes. Although it would be pointless to contest these assertions, the raid gave the major campaign against Ploesti the following year a vital head start in terms of reducing capacity and production, as will become clear later. This, in turn, aided the Allied and Red Army advances as German units experienced severe fuel shortages.[39]

By the time the evacuation and earlier transfers to the mainland ended on 17 August, Hube had overseen the escape of 60,000 German troops, 14,100 vehicles, 94 guns, 47 tanks, 1,100 tons of ammunition, 1,000 tons of fuel, and 21,000 tons of other equipment and supplies. Airpower had failed in its most critical task: preventing the escape of these assets. Ironically, the methodical ground advance had drawn in a huge number of Germans who were ripe for the taking after they ran out of maneuver room. Colonel General Heinrich von Vietinghoff, commander of Army Group 10, said Operation Lehrgang was of "decisive importance" because without it, the Germans would not have had the manpower resources to defend Italy south of Rome.[40]

Despite the nearly incomprehensible failure to stop the German evacuation, from mid-May to the end of the Sicilian campaign on 17 August, the Allies put air supremacy firmly in the bag, destroying more than 900 German aircraft and capturing another 482 abandoned on airfields, along with at least 618 Italian planes. During the latter part of the Sicily campaign, air forces went after enemy LOCs but rarely flew direct support missions given the enemy's ability to conceal, move at night, and take other countermeasures. Only twenty-eight calls for direct support went through British ASCs during the thirty-eight-day campaign.[41]

One of the most vital developments on Sicily, which extended from its early phases in North Africa, was the development of air operations posts (AOP) and their close coordination with air assets to pinpoint artillery. This proved extraordinarily valuable for maximizing the accuracy of Allied artillery fire and keeping German guns silent for the crews' fear of counterbattery fire. In the process, it tilted the advantages derived from the "queen of battle" heavily in the Allies' favor. The new P-51 was preferable to the Cub in certain instances, especially during extreme-range and indirection shoots in which Cubs were too vulnerable or lacked the dwell time. However, the Cub was a superb artillery spotting platform. By this point, red, white, and yellow smoke were available for air or artillery delivery or with hand grenades, making the effort even more effective.[42] The humble Cub worked with tactical reconnaissance fighters and the air control elements at each army headquarters to pinpoint targets for rapid air and artillery attacks. Just the presence of these aircraft silenced enemy artillery batteries—and they were often overhead.[43]

This combined-arms effort extended to heavy-bomber activity, with USAAF heavies striking Ploesti in the 1 August low-altitude raid and the

Messerschmitt Plant at Wiener Neustadt near Vienna two days later, both foreshadowing what heavy-bombers would soon do routinely from captured bases in Italy to support the CBO, Operation Pointblank, and advancing Grand Alliance armies. From 30 July to 17 August, heavies destroyed 263 German aircraft on the ground and damaged another 182. Meanwhile, air intelligence noted a major concentration of gliders, bombers, and other aircraft at Istres airfield in southern France. On 17 August, 182 B-17s attacked, destroying 106 aircraft and damaging 86.[44]

The bomber effort against LOCs during this period involved 8,981 sorties against targets in Sicily, Rome, Naples, and other locations. Although Axis records of the damage inflicted are sparse, Allied intelligence confirmed that the raids hampered (but by no means stopped) the movement of supplies and reinforcements. Kesselring noted that raids were systematically suppressing supply bases. On 13 August, a force of 91 B-17s, escorted by 45 P-38s, and another 102 B-26 and 66 B-25 mediums, escorted by 90 P-38s, attacked major marshaling yards in Rome, stopping through traffic to Naples for five days. Wellingtons dropped bombs and propaganda leaflets on alternating missions. Bomber Command shuttle raids also increased in intensity, with four in August by a total of 872 bombers on Milan, three on Turin by 359 bombers, and one by 72 bombers on Genoa. This effort to damage Italian industry and prompt the Italians to sue for peace received major emphasis.[45]

Despite these successes, Operation Husky highlighted several problems, the most severe of which was the failure to trap Axis troops on the island. Additionally, inadequate radio and landline communications, despite great efforts to date, demonstrated again how communications-intensive combined-arms warfare had become. The numbers of air headquarters exacerbated communications problems and C2. Combined with Sicily's rough terrain, the delays imposed by this system made close support of ground troops difficult. To address this problem, the British developed the first rover tentacles attached to leading brigades of leading divisions. Rover teams consisted of an armored scout car equipped with several radio sets for rapid and reliable communication with the army air support controls (AASCs) and air liaison officers (ALOs) at wing or group level who were responsible for coordinating air support. Rover VHF radios facilitated direct communication with aircraft over the target area. The crew consisted of an RAF ASC officer and an army AASC officer who advised the local army commander on requests for air support. He then contacted the ALO at the unit providing support. After aircraft arrived, the RAF member of the rover team directed them to target. The US Seventh Army and Twelfth ASC carried out similar and successful experiments with VHF sets mounted in jeeps. These techniques matured in northwestern Europe. The Allies also introduced an air-ground recognition system using luminous triangles issued to soldiers, red panels with white stars for platoons, and special pennants and markings for armored vehicles. These tools reduced fratricide and allowed for close-in direct support missions.[46]

The fighter-control system still had problems. Forward controllers' VHF radios overheated under heavy use. Ties between headquarters ships, main ground control parties, and forward ground controllers thus became tenuous at times. There were still seams in the radar and radio networks, but they continued to close. Nonetheless, the number of fighters in action often overwhelmed the VHF guard frequency. Planners had to assign an additional one in midcampaign. Fixing this problem, however, availed little without more effective shipboard and ground operations centers. The amount of information overwhelmed staffs and delayed the dispatch of aircraft. Landline parties also proved inadequate given the austere network on Sicily. Even with major improvements, officers had to control air assets using wireless radios for several days after the landings. However, the US use of a lateral ground-based reporting station promised better things to come. The RAF also adopted this practice before the Salerno landings.[47]

Radar coverage had room to improve, as did GCI, sea-controlled intercept (SCI) capabilities, and IFF. Radar sets were good but needed better mobility and operators. In Sicily's mountainous regions, they were often unable to track Luftwaffe aircraft reliably. There were significant problems with IFF, particularly regarding naval vessels, resulting in the loss of a number of Allied aircraft. Commanders gave a high priority to solving this problem.[48]

During the early days of Operation Husky, the Y Service proved exceptionally valuable. As in North Africa, however, Y Service sections continued to struggle with mobile battles. They needed better mobility and interconnectivity. Allied aircraft destroyed the majority of Axis radar sites before the invasion and suppressed the rest through electronic countermeasures (ECM). Once again, the Y Service played a key role in locating these targets.[49]

With a hard-fought and highly imperfect victory in hand, airmen and soldiers kept learning together. If they soft-pedaled the failure to stop the German evacuation, they had the satisfaction of knowing the Germans lost most of their heavy equipment. Very little of the Luftwaffe escaped, and this disaster, coming on the heels of extraordinarily heavy losses in North Africa since 1941, consigned German soldiers to another two years without air support. Even in Italy's excellent defensive terrain and weather, the pain they endured as a result was severe, and although it never broke them, it created massive operational difficulties. One clear indication of this is the fact that, despite their terrain and weather advantages and a number of excellent divisions, the Germans had far higher casualties in Italy on the defensive than did the Allies on the offensive.[50] This German misery in the Italian campaign, matched in similar ways by a corresponding Allied misery, became clear during and after the next two major Allied amphibious landings: Operations Baytown and Avalanche.

The Air Plan for Operations Baytown and Avalanche

Tedder's plan for Operation Baytown (the Eighth Army's move across the Straits of Messina onto the Italian toe on 3 September) and Operation Avalanche (the US Fifth Army's amphibious assault at Salerno on 9 September) was based on indivisibility given the geographical proximity of the two areas; the common targets required to assist with both, including airfields and LOCs; and the inherent flexibility of airpower, which could establish forward bases in Calabria without delay and flex to hit targets for either effort. However, the early phase of Salerno was more dangerous given its distance from air bases in Sicily, so the majority of effort would be to establish a permanent presence over the beachhead. Mediterranean Air Command (MAC) planners thus scheduled heavy preplanned support for Operation Avalanche and on-call support for Operation Baytown. The Allies had an 8-to-1 advantage in serviceable aircraft.[51] Based on Tedder's guidance, Carl Spaatz's staff set forth the following air objectives for NAAF: gain air superiority, cover the Operation Avalanche invasion fleet, isolate the battlefield from German reinforcements, transport and drop airborne forces, and provide support to Operation Baytown. The key problem was range: only P-38s, A-36s (a P-51 dive-bomber variant), and Spitfires with drop tanks could loiter over Salerno for more than a few minutes.[52]

The air chiefs also discussed opening a second aerial front against the Reich. Although doing so would detract from the key role heavy-bombers had thus far played in the interdiction and direct cooperation roles in Tunisia and Sicily, these deep raids would play a critical role in speeding Allied victory by hastening the destruction of the Luftwaffe, making the Normandy landings more secure, facilitating what turned out to be a crucially important amphibious assault on southern France in Operation Dragoon, speeding the Red Army's advance with a major transportation offensive in southeastern and central Europe, and rendering crucial assistance to Tito's Partisans. Key fighter factories at Regensburg and Wiener Neustadt were superb initial targets for the CBO and Operation Pointblank because they produced nearly half of all single-engine fighters, and *Luftflotte Reich* would defend them. Both came under increasing attack.[53]

The second aerial front would open southern Germany, Austria, Czechoslovakia, and the Balkan countries, including Romania and its Ploesti oil refineries, to attack. Given the number of aircraft factories, synthetic oil plants, transportation nodes, and other high-value targets in these locations, concerted bombing promised to speed the end of the war and lower its human and fiscal costs. Charles Portal's discussions with Eaker led to the 3 August raid on Wiener Neustadt by Twelfth Air Force heavies and the 17 August raid on Schweinfurt and Regensburg by the Eighth Bomber Command. If the Schweinfurt raid on ball-bearing factories was less than heartening, the other two did major damage to Me. 109 factories.[54]

Tedder was also enthusiastic about the effects that bombers were having on marshaling yards. He sent Solly Zuckerman's report on the effects of bombing in Sicily to Portal and said his chief scientist would be engaging in further examinations of the effects of transportation attacks as the Italian campaign developed. Tedder determined that it would be almost impossible to achieve a high level of success in Italy. Zuckerman arrived at the same conclusion, although both men saw clear value in a transportation campaign to aid Allied troops by reducing the flow of German reinforcements and supplies. The efficacy of attacks on marshaling yards proved relatively modest during the Italian campaign until greater concentration and persistence along with coordinated strikes on bridges, key stretches of railway, and road junctions came together in spring 1944. However, the learning that occurred during transportation campaigns in Italy proved crucial in the hammering of key rail centers around Normandy and in southeastern Europe, and of the *Reichsbahn* itself, which collapsed the German war economy.[55]

Although shocked about transport aircraft and glider losses in Sicily, and the poor accuracy of troop deliveries to their drop zones, Tedder rated the airborne operations a success in their primary objectives of securing the vital bridges south of Syracuse and Catania. Senior leaders all viewed airborne troops as exceptionally important even if the techniques of safe delivery required work. They had stopped German counterattacks and given Axis troops a false sense that 20,000 or more Allied troops had dropped from the sky. Airborne forces' value as combat troops, and agents of surprise and deception, were clear. The Allies employed major airborne landings for the rest of the war.[56]

Air Operations Begin

Tedder unleashed NASAF to attack airfields and LOCs throughout the summer. Every Luftwaffe base received a hammering along with supporting radar sites, as did marshaling yards and ports, especially Naples. The Desert Air Force (DAF, formerly WDAF) gave direct support to the Eighth Army's advance after it began. The Luftwaffe was largely inactive, conserving remaining assets for the next landing, but this was another losing proposition as bombers, cued by excellent intelligence, hounded the Luftwaffe and destroyed most aircraft before the landings.[57] However, efforts to keep German units from converging on Salerno in the weeks and days before the assault proved ineffective. This was in contrast to direct support during the battle's critical phase, which proved, along with naval gunfire and a determined defense on the ground, instrumental in forcing a German withdrawal. One German commander retreating in front of the Eighth Army characterized the air effort against columns moving by day as "astonishingly small." Others, having suffered constant air attacks in North Africa, marveled at the near absence of strike aircraft during their march to Salerno. Most of the air effort went to blocking the move south of stronger formations north of Rome.[58]

The air effort had followed established lines: air superiority, protection of the fleet, and attacks on enemy LOCs. After D-Day, the focus shifted to direct support of ground forces. With serviceability rates at 75 percent, the Allies could put 670 fighters over the beachhead at intervals. The USS *Ancon* was the director ship for all ground-based Allied sorties, with the USS *Hilary* focusing on fighters. Major General Edwin House, Twelfth ASC commander, directed operations from *Ancon*. Mediterranean Allied Photo Reconnaissance Wing (MAPRW) aircrews flew 670 sorties in July 1943 and 528 in August. Complete overhead and oblique coverage of the beaches and initial objectives went to headquarters down to the battalion level. The Twelfth ASC supplied USAAF units with fuel, bombs, and other supplies; built, improved, and maintained airfields; and repaired and maintained their aircraft. It prepared replacement aircraft for battle, built gliders, and engaged in salvage and repair work. In addition, the Twelfth ASC sent units ashore on D-Day, including airfield engineers, signal corps personnel, ordnance experts, and quartermaster corps troops. They built landing grounds, refurbished existing airfields, prepositioned fuel and munitions, and built a communications network.[59]

On the German side, Luftflotte 2 and Fliegerkorps II based most of their aircraft in northern Italy and would use air bases in southern Italy as advanced landing strips only when necessary. Allied intelligence confirmed this, so there were few worries about the landing in Calabria. Salerno was a different matter because the Luftwaffe had eight major airfields within 110 miles. Aircraft supporting Operation Avalanche had to fly between 175 and 215 miles to the beachhead. This gave a Spitfire V with auxiliary tanks twenty-five minutes over the beaches and a P-38 with extra tanks sixty minutes. The 106 Seafires aboard the four escort carriers and one aircraft repair carrier known collectively as Force V proved vital to success. This flotilla kept 22 Seafires over the beachhead continuously during the invasion's first week. However, even with the Seafires and overwhelming numerical superiority, aircrews faced limited loiter time, fewer sorties per day, and serious fatigue. Capturing airfields quickly was thus crucial, as was the occupation of airfields in the toe and up the boot of Italy as the Eighth Army advanced toward the US Fifth Army at Salerno. NASAF bomber attacks on airfields around Salerno would also be crucial before and during the landings. This Mediterranean pattern of employing heavies flexibly would once again prove valuable.[60]

Given the loss of most of its assets during Operation Husky, the Luftwaffe was not regularly active before Operation Avalanche. However, on 19 and 20 August, around sixty fighters attacked bombers raiding the airfields at Foggia. In two raids on Bizerte, a total of sixty German bombers sank or damaged four vessels and collapsed several oil tanks. These were among the last effective raids. Too few bombers and crews remained. The sporadic raids against Operation Baytown lacked concentration. By September, the Luftwaffe had lost what was left of the substantial forces it had built up to counter Operation Husky. From the end of the Sicilian campaign, when the Luftwaffe had 1,000 planes left in the Mediterranean, the total dropped to 600 on 1 September and dwindled from there as bombing raids against the Reich drew in most remaining fighters, and the deteriorating situation on the Eastern Front led to a transfer of most remaining bombers there.[61]

Operations Baytown and Avalanche Commence

On 3 September, Commonwealth forces landed on the Italian toe, encountering virtually no resistance. The Germans had anticipated an attack in the Naples area and placed most of their troops there. By 8 September, the Eighth Army was nearly beyond the toe and heading into open country. Italy surrendered the same day. Both of these developments boded well for a linkup with forces landing at Salerno. Continuing raids on the Foggia airfields brought up the last of the Germans' fighters, which suffered heavy losses. Axis airfields within range of Salerno and the Operation Baytown beachhead were so badly damaged that the Germans withdrew remaining aircraft to central Italy.[62]

As Operation Avalanche commenced, the Luftwaffe managed about thirty sorties over the beaches the morning of 9 September. An equal number appeared during the afternoon and evening, and a few more at night. They sank one small vessel and damaged another out of the 600-ship armada. As German Army counterattacks on the beachhead intensified, heavies and

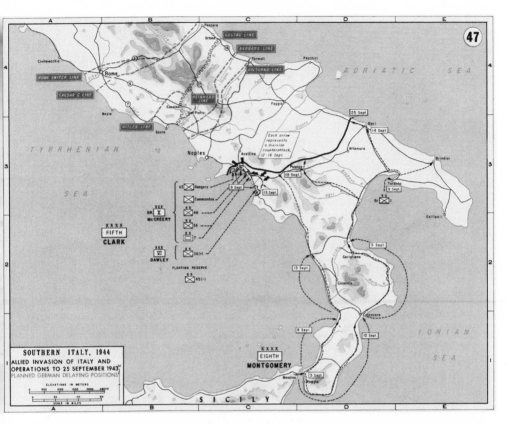

Operations Baytown (3 September 1943) and Avalanche (9 September 1943). Air action before these amphibious operations knocked out almost all that remained of the Luftwaffe in Italy; provided excellent reconnaissance for the invading forces; protected the landing fleets; and played a major role, along with naval gunnery, in saving the Salerno beachhead. After the crisis passed on 16 September, air support facilitated a rapid advance until the winter. (Maps Department of the US Military Academy, West Point)

mediums raided bridges and road junctions to delay the arrival of reinforcements. Allied aircraft flew 1,649 sorties on D-Day and continued this pace for the next several days as the fighting on the ground reached a culminating point.[63]

The key problem for the Allies at Salerno was German opposition just inland, which made it impossible to expand the beachhead and thus to unload enough men and supplies without creating gridlock. This was not something air or naval support could fix, and it nearly led to defeat. Given the

Destroyed Luftwaffe aircraft and fuel supplies in southern Italy, 1943. In what had become common practice, Allied aircraft hammered airfields in advance of the landings in Italy. Note the missing propellers, which were likely salvaged given endemic shortages. (Imperial War Museum Image No. NA 6704)

congestion on the beaches, any sort of credible Luftwaffe showing would have been a disaster. The Allies could not use Montecorvino airfield because, although friendly troops held it, artillery and mortar fire still ranged it. RAF ground parties began stocking fuel and ammunition there and placed a few ground crews on site for emergency landings and other requirements, but the airfield was not secure until 20 September. [64]

The Luftwaffe made a concerted effort against shipping with its remaining planes, including the Do 217 with radio-controlled bombs. Although they damaged the USS *Savannah* and HMS *Warspite*, the Germans had lost so many aircraft that any larger impact the radio-controlled bombs might

have had proved illusory. They averaged 100 sorties per day against shipping from 9 to 17 September and damaged thirty ships, most slightly. In the
process, they lost forty-four aircraft, with nearly the same number probably
destroyed, and their weight of effort declined markedly thereafter. However, Allied pilots were exhausted and were having more accidents as they
continued flying sorties from Sicily and the toe. German counterattacks on
the beachhead continued to deny the Allies captured airfields. Carrier-based
aircraft were losing their surge capability after more than a week of intensive operations. The USAAF aviation engineers completed their first landing
strip on 11 September, a second on 13 September, and a third on 16 September. Royal engineers had another runway ready on 11 September and a
second on 13 September. With these fields in action as fighters and fighter-
bombers arrived, the air situation improved. As the Eighth Army advanced
up the southern peninsula, German commanders evidently decided to attempt a holding action and a major reinforcement of the Salerno area from
the troops retreating out of the toe. The DAF had a field day as convoys
moved during daylight hours, destroying 90 vehicles and damaging 120
more on 9 September alone.[65]

On 12 September, as German Army counterattacks intensified, General
Mark Clark established Fifth Army Headquarters near the Albanella Railway Station, and House established Twelfth ASC headquarters several miles
northward to be able to move onto Montecorvino airfield. With the two
headquarters geographically separated and the intervening ground under
fire, landline and radio communications collapsed and, consequently, any
effective coordination of air support.[66]

Eisenhower saw a crisis coming, but insufficient naval transport and an
inflexible loading and delivery schedule slowed the arrival of reinforcements. This made expansion of the beachhead virtually impossible and created a massive logjam. German troops, in contrast, had relative freedom of
action despite heavy air attack, and their artillery ranged the beachhead.
As the situation on the ground reached crisis proportions by 12 September, air assets engaged in five days of extraordinarily intensive operations.
Tedder's aircrews flew more than 2,000 sorties during the twenty-four
hours of 14–15 September. Most operated at maximum ranges, but C2 difficulties made it very difficult for the armies and air forces to coordinate
airstrikes.[67]

Two squadrons of Spitfires arrived on the Italian mainland on 13 September at the new Tusciano landing strip. The US completion of landing strips
at Sele and Asa on 13 September allowed a USAAF P-40 wing to join Fifth
Army Headquarters at Albanella Railway Station, restoring coordination
of air support for ground forces. Finally, Tedder transferred the Ninth Air

Force (Lewis Brereton's command, first known as Middle East Air Force in Egypt) to NAAF, adding significant strength to NASAF's heavy-bomber force and allowing for greater concentration.[68]

During this critical period from 12 to 15 September, the Allies stopped an all-out German effort to overrun the beachheads with a combination of dogged resistance on the ground, air attacks, and naval bombardment. Spotting aircraft worked with specific ships to direct fire as fighter-bombers, mediums, and heavies hurled themselves against the enemy. Three short-notice airborne operations, ordered on 13 September and carried out on 14 September, put another 2,900 troops at Clark's disposal during the battle's most critical phase. They repelled attacks and then counterattacked, seizing critical points along the line.[69]

When Clark effused that "magnificent air support . . . contributed much to the success" of ground operations, he might better have said that it facilitated their success, along with naval gunfire and paratroopers, by averting a disaster.[70] Significantly, Clark also said there had been "no effective enemy attack since bombing [on the] 14th, which was probably decisive."[71] He credited the naval and air attacks with saving his troops' positions. Von Vietinghoff, Tenth Army commander, agreed, lamenting in his war diary, "The fact that attacks were unable to reach their objectives owing to the fire from naval guns and low flying aircraft, as well as the slow but steady approach of the Eighth Army, caused the Army Commandant to withdraw from the battle."[72] Eisenhower noted, "Without the concentrated use of Naval and Air strength we could hardly have kept from being driven back into the sea."[73] Further, he said,

> It appears to *me* important that one major lesson should never be lost sight of in future planning with the AVALANCHE landing now apparently secure against any major counter-attack that could seriously threaten us. This lesson is that during the critical stages of the landing operation every item of available force including land, sea and air, must be wholly concentrated in support of the landing until troops are in a position to take care of themselves. This most emphatically includes the so-called Strategic Air Force.[74]

This continuing plug to use heavies in interdiction and direct support roles, in which they had proven so effective, ensured that even with the CBO and Operation Pointblank pressing hard on available heavy-bomber assets, the Mediterranean theater would have a more varied use of them than occurred in northwestern Europe. Despite its failure to interdict the movement of German units to Salerno, airpower played a crucial role in preserving the beachhead and protecting the landing fleet at an extraordinarily low cost.

Of 1,870 sorties flown on 14 August, the Allies lost one plane—not an uncommon loss rate.[75] Mutually supporting combined-arms operations had salvaged a modest if real victory that put troops on *Festung Europa.*

If the landings were nearly a disaster, they were also, paradoxically, an example of Allied agility, flexibility, and improvisation. Every operational NAAF aircraft flew in the battle. Eisenhower had to convince the CCS to release three B-17 groups to carry out deep strikes on marshaling yards and other high-value targets between 21 September and 1 October, until the exhausted NAAF heavy-bomber crews could catch their breath and refurbish their machines. Eisenhower had relied so much more heavily than envisioned on airpower to turn the tide that he cabled General George C. Marshall with an urgent request for additional aircraft. His other problem was equally serious: good flying weather in the Mediterranean meant that aircrews completed their fifty required sorties in four to six months. Despite the utility of sending these airmen home to train their successors, too many were leaving and too few arriving. Ultimately, the solution lay in increasing replacement rates to give units sustainable aircrew numbers.[76]

From the victory in Sicily on 17 August to the end of the German counterattacks at Salerno on 15 September, Northwest African Tactical Air Forces (NATAF) flew 18,193 fighter and 2,259 fighter-bomber sorties. The Northwest African Coastal Air Force (NACAF) contributed another 3,571. NASAF and NATAF bombers made 8,000. This air armada destroyed 903 Luftwaffe aircraft at a cost of 205 planes. It was a telling victory and the last time the Luftwaffe made any kind of appearance over the battlefield. With southern Italy in Allied hands, the Germans evacuated Sardinia and moved their few remaining planes to Corsica.[77]

Moving Up the Boot

During the rest of September and October, Allied armies advanced up the peninsula under air cover. NAAF shifted assets from direct engagement of ground forces to concerted raids on airfields and LOCs. During September, they destroyed 510 planes, claimed another 50 probables, and damaged 110. They also destroyed nearly 1,200 vehicles and harried German troops retreating to the Gustav line. However, 29 September marked the first day when worsening weather grounded aircraft. Even the Wellingtons, which had operated on eighty-eight of the past ninety-two nights, could not fly. This problem hounded airmen for the next two falls and winters. Nonetheless, during September, in addition to their efforts over Salerno and during the pursuit, Allied airmen covered the unloading of more than 100,000 tons of supplies, 30,000 vehicles, and nearly 200,000 troops. As NACAF's antishipping campaign continued, another two large ships went to the bottom, two more suffered severe damage, and another nineteen sustained damage. Only the most carefully planned supply runs got through the aerial gauntlet, although the Germans became masters at this game.[78]

Italy's surrender forced a continuing and costly German commitment; eliminated twenty-six Italian divisions in the Balkans (most of their weapons fell into Partisan hands), forcing the Germans to garrison 7–12 of their own divisions there at any given point; brought the Italian Fleet almost entirely into Allied hands; and gave the Allies key airfields at Foggia from which to strike directly at the Reich and Ploesti (and the entire oil and transportation infrastructures in southeastern Europe). Hitler was forced to divert troops and other resources intended for the Eastern Front. These were not just to Italy and the Balkans but to the entire Mediterranean theater, where amphibious assaults now had OKW badly shaken. Finally, the successful lodgment in Italy proved vital to the success of Operation Overlord if only because the combined-arms capabilities Allied forces developed during four years in the Mediterranean went with them to France as senior commanders from the Mediterranean and Middle East became responsible for operations in northwestern Europe. It also freed a huge amount of shipping as the Mediterranean became the Allies' *Mare Nostrum* ("Our Sea"). By October 1943, ships transited with virtual impunity, and the ending of the pipeline around the Cape of Good Hope tripled available MVs to bring US troops, equipment, and supplies to the United Kingdom in preparation for Operation Overlord. To view the long series of campaigns in the Western Desert, Italian East Africa (IEA), Iraq, Syria, northwest Africa, Sicily, Salerno, and the larger Italian campaign from the traditional point of view—that they were at best a useful sideshow—is to miss the many ways in which exploiting the newly gained advantages they conferred helped to speed the end of the war and reduce the Grand Alliance's losses of men and materiel. In a fundamental way, because of this theater's distinctive geographic realities, and because of the steady growth in combined-arms acumen, Allied airpower was a cornerstone for these victories and the advantages they conveyed.[79]

However, the Germans would fight for Italy. The Allies thus planned to follow up their narrow victory at Salerno in four phases. The first included building up the Salerno beachhead; securing the ports of Taranto, Brindisi, and Bari; and building up air forces in the Italian heel. Second, US forces would secure Naples as the Eighth Army seized the vital airfields and the road and railroad networks around Foggia. Third, Allied armies would try to capture Rome. Fourth, they would continue north to seize the Port of Leghorn and perhaps Spezia along with the key cluster of airfields around Pisa. This fourth phase in the planned advance included capturing Florence and Arezzo, the gateways into the Po River Valley. As the Allies soon redis-covered, however, planning and doing were two different things, especially in Italy.[80]

Allied air forces attacked airfields and LOCs on 21 September prior to the ground advance. They tried to ring in German troops with roadblocks—a largely unsuccessful effort given German ingenuity. The day-bomber force was exhausted, and too few replacement crews were arriving. The terrain became steadily worse as troops moved north, and the habit of staying on main roads, rather than advancing overland with mule trains as Free French units did, resulted in a slow advance and higher casualties. Bad weather hampered air action, and the Germans were adept at moving during the night or in small groups during the day. A much more rapid advance on Fog-gia, when compared with the slog toward Naples, put the city and its associ-ated airfields in Allied hands on 27 September. The "two-front" bombing effort was set to begin.[81]

The buildup of air forces around Salerno continued, with twenty-four squadrons (380 aircraft) in place by 26 September. By 28 September Broad-hurst had moved nearly fifty squadrons (800 planes) onto the Italian heel. These moves increased responsiveness but could not solve the basic co-nundrum that terrain and weather had changed the character of the bat-tle, placing clear limits on what the combined-arms effort could achieve. Nonetheless, Luftwaffe losses prompted the Germans to preserve their few remaining aircraft for special missions or all-out support to ground forces at critical junctures. After 17 September, at most thirty Luftwaffe planes a day appeared over Allied lines. Night raids by Ju 88s and Do 217s had no success as air-intercept (AI) Beaufighters caused significant losses.[82]

Allied pursuit was a race against worsening weather with inevitable less-ening of the effectiveness of airpower. Still, the combined-arms team kept moving, with the armies pushing German troops mercilessly, heavy-bombers striking roads and rail lines, and fighter-bombers destroying more than 400 vehicles and damaging another 300 from 1 to 7 October.[83]

With airfields on the Italian heel and Foggia in Allied hands, heavy-bombers stepped up raids on the Reich. Missions against the aircraft indus-try had begun on 3 August with a major raid on Wiener Neustadt. Another followed on 1 October, doing serious damage. The Luftwaffe response to these raids was immediate and heavy. Three heavies fell to fighters, and flak damaged another fifty-two. However, from the outset in this theater, the bombers had fighter escorts (P-38s) all the way to and from targets. Con-sequently, German fighters paid dearly, losing twenty-five just in this first engagement. If Operation Big Week marked the point at which the entire might of the heavy-bomber force and its fighter escorts began destroying the Luftwaffe over the Reich, NASAF got a head start. Air intelligence demonstrated that bombing had forced large-scale dispersion of the German aircraft industry and destroyed substantial portions of it. As the industries dispersed, mostly to southern Germany, the Luftwaffe had too few propellers, engines, and tires and too little fuel. Ironically, this dispersal, begun as a result of RAF Bomber Command's Ruhr Offensive and USAAF raids on aircraft factories, prompted the building of satellite plants in NASAF's area of concentration just as the latter was starting to fly missions. Therefore, even dispersal left aircraft production in a vise and created a quality-control disaster.[84]

By the middle of October, heavy rains kept air units from giving regular support to troops. A mission against Wiener Neustadt on 24 October failed because of solid cloud cover over the target. October was thus a "light" month for NAAF, with 27,000 sorties and 12,000 tons of bombs delivered. However, the opening of a second aerial front against the Reich, with fighter escort from the outset, showed great promise. Aside from doing substantial damage to Wiener Neustadt, NASAF missions accounted for more than half of Luftwaffe aircraft losses in southern Europe. During October, Allied fighters destroyed 160 aircraft in the air with another 30 probable and 60 damaged. Bombers destroyed another 160 or so on the ground with another 40 probable and 80 damaged. NAAF lost 90 aircraft, most to flak.[85]

Exploiting the Sea Lines of Communication (SLOCs)

Another story of note in the Mediterranean was the huge revival in Allied shipping. Convoys sailed in relative safety before the end of 1942. By late 1943, the Mediterranean was the Allies' exclusive domain. After the Tunisian campaign, minesweepers cleared paths for MVs to traverse the Mediterranean. This freed up approximately 225 merchant ships for duty in other locations than the pipeline around the Cape of Good Hope—1.5 million additional tons of available shipping. An MV traveling from Liverpool to Egypt through the Mediterranean saved ninety days per roundtrip voyage. The Luftwaffe's failure to think adequately about maritime air warfare, if perhaps in part the result of being a Continental power, is incomprehensible given that Germany's principal enemies, Great Britain and the United States, relied heavily on SLOCs. Despite constant Kriegsmarine pleas for more aircraft and a greater focus on maritime airpower, Göring and Hitler refused.[86]

By April 1944, more than 300 Allied MVs per month delivered supplies to the Mediterranean and Middle East, with many supporting lend-lease efforts in Iran. German air attacks in April and May accounted for one destroyer and four MVs. U-boats sank another three MVs. This was one sinking for every 400 sailings. German tactics improved, but the Allies countered, and the Germans had far too few aircraft. Allied U-boat kills were nearly as numerous as Allied ship losses in large part because of air-sea ASW activity. The U-boat fleet soon disappeared.[87]

The reopening of the Mediterranean SLOCs to Allied shipping was one of the major contributions of the campaign to Allied victory. By drawing more than 50 U-boats and more than 300 torpedo aircraft from the Battle of the Atlantic at a critical time, and by freeing nearly 300 MVs from the cape run, Allied efforts in the Middle Sea sped victory in the Atlantic, the buildup of supplies for Operation Overlord, and supply deliveries to India as well as to Russia through Iran.

13

The Italian Campaign and the Invasion of Southern France

October 1943–August 1944

Hitler Fights for Italy

As ALLIED TROOPS STRUGGLED with worsening weather and rough terrain in Italy as they pushed north, Adolf Hitler issued a directive for the Italian theater on 4 October 1943. Delaying tactics were at the center of his guidance given his suspicion that the Allies might try to jump the Adriatic Sea and engage in a campaign with Josef "Tito" Broz's Partisans rather than fighting their way up the Italian peninsula. Hitler's plan to hold as much as possible of Italy was not an entirely unwelcome prospect for the Allies because they would hold a number of German divisions away from the Eastern Front. Between the 17 German divisions in Italy and the 13 in the Balkans, 11 percent of Germany's 277 total divisions were away from the Eastern Front. When combined with the 51 divisions in France, the Low Countries, Norway, and Denmark, 29 percent of the German Army was effectively immobilized by what the Allies might or might not do. With the 33 replacement and reserve divisions in the Reich, this total rose to 41 percent not fighting on the Eastern Front. Dwight Eisenhower's 8 November directive for the campaign in Italy was thus designed to keep German units in place and to secure Rome along with its nearby port facilities and airfields. He further directed that heavy-bombers would support the combined bomber offensive (CBO) but also be available to assist ground forces.[1]

Air forces quickly set up repair-and-salvage units (RSUs) in Sicily and Italy. The standard process was to repair every possible aircraft at forward fields. If a plane needed more serious work, it went to depots further south. By January 1944, depots were sending fully overhauled engines to forward air bases. Despite this outstanding logistical support, the air forces were hard pressed to carry out their full range of duties in October and November, not just because of the weather but also because of their wide range of tasks: air superiority, close support of troops, protection of convoys, aid to the Partisans, operations in the Dodecanese, and escorted heavy-bomber missions. In the battles around the Termoli in early October, the Sixteenth Panzer Division complained that air attacks were causing systemic fuel shortages, making movement to the battle area, maneuver, and withdrawal difficult. With Desert Air Force (DAF) assistance, the Eighth Army held despite heavy German attacks. On 2 October, aircraft began preparatory attacks on German positions north of the Volturno River. Mark Clark's Fifth Army jumped off on 5 October, but the weather was so bad that airmen were able to fly fewer than 100 sorties a day. Only a few raids on bridges and marshaling yards had any significant effect on German movement. When Allied troops occupied the major air bases around Naples, they captured 164 German and 101 Italian aircraft in various states of repair. This brought the bag for 3 September–15 October to 574 German and 523 Italian machines captured on the ground.[2]

However, air support to ground forces remained largely ineffectual because of weather, terrain, and German countermeasures. Although weather limited direct work with ground troops, heavy- and medium-bombers continued raiding marshaling yards, road junctions, and airfields. During this period, the Allies lost 326 aircraft and the Germans 267, of which 101 simply disappeared during operational flights. Although Allied action accounted for some of these, many crashed because of the increasingly serious defects in aircraft coming from German assembly lines (a result of the dispersal of factories and associated quality-control problems) and to the deteriorating quality of ground crews and maintenance as well as pilots.[3]

German commanders noted that troop movements in or near the battle area were virtually impossible by day. They also referred frequently to the demoralization among German soldiers. NATAF light-bombers met fighter opposition so rarely that they began flying without escorts. Even heavy- and medium-bombers had little to fear with the notable exception of flak. Allied logistics were virtually immune from air attack, and the sight of so many aircraft in the air, flying to and from their targets, was a boon to the troops' morale. Fighter-bombers flew relays over the front as often as weather permitted, ready to respond to calls for assistance or to pounce on any German movement. Night-bombers attacked the Germans as they moved, achieving occasional successes. Few specific air actions had major effects on tactical engagements, but the heavy-bomber effort had somewhat broader impacts over time. Heavies interrupted railroad movement in Italy and southern France, interdicted all other major lines of communication (LOCs), carried out occasional operations in the Balkans to aid the Partisans and weaken the will of the Axis Balkan governments, attacked all Luftwaffe airfields with aircraft in range of Allied positions in Italy, and supported the CBO. The air plan shifted to a concerted effort to cut as many railroad lines as possible with attacks on bridges and defiles. This began on 19 October with a B-24 raid near Ancona. The use of heavy-bombers to attack bridges soon proved inefficient and ineffective. Mediums and fighter-bombers took over with far better results. The objective was to hamper the forward movement of troops and supplies, cause maximum friction and chaos, and thus reduce the enemy's ability to fight or, just as important, to sustain a fight during a major battle for want of ammunition and fuel. The Germans turned increasingly to trucks and small boats to run in supplies, and they always brought in just enough to keep fighting—at least until Operations Strangle and Diadem the following spring. Nonetheless, the move to trucks and small vessels gave fighter-bombers a plethora of targets, and German losses climbed.[4]

From 26 November to 1 December, with help from the DAF, the Eighth Army had broken through the eastern reaches of the Bernhard line and might have exploited the advance had not weather, mud, and exhaustion combined to slow their progress. Allied tactical aircraft moved to "bomb carpets" in the bad weather, but with limited effectiveness.[5]

The Allies were discovering that airpower sometimes labored under severe limitations that precluded effective support to ground forces. Italy was a wretched place to fight a war of maneuver, especially in fall and winter. Mechanized warfare was all but impossible in the Italian mud, aircraft could not fly with any regularity, and German troops had ideal defensive terrain. In every effort leading up to this late 1943 slog toward Rome, airpower had played a major role in facilitating victory or forestalling defeat. Now, it was as powerless as the troops on the ground to exert any kind of major influence on the course of the battle. Allied planners thus turned to the possibilities inherent in another amphibious assault. Operation Shingle—the landings at Anzio—resulted. Meanwhile, however, the Allies continued to grapple with the disheartening realities of limited air and ground effectiveness in Italy.

Clark's Fifth Army winter offensive, designed to take Rome, started on 1 December, but weather once again prevented effective preparatory air attacks. After flying about 1,000 sorties on 1 and 2 December, Allied air forces were grounded with few exceptions until 8 December. There were eight good flying days the entire month. Air action did little to aid the ground offensive, which bogged down with heavy losses. German casualties now and during the entire campaign were higher, indicating that air attacks combined with ground action exacted a heavy toll, but this did not translate into a rapid advance up the peninsula. The Germans held, both sides bled, and the dreary Italian winter set in.[6]

Operation Shingle

In an effort to break the stalemate, Allied leaders agreed to attempt another amphibious outflanking of German positions at Anzio and a dash for Rome as the Eighth Army took Monte Cassino and broke through the Gustav line to link up with Clark's Fifth Army heading north from the beachhead. The Mediterranean Allied Air Forces (MAAF), formerly Mediterranean Air Command (MAC), had two primary objectives: further reduction of the Luftwaffe and heavy attacks on LOCs in northern Italy. Direct support of ground forces was also vital but expected to take up fewer assets now that the Germans were not counterattacking or moving during daytime.[7]

Every airfield around Naples was soon packed with aircraft. The first phase of the air effort, from 2 to 13 January 1944, began with attacks on northern Italy to distract the Germans while destroying aircraft on the ground. The second phase, from 13 January until the landings on 22 January, involved heavy raids on airfields, LOCs, headquarters, radar sites, and communications facilities. The third phase, from 22 January forward, included constant patrols over the beachhead and maximum assistance to ground forces. Diversionary air, sea, and land operations began on 2 January, with heavies and mediums hammering targets in northern Italy and army formations gathering with amphibious landing vessels on Corsica and Sardinia, leading the Germans to believe the landing would come further north.[8]

Direct support targets received prioritization through coordination between control ships and a control squadron on the beachhead. The process was better than the one in place for Operation Avalanche and, before that, Operation Husky. Night-fighters equipped with the latest radar sets and under capable ground-controlled-intercept (GCI) radar kept the few Axis aircraft away from the beaches. Airstrikes kept German reinforcements from massing around the beachhead for eighteen days, although Allied troops did too little to improve their positions. From 10 to 15 February, as German ground attacks became fierce, heavies provided direct support and played a crucial role in breaking up German attacks.[9]

Meanwhile, the Luftwaffe recalled a substantial number of bombers to Italy. Allied photoreconnaissance (PR) aircraft found them, and P-47s raided the bases on 30 January at treetop height, denying the Germans radar warning and giving them no time to engage in their customary fly-offs. The P-47s destroyed twenty-six bombers. Heavies followed immediately, wrecking another forty-two aircraft and effectively destroying the bomber force. Leading up to this disaster, the Germans had been using these Ju 88s along with Do 217s, stationed in southern France and carrying radio-controlled bombs, to attack the invasion fleet. Most raids occurred at dusk and involved fifty to sixty bombers. The Germans sank five ships. By 30 January, they were out of bombers.[10]

The Allies mustered 2,700 aircraft to support Operation Shingle within the new MAAF, which had recently succeeded MAC. (The establishment of MAAF receives detailed attention in the next chapter.) Unlike at Salerno or Sicily, these were close to the beaches on good airfields.[11] Major General John K. "Joe" Cannon, the Mediterranean Allied Tactical Air Forces (MATAF) commander, set clear and by now standardized priorities for air support of the Anzio landings. Destruction of the Luftwaffe, cover for the invasion fleet, close support of troops, and a steadily increasing number of raids on German LOCs topped the list. The bombing of LOCs accelerated between 15 and 21 January and again on 22 January (D-Day) and afterward. Air Marshal Hugh Lloyd, the Mediterranean Allied Coastal Air Force (MACAF) commander, provided heavy air patrols comprising antisubmarine warfare (ASW) aircraft and night-fighters to ensure safe passage for the fleet. There were always thirty-two day-fighters over the beaches. Eight Beaufighters replaced them at night, under the guidance of a fighter control ship and equipped with the latest air-intercept (AI) radar. The Sixty-fourth Fighter Wing would send its Forward Fighter Control Center ashore as soon as possible, this time equipped not only with a full array of radar and radio but now with a Y Service detachment to intercept tactical communications. This brought intelligence and operations side by side to an unprecedented degree.[12]

Albert Kesselring ordered units to be ready to move to the landing beaches without delay and organized heavy flak cover. The Germans moved at night to avoid air attacks. Engineering battalions were ready to make immediate road and railroad repairs. Hitler called for the utmost resistance.[13] Bad weather immediately after D-Day made it difficult for Allied aircraft to attack the six divisions converging on the battlefield to join another four already there. Exploiting these advantages as well as the painfully slow Allied buildup and weak efforts to advance, the Germans made their bid to take the beach. Despite heavy casualties, they came within a whisker of destroying the landing force. Heavy air strikes and naval bombardment made the difference. By the time German attacks ended, they had lost 10,511 men versus 3,472 at Salerno. The costs of mobile warfare and aggressive counterattacks had become prohibitive.[14]

One of the important effects of interdiction efforts as the Germans moved into position was the logistical constriction caused by heavy attacks on bridges and marshaling yards, which drove the railheads 50 to 100 miles from the beaches. This demotorized German divisions, which had to send their trucks north to collect fuel and ammunition, burning as much of the former as they brought back. These convoys also came under heavy fighter-bomber attacks. The two items in shortest supply were artillery ammunition and fuel. The former impeded German efforts to support their attack on the beachhead, especially because they lacked air support and naval gunfire, by denying them the opportunity to produce rolling barrages in support of advancing troops. Fuel shortages made it increasingly difficult for them to maneuver, especially after they were forced into mobile warfare. This pattern repeated itself in every subsequent major campaign in Italy and northwestern Europe and on the Eastern Front.[15]

Command issues on the Allied side were significant, but Ira Eaker defended the ground commanders to Henry "Hap" Arnold regarding their conduct during the battle if not before, noting that he and Carl Spaatz had both walked the battlefield and felt that trying to push north onto the high ground with two divisions would have allowed the Germans to cut them off from the other three, especially because they had ten divisions to the Allies' five, even if without any air cover. He also noted that a landing north of Rome would have been too far from air cover, giving the Germans an ideal opportunity to crush the effort. As with George Kenney's operations in the southwestern Pacific, airpower dictated the distance ground troops could leap from one amphibious assault to the next in Sicily and Italy.[16]

To counter events like the Anzio landings, the Germans developed a *Katastrophedienst* (catastrophe service) through which commanders could commandeer local troops and labor, along with the services of the Todt Organization, to repair marshaling yards with major damage. They had long since begun detraining their troops just outside of Italy or, in some cases, north of Florence, and then road-marching them to the battle area—resulting in tired troops who were under air attack by smaller aircraft much of the way, worn-out trucks, and gasoline shortages.[17]

As Operation Shingle went from "wildcat" to "beached whale" in Winston Churchill's words, the Cassino attack began in an effort to support Operation Shingle and break the Gustav line. Eaker told Arnold,

> Little useful purpose is served by blasting the opposition unless the Army does follow through. I am anxious that you do not set your heart on a great victory as a result of this operation. Personally, I do not feel it will throw the German out of his present position completely or entirely, or compel him to abandon the defensive role, if he decides and determines to hold on to the last man.[18]

Allied aircraft were tasked to bomb Cassino for three hours after the offensive began, which was delayed for three weeks by weather, with infantry advancing immediately on the heels of the last bomb detonations.[19]

With the Anzio attack a failure, the Eighth Army would move forward on its own rather than in tandem with the Fifth Army, against strong defenses and in atrocious weather. The air attack on Cassino, and the monastery at Monte Cassino, went forward on 15 March with 164 B-24s, 114 B-17s, 105 B-26s, and 72 B-25s—455 aircraft. Bomb plots confirmed that 47 percent of the weapons landed within a mile of the center of town and that the other 53 percent fell near the town or on Monte Cassino. Formations of eight fighter-bombers took to the skies every ten minutes, attacking every visible target below the overcast. A total of 319 fighter-bombers and day-bombers participated, with 312 fighters flying air cover and engaging ground targets when opportunities permitted. Artillery followed with an eight-hour barrage by all types of guns firing 195,969 rounds and then supporting advancing troops with a creeping barrage lasting 130 minutes. Targeted firing by more than 200 guns against known strongpoints in the ruins of Cassino followed. This blizzard of steel was unprecedented in the Mediterranean campaign. It was more like something from World War I in its ferocity as well as its ineffectiveness. Although this air and artillery onslaught killed or wounded half of the German defenders, it did not deter the rest, who held their ground and put immediate and heavy fire into advancing infantry.[20]

When Churchill asked on 20 March why the battle was not yet won, Harold Alexander noted the difficulties of weather and terrain and then emphasized, regarding the German First Parachute Division, "I doubt if there are any other troops in the world who could have stood up to it and then gone on fighting with the ferocity they have."[21] It might have been worth the Führer's time sending these troops against Malta in summer 1942 rather than bleeding them to death in Italy later.

Direct air-ground cooperation at Cassino was nearly impossible given the complex terrain and closeness of opposing troop formations. The Allies eventually occupied the town, but the price was too high, and they could not achieve a breakthrough. The decision to bomb Cassino and the abbey at Monte Cassino was taken within the letter of General Eisenhower's statement that the choice between destroying the monastery and reducing Allied casualties was clear. Nonetheless, it was a sad and ultimately ineffective operation not worthy of the brave men who fought and died in it. The fact that senior leaders persisted with the effort even after the failure of Operation Shingle is even more difficult to understand. Air assets had engaged in one of the largest bombing raids of the campaign, and it did no good. There were limits. The Allies had found and tried to overcome them without success.[22]

An immense amount of ink has been spilled on the Cassino attacks. After all the scholarly dust settles, we remain faced with the reality that no combination of air and ground action could have dislodged the Germans from these strong defensive positions, at this time of year, on a narrow and mountainous peninsula, using the technologies available at the time.[23]

Planning for Operations Strangle and Diadem

After the twin disasters at Anzio and Cassino, Allied commanders began planning for a combined air-ground-sea offensive to begin in May, when the weather was good and the roads dry, to take Rome and exploit the breakthrough from there. Operations Strangle and Diadem were thus born. The former was designed to maximize the disruption and dislocation of German logistics from the air, and the latter was the major combined-arms offensive designed to capitalize on the effects created by air attacks. Decisions at the first Quebec Conference the previous year had reconfirmed the continuing priority of Operation Overlord and called for "unremitting pressure" on German forces in Italy to keep them from reinforcing troops in France. The air and combined-arms efforts in Operations Strangle and Diadem must be viewed in this light. General Alexander believed he could do everything asked of him and more. In fact, he planned not just to destroy Army Group C (comprising the German Tenth and Fourteenth Armies) but to take Rome and then push north into the Po River Valley. He believed that airpower would be a major player in breaking the Gustav line and destroying the enemy south of Rome. Cannon collocated Headquarters MATAF with Alexander's headquarters to maximize air-ground coordination.[24]

The strategic considerations that General Henry Maitland Wilson, the new commander in chief (C in C) Mediterranean, faced at Armed Forces Headquarters (AFHQ) were complex. Despite the stalemate, twenty-eight Allied divisions were keeping twenty-four German divisions in Italy and away from both the Eastern Front and the defenses in France. Wilson and Alexander agreed that the Allies must keep the Germans from releasing divisions to other fighting fronts. Wilson too placed a high emphasis on the role of airpower in hampering the enemy's reinforcement and resupply. He also felt that taking Rome was vital not just for its political value but because of its airfields. Wilson knew the Germans would fight hard to keep Rome for symbolic reasons, and he sought to take advantage of this by drawing them into a major battle they could not win. On 26 February, the Combined Chiefs of Staff (CCS) declared at Wilson's urging that the Italian campaign would have priority over all other operations in the Mediterranean. The sole exception was Operation Anvil, the invasion of southern France, for which planning would proceed, and for which the CCS would allocate resources from the Italian front should that prove necessary.[25]

Within this larger strategic context, it is important to note that Wilson did not give final authority for Operation Diadem until after he received a 19 April 1944 directive from the CCS telling him that Operation Anvil (later renamed Dragoon for reasons of security), the amphibious assault on southern

France, would be postponed indefinitely and that he would proceed with the major offensive in Italy. By then Operation Strangle, the air effort to disrupt German LOCs to the point where the enemy could not resupply sufficiently to maintain a successful defense south of Rome, had been going on since 24 March.[26]

Within this larger set of developments, airmen knew Operation Strangle would have substantial effects but probably not to the point of forcing a German retreat exclusively through the use of airpower. A MAAF intelligence assessment of 22 November 1943 had concluded that just 5 percent of total rail and road capacity in Italy, along with seaborne supply deliveries, would allow the Germans to bring in enough supplies to hold their positions. Only a major combined-arms campaign would have any chance of dislodging them by forcing German troops to consume ammunition and supplies faster than they received them.[27] Eaker emphasized this when he said, "*The air phase* of the plan which will win Rome for us and *eventually* force the enemy back to the Pisa-Rimini line has been carefully worked out also. It calls for cutting lines of communication, road and rail, and the destruction of enemy coastal shipping to the point where he cannot possibly supply his 17–20 divisions South of that line."[28] Thus, although the stated objective of Operation Strangle was to "reduce the enemy's flow of supplies to a level which will make it impossible to maintain and operate his forces in central Italy," few senior officers failed to understand that only a major and well-orchestrated combined-arms offensive would succeed in forcing the Germans to retreat.[29]

Some airmen had their moments of excessive enthusiasm in the planning process. On 8 February 1944, Eaker issued MAAF Instruction No. 8, which sought, in the best case, to force a German withdrawal because of lack of supplies. During Operation Strangle, MAAF assets would attempt to make the German supply situation untenable over the long term, forcing the Germans to retreat. However, it is not clear from MAAF Instruction No. 8 or other sources whether they expected air attacks to do this alone, and it would have been utterly unrealistic to think that the Germans would simply leave their strong defenses as a result purely of air attack. It did in fact require a major combined-arms effort to drive back the Germans. Operation Diadem, coming hard on the heels of Operation Strangle, forced the Germans to use up their supplies, which air attack rendered insufficient for high-intensity combat, and then to retreat or be defeated in place.[30]

"It is evident," Eaker's 8 February order said, "that the communications system in Italy was uniquely susceptible to interdiction by air attack"[31] given its channelization along three rail lines and their branch lines. Prisoner of war (POW) reports made clear the seriousness of logistical problems, including a lack of fuel and especially artillery ammunition. German units, the report noted, "arrived in the battle area only after having suffered considerable casualties, losing large quantities of their motor transport and heavy equipment, and being dispersed to such an extent that they could not enter the battle as an integrated unit."[32] The air forces had eight weeks to accomplish their objectives—the absolute minimum time planners calculated would be necessary and another clear indication that a major ground offensive was part of the larger picture from the outset.

By 15 January, it was clear that Solly Zuckerman's call for all-out attacks on marshaling yards rather than rail lines and bridges would not work in Italy given its geographical realities, including its many small and redundant LOCs. The Germans continued bringing in enough supplies to keep fighting. Airmen turned to an interdiction campaign involving concerted strikes on specific stretches of railroads, key road junctions, bridges, and viaducts. Even this did not do the trick by itself, but it worked very well in concert with the ground offensive. This combination of attacks on key stretches of railway, bridges, and marshaling yards proved effective here and later in Normandy, the Reich, and southeastern Europe.[33]

Based on Arnold's efforts to use lessons learned across various theaters, he told Eaker that the new bridge bombing technique in the China-Burma-India theater was working very well. Mediums attacked from 300 feet of altitude, dropping delayed-action 1,000-pound bombs. These blew out the piers and rendered bridges virtually irreparable. He asked Eaker to look at this closely and try it. Airmen employed this tactic with great success during Operations Strangle and Diadem and in every other major campaign leading to Victory in Europe Day.[34]

Writing after the battle, Brigadier General Lauris Norstad said the Allies had done the right thing in employing air assets to prevent adequate reinforcements of supplies or personnel and also forcing abnormally large supply expenditures with the ground offensive. "This," he said, "was DIADEM—an operation designed to make the enemy 'burn both ends against the middle' . . . to make it impossible for the enemy to maintain his forces on his present line in Italy in the face of a combined Allied offensive."[35] This included maintaining air supremacy, going after every known logistics and supply target, and giving direct support to the armies.[36]

Operation Strangle

During the winter of 1943–1944, despite atrocious weather, air forces went after LOCs and defensive strongpoints without letup. This had not dislodged the Germans even in conjunction with a maximum effort on the ground around Cassino. Consequently, Wilson said, "a different employ-ment of our superiority in the air obviously was called for and the Air Forces responded promptly."[37] He noted that indirect support efforts reached ma-turity in spring 1944, producing what he said "has since been accepted as a masterpiece of tactical air procedure—the so-called 'Operation Strangle' to interdict enemy supply lines."[38] Rather than going after a few major marshaling yards with heavy-bombers, MAAF sent fighter-bombers and medium-bombers after bridges and key railroad lines to keep them out of commission.[39] Operation Strangle began officially on 24 March—nearly two months ahead of Operation Diadem. Wilson met with the senior air commanders—Eaker and John "Jack" Slessor—on 25 April after receiving his directive from the CCS and directed them to develop a modified bomb-ing plan by 28 April that Generals Alexander, Cannon, and Nathan Twining would agree was an effective contribution to Operation Diadem. The shift from an air to a combined-arms effort was under way.[40]

Up to 11 May, Operation Strangle involved 65,003 sorties delivering 33,104 tons of bombs on German LOCs. After Operation Diadem began on 12 May, MAAF kept flying, turning in another 72,946 sorties and drop-ping another 51,500 tons of bombs by 22 June. These attacks destroyed 6,577 motor vehicles, cut every rail line, and sank most small vessels trying to resupply German armies. On 4 April, Kesselring noted that the German Tenth Army, facing the US Fifth Army, and the German Fourteenth Army, facing the British Eighth Army, were receiving 1,357 tons of supplies each day rather than the 2,261-ton daily minimum for sustained operations.[41] The Allies needed to force the Germans to engage in battle and use up these insufficient supplies, which is what Operation Diadem did.

There is no question that Operation Strangle placed severe strains on the German logistical network, but it could not undermine the German supply effort on its own. In a memo to Charles Portal on 16 April, Slessor discussed why an air effort could not force a German withdrawal. "Perhaps most important of all," he said,

We are not forcing him to expend fuel or ammunition. I know it is necessary for the Army to rest, refit, regroup and train divisions for a major offensive, and that it is very difficult if not impossible with our present strength to maintain constant offensive pressure at the same time. But we must face the fact that by these long periods of inactivity the Army is automatically making its own task more difficult when it does not resume the offensive, by making it more dif-ficult for us now to reduce the enemy's capacity for resistance. As Joe Cannon said at Alexander's conference a fortnight ago, in this phase it is not a question of the Air supporting the Army—the Army must support the Air by making the Hun expend fuel and ammunition while we prevent him replenishing his supplies.[42]

On 28 April, with both Eaker and Slessor mindful of the inability to achieve any kind of victory without sustained combined-arms operations, the MAAF director of intelligence (A2) issued an updated assessment of German supply requirements: 4,000 tons per day. This was higher than Kesselring's 2,261-ton figure, which represented the absolute minimum requirement. Analysts assessed that 700 tons came by sea and 3,300 by rail and road. Attacks on rail lines and bridges would force a greater use of trucks to haul supplies around damaged chokepoints. This would exacerbate German truck and fuel shortages, starving them of these two vital commodities and hampering the delivery of ammunition and fuel to the front. The intelligence assessment estimated that German combat units had enough ammunition for perhaps thirty days of sustained operations and enough fuel for about ten. The start of the ground offensive would increase supply requirements to 5,500 tons per day (it in fact proved closer to 4,000 tons), of which 900 would have to come by sea. The bridge and rail campaign that began in February had already forced a much heavier use of trucks to carry supplies to the front; 1,500 fell prey to Allied planes in February and March. The A2 estimated, accurately, that the air-ground effort during Operation Diadem would keep the Germans at least 1,300 tons short of their daily supply requirements.[43]

Accordingly, as Slessor noted, Eaker's 28 April 1944 outline air plan for Operation Diadem reflected "our growing recognition that we have been unduly optimistic in our original hopes for *Strangle* in the directive of March 19."[44] As Slessor had recently told Portal, only a major combined-arms effort would work. The air portion of Operation Diadem thus had three elements, all of which built on efforts during Operation Strangle: neutralize remaining German Air Force aircraft, deprive German troops of adequate supplies, and provide direct support to Allied armies. These efforts would occur in three phases:

1. Preparatory: During 12 days prior to D-Day (attack airfields, LOCs, ports, and supply dumps with oil in top priority, followed by ammunition, and then other kinds of supplies; MASAF [Mediterranean Allied Strategic Air Forces] attacks as required to complement MATAF raids).

2. Assault: Duration not defined (continued LOC attacks, harrying a German retreat if and when that occurred, and providing CAS; MASAF ready to block major retreat/resupply routes).

3. Sustained Offensive: MATAF and MASAF provide any required assistance during pursuit.[45]

Meanwhile, MATAF began interdiction efforts. Medium-bomber crews were getting steadily better at knocking out bridges (from 200 tons of bombs per bridge in January to 62 tons by the middle of May). They also attacked marshaling yards and repair facilities. Fighter-bombers raided small marshaling yards, moving trains, key stretches of track, and motor transport. They also interfered with repairs at larger marshaling yards whenever weather grounded the mediums. Supply dumps near the front suffered severely whenever PR assets discovered them. However, most sorties went against targets more than 100 miles from the front to force the Germans to demotorize combat units by taking their trucks from these units and using them to bring supplies forward from the distant railheads, with the attendant difficulties and risks.[46]

On 11 April (a month before Eaker's formal directive for air action during Operation Diadem but more than a month into Operation Strangle), Kesselring appointed General Rudolf Wenninger, a Luftwaffe pilot and member of his staff, "General with special responsibility for the maintenance of rail communication in Italy." Railroad repair became a precise operation, with special engineering teams, equipment, and laborers moving to the points of greatest importance according to a process of centralized control and prioritization. Wenninger was a master of improvisation and used every means available to keep railroads open and supplies moving, but Allied airpower made his job extraordinarily difficult. Nonetheless, flexibility and repair prowess, lack of adequate Allied concentration and persistence, and relatively plentiful locomotives and rolling stock at the start of the air offensive kept just enough supplies moving forward to maintain a defensive posture on a stationary front. As capable as the Germans were, they could not overcome the high consumption of motor fuel and the steady attrition of motor transport by air attack and as a result of wear and tear. When Operation Diadem began, the wounds that Operation Strangle had inflicted became apparent. The Germans lacked the necessary transport to move the increased quantities of already insufficient ammunition supplies consumed in high-intensity combat operations, to bring forward reinforcements, and for tactical movement of troops. Slessor's fusion of the Zuckerman plan to attack marshaling yards and the MAAF A2 plan for interdiction of through lines, bridges, and other chokepoints thus worked as well as could reasonably be expected.[47]

On 10 May, two days before Operation Diadem began, Cannon issued a final directive. Brigadier General Gordon P. Saville, Twelfth Tactical Air Command (TAC) and Air Vice-Marshal W. F. Dickson, DAF, would maximize pressure on LOCs and focus their initial efforts on artillery and ammunition supplies. Rover David (Rover Joe for the United States) would provide immediate direct support using the Cabrank method of fighter-bomber formations. Cabrank kept a steady succession of six-aircraft fighter-bomber formations in the air to respond to immediate direct support requests and attack fleeting targets. A forward air controller (FAC) at a forward fighter control unit would radio the appropriate visual control post (when the FAC could see the target) or forward control post (when the FAC could not see it) to notify fellow FACs of the incoming aircraft. They provided terminal attack control. On 12 May, Allied air forces had 3,960 operational aircraft, excluding transports. The Germans had about 565. The Allies also had a massive artillery advantage and targeted every German gun, whether with artillery fire, airstrikes, or both.[48]

German intelligence officers said Cabrank was highly effective. A tank or armored car with very high-frequency (VHF) radio, and a ground station with height-finding (HF) radio, both located on high ground, established direct communications with fighter-bombers. These rover signal units controlled aircraft from a specific US Army Air Forces (USAAF) wing or Royal Air Force (RAF) group. Aircraft took up positions in a prearranged aerial waiting area from which the rover stations called them onto targets as needed. Because rover units had direct communications with ground combat units as well, they could orchestrate direct support missions with little delay. All three players—rover units, aircraft, and ground units—had identical maps and aerial photos with known enemy strongpoints and heavy weapons clearly marked. German Signal Intelligence Service (SIS) officers said that the USAAF led in the refinement of this deadly technique after 1943. It proved valuable during Operation Diadem and in the Normandy breakout as retreating German units became ideal targets.[49]

Operation Diadem

The Allies had two armies comprising seven corps and twenty-eight divisions for Operation Diadem, which lasted twenty-four days and was a major success. To keep the Germans guessing, there was no softening up of defenses with air or artillery attack prior to the 12 May jump-off.[50] On 11 May, as troops readied themselves, every heavy-bomber in MASAF overflew the front in a show of force en route to targets in Romania and the Reich as tactical air assets hammered defensive positions. This combination of attacks and shows of force did not cause large-scale panic among German troops, who were too well disciplined and used to constant air attacks. Still, the spectacle of nearly 1,000 heavies and escorting fighters parading overhead slowly and out of reach of flak, even as another 2,000-plus tactical aircraft hit everything they could find along the front, simply reminded the

Germans of the constant nightmare they labored under when it came to Allied control of the air.[51]

The offensive opened with 1,700 guns and howitzers pounding German positions, firing up to 420 rounds per gun during the offensive's first twelve hours. The Germans were confused by the barrage all along the front and further disoriented by heavy and successful air attacks on their headquarters. Sixty-five B-17s raided the headquarters of Army Group C, and sixty-eight others destroyed the headquarters of the Tenth Army. MATAF B-25s and B-26s raided the headquarters of the First Parachute Division, Forty-fourth Infantry Division, and Fifteenth Panzer Grenadier Division. To further compound the command-and-control (C2) disaster these raids caused, Colonel General Heinrich von Vietinghoff (Tenth Army commander) and General Fridolin von Senger und Etterlin (Fourteenth Panzer Corps commander) were away for two and five days, respectively, after Operation Diadem began. As air attacks intensified, Allied troops unleashed a sophisticated combination of mechanized and mountain warfare, in which the US and British troops moved up the roads and the Free French through the most difficult mountain passes, bypassing and isolating German units. Bostons from the Twelfth TAC dropped water, ammunition, and food to the French units. Whenever German troops moved to counter an advance, they came under immediate fighter-bomber and frequent night-fighter attack. On 18 May, the Poles secured Cassino and the entire surrounding area, including Monastery Hill, making any German effort to hold the Gustav line hopeless. Subsequent attacks on the Hitler line made clear that it would not hold either. General Siegfried Westphal, chief of staff, Army Group C, complained bitterly about "those damned French hanging round our necks" as they executed their flanking tactics.[52]

The Germans had seventeen divisions south of Rome and another seven to the north. They brought in three more but could not make up for high losses. "A factor of immense importance throughout the battle," Wilson said, "had been our great air superiority."[53] Attacks on LOCs and supply dumps had left the Germans unable to wage a protracted battle, and their large-scale movements during the day caused heavy casualties as fighter-bombers attacked. Between 12 and 31 May, MATAF destroyed 2,556 trucks and damaged another 2,236. In June, these figures increased to 3,318 and 3,302, respectively. After Rome fell on 4 June, the advance continued at seven miles per day. German troops were badly beaten, in the open, and under constant air attack.[54]

By 22 May, von Vietinghoff estimated that Tenth Army had suffered at least 10,000 casualties. Moreover, the Allies had broken through at several points, and efforts to reinforce and resupply were largely unsuccessful as a result of the constant air attacks. Clark's Fifth Army was supposed to break out from the Anzio beachhead and get astride the enemy's LOCs to Rome. Had Clark done so effectively, the Allies would have surrounded virtually all of Army Group C. Nonetheless, Allied deception efforts had led Kesselring to place his heavy divisions in the wrong locations, which worked in combination with airpower to hinder their approach to the battle area and then to escape from it.[55]

The Germans could not stabilize their front because Allied commanders, troops, and airmen had performed well with few exceptions (Clark's divergence from the plan being the main case in point). From 19 May to 5 June, MAAF bombers flew 9,226 sorties and dropped 5,673 tons of bombs (542 sorties and 334 tons of bombs every twenty-four hours). The "AOK 10 [Tenth Army] Withdrawal Triangle" was crucial, with tactical reconnaissance (Tac R) and fighter-bombers directed by rover teams forming a lethal combination. Tac R pilots led fighter-bombers directly to enemy formations.[56]

On 20 and 21 May, airstrikes destroyed all bridges over the Liri. Two reinforcing panzer grenadier divisions arrived late and in small packets because of concerted airstrikes. On 22 May von Vietinghoff lamented the near impossibility of bringing forward reinforcements over shattered LOCs and in the face of constant air attack. Fighter-bombers strafed key points in conjunction with artillery barrages, causing heavy casualties. In a 29 May report to Hitler, Kesselring praised his troops' excellence in the face of Allied air dominance. He also reported great delays in troop movements because of "road hunts" by night-bombers. On 30 May alone, Allied aircraft claimed 304 vehicles destroyed. The effects of air attack were cumulative and ultimately severe.[57]

Artillery and aerial bombardment caused heavy German casualties on 23 May and paved the way for the breakout to Rome. German estimates were 50–60 percent casualties in the two most affected divisions. German C2 also collapsed as a result of damage from artillery and air attacks. After German troops retreated, fighter-bombers swarmed the columns, destroying vehicles and creating huge impediments to travel over cratered roads choked with debris.[58]

Wilson said, "Throughout the second phase of the offensive, air power was a decisive factor on both the main front and beachhead [at Anzio] and much of the success was due to the finely coordinated team work of air and ground forces."[59] The weather was not good, but MATAF went after troop concentrations, gun emplacements, and ammunition and fuel dumps without letup. These attacks created havoc among German forces trying to counterattack or engaging in any kind of maneuver. The breakout from the Anzio beachhead on 23 May marked the beginning of the end of effective German resistance. By the time Operations Strangle and Diadem had achieved their combined purpose, the Germans had lost Rome and suffered 80,000 casualties. Twenty divisions were combat ineffective, they had lost 15,000 vehicles, and their supply situation was disastrous as they sought desperately to regroup.[60]

During Operation Strangle, the Germans lost about 20 trucks per day. During Operation Diadem, the number skyrocketed to more than 100. The Germans could not replace them, nor could they push through enough supplies by rail or sea given the heavy consumption rates for ammunition and fuel in particular. The result was a defeat followed by a rout. Ground and air assets working together and in a mutually supporting fashion facilitated this victory.[61]

Based on hundreds of POW interrogations, Allied Forces Headquarters (AFHQ) G2 concluded, "The almost uniform consensus of all the reports to date supports the idea that the largest factor in the collapse of enemy transport after D-Day was due to the breakdown of local distribution immediately behind the front."[62] These cascading effects began with the blockage of rail lines and the destruction of bridges leading to the front, which forced all available trucks to drive to the railheads and pick up supplies. They did so under constant threat of attack, at high risk of accidents driving over bad roads at night, and by using nearly as much fuel as they brought back. After the distribution of supplies broke down behind the lines, commanders had to send their remaining trucks to the supply points, causing major losses of vehicles and further demotorizing their units. After they had to retreat, the results were often disastrous as more mobile Allied formations caught them on the move and engaged them in conjunction with direct air support.[63]

In an effort they had now begun to perfect, Allied intelligence officers, planners, and operators kept very close track of repair efforts on rail lines, bridges, and other key targets. Aircrews would then restrike them as they were about to reopen. This agility in the intelligence-planning-operations cycle proved to be another key piece in the operation's success and would also carry forward into the war's remaining campaigns.[64]

During a visit to the battle area on 18 June, Arnold remarked,

> Took cars and drove to Viterbo. What a trip. Our Air Forces have destroyed 5,200 vehicles and damaged 5,000 more. I think I saw them all along the road to Viterbo. Tanks from the big Tiger with its 88 mm. down to the small baby size—trucks, large and small—half tracks—passenger cars—lorries, busses, rolling kitchens—wrecking trucks—winches—[fuel] tank[er] trucks—gasoline

drums—trucks loaded with everything—typewriters, food. Some burned—some half consumed—some just wrecked. In most cases the rubber tires were off the wheels. Most had been dragged from the road either by the Germans to aid in their escape or by our troops to open the road. These wrecks were along the road for over 40 miles and gave pictorial evidence of the desperate straits the Germans were in as they fled northward—with their divisions breaking up and disintegrating into small helpless bodies of men. This all caused by our Air Force striking with machine guns and bombs at the retreating columns.[65]

Given the nearly complete absence of the Luftwaffe during the offensive, Arnold concluded his note with the comment, "The general impression among the higher officers in the Allied Air Forces is that the high command of the G.A.F. has made one blunder after another, not only in their technique of employment but also in their grand strategy."[66]

Norstad emphasized that

DIADEM was a synchronized assault—an outstanding example of *mutual* support. . . . It is axiomatic that the success of combined operations depend [sic] upon complete coordination between the services. The means adopted in the Mediterranean to insure this degree of coordination and cooperation has [sic] served the theatre well. . . . This is the system, the formula for attack that grew up in the Desert victory. It has survived and sharpened its striking edge in the Battle for Italy.[67]

In his after-action report, Norstad delivered a commonsense summary of what airpower was able to do, and not to do, within the context of the Italian campaign. He emphasized first that airpower alone could not possibly have defeated a highly organized and disciplined army in ideal defensive terrain, even if the opposing army was entirely without air support. Nor could it force any kind of retreat on its own. Despite airmen's best efforts, they could not entirely prevent the movement of reserves to the front or tactical formations from one part of the front to another. They could slow them down and cause serious losses, but that was all. Airpower could not absolutely isolate the battle area from enemy supplies or reinforcements, nor could it stop all efforts at maneuver. However, what it could and did accomplish in Operation Diadem was to create an insuperable supply problem for the Germans in conjunction with the ground offensive, which forced them to expend their artillery ammunition stocks and then to maneuver and use up their limited fuel stocks as the Allies gained ground. After the battle became mobile, fighter-bombers and night-bombers swarmed the retreating Germans. Airpower thus played an important role in both the breakout and the rout afterward. Given the nature of the ground and the toughness of German soldiers, Allied ground forces could not have broken through had the Germans been able to mass, maneuver, reinforce, and resupply to anything approaching the same degree as Allied armies did.[68]

Air supremacy gave the armies freedom of action and denied it to the enemy. The Germans had to receive supplies and reinforcements from a Reich that had too few of either to give. What was available to keep them going had to make it through marshaling yards in northern Italy damaged severely by heavy-bombers. From there, logistical efforts relied on highly unpredictable rail transportation south, with multiple unloadings and reloadings. When within 50 to 100 miles of the front, depending on how far back the rail lines were cut, they relied on a worn-down motor transport force that suffered increasing attrition. After they were in place—if they made it at all—men and supplies went immediately into heavily camouflaged positions and became static. Efforts to supplement supply in Italy by sea proved exorbitantly costly and declined over time. It is always easy to point out what airpower was not capable of achieving, but in doing so its critics overlook the immense friction, attrition, and loss of maneuverability it created in the German ranks—and the opportunity this gave the Allies to craft and carry out a successful combined-arms operation that savaged the German Army, forced Hitler to send scarce additional divisions, and brought Allied bombers and escort fighters that much closer to key target sets in the Reich. Nonetheless, the Allied ground scheme of maneuver, which forced the great expenditure of fuel and ammunition and pushed German units back, slowly at first and then at an accelerating pace, was at the heart of this combined-arms success.[69]

Some airmen had been overly optimistic about what they could accomplish, Norstad continued, when they first set forth in February 1944 the objective of forcing the German Army to retreat, in the best case, without an associated ground campaign, and that was unwise from the perspective of cooperation as well as imprudent given recent experiences with airpower in Italy, Sicily, and North Africa. They realized their error in judgment as Operation Strangle unfolded, admitted this in their after-action report, and made it clear that it took not just the intensive air effort in the weeks and months leading up to Operation Diadem, but twenty days of intense air-ground-sea operations during Operation Diadem, from 12 May to 1 June, before the German Army broke and retreated. Field surveys and interrogations made it clear that the depletion of fuel and artillery ammunition, and the inability to rebuild stocks in the face of nonstop air attacks, proved decisive in conjunction with a major ground offensive in facilitating the breakthrough.[70]

Pursuit

Unfortunately, slow pursuit into early June and the lack of a large night strike force allowed the Germans to escape.[71] On 4 June, the day Rome fell, an aggravated Alexander said, "If only the country were more open we would make hay of the whole lot. However you may rest assured that both Armies will drive forward as fast as is physically possible."[72] Clark's decision to depart from the plan in an effort to speed the capture of Rome did the larger effort no good. Nonetheless, on 7 June Alexander effused, "A system of co-operation between my armies and M.A.T.A.F. has now been evolved whereby the full power of the air can be developed in support of my land operations. Any weakening of this combination would greatly decrease the value of this very powerful weapon of co-ordinated effort."[73] Clearly, he had seen the light since the early days in Tunisia.

Slessor was nevertheless astounded at the slow pursuit. He quoted an official letter from Clark—the only general he thought at all impressive in the theater—that in effect rehashed RAF and USAAF doctrine developed over the past three years. "That," Slessor said of the excerpt, "might well be an extract from a lecture at ANDOVER [the Army Staff College, where he taught] in 1924."[74] Slessor's publication of the prescient and timeless *Air Power and Armies* in 1937 underscored his understanding of the importance of effective combined-arms efforts. On a positive note, he said Eaker was a "first-class little man and I think is doing very well."[75]

Slessor was also deeply concerned about the geographical separation of key headquarters and believed this had also played a role in allowing the remnants of Army Group C to escape. "It is obviously undesirable," he said, "for A.F.H.Q., the Naval C.-in-C., and Air C.-in-C. to be widely separated. It is quite impossible to exercise effective control over air operations from ALGIERS—one might just as well do it from LONDON."[76] This division of the high command into three separate locations was an "infernal business" in need of immediate repair, and the planned move of AFHQ to Caserta should happen as quickly as possible. As it turned out, AFHQ did not move to Italy until April 1945, confirming Slessor's fears that the command might stay geographically divided, with the negative repercussions that followed for the campaign in Italy.[77]

Twelve days after the start of Operation Overlord, Slessor wrote to Arnold, at the latter's request, summarizing the most important things he, Eaker, and the air-ground team had learned during Operations Strangle and Diadem. Slessor began by making clear that airpower could not by itself defeat a highly organized and disciplined army even if it had no air support of its own. Nor could it force a major withdrawal just by slowing the flow of supplies. Ground and air forces had to support each other with maximum effect to bring about the enemy's defeat, in this case by slowing the flow of supplies from the air while forcing their consumption on the ground. In conjunction with the ground offensive, this made it impossible for the Germans to offer prolonged resistance even though the ground war was ideally suited to their defense. It also turned the German retreat into a rout and played a vital role in the elimination of most German divisions as effective combat units.[78]

The converse, he continued, was equally true of the Allied armies: "No-one who is familiar with Italy with the vast supply dumps and camps—laid out without regard to dispersion—the crowded shipping in Naples Harbour, the endless columns of vehicles almost nose to tail on the roads right up almost to the front line, the railways working to capacity day and night, the packed airfields, can adequately appreciate the appalling difficulty of supplying, maintaining and moving a great Army in the conditions prevailing on the other side of the line. It is hardly an exaggeration to say that the Army can safely disregard, and has virtually disregarded in Italy, the existence of an enemy air force."[79] Slessor also pointed to the wealth of Allied reconnaissance, in contrast with the German inability to gather any, and to the superb interactions between Allied spotting aircraft and artillery.

He noted that even with major LOC strikes before Operation Diadem, the ground battle was very hard fought from 12 May to 1 June, when the heavy use of artillery ammunition, fuel, and other consumables made it impossible for the Germans to resupply in the face of a combined-arms offensive. Slessor said the army played the pivotal role, but that "there is not the slightest doubt the break *through* would have been impossible but for the air."[80] In addition, even though airpower could not keep German reserve units from reaching the line, it caused such long delays and heavy losses that Kesselring was unable to engage in the flexible tactics he preferred. The Hermann Göring Division's inability to slow, much less stop, the Allied breakthrough around Valmontone was a key case in point.

Slessor concluded with Trenchard's assertion that "all land battles are confusion and muddle, and the job of the air is to accentuate that confusion and muddle in the enemy's Army to a point when it gets beyond the capacity of anyone to control."[81] This is exactly what the air attacks did to the German Army in Italy in the critical last days of May and first days of June. As Slessor emphasized,

> Roads were cratered and blocked by destroyed vehicles, telecommunications were cut, villages became a mass of rubble barring through movement, local

reserves could not be moved because there was no petrol available, forward troops were out of ammunition and out of touch with their controlling head-quarters, nobody knew for certain where anyone else was, and the troops were hungry, thirsty, tired and demoralized by constant attacks from the air. Above all, perhaps, the enemy was deprived, by the impossibility of rapid and coherent movement, of that tactical flexibility which has always been such an admirable quality in German defensive fighting.[82]

In a note he wrote to the RAF's Air Historical Branch nearly thirty years after Operations Strangle and Diadem, Slessor discussed a memorandum he had penned in April 1944, in which he put forward principles for the most effective employment of tactical airpower for the offensive. He recognized Operation Strangle's shortcomings from the soldier's point of view but also emphasized that the plan was never intended to force a German withdrawal on its own, despite some airmen's hopes that it would do so, and as some people later claimed was its objective. Slessor said,

> We thought no such thing. . . . Indeed we were constantly emphasizing that the Army *must* co-operate with the Air Force by maintaining pressure, to force the enemy to *expend* ammunition, supplies, etc. There is certainly nothing in the April memorandum to encourage the idea of independent action by the air. . . . Who could seriously imagine that nineteen German Divisions, after some weeks of bombing of their supply lines (in Italian weather) would wake up one morning and say "by Jove, we have not got any supplies of ammunition or petrol or rations left—come on chaps" and just pack up and move North, without any pressure by the Allied armies on the ground? Of course not.[83]

He emphasized that the role of tactical air assets was to exhaust the Germans and their supplies in conjunction with ground forces after Operation Diadem began, allowing for a more rapid breakout with lower losses. He did acknowledge that the air forces should have focused more effort on delaying the movement of German troops and reinforcements. In his last comment about the April memo, Slessor opined that too few senior officers had read it and fairly laughed when he said, "I would not mind having a small bet that Alexander never read it at all!"[84] He closed by stating that the Fifth and Eighth Armies "would not have had a *hope* of breaking the Gustav Line without M.A.T.A.F.'s air cover and support."[85]

The Germans corroborated what Slessor and other commanders said about the role of airpower in the battle. The level of friction and outright chaos these nonstop air attacks caused is impossible to imagine without read-ing carefully the 200 or so POW interrogation reports about the casualties, vehicle losses, immobility, abject frustration and despair that accompanied the raids, and lack of warning when fighter-bombers strafed. German soldiers were tough and got on with their jobs, but no amount of will could make up for the massive drain on efficiency and effectiveness resulting from airstrikes. The 577th Panzer Grenadier Regiment, for instance, lost more than two-thirds of its vehicles in the first two weeks of the offensive—the norm for German units. Most were loaded with supplies. The retreat to Rome occurred by night and with commandeered Italian trucks. Another soldier said aircraft attacked convoys every ten minutes after they located them. Vehicles moved only at night, and even then only in case of a logistical or other emergency. For its part, the 525th Assault Gun Battery dispatched fifty trucks weighing three tons each that made the mistake of moving during the day. Fighter-bombers destroyed every one.[86]

After it became clear to Wilson on 22 May that Operation Anvil would likely proceed, he pushed Alexander to continue destruction of German forces in Italy and capture Ancona with its port and airfields. On 7 June, Alexander authored a review in which he reemphasized his focus on causing attrition of German units in Italy, forcing the Oberkommando der Wehrmacht (OKW) to send reserves there rather than to reinforce Normandy. He assessed that the twenty German divisions involved in Operation Diadem now had the effective fighting power of six. Even with the commitment of remaining reserves, the Germans could not man the Pisa-Rimini line with any more than ten division equivalents—two short of what his intelligence staff considered the bare minimum. Alexander felt he could breach it and be into the Po River Valley by late summer. He wanted to force the Germans to keep fighting a mobile battle as they retreated, which exposed them to very heavy losses from air attack and caused them to run out of fuel. Therefore, Alexander proposed a continuing offensive to keep the Germans off balance, drive them over the Apennine Mountains, take the Bologna-Modena area with its major airfields, deny the Po River Valley's agricultural resources to German armies, and allow for exploitation into France, Austria, or both. However, Alexander insisted that such a campaign would only be possible without diversion of land or air forces to an amphibious assault on southern France. Wilson thought Alexander would be able to continue the offensive into July before the CCS made any decision on Operation Anvil (later renamed Dragoon), but on 14 June it signaled Wilson to make preparations for the release of the US and Free French divisions. Efforts to keep Alexander's units in place and his offensive rolling failed. George C. Marshall and the Joint Chiefs of Staff (JCS) refused to consider operations in the Balkans, and Marshall felt that any effort to exploit beyond the Pisa-Rimini line would be largely ineffectual in terms of speeding Germany's defeat. Despite his stated support for Operation Dragoon, Wilson felt that the invasion could only help to ensure the war's end by spring 1945, whereas aggressive action in Italy with a breakout into the Ljubljana Plain or (even more fancifully) over the Alps might produce an Allied victory in 1944. Consequently, he backed Alexander's plan (unsuccessfully) for a continuing, major offensive in Italy as the best way to force German diversion of troops from other theaters and hold open the possibility of exploitation into Austria and southern France.[87]

However, by 21 June, German resistance was already hardening. By the middle of July, after US troops had come out of the line for Operation Dragoon, the Allied advance was grinding to a halt. The two French divisions left the line on 21 July. One more major push resulted in the capture of Leghorn and Ancona, but remaining Allied units were exhausted by the time they reached the Pisa-Rimini line, where German units were now well established in strong defensive works. Wilson concluded that the 1944 summer campaign had been very successful.[88] Yet, he continued, "The campaign was destined to be limited to tactical rather than strategic achievements however, because of the heavy diversions of troops from Italy for employment elsewhere, and its full promise therefore was not to be realised in 1944."[89] Nonetheless, dry ground, superb air support, and a three-to-one superiority in ground strength had allowed the Allies to pin the Germans, run them short of supplies, force them to move against their will, and then cause heavy losses during Operation Diadem. Wilson said of the air forces, "The employment of our air power opened a new chapter in the history of the effectiveness of this arm, and much of the success of the Italian campaign can be attributed to the punishment meted out to the enemy by the aircraft of the Mediterranean Allied Air force [*sic*] and to the protection given our own forces from the waning German Air Force."[90] By the time the advance ended at the River Arno, Allied armies had moved 250 miles north from their jump-off points.

After the battle, it became clear how heavily air assets had affected German troops and their ability to operate. A captured order issued during Operation Diadem noted that

> aerial reconnaissance detects our every movement, every concentration, every weapon, and immediately after detection smashes every one of these objectives. This is accomplished by a close coordination of Air Force and Artillery. Every soldier must be made to realize that the enemy's present superiority in the air is not of temporary duration—subject to time and location—but rather that it is a part of a permanent set of conditions that must be faced by our troops. Experience demonstrates that the enemy knows how to reconnoiter and destroy our every concentration. Every weapon detected by the enemy is destroyed

by coordinated fire, directed by his OP's. The enemy can conduct his artillery reconnaissance observation completely unhindered. Every vehicle must post an air look-out. The best means of locomotion for individuals, as well as whole units, is the bicycle. To find appropriate cover with lightning speed, practice is necessary. More than ever before, units must be trained fully to master the art of camouflage. It is not enough that soldiers know how to attach grass and twigs to their helmets. Complete camouflage of men and material is required.[91]

A captured artilleryman lamented, "With all your tanks against us, it is a sad state of affairs that we have to come along with our horse-drawn guns and six rounds a day to be fired only on the authority of the divisional commander."[92] The Germans had indeed run out of artillery ammunition and were still drastically short of it a month after the fall of Rome. A soldier from a German antitank company told his captors, "Our greatest fear here in Italy was your aircraft; it was impossible to move about behind the front during the daytime. We could only do so at night."[93] When asked about the combination of artillery and air spotting, he said that firing at an Allied plane was nearly suicidal. "If we had shot at them," he said, "it would have been the end of us."[94] Finally, he said, the effect of air attacks on vehicle movements was catastrophic during the day and substantial even at night: "If one of our vehicles so much as moved it would be shot to smithereens."[95]

Two artillery officers captured at Rimini on 2 September 1944 remarked that Allied artillery fire was very accurate, even at night. When the interrogator asked whether the Germans had maps with US artillery positions, one of them replied, "Yes, we've got maps all right—but no ammunition!"[96] They had twenty-six rounds for all the guns and had to move using oxen to pull the guns with the enlisted men sitting on them, as the other POW remarked, "just like Italians."[97] This shortage of ammunition was a key reason for the collapse of German defenses during the offensive and their inability to stabilize the line afterward.[98] A private, thinking back on the Italian campaign in December 1944, asked, "Isn't it interesting how the Luftwaffe just suddenly disappeared [after the Sicilian campaign]?"[99] Finally, and perhaps most instructively, a German corporal, contemplating a question about the role of airpower in the campaign, simply asked, "I wonder what would happen if the Germans had the Allied Air Force?"[100]

From April to September 1944, the intensity, reach, roles, and tasks of Allied aircraft in the Italian campaign reached their zenith. Although the scholarly debate about the campaign's necessity will continue ad infinitum, it is clear that air supremacy facilitated major combined-arms victories during good-weather months; took a fearful toll on German logistics, C2, and operational capabilities; and killed a very large number of Germans. The fact that many more German troops than Allied ones died, were wounded, or went into POW camps in a campaign suited ideally to a defensive effort waged by excellent troops underscores the role airpower played in this effort. As flawed as it was in many ways, from planning to execution, and however many detractors it has, the Italian campaign held from seventeen to twenty-eight German divisions in place at any given time and bled them white. It also netted the Foggia airfield complex and allowed for the "second aerial front" against the Reich. It was, at the end of the day, one part of a much larger, complex, and successful effort to defeat Nazi Germany and Japan in a global and always costly total war of attrition.

The Italian campaign, though important in tying down substantial numbers of German divisions until Victory in Europe Day, had played its major role by the end of the 1944 summer offensive. The air-ground focus shifted to Operation Dragoon in southern France, and heavy-bombers engaged in highly effective raids on German oil and transportation assets to demotorize the Wehrmacht on all fronts and speed the advance of the Grand Alliance's armies—and in particular the Red Army. Assistance to Tito's Partisans in the form of the Balkan Air Force also aided substantially in the latter effort and produced political results by helping to give Tito the credibility and power he needed to stay out of Stalin's camp. We now turn to these crucial aspects of airpower exploitation in the war's closing year.

Operation Dragoon

Discussions about the viability of an invasion of southern France began during the Trident Conference in May 1943. Allied leaders convened in Washington, DC, to discuss their actions after North Africa and determine long-term strategy.[101] The Allies originally timed Operation Anvil (renamed Dragoon) to occur simultaneously with Operation Overlord in order to prevent German troops in southern France from reinforcing units in Normandy. US planners believed Operation Anvil was critical to the success of the Normandy landings and victory in Europe. Aside from pinning German forces in southern France, it would allow the Allies to seize major ports there, which would supplement supply deliveries to Normandy and Brittany.

Although support for Operation Anvil waxed and waned during the next year, the near disaster at Anzio revived its prospects. Isolation of the Anzio beachhead and the requirement for the Fifth Army to assist with the capture of Rome during Operation Diadem ensured that the troops in Italy would not be available for any version of Operation Anvil that coincided with Operation Overlord. However, when the Allies seized Rome earlier than expected, the debate about Operation Anvil reopened. Wilson told the CCS that he could start to release troops for another action in the Mediterranean, suggesting a small invasion of southern France. Support for Operation Anvil increased after the Germans wrecked the ports in northern France. Eisenhower saw it as the best way to secure an additional supply line. US planners determined that an invasion during the second half of August was practical. On 2 July 1944, the CCS directed Wilson to plan for a three-division invasion of southern France on 15 August. Severe Anglo-American arguments followed. The British opposed Operation Anvil on the grounds that it would hollow out the force structure in Italy and thus rob that effort of its potential to break into the Po River Valley and across it to the Alps and the southern Reich. Eisenhower supported his countrymen. The United States ultimately won the argument after a hard-fought battle.[102]

Eaker and Slessor claimed that MAAF could not support operations in Italy and southern France simultaneously. Nonetheless, the CCS directive of 2 July 1944 to Wilson ordered him to launch Operation Anvil no later than 15 August. It would receive first priority for resources, with the remainder going to the Italian campaign. The CCS recognized that Commonwealth troops in Italy depended heavily on US air forces and told Wilson to do what he could here.[103]

Planning the Assault

The MAAF Plans Division under Air Commodore Leonard Pankhurst planned the air aspects of the operation after MAAF received orders in January 1944 to prepare the outline air plan. Operation Dragoon's objectives were fourfold: to assist Operation Overlord, to capture major ports through which to bring in reinforcements and supplies, to liberate southern France, and to join up with Allied troops from Normandy in an effort to bring about the collapse of German armies in the west. Air forces were to destroy Luftwaffe aircraft, cover the assault and all following engagements, prevent movements of German reinforcements into the area, deliver airborne units, and support Maquis irregulars (eventually brought together as the French Forces of the Interior, or FFI).[104]

The storm of 19 to 22 June in the English Channel, which destroyed one of the Mulberries transloading supplies to the beach, gave the final push to Operation Dragoon. This compounded the massive German-produced wreckage at the Port of Cherbourg and other logistical hubs and the insufficient number of smaller ports to facilitate the buildup of forces for the breakout. Allied occupation of Marseilles and Toulon would produce a major increase in reinforcements and supplies. Despite the fact that the breakout from Normandy occurred before Operation Dragoon, the latter operation gave the

Allies a major logistical boost; sped the rout of German troops from France; and helped set the stage, along with the oil and transportation campaigns, for the rapid collapse of German forces in 1945. Allied air assets played a central role in this operation.[105]

The shift of combat squadrons for Operation Dragoon, and to the United Kingdom for Operation Overlord, accelerated. The US Twelfth TAC and three British wings, along with Headquarters MATAF and all its medium-bombers, were in place on Corsica and Sicily by 19 July. Cannon ordered Air Vice-Marshal William F. "Tommy" Dickson, air officer commanding (AOC) DAF, to provide air support for both the Fifth and Eighth Armies in Italy. Dickson had thirty-six and one-half squadrons. Operations in Italy suffered as a result, but Allied troops were still able to advance until the weather worsened. They resumed their advance in spring 1945 with limited but highly effective air assets.[106]

Air-Ground Coordination

The final operations plan of 16 July incorporated air-ground capabilities and coordination developed throughout the war in the Mediterranean. The army staff had an army air support control (AASC) comprising sixteen officers to ensure a strong liaison function. These officers worked in operations (G3) to process and coordinate all army requests for air support. They also dealt with all matters concerning potential air targets and provided advice on target suitability. Army liaison officers also worked at each aircraft group within the Twelfth TAC. At the corps level, an air section consisting of an assistant G3 for air, and an assistant G2 (intelligence) for Air, handled all air intelligence matters through the army G2. He arranged strikes in support of the corps through the AASC. Division staffs were structured in the same fashion.[107]

Forward army elements also coordinated air support by keeping the AASC informed of the ground situation. Based on what they observed, they would submit requests for airstrikes, designate and prioritize targets, and state the expected results. Ground commanders used these reports to adjust the air plan and change the bomb line to minimize fratricide. Forward army elements also disseminated the staff's air plan and kept ground units informed of changing air plans, activities, and effects.[108]

As these processes took final form, Operation Anvil became Operation Dragoon officially on 1 August. Lieutenant General Alexander M. Patch, who had a strong record with amphibious assaults in the Pacific, received command of the Seventh Army—an indication that Marshall and Eisenhower, like Arnold, were putting lessons learned and successful commanders from other theaters in place for analogous operations. The deception plan, code-named Ferdinand, included a major and effective air component along with a major ground push in Italy that focused on OKW there and covered preparations for Operation Dragoon.[109]

The air plan had four phases: preliminary operations from 17 July to 9 August (D-6), second phase from 10 August to 3:50 a.m. on D-Day (15 August), third phase from 3:50 a.m. to H-Hour (8:00 a.m.), and a fourth phase of heavy and persistent air attack from H-Hour until the fall of Toulon. Each phase emphasized deception. The preliminary air phase involved countering and suppressing Luftwaffe operations, interdicting German Army communications, and neutralizing U-boats. The third task faded in importance after the 5 July and 6 August MASAF raids on Toulon reduced operational U-boats in the Mediterranean from eight to one. Interdiction remained a major effort, with 2,188 sorties flying against targets in Italy and France to confuse the Germans. Railways were the key focus of this effort. The second phase focused on destroying coastal defense batteries and radars, and also degrading German morale, along a stretch of coast from west of the actual landing area to Genoa. Another 2,350 sorties flew during this period, hammering coastal batteries and artillery positions. In addition, 1,047 more sorties went after Italian coastal targets in accordance with deception plan Ferdinand. This entire range of air operations also included opening attacks on German Army Headquarters, playing havoc with communications and, thus, command and control. The Nineteenth Army and Army Group G struggled from this point forward to maintain even a foggy view of the battle space. During the third phase, air attacks and naval gunfire destroyed coastal defenses and artillery positions, relying heavily on PR and photographic intelligence (PI). The last phase transitioned the air forces to direct support of the invasion force. Heavy-bombers attacked key rail centers and bridges to prevent reinforcement, resupply, and maneuver.[110]

The Airborne Task Force was set to begin with pathfinder drops at 3:23 a.m. and the main drops at 4:23 a.m. and 6:10 a.m. employing 500 aircraft towing 480 gliders. Resupply on D+1 involved a further 100 aircraft. Careful planning and incorporation of lessons learned from the Sicily disaster through to Normandy made this the smoothest airborne operation to date.[111]

By the time Operation Dragoon began on 15 August, it was the Grand Alliance's fourth major offensive in as many months. Operations Diadem, Overlord, and Dragoon were draining the life out of German armies in the west, and the massive Red Army offensive during the same period destroyed Army Group Center and carried the Russians deep into Poland and the Balkans, causing Bulgaria and Romania to defect. The Seventh Army, with ten

US and Free French divisions, would make the assault under heavy air cover and with support from 450 warships.[112]

From 28 April to the invasion on 15 August, Allied planes attacked a variety of targets within and outside of the landing area to obscure the real target and cause real damage. Heavies attacked all marshaling yards in southeastern France to hamper the movement of reinforcements and supplies, and they raided Luftwaffe bases. These targets absorbed about 75 percent of the total effort, and the other 25 percent went against coastal gun emplacements, radar sites, and the other targets slated for destruction in advance of Operation Dragoon. These came under heavy attack on 11 August or later to give the Germans minimum time to realize the raids were associated with an amphibious assault. Further deception activities included heavy raids around Genoa, which fooled the Germans into thinking it was the target, strafing and bombing of radar sites on D-1, radar jamming by Mandrel-equipped Liberators and ground stations on Corsica, and demonstrations by naval and air forces west of the landing beaches to draw German attention there. By D-Day, there was one bridge left standing, and it soon fell. Allied air forces had effectively sealed the battle area while deceiving the Germans and causing them attrition.[113]

The intelligence effort was also effective. PR and PI work resulted in oblique photos that facilitated highly effective raids against gun positions. Intelligence personnel distributed a full complement of basic photographic coverage, mosaics, lithographs, overhead and oblique target and target-approach photos, flak maps, and intelligence assessments of various kinds. Damage assessment was excellent, with turnaround times under twenty-four hours for bomb damage assessment (BDA) reports to determine the need for restrikes. Aircrews guided by first-rate intelligence struck and restruck targets with impunity.[114]

German commanders led by Colonel General Johannes Blaskowitz, commander of Army Group G in southern France, had extensive experience but not much of an army to lead. Blaskowitz had fallen out of favor with Hitler for trying to stop Heinrich Himmler's *Einsatzgruppen* as they murdered "undesirables" after the conquest of Poland. Blaskowitz put General Friedrich Wiese in charge of the Nineteenth Army, which would defend the beaches against the Allied invasion. Wiese had spent three years fighting on the Eastern Front and was noted for his tactical skill.[115]

Regarding Blaskowitz's troops, an Allied assessment noted that "the 338th, 244th, 242d, and 148th Divisions, which were the only ones in the immediate vicinity of the assault area, were all Limited Employment Divisions, generally of poor caliber, and containing large numbers of non-Germans. Total strength of each was not more than 8,500 men. Equipment was only partly mobile, and of inferior quality."[116] And these were the best divisions. The Kriegsmarine had five destroyers, twenty-eight torpedo boats, nine U-boats, and fifteen patrol craft. After heavy bombing of Toulon, it had one operational destroyer, three submarines, and a handful of torpedo boats.[117] The Luftwaffe was in the worst shape with thirty Me. 109s, sixty-five Ju 88s, thirty-five reconnaissance aircraft, fifteen He 177s, and about thirty obsolete Do 217 bombers. Another fifty fighters in northern Italy were available for transfer.[118] These forces faced a 20:1 ratio against the MAAF's 4,056 aircraft.[119]

The Luftwaffe was unable to interfere with the landings. Allied naval and air bombardment of coastal batteries and defensive positions was so effective that fewer than 100 enemy artillery and mortar rounds landed among the assault craft, doing no damage. Heavies and mediums, whose crews had drawn the right conclusions from preassault bombings at Utah and Omaha Beaches, used the approach from the former, flying down the beaches and striking predesignated targets. Concussions and direct hits incapacitated or killed most of the gun crews. Once again, heavies played a major tactical role, as they had since summer 1942. The army's report on the beach bombardments was effusive. Bombs breached the barbed wire and killed or stunned many of the defenders, but the beachheads were not cratered to the extent of impeding the advance of Allied troops.[120]

The bombing of the landing beaches was highly accurate, especially because the crews had bombed parallel to the coast and on cue from the best H2X Pathfinder crews. The distinction between water and land was clear on the operators' scopes, making the beaches ideal targets. Aviation engineers had a fighter-bomber airfield operational on French soil four days after the landing. Three more were operational within the next week. They had 6,000-foot runways and could thus accommodate fighter-bombers and mediums. Saville's aircraft thus responded very rapidly to army calls for assistance as the aviation engineer units kept up with the advance.[121]

By D+30 the invasion force was where the plans called for it to be on D+120. This rapid advance caused logistical and air support problems.[122] It also drove the bomb line well in front of the ground troops to ensure they were relatively safe. This worked well as long as the troops were moving, but when they stopped to fight, the bomb line tended to be in the wrong place for several critical hours. Only someone on the ground could drive home the need to adjust it. The plan for Operation Dragoon should have provided a framework for this coordination, but it did not do so adequately, as the Battle of Montelimar demonstrated. The Germans took heavy losses there but managed to break through US units and reach the borders of the Reich.

The Battle of Montelimar

US troops sought to stop the retreating German Nineteenth Army near the town of Montelimar along Highway N-7, which followed the Rhone River north to the borders of the Reich. A very fluid campaign suddenly produced a pitched battle. After initial skirmishes on 21 and 22 August, the United States continued shelling the Germans as they moved north. They also cut off Highway N-7 with roadblocks, but the Germans broke through. The battle, which ended on 28 August, was difficult and bloody, but 80 percent of the Germans escaped, mostly without heavy equipment.[123]

Airpower was almost completely absent. The US Army's official historians said that "Perhaps the most remarkable feature of air operations in the southern France campaign was the total absence of normal close air support activities involving the use of air-ground liaison teams . . . the problem was brought home most forcefully during the battle at Montelimar."[124] At the very moment it reached its zenith, the air-ground system failed in this instance. The rapid advance and the bomb line were at the heart of the problem. During the battle, it was well forward of the troops because of the fluid situation and drove the Twelfth TAC to focus on vehicle convoys outside of the immediate battle area. Ground commanders called for airstrikes against German troops around Montelimar, but given the air component's concentration on attacking vehicle convoys outside the developing battle area and the fact that forward airfields could perform only basic maintenance, most of the Germans escaped.[125]

Evidently, the Twelfth TAC thought the placement of the bomb line was appropriate. The US Army agreed until Montelimar. During the battle, Twelfth TAC assets tried to interdict German forces before they reached Montelimar, rather than flying direct support missions. The Twelfth TAC claimed it played a successful role in the battle, noting, "An enemy column near Montelimar blocked by bombed bridges and an encircling ground force maneuver was almost totally destroyed by attacking fighters and fighter-bombers. This was again conclusive proof that under conditions of air superiority the air arm can deny the use of roads to the enemy in daylight and restrict his mobility."[126] However, there was almost no direct support. Although air assets destroyed a great deal of German equipment away from the fight, ground commanders needed them immediately over the battlefield. Perhaps the Combined Operations Headquarters provided the best assessment of air-ground coordination in its report on the battle: "Through laying down a bomb line for considerable periods, it became necessary to put it well forward and the armed reconnaissance line even further forward. As a result, it was felt that these precautions at times erred too much on the side of safety."[127] This helps us to understand various coordination issues during Montelimar, but it leaves one to wonder how airmen and soldiers allowed this to happen.

Although many of the commanders in Operation Dragoon talked to each other during planning, they did not adequately take into account lessons learned by commanders in other theaters. Major General Pete Quesada, commander of the Ninth TAC, had experienced problems with the bomb line and fratricide during the breakout from Normandy. During the planning for Operation Overlord, Quesada communicated with Cannon, asking about air-ground coordination in Italy. He then sent more than 200 officers to Italy to collect lessons learned and apply them to the campaign in Normandy.[128] Quesada solved the problem of coordinating the bomb line by putting radios in tanks with the forward troops (as had already occurred in Italy) so that they could talk to aircraft overhead.[129] However, there is no indication in existing sources that Quesada spoke with Saville about this. He tended to communicate with Cannon. Nor is there any indication that Cannon discussed Quesada's bomb line concerns with Saville, who needed to know about them. Finally, there is no indication that Saville addressed these issues in the same effective way as Quesada. Lessons learned from Operation Cobra in Normandy should have flowed to Operation Dragoon air commanders.

However, even with a perfect bomb line, US forces would have been hard pressed to ensnare the entire Nineteenth Army. They were short on troops, artillery ammunition, and airpower. In the latter case, aircraft on Corsica were out of range, and there were only four airfields operational on the mainland, all more than 400 miles away. Despite losing huge numbers of vehicles, the Germans fought their way to the frontiers of the Reich, where they formed the southern anchor of a defensive line. Based in part on the battle of Montelimar, the official historians said there was no denying Operation Dragoon's tactical success, but they disputed Eisenhower's claim that the operation had decisive effects in speeding the defeat of Nazi Germany. They also noted that only six US divisions disembarked at Marseilles and Toulon, three in October and three in December.[130]

What they did not discuss, however, was the vital importance of taking two major ports intact, through which 40 percent of all supplies for the campaign in northwestern Europe flowed. Allied forces needed the supplies more than additional divisions, which is why only six came ashore along with the mountain of supplies. The operation also caught two German armies (the Nineteenth Army and elements of the First Army) in a vise, resulting in the capture of nearly 200,000 men. Finally, after the six divisions joined

Patch, his Seventh Army advanced well into Czechoslovakia after the break-out in spring 1945.

We might also ask what the other options were at this point. When they heard that the Twelfth Air Force would move to southern France, under the command of Eisenhower's Supreme Headquarters Allied Expeditionary Force (SHAEF), and work directly with Patch's Seventh Army, Wilson and Alexander protested that they would lose most of their medium-bomber support. Eaker discussed the matter with Portal and Slessor, noting that Portal, as always, "took the broad, high-minded, long-range view."[131] Portal asked the army commanders what their plans were for offensive operations in Italy. They said the campaigning season was nearly over, the troops were tired, and they could not force the Ljubljana Gap until the spring. Portal said it would be a waste to leave an entire numbered air force in Italy to support troops that were not advancing while those in southern France were doing so quickly. This pragmatic response underscored the degree to which Operation Dragoon surpassed any of the available options in Italy or elsewhere in the Mediterranean. It also reminds us that it served a larger purpose by providing critical logistical and operational support to the advance of troops through northwestern Europe and into the Reich.

14

The Second Aerial Front
November 1943–January 1945

Grand-Strategic Developments

THE TEHERAN CONFERENCE OF 28 November to 1 December 1943 largely settled grand and military strategy for the remainder of the war, although there were many further arguments about the details. The Allies would establish a second front in 1944, and the Russians would join the war against Japan as soon as Germany collapsed. By this time, the United States and USSR were at the helm, with British influence fading. Despite Winston Churchill's continuing emphasis on actions in the Mediterranean, even with the recent disaster in the Dodecanese Islands (Operation Accolade), Josef Stalin and Franklin Roosevelt objected. Both were entirely in favor of Operation Overlord and believed a cross-channel invasion was both overdue and far preferable to any continuing major effort in the Mediterranean. Despite Stalin's limited understanding of sea and airpower, he appears to have understood the potential value of heavy-bomber attacks from bases in the Mediterranean to assist the Red Army's advance. (This subject receives attention later.) The Allies already had airfields in Italy from which to launch their raids, so Stalin viewed further efforts there as a waste of effort. Accordingly, Stalin also sided with Roosevelt regarding the use of forces currently in Italy for Operation Dragoon. Although the Mediterranean's position as a major theater of war was in relative decline, its importance in speeding Allied victory, especially through the effective use of heavy-bombers, was increasing.[1]

As the heads of state and their military chiefs talked, Henry Maitland Wilson worked with his airmen to activate the Mediterranean Allied Air Forces (MAAF). As the command was established, a number of major personnel changes occurred. Dwight Eisenhower was named supreme commander allied expeditionary force (SCAEF) for the planned Normandy invasion. Arthur Tedder became Eisenhower's deputy SCAEF. Ira Eaker succeeded him as MAAF commander. Arthur Coningham became air officer commanding in chief (AOC in C) of the Second Tactical Air Force (TAF). Carl Spaatz became US Strategic Air Forces in Europe (USSAFE) commander. John "Jack" Slessor became Eaker's deputy and commander of all Royal Air Force (RAF) units in theater. Major General Nathan Twining would command the Fifteenth Air Force, John K. "Joe" Cannon the Twelfth Air Force, and Air-Vice Marshal William "Tommy" Dickson the Desert Air Force (DAF). Bernard Law Montgomery turned over the Eighth Army to Oliver Leese as he took command of the Twenty-first Army Group for Operation Overlord.[2]

MAAF Is Activated

With the sea lines of communication (SLOCs) secure, the Allies expanded their second aerial front against the Reich from the Mediterranean. This was now possible given air supremacy, the rapid evolution of Allied air forces—especially the US Army Air Forces (USAAF)—in the period since Operation Torch began, an increasing concentration of Luftwaffe fighters in the Reich, and the Allies' ability to strike at aircraft factories, synthetic oil plants and crude-oil refineries, and transportation assets throughout nearly all of occupied Europe. These new realities drove the development of MAAF, which grew out of the Mediterranean Air Command (MAC), and the incorporation of its USAAF heavy-bombers into USSAFE, which became active under General Spaatz's command on 1 January 1944. His mission was to employ the Eighth Air Force under Major General James "Jimmy" Doolittle and Major General Twining's Fifteenth Air Force (assigned to Lieutenant General Eaker within MAAF) to destroy the Luftwaffe; ensure air supremacy during Operation Overlord; and destroy Germany's aircraft industry, transportation infrastructure, and oil production. This would undermine the Wehrmacht's ability to mass, maneuver, and fight at anything approaching parity.

On 31 October, Tedder stated his priorities for MAAF: assist Allied armies in Italy, contribute to the combined bomber offensive (CBO), and weaken the German hold on the Balkans and the Aegean. He added the destruction of the Luftwaffe wherever it could be found in support of Operation Pointblank. Tedder drove home the value of a second aerial front in terms of reaching all the key targets in the Reich, and of forcing *Luftflotte Reich* to divide its forces.[3]

The month before Tedder's formal statement of priorities, efforts to set up MAAF were in high gear. Eisenhower told George C. Marshall there were five major advantages in flying heavy-bomber raids from Italy:

1. The heavies could reach targets out of range of those stationed in the United Kingdom.
2. Flying weather would be better in Italy, even in the winter and especially after the heavies climbed through clouds into clear skies.
3. There would be fewer Luftwaffe fighters and flak at the outset.

4. A heavy series of raids would soon force the Germans to split these forces and send more to the southern Reich, Ploesti, and other key target areas.
5. There would be greater flexibility in the targeting process because many of the key factories, oil refineries, marshaling yards, and other targets were in range of heavies from both the United Kingdom and Italy.

This effort would mean lower aggregate bomber losses, especially after the Eighth Air Force started flying missions with escort fighters. Henry "Hap" Arnold was an avid supporter of this plan and noted that in addition to its other advantages, it opened the door to shuttle bombing and even to coordinated raids that would confuse, mislead, and overtax German air defenses.[4]

A key reason for developing MAAF was in fact to bring Twining's Fifteenth Air Force under USSAFE, giving Spaatz operational control of all USAAF heavies in time for Operation Big Week, which was designed to give the Allies air supremacy for Operation Overlord. The Fifteenth Air Force's heavies could reach two of Germany's largest aircraft factories, which accounted for 60 percent of single-engine fighter production, the critical oil refineries at Ploesti, several major synthetic oil plants, the Danube (which No. 205 Group became adept at mining), and the entire transportation network in southeastern Europe. Mediterranean Allied Strategic Air Forces' (MASAF's) contribution to victory, much underrated when compared with that of the Eighth Air Force, became evident during the course of concerted raids on these target sets.[5]

When the Fifteenth Air Force was established on 1 November 1943, it included six heavy-bomber groups and two of long-range fighters. By 31 March 1944, it had twenty-one heavy-bomber groups, seven of long-range

(*Opposite*) US Army Air Forces (USAAF) heavy-bomber combat-radius rings from Royal Air Force Alconbury and Foggia, October 1943–May 1945. After the second aerial front against the Reich opened, and the USAAF defeated the Luftwaffe over the Reich, the oil and transportation offensives robbed the Wehrmacht of fuel, logistical support, mobility, and thus the capacity to resist the Grand Alliance advance on Germany with anything approaching what it might have been. The 600-mile radius depicts maximum internal bomb loads of 8,000–12,000 pounds depending on bomber type and model. (Originally published in Stephen L. McFarland and Wesley Phillips Newton, "The American Strategic Air Offensive against Germany in World War II," in *Case Studies in Strategic Bombardment,* edited by R. Cargill Hall [Washington, DC: Air Force History and Museums Program, 1998], 199, now public domain)

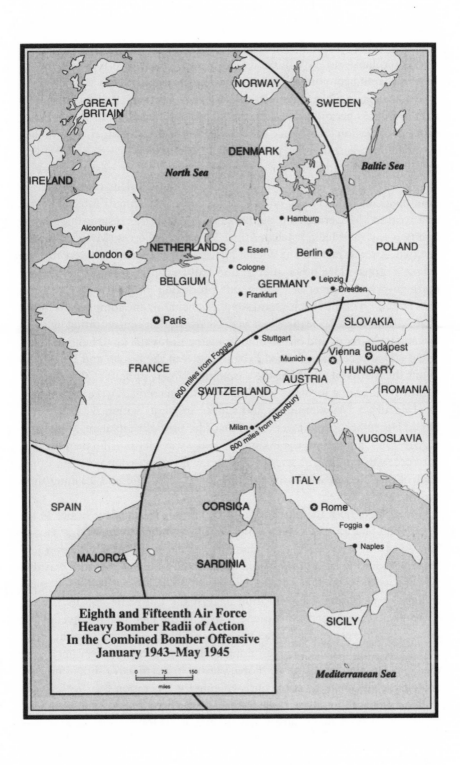

**Eighth and Fifteenth Air Force
Heavy Bomber Radii of Action
In the Combined Bomber Offensive
January 1943–May 1945**

0   75   150
miles

fighters, and one reconnaissance group. Its mission was to assist in Operation Pointblank and the CBO and to coordinate closely with the Eighth Air Force. It would support Allied armies on an emergency basis.[6]

Tedder, Spaatz, Eaker, and Doolittle met at Gibraltar on 8 and 9 November to coordinate operations between the Eighth and Fifteenth Air Forces. They divided Operation Pointblank and other target sets according to factors such as range to target, weather patterns, available fighter escorts, and requirements for H2X radar-equipped bombers. MAAF's four missions were to destroy the Luftwaffe in the air and on the ground, participate in Operation Pointblank including raids on fighter factories, aid ground forces in Italy, and weaken the Germans in the Balkans. MASAF was responsible for the first two and the Mediterranean Allied Tactical Air Forces (MATAF) for the last two. Aircraft factories at Wiener Neustadt and Regensburg were the two primary targets. Five other aircraft and ball-bearing factories rounded out the high-priority target list. Oil plants producing 76 percent (11,825,000 tons annually) of the Reich's total supplies, including the complex at Ploesti, were within range. US leaders believed, correctly, that concentrating on these target sets—aircraft and oil—would bring the Luftwaffe up to fight and die.[7]

On 4 December, after heated debates between the British and US leaders about the merits of USSAFE, the Combined Chiefs of Staff (CCS) released a directive agreeing formally with the Joint Chiefs of Staff's (JCS's) decision to place all US heavy-bombers and supporting units within the command. On 20 December, Eisenhower issued an order officially replacing MAC with MAAF. On 5 January, as the Eighth Air Force was refitting and the Fifteenth Air Force was building its strength, Spaatz, who exercised direct operational control of Fifteenth Air Force through Eaker, gave him and Twining their orders.[8]

Arnold gave both the Fifteenth and Eighth Air Forces clear notice in his "year's end" message, which concluded, "Therefore, my personal message to you—and this is a *MUST*—is to destroy the enemy air forces wherever you find them, in the air, on the ground and in the factories."[9] Arnold was dissatisfied with the number of what he called "diversionary" and small raids, calling for maximum concentration against the Luftwaffe and its sources of production and supply. He pledged to send the Eighth and Fifteenth Air Forces each five groups of heavies per month with 100 percent combat crews and 50 percent aircraft reserves along with more escort fighters, including the P-51. Despite the Eighth Air Force's heavy losses to date, both numbered air forces (NAFs) would come out swinging for Operation Big Week.[10]

The Fifteenth Air Force thus completed the aerial encirclement of Germany, laying every major target open to attack. Its top-priority mission was the destruction of the Luftwaffe during Operation Pointblank. Following this, attacks would fall on all oil refineries and oil supply installations within range. However, the nonstop transportation campaign in the Balkans absorbed by far the greatest effort. This was intended "to disrupt and interdict completely . . . communications and supply lines, and to advance to the greatest possible extent the ground operations of Allied [and especially Soviet] Armies."[11]

The Fifteenth Air Force reached maturity in April 1944 with an average strength of 1,427 heavies and 632 escort fighters. From its first official mission on 2 November 1943 against Wiener Neustadt to its last on 1 May 1945 against the main marshaling yard in Salzburg, its aircrews flew 152,542 bomber sorties and 89,835 fighter sorties, dropping 309,278 tons of bombs. Aside from destroying huge numbers of enemy aircraft on the ground and in factory raids, the Fifteenth Air Force scored 3,946 aerial victories. The cost was 3,410 aircraft destroyed and 14,181 damaged. Manpower losses were 2,703 killed, 12,359 missing, and 2,553 wounded. Of the total bomb tonnage delivered, 42,156 went against aircraft factories, 14,976 against other industries, 59,854 against oil targets, 25,374 toward air-ground cooperation, and a stunning 159,671 against lines of communication (LOCs). The latter figure represented 52 percent of all bombs dropped. Heavies and their escorts penetrated nearly 800 miles into occupied Europe, bombing targets within an arc from eastern France to southern Poland and the southern Ukraine. Accuracy was very good given the technologies of the time. More than 80 percent of bombs struck within 2,000 feet of the desired mean point of impact—60 percent within 1,000 feet and 36 percent within 600 feet. The Fifteenth Air Force's accuracy in visual bombing and H2X proved superior to that of the Eighth Air Force.[12]

One of the key advantages the Fifteenth Air Force had over the Eighth Air Force was its ability to fly parallel to and thus between weather fronts, whereas the Eighth Air Force and RAF Bomber Command had to fly through them. The successful raid on 3 November against Wiener Neustadt drove home this fact. The 112 heavies did severe damage to several assembly shops and hangars, and the 72 escorting P-38s fought off attacks by between 130 and 160 fighters. The United States lost 11 bombers, mostly to flak, and destroyed more than 50 enemy fighters. Intelligence estimated that the raid reduced Me. 109 output by 250 fighters per month for at least two months. Operation Pointblank's objective was the destruction of *Luftflotte Reich* and the larger Luftwaffe. The Fifteenth Air Force did so from the outset by achieving air supremacy using counter-air-force operations and destroying enemy aircraft production.[13]

The key problem in the Fifteenth Air Force's early days was a critical shortage in ground and aircrews. Aircrews could fly only eight long-duration missions per month due to physiological and psychological factors, but heavy-bombers could make sixteen based on serviceability rates. This required a 2:1 ratio of aircrews to planes. The Mediterranean theater's better flying weather allowed crews to complete their tours more quickly, leading to chronic shortages. Arnold authorized an immediate increase in the replacement rate from 0.5 to 2.5 percent per month. During the first week of November, the Fifteenth Air Force had 20,499 men and 931 operational aircraft. Arnold promised three new B-24 groups each month from November 1943 to January 1944, and three B-24 groups plus another of B-17s after that—a promise he kept. Even more importantly, ground and aircrews started arriving in large numbers.[14]

Unfortunately, the number of fighter groups did not expand until April 1944 because so many headed to the United Kingdom as drop tanks became available. Although there were fourteen fighter groups in theater, there were just three of P-38s—the only long-range escort available. The five P-40 groups would re-equip with P-47s, but the need for Thunderbolts in the United Kingdom delayed this upgrade. Also, as part of the larger arrangements for the activation of MASAF and the Fifteenth Air Force, all mediums moved to MATAF.[15]

After Eaker and Slessor arrived as MAAF commander and deputy commander in January, one of their first duties was the movement of all remaining logistical assets and organizations to Italy from North Africa. Eaker and Slessor thus had to simultaneously build a larger command structure, make the logistical moves, maximize air effectiveness in the difficult combined-arms effort in Italy, and engage in the heavy-bomber campaigns.[16]

The command consisted of MASAF (the Fifteenth Air Force and No. 205 Group), MATAF (the Twelfth Air Force and DAF), Mediterranean Allied Coastal Air Force (MACAF), and Royal Air Force Middle East (RAFME), with about 12,500 aircraft and 325,000 personnel. The combined plans and operations staff came together quickly at Headquarters MAAF in Caserta, Italy. The Fifteenth Air Force became part of USSAFE on 1 January 1944, facilitating coordinated raids on vital target sets.[17] MAAF's guiding principle was to have combined USAAF/RAF operational staffs but parallel administrative staffs. This represented a major change from MAC and the Allied Air Forces, Northwest Africa (known as NAAF) but worked well because Eaker and Slessor drove maximum effectiveness both in terms of coordination between the national staffs and the combined plans and operations staff. This system maximized flexibility in three ways: by allowing national contingents to operate as they were used to doing, by facilitating effective combined operations, and by producing maximum flexibility in air and combined-arms operations. Only the joint staff, which included the Signals, Operations, and Intelligence Sections, took orders directly from Eaker and Slessor. The national administrative chains of command facilitated the myriad support requirements for waging war.[18]

Eaker had high standards for MASAF. However, in a 21 March note to Arnold, he said of the Fifteenth Air Force that by Eighth Air Force standards it was a "pretty disorganized mob" in need of shaping up—a task he took on with great but seldom recognized success. Part of the problem was that its units had been engaged mostly in tactical missions supporting combined-arms operations and could not yet accurately bomb point targets from high altitude. Twining was putting that straight. "We shall spare no pains," Eaker said, "in order to leave nothing undone and to build the Fifteenth up to my standards as quickly as possible."[19]

Operation Pointblank missions included coordinated operations against aircraft factories and ball-bearing plants. German fighters had to defend these targets, so the process of attrition went into high gear as USSAFE ultimately destroyed the Luftwaffe. The Fifteenth Air Force's contribution was immense but has received relatively little attention. Part of this has to do with the fact that the Luftwaffe proved a cagy opponent, often going for the smaller of the two forces. On 25 February, the final day of Operation Big Week, about 200 German fighters attacked the Fifteenth Air Force's 342 aircraft but sent only 59 against the Eighth Air Force's 1,490. The Fifteenth Air Force contingent shot down more than 80 aircraft but lost 32—18 percent of the attacking force. However, the fact that the Luftwaffe was shying away from large formations and going for smaller ones was a clear indication that *Luftflotte Reich* was losing the battle. By forcing the dissipation of fighter assets against major formations on two aerial fronts, Operation Big Week paid huge dividends. The Fifteenth Air Force faced strong resistance throughout the campaign, encountering 1 enemy fighter for every 3.8 bombers dispatched, whereas the Eighth Air Force saw 1 for every 8.1 bombers. Fortunately, it had more fighter protection than the Eighth Air Force, with 1.4 fighters escorting each bomber as opposed to 1.2. However, the increasing ferocity of Luftwaffe attacks on the Fifteenth Air Force as the Eighth Air Force became too powerful to engage meant that it had a 3.3 percent average loss rate per mission as opposed to the Eighth Air Force's 2.2 percent. This was also in part a result of heavy flak and air defenses around Ploesti, Wiener Neustadt, and the Brenner Pass.[20]

Despite the Fifteenth Air Force's steady improvements to date, Eaker continued, it still had some distance to go. He added, however, that many of the senior officers and crews were outstanding. Twining was teaching the crews in his twenty-one groups—up from three just three months earlier—to bomb from high altitude. The seven fighter groups attached to the Fifteenth Air

Force were also upgrading quickly to P-47s and P-51s from P-38s, which were moving increasingly into the fighter-bomber role. Thirty pathfinder crews and aircraft equipped with H2X radar had also arrived, facilitating bombing through clouds, although it was often inaccurate despite the high quality of the crews.[21]

On 5 April, the offensive against Ploesti began. A month later, Fifteenth Air Force heavies began raiding marshaling yards in southern France as part of the larger transportation campaign to hinder the movement of German reinforcements and supplies to Normandy. This proved exceptionally useful after the northeastern and eastern regions of the French National Railway collapsed and the Germans had to divert trains through southern France, where railroads were also heavily damaged.[22]

On 7 May, Eaker sent Arnold an update, noting that the past month had seen the most intensive bombing to date: more than 40,000 sorties dropping 30,000 tons of bombs. The Fifteenth Air Force had flown twenty-six major raids on Balkan marshaling yards and other LOCs to keep the Germans from moving supplies and reinforcements to the Eastern Front. Another twenty-three raids went against aircraft factories and ball-bearing plants. Ten aircraft factories were destroyed or seriously damaged. Ploesti, Bucharest, Sofia, Belgrade, and nine other key marshaling yards also suffered severe damage. Bombers caught a number of trains at these locations, destroying them and most of their contents. No. 205 Group added to the disruption by mining the Danube and bringing barge traffic to a near halt. The average distance covered during a sortie had increased from under 100 miles to more than 400 in an area with an arc of more than 1,200 miles. Average missions per month at the group level increased from six to eight, and the bomb tonnage delivered from five to eight. MASAF was flying about 1,000 sorties per day, mostly against LOCs. The H2X crews often out-performed crews using conventional methods. Eaker made full use of this key capability, ultimately giving each heavy-bomber squadron a primary and alternate H2X Pathfinder Force (PFF) crew and aircraft.[23]

By the time it matured in summer 1944, the Fifteenth Air Force was a superb command. German Signal Intelligence Service (SIS) officers were impressed, noting that its superb weather reconnaissance gave the SIS fits because the aircraft reported weather over large areas rather than target localities, making it very difficult for the Luftwaffe to determine where and when raids would occur. Intercepts were correspondingly rare. They also noted that they could not keep pace with the strain imposed on their intercept efforts by frequent and effective code changes. Tactics were also outstanding, largely because of the leadership, which the SIS officers rated as "impressively superior." They said, "This was evinced by the nature of its operational planning, which resulted in carefully designed attacks on high priority targets. . . . The refined technique underlying the assignment of call-signs and frequencies to combat units, as well as the employment of R/T interceptors, was a high tribute to the acumen of the signal officers."[24] They had equally high praise for the command's training efforts and rated every element of its operations superior to those of the Eighth Air Force—quite a tribute to a group of men Eaker had referred to as a "disorganized mob" three months earlier.

SIS officers were also stunned that the Fifteenth Air Force flew a major mission every three days on average, each comprising 400–500 heavies. During good flying weather, raids on marshaling yards and other LOCs in the Balkans averaged 600–700 aircraft per day. And as always, they were most deeply impressed with No. 205 Group. After this unit moved to Italy in early 1944, it had two Liberator wings along with the Wellingtons. These were H2S equipped, and their radar operators were outstanding. They often served as pathfinders for the Wellingtons and were highly proficient in attacking targets through clouds in their own right. They went after Graz and Marburg, the two most important LOCs for German troops in the Balkans during late 1944, hitting them whenever they showed signs of recovery and with a level of accuracy rarely seen from heavies. They also mined the Danube so effectively that they virtually halted barge traffic from April 1944 until Romania's defection in August. Radio and radar discipline were excellent, as was the ability to avoid flak and night-fighters. The SIS concluded, "The results which it achieved, as the only long-range night bomber unit in the South, far exceeded those which might ordinarily be expected from a unit its size."[25]

In another sign of MASAF's tactical acumen, Eaker told Arnold,

> I think we have done nothing better in this war than the working out of this technique of passing heavy-bomber formations from group of fighters to group of fighters, affording the bombers thorough and efficient protection on deep penetrations into enemy territory. All Groups yesterday afternoon reported a definite reluctance on the part of the German fighters to close. At each Group where we talked to the pilots and leaders who had just returned they said, "We saw a few German fighters, but always diving away."[26]

Eaker had superb air units, and he also profited greatly from MAAF's intelligence organization. His director of intelligence and operations (A2/A3), Brigadier General Charles Cabell, was a brilliant planner and analyst who would eventually become deputy director of the CIA. The Mediterranean Allied Photographic Reconnaissance Wing (MAPRW) and all of its components, from aircraft and pilots to photointerpreters and the products they sent planners and aircrews, came under his authority. He also had a special photo section tied directly to Headquarters MAAF for immediate delivery of raw and annotated imagery from heavy-bomber raids. The Signal Section handled 6,600,000 cypher traffic groups in December 1943 and 14,020,000 in July 1944—a stunning increase. Use of the combined cypher machine, which the Germans never broke, facilitated secure communications. Cabell's directorate was the hub of air operations in the Mediterranean. This combined staff engaged in every aspect of operations planning and execution. The Intelligence, Plans, and Combat Operations Sections were at its core. The initial complement of 118 officers—69 British and 49 American—changed little.[27]

When the Mediterranean Photographic Interpretation Center (MPIC) was established on 1 April 1944 to play the same general role as the Joint Photographic Reconnaissance Center (JPRC) in the United Kingdom, MAPRW was able to focus entirely on the photoreconnaissance (PR) and photographic intelligence (PI) missions, giving MPIC the responsibility for prioritizing and disseminating imagery to all customers in the theater. MPIC built an effective means for generating, organizing, and keeping records. Its tasking system, developed in close coordination with other elements of the staff, functioned like the JPRC's. In June 1944, MPIC processed 2,453,000 print orders with distribution lists that often included more than 100 recipients. By this time, "the use of Photo Intelligence material had become so widespread that it could nearly be said that no operations either land or air took place without recourse to this intelligence medium."[28]

The MAAF A2's Target and Analysis Subsection was particularly crucial because it picked targets; assessed damage in coordination with the MPIC, Air Ministry A.I. 3c(1), USSAFE, Office of Special Services (OSS), Enemy Objectives Unit (EOU), and other organizations; determined restrike timing; and sent bomb damage assessment (BDA) teams to inspect targets captured by ground forces. This section also worked with the combat intelligence personnel in the German Air Force (GAF) Section to determine levels of threat and the best countertactics. Target recommendations went to the Combined Strategic Targets Committee (CSTC) and its predecessors for prioritization and inclusion in master target lists for oil, transportation, aircraft factories, and other target sets. This effort proved central to the execution of effective campaigns from Ploesti to the transportation attacks in southeastern and central Europe. These two heavy-bomber efforts played an important role in speeding the Grand Alliance's victory.[29]

The Ploesti Campaign

Ploesti had been a location of great interest both to Adolf Hitler and to his opponents since before the war. With its major refining complex and massive oil output, it had immense strategic importance. Much of Hitler's planning for the Balkans campaign was predicated on keeping bombers out of range of Ploesti. Churchill, Charles Portal, and Tedder kept it perennially in their sights and determined to go after it the minute bombers could reach it. Arnold, Spaatz, and Eaker joined them in this objective.

As the oil offensive commenced, a delegation of US officers visited their Soviet counterparts from 10 to 15 May to discuss how bombing could best support the Red Army's advance. The Russians made one insistent request: attack Ploesti. The timing of the visit was significant, coming during the opening effort against synthetic oil plants and just more than a month into the push against Ploesti. Russian emphasis on oil attacks confirmed their importance and probably gave Spaatz leverage with Eisenhower for approval of a major oil offensive.[30] During a meeting on 13 May, General Vladimir Grendal, the Red Army's senior intelligence officer, emphasized that oil was the key target. He said the Germans still had enough gasoline to use motor transport for their troops often and that the "latest information available from POW's and other sources showed on the whole Soviet front from the Baltic to Roumania there was no shortage whatever of oil. That very large oil dumps have been built up and are kept well supplied."[31] Although German fuel supplies were nowhere near as robust as Grendal claimed, his statement was a thinly veiled request for assistance in creating an oil crisis for German armies on the Eastern Front.[32]

Major General John Deane, head of the US military mission to Moscow, later noted the early effects of the oil offensive during his tour of the battlefield outside of Vilna after the Russians retook the city in July 1944. "It was apparent," he said, "that the Russian victories were won by superior mobility. The combined bomber offensive of the Western Allies was taking its toll on German oil, and the German artillery and much of the transport we saw was mostly horse-drawn. The Russians with their preponderance of motorized equipment were thus able to outmaneuver the Germans."[33]

After Ploesti came under heavy attack, fuel shortages created massive problems for the Wehrmacht, to the direct benefit of the Red Army. The interconnections between the oil and transportation campaigns in the Balkans and central Europe produced an operational cataclysm for German units that destroyed their mobility, starved them of ammunition, and made them far less able to defend the borders of the Reich. The Germans were already losing the war, but the bombing of these target sets accelerated the process and reduced its costs.

The Offensive Begins

Ploesti's eleven major crude oil refineries provided Germany 6.5 million tons of petroleum, oil, and lubricants (POL) products annually—nearly 30

MASAF B-24s strike the Ploesti refineries. The concerted effort against this key target, combined with No. 205 Group aerial mining of the Danube and simultaneous attacks on all transportation nodes between Ploesti and the Eastern Front, reduced the delivery of POL products from Ploesti by more than 95 percent. This in turn created major fuel shortages for German troops in Army Group South Ukraine and further north along the Eastern Front. When combined with the destruction of southeastern and central Europe's transportation infrastructure, it helped to create an operational cataclysm for the Wehrmacht. (National Museum of the US Air Force Photo No. 050616-F-1234P-017)

percent of the Reich's total supplies.[34] The first major effort, Operation Tidal-wave, was a low-level attack by Ninth Air Force B-24s on 1 August 1943. The US forces suffered heavy losses but did severe damage, decreasing annual refining capacity from 9.2 million to 5.3 million tons—a 42.5 percent decline.[35] Luftwaffe Lieutenant General Alfred Gesternberg, in charge of Ploesti's air defenses, built a ring pipeline that facilitated the movement of crude oil from damaged to operational refineries, but he could not convince the Romanians to repair all their refineries.[36] Because they had excess refining capacity, they viewed this as wasted effort.[37] Instead, they wrote off two refineries, reducing their annual capacity to 8,000,000 tons.[38] Nor did they build protective concrete walls around key equipment and storage tanks. Consequently, the April–August 1944 MAAF campaign had rapid and severe effects.

Increased Romanian deliveries in November 1943 signaled production was back on track, although it now took nearly all of Romania's output to meet Germany's demands. One key development was the increase of oil shipments on the Danube from 2,260 tons in November 1943 to 3,850 in December. This required 14.7 oil trains per day, each with thirty cars holding 13.6 tons of oil per car, unloading at Giurgui.[39] Spaatz and Eaker pushed hard for authority to attack but did not receive CCS permission to mine the Danube until early April, and it took nearly two months more for permission to attack Ploesti. However, in an act of calculated insubordination, Spaatz and Eaker ordered the first raids on 5 April by targeting the adjacent marshaling yards and moving the aimpoints just slightly to the refineries. In fact, two weeks *before* the CCS gave permission, Eaker told Spaatz, "We mean to *finish off this job* in the Ploesti area with the first favorable weather."[40]

When it had permission to attack targets on the Danube, the MAAF Strategic Targets Committee quickly exploited vulnerabilities in Romania's oil-shipping activities, including the rail-to-barge transshipment process and barges, the former with bombing raids and the latter with aerial mining. By May 1944, mining caused severe difficulties because the Germans lacked equipment to sweep the advanced acoustic mines.[41] No. 205 Group began its efforts on the night of 8–9 April 1944. Its crews mined at 200 feet of altitude in high-moonlight conditions, completing eighteen operations by early October 1944 and laying 1,382 mines of 1,000 pounds each. This brought river transport to a near standstill.[42] By August,

> the success of the minelaying of the Danube was becoming a legend. This vast river had become a very important lifeline for the German forces [by] connecting the oilfields of Rumania and the grain fields of Hungary with the German war machine. One barge could carry a load equivalent to that carried by a hundred large railway wagons, and hundreds did so. Because of the Allied air attacks on the road and rail links, the river had been used by the Germans extensively, but now it had been put out of action for months—a most serious blow to the enemy.[43]

Raids on the Danube reduced shipments 70 percent by August. During July, the Kriegsmarine cleared 510 mines, but 202 barges sank. Another 172 were damaged. Renewed minelaying on 1 and 2 July along with air attacks on ports and transshipment installations caused severe damage. No. 205 Group flew 127 sorties during July, laying another 428 mines, and the Fifteenth Air Force flew 145 sorties against oil storage facilities, dock gates, and oil transshipment and loading points. Beaufighters escorting the Wellingtons sank another 8 oil barges and damaged 102. Meanwhile, August brought another 37 mine-laying missions, and USAAF heavies flew 65 sorties against oil-related infrastructure. New magnetic mines, immune to detection, produced another huge increase in sinkings. The Germans floated damaged barges in front of operational vessels in an effort to detonate the mines. Minesweepers became ineffective. When the Russians occupied the Ploesti oil fields at the end of August, bombing and mining had long since rendered them, the oil-pumping stations on the Danube, and barge traffic nearly inoperable and unable to move more than a trickle of oil.[44]

The director of Commercial Marine, a major shipping company, and a tugboat pilot familiar with the Danube, said mining before Romania's defection to the Allies was significant, with more than 200 barges and 60–70 tugboats sunk. After the defection, Romanian artillery and gunboats destroyed more than 100 German boats, and the Germans themselves sank several more to block river navigation. Albert Speer noted in a postwar interrogation that mining the Danube had been even more harmful than attacks on Ploesti's refineries.[45]

A report from human intelligence sources, corroborated by air reconnaissance, confirmed that serious damage to Ploesti's marshaling yards had caused major reductions in the movement of fuel from there to Germany and the Eastern Front. Damage to pumping stations also prevented the shipment of oil up the Danube. Consequently, because surplus fuel and oil production could not be shipped, the Romanians shut down two-thirds of oil field production while awaiting repairs. The report closed with advice for MAAF to keep pumping stations and loading terminals out of action, continue mining the Danube, and carry out attacks on the tugboat fleet. To hamper rail shipments, MAAF bombers continued attacking marshaling yards and interdicting bridges between Ploesti and the Eastern Front.[46]

As mining took its toll, attacks on Ploesti went into high gear. MAAF bombers flew twenty missions from 5 April through 19 August. These destroyed 89 percent of Ploesti's productive capacity (from 709,000 tons per month to 77,000) and reduced its gasoline and aviation fuel output by 91 percent (from 177,000 tons per month to 15,400). This was highly significant because most fuel went to the Eastern Front. Combined with continuing transportation attacks, this resulted in major operational problems for German forces in southern Russia.[47]

Heavy attacks during the first week of August, coupled with Romania's defection to the Allies, proved disastrous for German forces. With the loss of Ploesti, total monthly output of finished products in the Reich, including gasoline and aviation fuel, was 539,000 tons, only 40 percent of the preattack total.[48] By late May, attacks on Ploesti and German oil plants were already affecting German military operations and training. The German Army joined the Luftwaffe and Kriegsmarine in a steadily diminishing fuel quota system. By early July, the Germans had tapped strategic reserves on all fronts and were experiencing severe distribution problems. To deal with this, Oberkommando der Wehrmacht (OKW) cut fuel quotas by 20 percent for all services. Continuing attacks on oil refineries and oil supply installations along with intensive military operations on three fronts were depleting supplies so rapidly they soon fell 35 percent short of the minimum needed to supply all combat divisions with adequate fuel.[49]

By the time his unit tried to flee Greece in October, a dispirited POW said, "I wonder where they'll get the petrol now that we've lost Rumania?" He complained that there was no gasoline left, just a 95-percent-alcohol-to-5-percent-gasoline mixture that often would not start vehicle engines and destroyed them after a brief period.[50]

Ultra signals intelligence decrypts disclosed that fuel shortages were having a major impact on German operations on the Eastern Front. On the southern Russian front, Luftflotte 4 reported serious fuel shortages and ordered Fliegerkorps I to operate with minimum required forces.[51] On 13 October, Luftgau VII in Germany said the strictest fuel economy measures were necessary because no more fuel would be available *until the end of November.*[52] Equally troubling, Army Group North in Russia said, "Since August 1944 there has been an acute shortage of liquid fuel of every description. . . . As a result, operations by Panzer Troops and Air Forces have been temporarily reduced to a minimum. Only some 100 aircraft have been operating daily . . . during this period."[53]

German POWs who served on the Eastern Front in summer and early fall 1944 and were transferred to and captured in the West highlighted the disastrous fuel situation in Russia. One said, "The Russians encircled the Germans again and again, moving ahead faster than the Germans could retreat. . . . Supplies were sent everywhere except where most needed. General chaos existed."[54] Another said, "Gasoline is particularly lacking. In most instances supplies and artillery are moved by horse-power, therefore, very slowly. In addition to the gasoline shortage, bombing of rail lines has further disrupted supply lines."[55] Ultra intercepts and POW interrogations corroborated each other increasingly as the German fuel position worsened.

After the campaign, the Romanian minister of economics told Eaker, who was touring Ploesti, how effective the Ploesti oil and transportation campaigns had been. "Prior to your first bombing attack," he said,

our production from all the refineries was 26,000 tons a day of fuel products of all types. The day the Germans moved out, the production from all of these establishments was 3,000 tons a day, or between 10% and 15%. However, this does not tell the full story. As you will see, we have a surfeit of motor gasoline and petroleum products in this area. Even the 10% the Ploesti refineries were down to provided more than the German could get out. You had destroyed their tank cars, broken their bridges, knocked out their rail lines, attacked the marshaling yards and mined the Danube. I do not believe, therefore, that the German was getting out more than two to four percent of the fuel products being manufactured at Ploesti at the time he was forced to relinquish the area.[56]

By June, heavies were also raiding oil fields in Austria and Hungary and the refineries serving them in Czechoslovakia, Poland, and Germany. Brüx, Blechhammer North and South, and two smaller synthetic oil plants rounded out the target list. All suffered crippling blows. Eaker noted that from 16 June through 20 June, good weather had allowed his heavies to hammer Blechhammer North and South twice, doing immense damage.[57]

When the Germans tried to disperse oil production or build new plants, photointerpreters detected their efforts. Intelligence specialists confirmed that the oil refinery at Moosbierbaum was the same one dismantled at Port Jerome in France for relocation to the Caucasus. It refined crude oil and produced a high-grade distillate to boost aviation fuel octane. Estimated output was 150,000 tons per year. Within days, Fifteenth Air Force bombers permanently destroyed the plant.[58]

With the successful campaign against Ploesti, MAAF played a major role in the attack on oil, the weakest point in the German war economy. Although it did not have the strength required to attack oil targets decisively until spring 1944, the effort shortened the war. The cost in blood and treasure would have been higher had the Wehrmacht been able to defend the borders of the Reich with more substantial fuel supplies, a functioning transportation network, and the related ability to engage in *Bewegungskrieg* (mobile warfare) at even a modest level. MAAF knocked out the forty-six crude-oil refineries (including eleven at Ploesti) and five synthetic oil plants within its range. These installations accounted for about 60 percent of German oil supplies including a very high proportion of aviation fuel.[59]

As painful as it was for the Wehrmacht, the oil offensive's interactions with the Fifteenth Air Force's massive transportation offensive in southeastern Europe created a series of cascading effects that caused the Wehrmacht to run out of fuel, undermined its maneuver by road and train, saddled it with severe ammunition shortages, and largely paralyzed it just as it reached more defensible terrain near the Reich. Coordinated oil and transportation attacks collapsed the German-Romanian Army Group South Ukraine in August 1944, allowed for rapid advances on Budapest and Vienna, and had impacts throughout the faltering German defenses.

The campaign against Ploesti reduced Romanian rail capacity by 45 percent in addition to destroying oil refineries. When the Red Army offensive in southern Russia and the Balkans began on 19 August, the Germans tried to evacuate eight motorized divisions by rail. This required the entire remaining Romanian rail capacity of 500 trains, but routes were so congested that the Germans had to retreat by road, using up their fuel reserves even as trains transporting oil and fuel dropped from twenty-four per day to five. The two rail lines carrying troops to the front could handle twelve military trains a day, but the capacity often dropped to zero as heavies struck marshaling yards, key sections of lines, and the nineteen railroad bridges facilitating movement of traffic eastward.[60] These developments formed part of a much larger transportation offensive to support the Red Army's advance and bring a more rapid end to the war with lower Grand Alliance casualties.[61]

Bombing in Support of the Red Army's Advance

The geography in the southern portion of the German lines facilitated an effective effort to cut off fuel supplies and avenues of retreat through Romania and the Carpathian Mountains. This was a crucial location where the complex interconnections between oil and transportation attacks became evident. Group Captain J. C. E. Luard, chief of the Transportation Section within MAAF's Intelligence Division, produced a paper examining the "possibilities of aid to the Russians through attacks on the Balkan communication system by Italian-based long range heavy bombers."[62] The Russian capture of Cernauti had severed the main German supply lines. This action threw practically the entire burden of supplying troops on the Hungarian and Romanian rail systems, confined between the Carpathians on the north and the Danube on the south. In between, Luard observed, trains moved through multiple bottlenecks before emerging onto flatlands to the east. This meant that troops along the southern portion of the front were entirely dependent upon Romanian fuel for mobility. Luard identified three vital bottlenecks: Bucharest, Brasov, and Ploesti. The daily capacity of the two main rail lines leading through Ploesti was 33,000–36,000 tons. The Danube was no longer an alternative because of mining.[63]

Luard then estimated German supply requirements along the southern portion of the Eastern Front at 12,000 tons per day. This yielded a 3:1 rail capacity to supplies ratio, compared with 10:1 in Italy, making the three Romanian rail centers and the bottlenecks between them prime targets. Luard believed concerted raids would speed the Russian advance, causing the defection of Germany's Balkan satellites. Raids would also undermine German mobility, setting the stage for a follow-up effort to hinder the escape of German units from Greece and Yugoslavia. "The main purpose," Luard concluded, "is not to attack transportation per se, or to destroy freight cars and locomotives, but to reduce the rail capacity to the Eastern Front to the point where the enemy is no longer able to supply his troops opposing the Russian Army."[64] The raids that followed accomplished this, underscoring the vital interactions between forward-looking intelligence and good operations plans and the complementary nature of transportation and oil attacks.[65] As one scholar has noted, "In the eyes of some Allied commanders, especially Portal and Wilson, this [the tenuous LOCs for more than forty German divisions of Army Group South Ukraine and the resulting Luard plan to exploit their vulnerability] presented the western Allies with a golden opportunity to attack the Germans at a vulnerable point and aid the Soviets."[66]

The combination of a rapid Soviet advance, loss of and damage to key rail junctions, and rapid reduction in oil shipments sped the collapse of German resistance along the southern portion of the Eastern Front. The effects of transportation raids against Romania made themselves felt in conjunction with the renewed Soviet offensive on 20 August. Rail centers were so damaged and overloaded that reinforcing divisions had to move by road, using up their *Versorgunzsetze* (fuel increments) as they went. This lack of mobility was particularly dangerous in the East, where Soviet armored and motorized divisions, with their copious lend-lease vehicles and fuel supplies, already outmaneuvered the Germans.[67]

German Resistance before the Bombing

To understand how devastating the transportation offensive was to German operations in southeastern and central Europe, we must assess how well their units fought while they could still engage in *Bewegungskrieg*. Ever since its victory at Kursk, the Red Army had driven back the Wehrmacht relentlessly, exploiting its superior mobility, numbers, logistics, and increasing operational acumen. One of its major but until recently little known offensives was against Army Group South Ukraine in April and May 1944, with the triple objective of knocking Romania out of the war, capturing Ploesti, and inflicting another disaster on the reconstituted German Sixth Army. The Russians failed in large part because German units had good defensive terrain, good air support, and above all good rail and road networks on which to move oil, other supplies, and reinforcements. The Russians were also, uncharacteristically, in a relatively weak logistical position after their rapid advance.

Given the new geographical and logistical situation, which favored the

Axis, Marshal Ivan Konev's second Ukrainian front received an array of specialized units, including twenty-four combat engineer battalions, to help deal with the strong defensive positions in Romania's hilly and still muddy terrain. For the first time in nearly three years, the Germans were out of the steppes, in good defensive terrain, and backed by a relatively dense logistical network. Konev aimed to break through quickly and rob the Germans of these advantages. Army Group South Ukraine had some of the best army and *Waffen SS* divisions, and it had received substantial troop and tank replacements. From 8 to 11 April, Konev's troops battered against the German lines. To their dismay, German units counterattacked with great effect because they had good equipment, copious fuel and ammunition supplies, and heavy air support. On 12 April, the Russians called off the attack. Lieutenant General Hasso von Manteuffel, commander of the division *Grossdeutschland,* and other commanders reported that the Russian effort had failed. Like his fellow commanders, von Manteuffel had organized *Grossdeutschland* into *Kampfgruppen* (regimental combat groups) that could move quickly to key points and fight with exceptional effectiveness. This tactical acumen, combined with logistical and positional advantages and the last good air support of the war, savaged the Russians. It was the first major German victory since Kharkov a year earlier and would be the last.[68]

German operations demonstrated what was possible when conditions for *Bewegungskrieg* were still present. Every time the Russians advanced, the Germans shifted troops rapidly and made savage counterattacks. Their movements were so rapid that Konev's intelligence staff, long used to knowing where their adversaries were, could not keep track, resulting in multiple surprises and heavy losses. During late April, the Russians tried and failed again. Every time they tried to regroup, which the Russians had done at will since Kursk, German spoiling attacks from the hilly defensive positions caused their formations attrition.[69]

No longer used to these kinds of trouncings, Stalin and the Stavka (High Command) ordered the offensive to continue into the first week of May—a serious error given the casualties already sustained and the dim operational prospects. Six more days of intense fighting left the Red Army with extraordinarily heavy losses and no gains. Effective and frequent German air attacks, supplied with aviation fuel from Ploesti, played a key role in stopping Soviet advances and aiding German ones. Luftwaffe 88 mm guns knocked out dozens of Soviet tanks at critical positions. German counterattacks erased Soviet gains by 7 May. Further counterattacks along the Dniester River through 9 June were not entirely successful but left the Red Army unable to attack again for two months.[70]

After the battle, von Manteuffel called the German victory a "classic example of mobile defense. . . . It shows how successful even depleted units can be against a numerically far superior enemy, if their morale is high, if they are led with resource and skill, and react quickly and with dash and determination."[71] These observations were important because *Bewegungskrieg* as von Manteuffel described and practiced it depended on a steady flow of fuel, ammunition, and reinforcements. Without fuel in tactical situations and trains to move large units from place to place during lulls in the action, *Bewegungskrieg* was impossible. The oil and transportation offensives, which began having a major effect on fuel supplies and rail mobility by June, account to a substantial degree for Army Group South Ukraine's rapid collapse in August with the renewed Red Army offensive, an abject disaster utterly unlike the spirited defense of the spring.

The Russian defeat in May was so heavy that Soviet official histories did not mention it. Only after the end of the Cold War, as the archives opened and a new four-volume official history appeared, did the truth emerge.[72] The scholar who brought this to light has asserted convincingly,

> The "discovery" of this forgotten offensive fundamentally alters longstanding perception of the military strategy Stalin and his Stavka pursued during 1944, if not the entire final two years of the war, by revealing what should be properly termed Stalin's "Balkans strategy." . . . Simply stated, vital military, economic, and political factors prompted Stalin to order his Red Army to mount a major offensive of immense potential strategic significance into Rumania between mid-April and late May 1944.[73]

Just as important was the German Army's exceptionally effective resistance. This had much to do with the excellence of troops and commanders but even more with good defensive terrain, abundant supplies, intact transportation networks, and effective air support. At a cost of approximately 45,000 men, the Germans inflicted about 150,000 casualties and destroyed more than 300 of the 500 Soviet tanks engaged. Their own tank losses were not entirely clear but less than 50. The Russians also lost nearly 1,000 artillery pieces and antitank guns.[74] Stalin relented and ordered Konev to prepare for a major offensive in August, after the logistical situation improved and reinforcements arrived. Although we cannot know with any certainty how well Stalin understood the effects of Allied bombing on German oil and transportation assets, he was aware that the Allies were doing everything in their power to hasten the Red Army's advance by employing heavy-bombers against these target sets, and his intelligence assets made the extent of the damage after May 1944 clear to him. By the time Konev's troops advanced in August, the Germans faced severe fuel shortages, were

unable to use the Romanian railroad network, and could not resupply—and
the Russians knew it. The resulting Red Army offensive, carried out with
its usual skill and ferocity, thus encountered much weaker resistance than
in April and May. The victory propelled Russian troops rapidly through
Romania and into Yugoslavia, Hungary, and ultimately Austria and eastern
Czechoslovakia.

The Transportation Campaign Intensifies

From April to August 1944, MASAF flew 7,579 sorties against transportation targets in southeastern and central Europe.[75] On 31 August, Eaker
wrote Arnold:

> All goes well with us, but we are pretty tired, having waged in the month of Au-
> gust the greatest air offensive ever put on by any air force. Our strategic bomb-
> ers operated in force on 30 days of the 31, and our Tactical Air Force operated
> every day. We put out over 51,000 sorties. . . . I think it will be found, when
> air history is compiled, that no air force of a similar size has ever brought such
> pressure on the enemy and his material resources in one month as the Mediter-
> ranean forces have in the past month of August, in addition to supporting two
> armies in Italy and another Army in the invasion of Southern France.[76]

Transportation raids in southeastern Europe intensified between 20 August and 25 October. The Fifteenth Air Force flew 6,223 heavy-bomber
sorties during this period. This disorganized and slowed German troop
movements and kept more than one-half of the German units in Greece and
Yugoslavia from escaping northward. The devastation of railways forced
the Germans to move largely by road, but they had insufficient fuel. Losses
in Romania and Hungary were particularly heavy as the combination of
rail crises, destruction of bridges, and fuel shortages made withdrawals and
maneuver extraordinarily difficult. The Germans tried to throw together a
defensive line in the Carpathians, but the Red Army got there too fast.[77]

When Bulgaria declared war on Germany on 8 September, Field Marshal
Maximillian von Weichs decided it was time to get his troops off of the
Aegean Islands and out of Greece. He had 57,000 Germans and 13,850
Italians to evacuate using 100 Ju 52 transports and a dwindling number of
Kriegsmarine vessels. As a result of heavy Allied bombing raids on transportation nodes from Greece to Hungary and the disruptions these caused
to fuel deliveries as well as the movement of troops by rail and road, he had
to leave 26,495 Germans and 11,000 Italians behind after his transports,
vessels, and most vehicles ran out of fuel.[78]

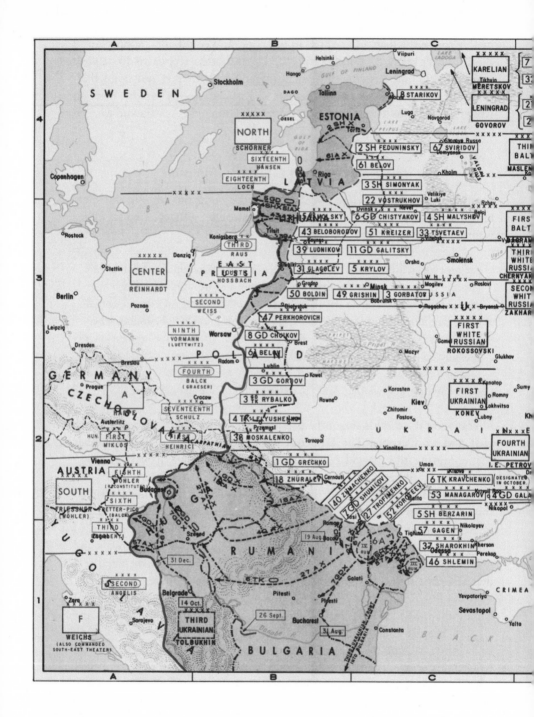

EASTERN EUROPE, 1941
RUSSIAN BALKAN AND
BALTIC CAMPAIGNS

Operations, 19 August - 31 December 1944

31

SCALE OF MILES
0   100   200

GERMANY'S ALLIES

25 AUG RUMANIA DECLARED
        WAR ON GERMANY.

8 SEPT. BULGARIA DECLARED
        WAR ON GERMANY.

15 OCT HUNGARY ANNOUNCED
        END OF HOSTILITIES.

18 OCT GEN MIKLOS (FIRST
        HUNGARIAN ARMY)
        JOINED THE SOVIETS.

The Russian offensive in the Balkans, 19 August–31 December 1944. This major offensive followed a failed one in April–May 1944, during which well-supplied German troops, still able to conduct mobile warfare with reliable logistics and good air support, demonstrated how much fight the Wehrmacht had left in it as it reached more defensible terrain closer to the Reich. Heavy-bomber raids on Ploesti and marshaling yards throughout the Balkans undermined German mobility and resistance, facilitating a rapid and massive Red Army breakout and advance in August. The oil and transportation offensives proved very effective in speeding the advance of the Grand Alliance's soldiers to the Reich. (Maps Department of the US Military Academy, West Point)

Heavy-bomber raid on marshaling yard, 1944. The massive transportation
campaign in southeastern and central Europe produced innumerable scenes
like this one. The Wehrmacht's fuel and ammunition supplies, mobility, and
communications largely collapsed, allowing the Red Army to advance with great
speed even in what would otherwise have been highly defensible terrain. By late
1944, German units on the Eastern Front were often immobile for days. (National
Museum of the US Air Force Photo No. 091002-F-1234S-019)

From 25 October 1944 to 3 January 1945, raids on Yugoslavia's Vardar
Valley, through which German units fleeing Greece were retreating, were
heavy. Most of the units escaped, but with heavy losses of men and equipment as frequent delays made them vulnerable to Partisan attacks. There
were nearly 200,000 Germans retreating along this route. A heavier air effort would have produced greater results, especially if the Allies had also attacked the route that the relief column from the north was using and the few
trains bringing fuel to retreating troops. Nonetheless, the Germans lost most
of their vehicles and heavy weapons. By the time Army Group E escaped, it
was severely weakened.[79]

Transportation attacks continued during the final German counteroffen-
sives in the East. Between 4 January and 14 March 1945, the Fifteenth
Air Force flew 8,411 sorties against rail targets. The chaos brought on by

bombing is difficult to overstate. Every main and alternate route closed, mostly permanently, to anything but small trains rushed through after repair crews made the lines operational until the next raids, which were never long in coming because PR and PI told targeteers when the lines would reopen. The German counteroffensive around Lake Balaton stalled for lack of fuel, surviving formations were unable to retreat with any order because of rail and fuel shortages, and the Soviet offensive against Vienna proceeded quickly as a result. During the final period of Fifteenth Air Force transportation attacks, from 15 March to 6 May 1945, heavies flew another 5,616 sorties. The Soviet capture of Vienna netted 130,000 POWs and destroyed eleven panzer divisions fighting from static positions because of lack of fuel.[80]

Eaker was pleased with the Fifteenth Air Force's huge strides, from ground crews to bombing. Twining was employing a new bombing technique in which he used his 100 pathfinder (H2X-equipped) bombers individually during bad weather to attack a variety of high-value targets. Their tactic was to stay above the clouds and German fighters the entire way. These were the best H2X operators in the USAAF and could hit targets with reasonable accuracy. Eaker also emphasized to Arnold that MAAF had a liaison team with the Second Ukrainian Army in Bucharest. This comment was immensely important because it confirms that Fifteenth Air Force liaisons were posted to Red Army headquarters and units to coordinate bombing raids with the Soviet advance.[81]

Equally significant, German SIS officers noted, "The U.S. Strategic Air Force was pre-eminently responsible for the eventual collapse of the Balkan countries, thereby contributing in no small measure to final Allied victory."[82] They further noted, "Building upon its experiences of the two bomber commands in Africa, the Fifteenth USAAF began to develop completely individual and flexible tactics. These tactics differed in essentials from those employed by the puissant, though comparatively primitive, procedures of its counterpart in Great Britain, the Eighth USAAF."[83] German fighter units had trouble reacting to Fifteenth Air Force raids because of the huge area to cover and their relatively small number of fighters. The first half of 1944, the SIS officers note, was dedicated to knocking the Axis satellite states in the Balkans out of the war. "In this connection," they said, "co-operation with the Russians was clearly noted by the Luftwaffe SIS in the Russian air raid warning networks."[84] The Fifteenth Air Force was the Red Army's de facto heavy-bomber force on the Ukrainian front, carrying out huge numbers of raids against LOCs and supply dumps. Attacks on marshaling yards in the eastern Balkans, which were "of such decisive importance in the East,"[85] sped the advance as German units struggled to find enough fuel and ammunition to make a stand.

Heavy-bombers kept German forces in northern Yugoslavia and western Hungary from reaching Vienna in time to defend it. Severe fuel shortages virtually grounded remaining Luftwaffe bomber units by fall 1944, even as intense Fifteenth Air Force raids on LOCs made ground maneuver extraordinarily difficult and costly. After the Germans suffered catastrophe in Vienna, there were virtually no forces left to counter further Soviet advances. The hammering of marshaling yards, fuel dumps, and other high-value targets demotorized the remaining panzer divisions, made movement virtually impossible, and precluded maneuver as units ran almost entirely out of fuel. The Germans pushed trains through where they could, but the Wehrmacht's heavy reliance on the *Reichsbahn* for operational and strategic mobility undermined the foundational elements of German war making as the rail system collapsed. In Yugoslavia, where fighter-bombers could reach the trains, the Germans simply gave up rail travel and moved by road at night. Partisans took full advantage of this.[86]

The concerted raids during this little known but massive transportation offensive had a number of important effects. The sheer size of the effort—308,532 bomber sorties and 69,040 tons of bombs delivered at a cost of 340 aircraft—makes it clear that the Western Allies were in deadly earnest about assisting the Soviet advance and thus bringing a quicker and less costly end to the war. This campaign involved 43.6 percent of the total air effort against transportation targets in the greater Reich—a stunning figure. The oil offensive, combined with poor German logistical practices, had already left the Wehrmacht exceptionally vulnerable to the kinds of deep, narrow, and rapid Soviet penetrations of their lines and the inevitable exploitation that followed after German units fighting at the front had used up their fuel reserves and could not retreat in good order or at all. The Fifteenth Air Force official history simply noted, "The effort expended in facilitating the liberation of the entire area of Southeastern Europe formed a strikingly impressive part of the total Air Force effort."[87]

German Logistics: Still an Achilles' Heel

The German Army's supply system, which suffered from a bifurcation of operational and supply responsibilities, magnified the effects of bombing and Russian advances. Poor placement of depots combined with decreasing deliveries and deep Soviet penetrations to produce the operational cataclysms already addressed. The cavalier German attitude toward logistics was reflected in the fact there was no supply expert at the Führer's daily meetings. Consequently, logistical difficulties received insufficient consideration. Ironically, oil and transportation attacks drove a centralization of supply functions at OKW. The result was a shift from a supply system in which nobody paid attention to details to one in which rigid centralization produced inflexibility. Additionally, the OKW quartermaster general had no control over each service's quartermaster until late 1944, leading to wasted resources.[88]

Equally serious, supply channels ran directly from the quartermaster general of the army to the chief supply officers of the field armies. Army groups and corps headquarters were *not* part of official supply channels until December 1944. Nor was there any centralized control of supply. Field armies carried and distributed supplies, in theory, through army supply distribution groups. These had too few trucks, so combat divisions used their own but had too few to build large reserves. Finally, and most injurious in terms of fuel supplies, was the policy on the Eastern Front of placing large depots at the field army level, about sixty miles behind the front. This forced German divisions to send trucks on supply runs rather than using them to shift troops to key areas.[89]

The irony here is that German armies in the East, unlike their counterparts in the West, did not have to do this. The threat of air attack was much lower, so they could have placed large supply dumps—or a large number of smaller ones—closer, where truck convoys from divisions could load up and be back within hours rather than days. In certain instances, such as the effective operations in Romania during April and May 1944, German logisticians did just this, backed as they were by a capable transportation network. However, after bombing caused the rail network to fail, these resupply capabilities disappeared even where logisticians understood how to leverage them, as the rapid defeat in Romania in August 1944 underscored.

The combination of Red Army tactics and the gradual failure of the rail network meant that divisions fought too far from field army depots. Because of the oil offensive, they often had too little gasoline to retrieve supplies. Consequently, after Soviet armored spearheads penetrated the front to a depth sufficient to cut off divisions from their depots, disaster ensued. The huge expanses of the East meant trucks were often the only means for getting supplies to the front. When depots were overrun and the fuel pipeline disrupted, combat units relied on fuel they already had or retreated sufficient distances to receive fuel supplies at the railheads. As marshaling yards in southeastern and central Europe were savaged starting in summer 1944, even when units arrived at railheads there was often nothing there because of severe traffic disruptions. Speer said, "Fuel was only obtainable by chance and not at the place where it would be needed."[90] To make the disaster complete, Hitler often forced units to stand their ground without resupply.[91]

Until June 1944, the normal fuel issue to an armored division was five consumption units.[92] Army units continued to receive fuel allotments until July 1944, at which point bombing-induced shortages led to the issuance of fuel quotas every ten days, then every five. Army units lived hand to mouth on these small and unreliable allotments. Given the tempo of oil and transportation attacks, allocations dropped in fall 1944 from five consumption units to two or three for heavy divisions.[93] Tanks also used three times as much fuel in muddy conditions as in dry ones, and the winter of 1945 in eastern Europe was particularly muddy as a major surprise Russian winter offensive began on 12 January.

The upshot of these logistical woes was a massive fuel shortage resulting in the immobilization of 1,200 German tanks and assault guns at the Baranow bridgehead and other locations—65 percent of the 1,850 they had on the entire Eastern Front—in the first few days of the offensive.[94] Speer commented on the enormity of this disaster:

> We had utilized here approximately 1,200 tanks for the defense and here it was for the first time that the troops had only one or two sets of supplies for fuel [fuel increments] so that the tanks were practically unable to move when the Russian attack started. To the best of my knowledge tanks were not utilized again [in large quantities] in battle.[95]

The shortage of fuel Speer referred to was a product not only of the oil offensive but of transportation attacks that made fuel and ammunition deliveries unreliable.

Equally problematic was the proportion of fuel in the "pipeline" stretching from oil plants and refineries to troops on the line. This included all fuel in circulation and was the minimum needed to keep combat units supplied. About 10 percent of all fuel was in transport at any given time, making fuel supplies very susceptible to transportation attacks.[96] As a US Strategic Bombing Survey (USSBS) study noted, "The bombing of oil targets was concomitant with the attacks on transport in bringing about a paralysis of the German logistics system. This blocking of logistic circulation, together with the draining through battle wastage of the supplies remaining in the system, was the immediate cause of the collapse of the German armies at the time it occurred."[97]

MASAF assets played a vital role in speeding the Red Army's advance and helping to bring the war in Europe to an earlier and less costly end. There were several reasons for this. First, unlike the Eighth Air Force and Bomber Command, the Fifteenth Air Force specialized in striking oil and transportation assets even as it hit a number of fighter factories and other high-value targets. Second, because it had focused on "point" targets almost exclusively, its bombing was more accurate. Third, it hit the Wehrmacht's means of maneuver and supply at the very moment that German divisions, many of them still highly effective, were reaching more defensible terrain. This included not only raids on marshaling yards but also No. 205 Group's highly successful mining of the Danube. As one scholar has noted, this "had a decisive and negative impact on the overall German oil position . . . transportation system attacks . . . interfered with the movement of Axis troops in the Balkans and with the shipment of raw materials to the Reich."[98] The loss of fuel and transportation assets, and thus of the ability to maneuver, led to operational calamities throughout the East. MASAF assisted the Russian advance in crucial ways, helping to shorten the war and reduce the Grand Alliance's casualties.

15

The Balkan Air Force
January 1944–April 1945

Supporting Irregular War

BY 1944, THE ALLIES HAD EMPLOYED airpower effectively in all of its major roles, including air superiority, interdiction, direct support, heavy-bombing of vital target sets, air transport, photoreconnaissance, and aero-medical evacuation. They had learned to use these diverse capabilities in a carefully coordinated fashion in all three domains—air, land, and sea. Cooperation with the other two services had developed steadily and reached a high pitch of combined-arms effectiveness. The Allies were about to put all of this expertise to work in a very different context—irregular war—as they supported the Yugoslavs in their struggle against Axis occupying forces. The larger Allied intent was ultimately to ensure that the postwar Yugoslav government did not become a Soviet satellite state.[1]

The first British liaison officers reached General Dragojul Mihailovich's Chetnik headquarters in September 1941 and soon found the former very difficult to work with and utterly unwilling to cooperate with the other emerging resistance leader: the Croat Josef "Tito" Broz. By the time Henry Maitland Wilson replaced Dwight Eisenhower as commander in chief (C in C) Mediterranean theater, Tito's Partisans had become far more effective fighters than the Chetniks and more willing to engage occupying forces. When the Germans launched their sixth offensive against Tito's troops in December 1943, Wilson had determined that supporting them would hold Axis divisions in place and create serious problems for the extraction of natural resources to support the German war economy. The first British liaison officer had arrived at Tito's headquarters in May 1943. In September, a liaison staff with substantial signal capability joined Tito. Numerous US Office of Special Services (OSS) and later Jedburgh teams also arrived. The Allies had to tread carefully as they shifted the bulk of their support from Mihailovich to Tito while maintaining diplomatic support for the government of King Peter, but they were determined to work with Yugoslav irregulars to create maximum problems for the Germans and gain leverage against the Russians.[2]

Allied support increased in October 1943 as torpedo boats and aircraft began delivering supplies. The totals were minuscule—twenty-three tons for the Chetniks and six tons for the Partisans between June 1941 and June 1943—but increased dramatically after the Allies had adequate basing facilities and air-sea assets in October. By this time the Partisans had more than 220,000 men and women under arms, organized in twenty-six divisions, and controlled more than half of Yugoslavia.[3]

Given the political and operational complexities involved, Wilson took direct control. From the outset, air, naval, and special operations assets played predominant roles in supporting Tito with supplies and in direct-action missions. In January and February, Allied aircraft had dropped 130 tons of supplies to Tito's forces, and ships had brought in even more through the Island of Vis, which was so heavily defended and had such strong air and naval support that the Germans did not try to take it. The Mediterranean Allied Tactical Air Force (MATAF) and Mediterranean Allied Coastal Air Force (MACAF) played key roles by driving the Luftwaffe out of the area and then engaging in numerous attacks against land and sea targets. On 12 April 1944, Force 266 became the chief operational organization in Yugoslavia and Albania. The Russians had established their own liaison group alongside the Allies' in late 1943, and in March 1944 they requested and received permission to use the major airfield at Bari for twelve Dakotas and twelve fighters, with the former dropping supplies to the Partisans. The fact that the Russians used lend-lease aircraft with lend-lease air-drop equipment to drop lend-lease supplies to the Partisans must have been amusing to those with any time to notice.[4]

Air drops increased to 183 tons in March 1944, and with the Partisans liberating increasingly large areas, aircraft now landed supplies directly with the help of the Allied flying control and unloading party at Medino Polje. Aircraft evacuated wounded Partisans to Italy, giving Tito's forces maximum mobility and effectiveness because they no longer had to defend field hospitals. On 25 May the Germans launched a major offensive aimed at destroying the Partisan hold in western Bosnia. Wilson ordered every available Allied aircraft into the fray, driving off the Luftwaffe and bringing troops under concerted attack. The German offensive failed. By this time, Wilson had concluded that he needed an intermediate level of command to oversee all operations, but especially the burgeoning air effort. The Balkan Air Force (BAF) was about to enter the fray.[5]

German Signal Intelligence Service (SIS) officers noted that the British took advantage of guerrilla movements in the Balkans as soon as possible, flying six to ten sorties every night to drop weapons. The area was too rough and the lines of communication (LOCs) too poor to allow Axis forces to exert permanent control. By summer 1943, forty-five British liaison teams in Yugoslavia, and more elsewhere, provided radio and beacon guidance for transport aircraft that often landed at captured airfields to discharge cargo. By 1944, about seventy B-24s and Halifaxes were delivering supplies. The SIS's counterparts in the Army Radio Defense Corps picked up most of this radio traffic, but getting troops to the drop zones without being ambushed was difficult. In addition to air transport assets, two Royal Air Force (RAF) fighter-bomber wings, another of bombers, and units of the Italian Allied Air Force were flying strike and direct support missions for the Partisans.[6]

Forming the Balkan Air Force

As the SIS tracked the growing effort to help the Partisans, Allied senior officers knew that they had to do more if they wanted Tito's forces to continue tying down large numbers of Axis troops. The Germans' sixth offensive also drove home the possibility that they might kill Tito, creating a disaster for Yugoslav resistance efforts. Air Commodore Leonard Pankhurst, chief of the Mediterranean Allied Air Forces' (MAAF's) Plans Division, led the effort to form the BAF. John "Jack" Slessor told Wilson that the existing organization was unsatisfactory even if effective in the aggregate. A clear command structure and relationships would make it much better. Slessor proposed that Wilson appoint a task force commander or a permanent commander subordinate directly to Allied Forces Headquarters (AFHQ). Ira Eaker agreed. In fact, Slessor commended Eaker after the war for going "to the very limits of his authority to help us in the Balkans" even though he was not authorized to engage officially in the development of Balkan strategy, which the Joint Chiefs of Staff (JCS) had long since made clear was a British affair.[7] With Wilson's and Eaker's support, Slessor suggested further, given the preponderance of air involvement in the Balkans, that the commander be an airman with operational control over all combined-arms efforts. Accordingly, the draft directive to the BAF commander said,

> You will be responsible for supporting them [the Partisans] to the greatest practicable extent by increasing the supply of arms and equipment, clothing, medical stores, food and such other supplies as they may require. You should also support them by commando operations and by furnishing such air support as you may consider advisable in the light of the general situation.[8]

Meanwhile, planners determined that available aircraft could deliver 1,875 tons of supplies per month and evacuate 11,250 wounded. Planners warned that the Partisans were becoming frustrated because deliveries had not equaled promised amounts—an operational as well as a political issue because Tito appeared likely to lead Yugoslavia after the war. Pankhurst warned that the Allies had to deliver more and not promise more than they could deliver.[9]

To address these problems, Slessor hosted General Vladimir Velebit, Tito's representative, for discussions on 21 April. Velebit said the Partisans could hold perimeters around all four available airfields long enough to allow for major deliveries of heavy weapons, ammunition, food, and other supplies. He advised against building runways because they would alert the Germans. Each existing airfield already had a Partisan commander responsible for its security and for coordination with the Allies through liaison officers. Slessor agreed to provide flares immediately so the Partisans could mark those runways for night deliveries. Many of the 10,000 wounded were located around one of the airfields and ready for evacuation. General Velebit asked how many wounded the Allies could evacuate each week, so he could stage them to the airfield. The use of fighter-bombers presented a more complex problem because they would have to base at these airfields with maintenance personnel, munitions, and all the other required assets. Pankhurst suggested placing external fuel tanks at the airfields so fighter-bombers could fly operations against German forces, land at the airfields, have maintenance personnel attach the tanks, and fly back to Italy. Wireless radio operators would accompany the refueling teams to ascertain aircraft arrival times and minimize their time on the ground. These agreements brought the BAF one step closer to formation.[10]

The outcomes of the meeting were driven by realities on the ground. The Germans were having increasing difficulties controlling Tito's 300,000 troops. These were poorly armed and fed but holding nearly twenty German divisions in place, sabotaging railways, and inflicting heavy casualties. However, the Partisans lacked the heavy weapons to keep the Germans from moving largely at will. The BAF's key functions, therefore, would be to transport heavy weapons and trainers to the Partisans, evacuate Partisan wounded, and develop airfields for transports and fighter-bombers to support Tito's operations.[11]

MAAF planners believed that anything other than immediate action would mean a lost opportunity of strategic importance, with a corresponding loss of Allied prestige just as the Russians were approaching Yugoslavia. The planners also wanted to isolate German troops in Greece and the Aegean by denying them the LOCs in Yugoslavia upon which to retreat. Transports delivered the vast majority of supplies, although naval craft also brought in some. Vis provided an ideal advanced airbase. In concert with developing and defending Vis, the Partisans would use heavy weapons delivered there to drive the Germans away from the coast wherever possible. More airfields

would go into operation as portions of the coast became secure. No. 334
Wing would lead the transport effort, while additional fighter-bomber units
provided direct support. A large number of RAF air stores parks (ASPs) and
other liaisons were in place.[12]

Slessor signaled Charles Portal in his usual fiery way that close coordination with army, navy, special operations executive (SOE)/Special Operations
Service (SOS), and Partisan forces would be crucial:

> The operations in BOSNIA have turned out well and look like having resulted
> in a complete failure on the part of the Hun and something of a triumph for the
> Air Force. Anyway the Partisans' tails are sky-high and our stocks have soared.
> Even the RUSSIANS have been complimentary on air support. But this week
> has impressed on me more than ever the essential need for an H.Q. running the
> show over there.[13]

Eaker also weighed in on the importance of establishing the BAF, telling
Henry "Hap" Arnold that Tito's Partisans had rescued many combat crews
downed in the Balkans. He was stunned by the kindness and assistance they
rendered, often at great risk. The Partisans gave aircrews scarce clothing
and food as they saw them to safety. Randolph Churchill had dined with
Eaker the night before after spending several months supporting Partisan
operations. He said the Partisans cheered every time the Fifteenth Air Force's
heavies flew over. These aerial armadas encouraged Tito's men. US aviation
engineers had just completed an all-weather airfield on Vis, placed a radar
site there, and thus facilitated very effective MACAF attacks on Axis shipping as well as fighter-bomber support to the Partisans. Eaker closed by
saying, "I feel that our support of the Partisans and our effort in the Balkans
has far-reaching effects on the overall view of the war. It is pinning down a
large number of German divisions, it is destroying a lot of German supplies
and it is killing many Germans."[14]

Eaker, Slessor, and Wilson agreed that it was high time to appoint a commander for the BAF and activate it. This RAF officer would have a dual
role: commander of No. 334 Wing, which supported Tito's Partisans, and
coordinating and planning all operations in Yugoslavia. The BAF's mission
was to contain and destroy as many enemy forces as possible in close cooperation with Tito. A senior RAF liaison officer would join Brigadier Fitzroy
MacLean's mission at Tito's headquarters to maximize BAF effectiveness.
The BAF commander would also be responsible for all antishipping missions
in the seas around Greece, special operations missions, and coordination
with Russian aircraft and troops as they began appearing. He would also
plan and execute all amphibious operations and construction of forward

airfields in the Balkans. Finally, he would coordinate with the SOE, MACAF, and all other organizations with overlapping responsibilities in his area of responsibility.[15]

To drive home the importance of establishing the BAF, Slessor penned a note in which he observed that the Germans still had twelve divisions in Yugoslavia and Albania. The Partisans could not prevent them from retreating. Wilson thus directed that the BAF find good locations for building landing grounds from which fighter-bombers could operate in conjunction with Partisan military actions. Transport aircraft would deliver the men and materials. The BAF was thus evolving even before it became official. From assisting Tito's Partisans, it was now moving toward an interdiction role designed, in concert with Partisan ground actions, to prevent the escape of German units from the Balkans. Given the timing of Slessor's note, just as the oil and transportation offensives got into high gear, his thoughts were clearly linked to the larger prospects of demotorizing German units and destroying them. Finally, the political importance of keeping Tito from falling under Stalin's sway was always clear. During the week prior to the BAF's official activation on 1 June, Allied aircraft flew 982 combat sorties in support of the Partisans, helping to stop the Germans' seventh offensive. Tito's decision to accept refuge in Bari early in the German offensive, when airborne troops nearly captured him, proved unnecessary from an operational standpoint but fortuitous from a planning one. During meetings at Bari with Slessor and other senior Allied officers, Tito asked for great efforts to defeat the Luftwaffe in the Balkans, to step up the supply of arms and equipment, and to work harder to evacuate Partisan wounded. The Allies made good on all three requests. Two weeks later, Tito effused, "British and American aircraft have driven the *Luftwaffe* from Yugoslavian skies, and considerable damage has been inflicted on German forces. As a result Partisan morale and Anglo/American prestige have soared."[16]

The BAF Is Activated

Eaker's directive to Air Vice-Marshal William Elliot, the new BAF commander, said, "You will be responsible for co-ordinating both planning and day to day conduct of combined amphibious operations and raids by Allied air, sea and land forces on the islands and Eastern shores of the ADRIATIC and IONIAN Seas. The system of joint command will be followed in trans-ADRIATIC operations."[17] Elliot would have a combined operations room, intelligence center, and communications center at his headquarters to ensure maximum effectiveness for the complex and fast-paced operations that typified support to Tito's Partisans. Ideally, further successes would facilitate large-scale supply of Tito's forces. In addition to controlling all air and joint operations, he also coordinated with Brigadier Fitzroy MacLean at Tito's headquarters.

Although US senior leaders did not approve the direct employment of US ground units in the Balkans, they provided substantial airlift capability. Employment of MAAF assets for operations in the Balkans brought them under Elliot's tactical control. Air transports delivered eighty-four 75 mm mountain howitzers to Tito's units. During June, with the BAF increasingly active, the Partisans extended their operations into Serbia and interdicted rail lines. BAF transports and No. 205 Group bombers, with their crews' expert night-flying and navigation capabilities, delivered 312 tons of supplies—a new monthly record. As Tito's victories increased, so did his recruits. By the end of August, 498 tons of additional supplies were in Partisan hands—309 landed by aircraft and another 189 airdropped or brought in by sea. Nearly 1,700 Partisan wounded headed to Italy for treatment. The Russian aircraft at Bari delivered another 321 tons of supplies and evacuated another 625 Partisan wounded. By September, the strategic picture in the Balkans was bright as the Partisans scored more successes. The German loss of Bulgaria forced a major withdrawal from Greece and Albania along with most of Yugoslavia, a process with which the Partisans assisted gleefully, making widespread attacks on rail lines and roads. However, as the Russian advance into Yugoslavia began, Tito and the Partisans cooperated less with Allied officers, and Tito made it clear that he would soon move his headquarters from Vis to Belgrade.[18]

By the end of August, the BAF comprised sixteen squadrons. The greatest threat was the flak that the Germans deployed liberally along major LOCs. Fighter-bombers strafing targets in the narrow valleys often found themselves fired upon by flak positions *above* their altitude as well as those below—not a happy or safe set of circumstances. The BAF flew out more than 10,000 wounded troops from June through September, increasing the Partisans' mobility and giving them assurance that their comrades would not be captured and executed. However, operational problems paled in comparison with the political problems surrounding BAF coordination with Tito's Yugoslav Army of National Liberation (YANL). Wilson and Tito resolved many of these in their August meetings, and Operation Ratweek ensued, during which the Partisans and BAF assets engaged in generally successful efforts to impede the retreat of German troops through Yugoslavia. The BAF flew 38,340 sorties in support of the Partisans, dropping 6,650 tons of bombs and carrying out large numbers of direct support missions. Despite these challenges, Wilson and Tito were pleased with each side's handiwork when they met at Caserta on 6 and 10 August to discuss their next steps.[19]

Elliot formed the Air Liaison Organization, Balkan Air Force, to work directly with Tito's forces in the provision of air support. The organization had air support commands (ASCs) for employment with fighter-bombers at forward airfields, air liaison officers (ALOs) to work with the Partisans, and very effective signals and maintenance capabilities. The latter were particularly versatile given the variety of aircraft to maintain. The Partisans valued airpower and worked well with their RAF counterparts. Headquarters BAF included staff elements from all three services and was in constant communication with the military mission to the Partisans. An advanced landing ground at Vis had ALOs who briefed all aircrews on mission details. RAF liaison officers attached to Partisan units supervised air drops, transport aircraft deliveries, and direct support missions. By the end of 1944, the BAF had played a vital role in clearing southern and central Yugoslavia of German troops. During the last three months of the war, rover tentacles moved forward with the Partisans' Fourth Army as it cleared German troops from the remainder of the country. From 19 March to 3 May, BAF aircraft destroyed more than 900 vehicles, 126 locomotives, large amounts of rolling stock, and 46 surface vessels.[20]

As these efforts unfolded, Wilson proposed Operation Baffle to Tito, which would allow 6,000 Allied personnel to establish a major air base at Zadar from which the BAF could attack retreating German forces. Mediterranean Allied Strategic Air Forces (MASAF) heavies would have the option of diverting there. With large German formations still posing a threat to this air base, the Allies scaled it back to an advanced landing ground with a much smaller footprint. Nonetheless, the airfield saw extensive action with its capacity to handle 200 fighter-bomber sorties and many bomber landings every week. In fact, heavy-bombers used it extensively, with as many as seventy there at any given time.[21]

Eaker visited the BAF in November and was highly impressed. He referred to it as

the most heterogeneous collection of personnel anywhere in the world, there being units of nine different nations represented there, and it is an up and coming organization, full of enthusiasm and is really paying dividends in chewing on the retreating columns of Germans coming out of Greece, Albania and southern Yugoslavia and wearily fighting their way through partisan and guerilla [*sic*] bands along the limited road network on the other side of the Adriatic.[22]

Eaker was confident the combined BAF and Partisan actions would delay the German retreats until winter, when the weather would make their plight even more miserable. He proved correct, and these unfortunate Germans simply moved from one crisis to another as they found themselves in the middle of MASAF's major oil and transportation offensives. Less than half of them made it back to the rapidly contracting German lines.

As BAF operations wound down by spring 1945, the command had performed its mission exceptionally well, increasing Partisan combat power tremendously and making German operations in and the withdrawal from the Balkans very costly. The command's diplomatic benefits did not go unnoticed either. Despite Tito's pragmatic requirements to deal with the Russians, who were in his country in large numbers, his long association with the Allies gave him bargaining power none of the other Balkan states had. His huge military force, now numbering more than 300,000 and with a demonstrated ability to engage in sustained irregular warfare, did not hurt either. All of these factors, and several others, facilitated Tito's ability to avoid either Soviet domination and occupation or too much influence from the Western camp. The BAF had done its wartime job and, in the process, helped set the terms for the peace.

PART IV

Retrospective

16

Taking Stock

Crucial Theater, Crucial Weapon, Crucial Victory

BY 1945, THE INTENSITY, REACH, ROLES, and tasks of Allied aircraft reached their zenith in the Mediterranean theater. Although the battle for its control had long since turned in the Allies' favor, the role airpower played in these events, and as a vital partner in the combined-arms arena, continued to grow. When Air Chief Marshal Arthur Longmore took command of the Royal Air Force (RAF) in the Middle East on 13 May 1940, he had 205 aircraft in Egypt and 165 in East Africa. This made it more important than ever for Longmore to concentrate all air assets under his control. This concept proved vital to victory in the Mediterranean. When the Allies captured Tunis on 13 May 1943, exactly three years after Longmore took command, Air Chief Marshal Arthur Tedder, who succeeded him on 1 June 1941 and was by this time both air commander in chief (AOC in C) Mediterranean and Dwight Eisenhower's air deputy, had 3,500 modern combat aircraft at his disposal. Whereas Longmore had freedom of action but too few resources, Tedder had an abundance of both. A year later, Lieutenant General Ira Eaker's Mediterranean Allied Air Forces (MAAF) had 5,000 modern combat aircraft and could range the entire Mediterranean, much of the southern Reich, and all of eastern Europe. As the official historians note,

> Much less would have been achieved had all the Allied air resources not been concentrated in the hands of one man as it [*sic*] had been since Longmore's day, with the exception of the short but confused period during the earlier part of the Tunisian campaign before Tedder was given overall command. Allied airpower could always be applied where it was needed most at any given time, and yet the never-ending run of the mill air commitments could still be met. In brief, unified control did make it possible not to squander these vast resources.[1]

Airpower played a central role in the Mediterranean and Middle East, and its strategic effectiveness there exercised a major influence on the larger war. Within the theater, given the huge distances involved, the complex geography, and the opportunities and vulnerabilities these created, land-based airpower proved an indispensable partner with the other fighting services. This became evident by spring 1941. Without control of the air, naval and ground operations were extraordinarily vulnerable to direct attack and, more importantly, the indirect effects brought on by the battle for the sea lines of communications (SLOCs). Superior Allied employment of airpower, despite the struggles associated with numerical and technological inferiority during the campaign's early years, ultimately gave the United Kingdom and then the United States the full range of advantages associated with air supremacy. In this sense, they created a balanced air instrument and employed a general air strategy. They then used this superior air organization to help secure combined-arms victories in the Mediterranean and elsewhere *before* Allied war production got into high gear. In short, RAF airpower employment was substantially more effective than that of the Axis air forces and immensely consequential in a complex theater where land-based airpower of all kinds played vital and coordinated roles in the larger combined-arms effort. The United States followed suit.[2]

These military-strategic and operational advantages translated directly into grand-strategic ones. In *A World at Arms,* Gerhard Weinberg focused on the grand-strategic impacts of Allied victory in the Mediterranean and Middle East. Among the most important of these was keeping the Axis powers away from oil assets in the Middle East, the Suez Canal and, even more important, preventing a linkup with the Japanese had Egypt fallen and the Allied position in the Middle East collapsed. Weinberg was the first one to remind us that major diversions of U-boats and torpedo-bombers to the Mediterranean played a major role in Allied victory in the Battle of the Atlantic. Additionally, victory in the Mediterranean released a huge amount of merchant shipping from the cape pipeline, allowing for its use to build up the human and material resources necessary for Operation Overlord and the campaign in northwestern Europe.[3]

For their part, the Axis air forces employed what Richard Overy has referred to as a truncated air strategy, focusing on direct support of ground forces and only occasionally and sporadically going after the most important target sets in the theater—the RAF, the Suez Canal, and the occasionally very vulnerable Eighth Army. With few exceptions, the Luftwaffe and Italian Air Force (IAF) displayed far too little concentration of effort or persistence. In fact, dissipation of effort proved one of the greatest Axis weaknesses in the air war. The irony is that the Luftwaffe could and should have achieved air superiority with the assets already in theater, even when Luftflotte 2 was on the Eastern Front. The Germans and Italians had more aircraft and, in the Luftwaffe's case, better ones and better crews until summer 1942. However, ineffective command relationships and the gaping seams they produced within the larger effort doomed the Axis air forces to steady and heavy bleeding, ensured RAF air superiority in almost all cases, and ultimately led to defeat in the air, at sea, and on the ground. The convoluted command-and-control (C2) arrangements in the Luftwaffe and IAF, not to mention the nearly complete absence of cooperation between them, set the course for disaster.[4]

After the RAF and then the US Army Air Forces (USAAF) gained air superiority and held it, the Allies began winning key battles in the theater. The most important of these was the fight for the SLOCs, which often hung in the balance given the episodic but major Luftwaffe raids on Malta, but which the British ultimately won by holding Malta and employing an increasingly deadly No. 201 Naval Co-operation Group to savage Axis convoys. Conversely, the Axis failure to keep Malta down and out, or to develop an equally effective ship attack capability, placed it at a serious disadvantage. The only way to resolve the issue was to take Malta, and there is little question, after an exhaustive review of available source material, that Germany and Italy would have done so had they invaded. The Royal Navy was barely in existence by spring 1942, the Axis had air superiority over the central Mediterranean, and the plan for taking the island was sophisticated and sound. Given the mauling that British vessels took at Crete and every time they approached Malta, it is reasonable to think that the same thing would have happened in May or even July 1942 to any relief flotilla. With the vast majority of the RAF out of range at the time given Erwin Rommel's advance into Egypt and Axis control of the airfields in Cyrenaica and well into Egypt, it would have been impossible for the C in Cs to provide any meaningful air assistance. Malta's garrison was emaciated and low on everything, the RAF contingent could muster fewer than ten sorties a day, and the Luftwaffe controlled the central Mediterranean. This failure to take Malta, or even to engage in the calculated risk, was the single most disastrous Axis decision of the campaign and underscored the lack of grand- and military-strategic vision regarding the Mediterranean and Middle East.

The argument that Adolf Hitler and the Oberkommando der Wehrmacht (OKW) had to choose between Russia and the Mediterranean is patently false. Had the Germans effectively employed just the assets they had in the Mediterranean, they could and likely would have defeated the Eighth Army and exploited their success by taking Egypt and the oil fields beyond, and perhaps even linking up with the Japanese after the British shipping position in the Indian Ocean collapsed as a result of the Axis ability to bring air assets and U-boats into action there. It was not an either-or proposition. It was, rather, a failure to understand that the immense advantages of victory in the Mediterranean and Middle East were there for the taking without an additional commitment of forces beyond the episodic Luftflotte

2 engagements. One telling example of the waste of assets was the tendency for Fliegerkorps X bombers in Greece and Crete to engage partially or not at all during the campaign's most critical phases. In fact, Axis air forces held back or underused nearly half of their total strength in theater during every major operation from Compass to El Alamein. Although the British were guilty of this as well in a different sense by keeping their best aircraft in the United Kingdom for far too long, Longmore and then Tedder used the older models they received, and newer US ones, with maximum effect. They held nothing back.

The result was decisive in the air and crucial in terms of increasingly successful combined-arms warfare. With air superiority—the enabler for all subsequent operations—the British exploited their advantage by engaging in a nonstop effort against enemy airfields to keep the Luftwaffe and IAF weak and off balance. From there, they delivered immense assistance to the army by attacking ship convoys and then vehicle convoys, starving Panzerarmee Afrika of supplies and protecting the Suez Canal from a very real Axis air threat that never took serious form. The destruction of the Italian Merchant Marine was extraordinarily important because it was Rommel's logistical lifeline. Even earlier, the key role of airpower in the Italian East Africa (IEA) campaign opened the Red Sea to US shipping—a huge if seldom recognized strategic victory. The RAF's role in "closing and bolting" the back door by helping to secure Iraq, Syria, and Iran also paid major dividends for the larger Allied war effort. In the case of Iran, more than one-third of lend-lease aid to Russia flowed north from the Persian Gulf. In terms of assisting the ground forces, army commanders and staff officers said on numerous occasions that Rommel would have annihilated the Eighth Army had the RAF not kept his advancing columns at bay. The basic fact that the army lived to fight another day—on three separate occasions—and thus had time to learn how to defeat Rommel on his own terms, was in and of itself of fundamental importance. Without ground forces, there is no victory in a major war. The fact that the Eighth Army survived and ultimately defeated Rommel in coordination with the RAF and Royal Navy ensured the Allied victory in North Africa and the many advantages that came with it in terms of follow-up amphibious, airborne, and ground operations.

As the war continued and Allied air superiority became air supremacy by 1944, an entire new range of air actions exploited earlier successes in the Mediterranean and Middle East, bringing the war directly to the Reich, helping to destroy the Luftwaffe, creating cataclysmic maneuver and supply problems for the Wehrmacht during the oil and transportation offensives in southeastern and central Europe, and giving aid to Tito's 300,000 Partisans as they took their toll on German forces trying first to control them and then to flee to the Reich. This exploitation of the air advantage allowed the Allies to pursue their grand-strategic aims through a highly effective collection of military-strategic actions involving combined-arms efforts such as the landings from Sicily to Anzio, the breakout during Operation Diadem, and Operation Dragoon, along with purely air efforts in others, including the defeat of the Luftwaffe and the crucial oil and transportation offensives.

Airpower supported Allied grand strategy in other ways. Italy's surrender was in large part the result of a long effort, spearheaded by land-based airpower, that starved Axis troops of supplies and, in tandem with Operation Torch and its ultimately effective air operations, resulted in the victory in Tunisia. This Axis disaster set the stage for Operations Husky and Baytown and for Italy's surrender. The Italian surrender and Allied advance in turn led to the capture of the Foggia airfields, facilitating heavy-bomber missions during the exploitation phase of the campaign. By drawing U-boats and torpedo-bombing assets to the theater in an effort to hold the SLOCs, land-based airpower robbed the Germans of critical assets for the Battle of the Atlantic just as that most important of efforts was entering its crucial stage. After aircraft and naval assets had dispatched most of these threats, amphibious operations in Italy and France proceeded, with major logistical benefits in the latter case.

Although there were substantial disappointments during the course of the Mediterranean and Middle East campaign, beginning with Rommel's major counteroffensive at Sollum and ending with the twin disasters at Anzio and Cassino, with multiple potential disasters in between to include periodic Axis control of the SLOCs and the subsequent offensives this allowed Rommel to launch, these failures did not detract in any fundamental way from the larger advantages the Allies reaped during the campaign. They were painful and sometimes unnecessary (Cassino and the disaster during Operation Accolade in the Dodecanese come to mind), but they did not derail the grand- and military-strategic advantages the campaign conferred on the Allies.

The final series of major air campaigns, flown by Mediterranean Allied Strategic Air Force (MASAF) heavy-bombers, helped to destroy the Luftwaffe; robbed the Reich of more than half of its oil supplies; and created insuperable logistical, maneuver, and other problems for German armies on the Eastern Front, especially from the Ukrainian front south into the Balkans. Although we cannot know with any precision how much these attacks sped the Red Army's advance (and those of the Western Allies), how many men they saved, or how much they shortened the war, there is no question that they did all three. Given the more vulnerable transportation infrastructure in southeastern and central Europe, with its relatively few and highly vulnerable marshaling yards, raids on these facilities, combined with the oil offensive, interacted to create manifold disasters for the Wehrmacht in the war's final year. The terrain closer to and in the Reich was much more defensible than that in Russia, and German troops were still quite willing to fight and die for the fatherland. We must consider how much more effectively they would have done so had the oil and transportation offensives not demotorized them, robbed them of ammunition, undermined their ability to maneuver, and kept them from concentrating to meet major Red Army advances, much less counterattack with any effect. The Germans were going to lose in any case, but as the Grand Alliance's casualties reached a peak during the battles along the Reich's borders fighting an army without mobility or enough ammunition, how much worse would it have been had the oil and transportation offensives not exerted their major effects?[5]

Douglas Porch's assertion that "the beleaguered in the Mediterranean would soon cease to be the British and become the Germans, who had committed themselves to fight in a geographically complex theater with inadequate resources, a poorly adapted force structure, and an undependable ally"[6] is one of the most important points any historian has made about the campaign. In what was supposed to be a primarily supporting action to keep the Italians fighting and the Vichy territories safe, and only then an offensive effort either before or after Operation Barbarossa, Hitler's war in the Mediterranean became, as Carl von Clausewitz warned, something alien to its nature, drawing in immense air assets along with naval vessels, submarines, and merchant vessels (MVs), consigning them to a defeat in detail as the Axis High Command fed them into the giant Mediterranean maw piecemeal with no larger sense of strategic purpose. Axis losses in the campaign, aside from the more than 730,000 Germans killed, wounded, or captured and the nearly 400,000 Italians, were staggering in a theater where the Germans had simply planned to assist their Italian allies but could and should have seized the strategic initiative.

Total Axis air losses from June 1940 to May 1943 were 5,211 aircraft destroyed outright, 1,420 probables, and 3,320 damaged. Within each category, 2,494, 918, and 2,410, respectively, were German. The remainder—2,717, 502, and 910, respectively—were Italian. The Luftwaffe lost more than 1,500 planes in Sicily and Italy during the rest of 1943 and another 140 in 1944. The IAF lost 700 in Sicily in 1943. This does not count the more than 4,000 IAF aircraft captured in various states of disrepair after Italy surrendered. Because the majority of probables and damaged aircraft fell prey to advancing Allied armies as a result of perennial supply shortages and the inability to move damaged airframes, between 70 and 80 percent of the total machines noted above were write-offs. The total lost, then, was on the order of 9,700 aircraft. When we add the 7,849 fighters and other planes destroyed engaging Fifteenth Air Force raids or other Allied aircraft in the air on the "second aerial front," or falling prey to them on the ground, the total grows to around 17,550. MAAF assets damaged another 2,407 *Luftflotte Reich* assets in the air. The Mediterranean campaign thus became an aerial cataclysm for the Luftwaffe and killed the IAF. These losses compare with 20,419 in the West and during the air defense of the Reich, and perhaps 11,000 on the Eastern Front. Axis aircrews were even more irreplaceable than their aircraft given Luftwaffe and IAF shortcomings in the training arena and represented a huge loss neither Axis partner could make good.[7]

Regarding Axis shipping losses in the Mediterranean, the Headquarters Royal Air Force Middle East (RAFME) Tactics Assessment Office concluded its major report on antishipping operations with a note that from the start of the war to 30 June 1943, the Axis had lost 1,659 merchant ships—897 in all areas outside the Mediterranean and 762 within it—46 percent of the total. Of the 5.7 million gross tons sunk, 2.4 million—42 percent of the total —went to the bottom of the Mediterranean. During the course of the entire war, Commonwealth forces sank 3,679 ships of the European Axis totaling 10.1 million gross tons. Of these, 1,544 totaling 4.2 million gross tons (42 percent of the total in both cases) were in the Mediterranean. This theater thus became a graveyard for Axis merchant shipping as well as its air forces—another sobering reminder that "subsidiary" theaters can become anything but that without the proper level of grand-strategic insight and focus.[8]

We return, then, to assertions that the Mediterranean and Middle East theater of war was a sideshow, a "mere byplay" somehow less important than other theaters. Given the evidence at hand, we must reject this assertion. The theater was, in fact, vital to Allied victory. It was also the one place where the British and then the Americans could come directly to grips with the Germans and Italians. Entirely aside from the political importance of Operation Torch, for instance, it resulted in a major military victory rivaling Stalingrad. Although Josef Stalin disparaged the "inadequate" Allied effort to assist his troops in surviving and then going over to the offensive, he ignored several vital factors. On the air front, the constant movement of Luftflotte 2 from the East to the Mediterranean was in and of itself a help to the Russians, and the Axis loss of around 17,550 aircraft and most of their crews around the Middle Sea, in the southern Reich, and the southern reaches of occupied Europe meant they were not present on the Eastern Front. The safe route for lend-lease materiel through Iran was another massive force-multiplier for the Red Army, which became much more mobile than the German Army precisely because of the more than 500,000 US-made vehicles, 420,000 radios and field phones, 1 million miles of telephone wire, and other equipment and food it received, mostly by this route. These assets allowed the Red Army first to match and then to beat the Germans at their own game.[9] Allied armies also held between seventeen and twenty-five first-line German divisions in Italy at any given time. They would have been useful on the Eastern Front along with the many others stuck in the Balkans or guarding other parts of the Mediterranean Basin. So, too, would have been Rommel and his very capable subordinate commanders. Placed at the right level within the command structure in Russia, he would likely have been one of the best leaders there, along with Ludwig Crüwell, Walther Nehring, Georg von Bismarck, Wilhelm Ritter von Thoma, and several others. Albert Kesselring had also done well in Russia before he left for the Mediterranean. The insidious but huge drain of materiel and manpower, and of the intellectual capital inherent in the best of the German leaders in the Mediterranean, put the lie to Stalin's assertions and undermine those of historians who characterize the Mediterranean as anything other than a major theater of war and, among other things, a key to the survival of the British Empire and its ability to continue in the war as a major partner in the Grand Alliance.

Context, Understanding, and Victory

In every war and every campaign, the degree to which senior leaders and their troops understand the contextual realities, and use them to their advantage whenever possible, often decides the issue. Cultural factors also come into play, often with great force, as they did in the Mediterranean with the abject failures of German-Italian cooperation and the many Anglo-American successes (despite plenty of friction) in this arena. Clausewitz, B. H. Liddell Hart, Colin Gray, and a host of other theorists have emphasized this. The Allies proved far superior in their efforts to apply a clear grand strategy, a "general air strategy," and combined-arms operations attuned to the contextual realities of the theater. This ability to read context and frame one's actions, from the level of grand strategy to that of tactics, is an abiding one. Today's and tomorrow's policy makers and military leaders would be wise to pay heed to this. Senior military officers would also profit from understanding that the Mediterranean campaign was a model (if an imperfect one, as all things are) of a combined-arms effort in which ground, sea, and air assets—and for the most part their commanders—supported each other. In the Mediterranean and Middle East, it was misguided to refer to "supported" and "supporting" services. Although this applied with special force given the theater's contextual realities, "learning organizations" today and tomorrow will take note and focus on the reality that all warfare depends on mutually supporting services and assets. After the war, John "Jack" Slessor became disillusioned with the re-emergence of severe interservice rivalries not just in the United Kingdom but also in the United States. It was sad, he said of a number of statements made in US postwar official histories, that "the inter-Service feeling that is such an unhappy feature of the American scene should have been allowed to run away with historical objectivity."[10] Slessor's warning to his country and his allies still carries great weight nearly sixty years after he penned it.

The Mediterranean campaign also brings us face to face with another complex reality—the interaction of military forces within and between various domains. For the Romans and their *Mare Nostrum* ("Our Sea"), this involved two domains. During World War II, it involved three. Now, outer space and cyberspace have joined the fray, and there is much argument about whether the latter's unique characteristics make it a domain or a parallel "theater" of sorts that exerts an enormous impact on all other domains. Air superiority facilitated success in the Mediterranean, although only in close coordination with land and sea assets. The Mediterranean theater was a vital "laboratory" in which the Allies learned combined-arms warfare prior to Operation Overlord. Will the cyber "domain" be analogous to air superiority in future conflicts—and equally vital? If so, how do we prepare for that and harness its advantages while minimizing the threats it poses, and how do we incorporate it fully and properly into a combined-arms effort that produces maximum combined effects and effectiveness? The Allied and Axis experiences in the Mediterranean provide some useful insights into this question.[11]

Victory in the Middle Sea was, above all, an example of the interactions between contextual understanding, the ensuing formulation of grand strategy, and the degree to which each side focused its military power to achieve the governments' larger aims. The Allies ultimately proved superior at making these choices and engaging in a particularly effective series of combined-arms operations that led to victory in the theater and major grand- and military-strategic benefits beyond it. Air operations around and over the Mediterranean Basin were of central importance here. Allied leadership leveraged these, and the combined-arms operations in which they played their many roles, to secure one of World War II's most important victories. Making the Middle Sea the Allies' *Mare Nostrum* facilitated further victories in other theaters and a more rapid and less costly end to the war.

NOTES

Preface and Acknowledgments

1. Richard Overy, *The Air War, 1939–1945* (New York: Stein and Day, 1980), 107; Arthur Tedder, *Air Power and War* (London: Hodder and Stoughton, 1947), 29–30, 89.

2. John Slessor to C. R. Cox (Army Co-operation Command, RAF), 25 August 1942, AIR 75/43, National Archives, 1.

3. Ibid., 4.

4. Charles Portal, "Command in the Middle East: Note by the Chief of the Air Staff," 11 September 1942, AIR 75/43, National Archives.

Chapter One. The Approach to War

1. John Ellis, *Brute Force: Allied Strategy and Tactics in the Second World War* (New York: Viking, 1990), 292.

2. Corelli Barnett, *Engage the Enemy More Closely: The Royal Navy in the Second World War* (London: Hodder and Stoughton, 1991), 689–692.

3. Vincent P. O'Hara, *Struggle for the Middle Sea: The Great Navies at War in the Mediterranean, 1940–1945* (Annapolis, MD: Naval Institute Press, 2009), xvi.

4. John B. Hattendorf, "The Contexts of Mediterranean Seapower," in *Naval Policy and Strategy in the Mediterranean: Past, Present, and Future,* ed. John B. Hattendorf (London: Cass, 2000), 420.

5. Williamson Murray and Alan R. Millett, *A War to Be Won: Fighting the Second World War* (Cambridge, MA: Belknap Press, 2000), 108.

6. Carlo D'Este, *World War II in the Mediterranean, 1942–1945* (Chapel Hill, NC: Algonquin, 1990), xiv.

7. Ibid., xxi.

8. Douglas Porch, *The Path to Victory: The Mediterranean Theater in World War II* (New York: Farrar, Straus, and Giroux, 2004), 10, 12.

9. Gerhard Weinberg, *A World at Arms: A Global History of World War II* (Cambridge, UK: Cambridge University Press, 1994), 307–348.

10. Porch, *Path to Victory,* 3.

11. Michael Simpson, "Superhighway to the World Wide Web: The Mediterranean in British Imperial Strategy, 1900–45," in *Naval Policy and Strategy in the Mediterranean: Past, Present, and Future,* ed. John B. Hattendorf (London: Cass, 2000), 51. Ironically, Simpson too characterized the Mediterranean campaign as a sideshow and a waste of resources (68–70).

12. John Strawson, "The Shape and Course of the Mediterranean War, 1940–43," in Royal Air Force Historical Society, *The End of the Beginning: Bracknell Paper No. 3, A Symposium on the Land/Air Cooperation in the Mediterranean War, 1940–43,* ed. Derek Wood (Bracknell, UK: Royal Air Force Historical Society), 20 March 1992, 14.

13. Ibid.

14. Quoted in Porch, *Path to Victory,* 107.

15. Strawson, "Shape and Course of the Mediterranean War," 14.

16. Simpson, "Superhighway," 57–59. Admiral Andrew Cunningham, commander of the Mediterranean Fleet from June 1939 to October 1943, was among these officers.

17. I. S. O. Playfair, *The Mediterranean and Middle East,* vol. 1, *The Early Successes against Italy [to May 1941], History of the Second World War: United Kingdom Military Series* (Uckfield, UK: Naval and Military Press, 2004), 1–3. This and all subsequent volumes of British official history are referred to as "BOH" with volume number.

18. BOH 1, 17–21.

19. Humphrey Wynn, "The RAF in the Mediterranean Theatre," in *The End of the Beginning: Bracknell Paper No. 3, A Symposium on the Land/Air Cooperation in the Mediterranean War, 1940–43,* ed. Derek Wood (Bracknell, UK: Royal Air Force Historical Society), 20 March 1992, 21.

20. Ibid.

21. BOH 1, 17–21.

22. There are many works regarding Italian ambitions, policy issues, and military shortcomings. See Reynolds M. Salerno, *Vital Crossroads: Mediterranean Origins of the Second World War, 1935–1940* (Ithaca, NY: Cornell University Press, 2002). Another crucial source is Gerhard Schreiber, Bernd Stegemann, and Detlef Vogel, *Germany and the Second World War,* vol. 3, *The Mediterranean, Southeast Europe, and North Africa, 1939–1941,* trans. Dean S. McMurry, Ewald Ossers, and Louise Willmot (Oxford, UK: Clarendon, 1995), 1–126. This work, the German official history, is referred to as "GOH" in subsequent notes. For Mussolini's longer-range ambitions, see Gerhard Weinberg, *Visions of Victory: The Hopes of Eight World War II Leaders* (Cambridge, UK: Cambridge University Press, 2005), 41–55.

23. Dennis Mack Smith, *Mussolini: A Biography* (New York: Knopf, 1982), 194–220; GOH 3, 50–62.

24. See Norman J. W. Goda, *Tomorrow the World: Hitler, Northwest Africa, and the Path toward North America* (College Station: Texas A&M University Press, 1998).

25. See Jak P. Mallmann Showell, Foreword, *Fuehrer Conferences on Naval Affairs, 1939–1945* (London: Chatham, 2005), abbreviated as *FCNA* in subsequent citations. Raeder's conferences with Hitler from 26 September 1940 to 18 March 1941 (141–186) are instructive regarding his arguments for a moderate but serious and sustained level of involvement in the Mediterranean to destroy the British position there, turn it into an Axis lake, and make it a springboard for colonial conquests stretching from Egypt to India and south into East Africa to conquer the Red Sea and link up with the Italians. For colonial aspirations, see GOH 3, 279–301.

26. Klaus Schmider, "The Mediterranean in 1940–1941: Crossroads of Lost Opportunities?" *War and Society* 15, no. 2 (October 1997): 19.

27. Walter Ansel, *Hitler and the Middle Sea* (Durham, NC: Duke University Press, 1972), 3.

28. Norman Rich, *Hitler's War Aims: Ideology, the Nazi State, and the Course of Expansion* (New York: Norton, 1992), p. 177. Italics in original.

29. GOH 3, 758.

30. See *FCNA,* 141–146, for a realistic and achievable alternative strategy, even with Operation Barbarossa claiming pride of place, had the Germans and Italians used available resources with greater effect and clearer purpose.

31. D. A. Farnie, *East and West of Suez: The Suez Canal in History, 1854–1956* (Oxford, UK: Clarendon, 1969), 601; BOH 1, 27, 75, 79. For Romanian production, see Gavriil Preda, "German Foreign Policy towards the Romanian Oil during 1938–1940," *International Journal of Social Science and Humanity* 3, no. 3 (May 2013): 327. Production in the Middle East came mostly from Iran at 14 million barrels in 1941. The other 1.6 million came from Iraq. See Steven Morewood, *The British Defence of Egypt: Conflict and Crisis in the Eastern Mediterranean* (London: Routledge, 2004), 23, for Iranian production. See Joseph Sassoon, *Economic Policy in Iraq, 1932–1950* (London: Cass, 1987), 248, for Iraqi output. Capturing Middle East oil would have nearly doubled German annual supplies.

32. By 1941, the Russians were producing 62 million barrels of oil a year at Baku, 16 million in the northern Caucasus, and 49.9 million barrels elsewhere, including the new and rapidly growing Volga–southern Urals region. See Michael P. Croissant and Bülent Aras, eds., *Oil and Geopolitics in the Caspian Sea Region* (Westport, CT: Greenwood, 1999), 12. For comparative purposes, the United States produced 1.35 *billion* barrels in 1940—65 percent of the world's production—a reminder of how short of petroleum, oil, and lubricants (POL) products the Germans were even with Ploesti.

33. Michael Howard, *Mediterranean Strategy in the Second World War* (London: Greenhill, 1993), 9.

34. Gerhard Weinberg, *Visions of Victory,* 7–17. The map on page xxiv is a useful reference for understanding these larger territorial ambitions.

35. This "lost year" for Axis fortunes in the Mediterranean and beyond receives greater attention later. Several recent works highlight its importance. GOH 3 is the most comprehensive reference (see 205 regarding Hitler's desire to remove "Britain's sword on the Continent" by destroying the USSR). Also, see Salerno, *Vital Crossroads,* for the antecedents to the "lost year." For an excellent look at Hitler's western Mediterranean ambitions and their impediments at the hands of Pétain, Franco, and Mussolini, see Goda, *Tomorrow the World.* Citino defined the problem succinctly when he noted that the rapidity of the German military victories caught the Axis unprepared, as did Britain's refusal to come to terms. He said, "The Wehrmacht, in a sense, had conquered its way into a strategic impasse." See Robert Citino, *Death of the Wehrmacht: The German Campaigns of 1942* (Lawrence: University Press of Kansas, 2007), 33.

36. GOH 3, 9.

37. Ibid., 39.

38. Ibid., 9, 17.

39. Karl Gundelach, *Die deutsche Luftwaffe im Mittelmeer,* vol. 1 (Frankfurt am Main: Lang, 1981), 15–16 (referred to from this point forward as *DLM*); GOH 3, 25–36.

40. GOH 3, 35, 46–49.

41. Porch, *Path to Victory,* 28–29; GOH 3, 7–50.

42. GOH 3, 62.

43. Ibid., 61–62.

44. Weinberg, *World at Arms,* 65–67; F. H. Hinsley et al., *British Intelligence*

in the Second World War, vol. 1 (London: Her Majesty's Stationery Office, 1979), 199–203.

45. Howard, *Mediterranean Strategy,* 9.

46. Brian Bond, ed., *Chief of Staff: The Diaries of Lt. Gen. Sir Henry Pownall,* vol. 2 (London: Cooper, 1974), 95.

47. Howard, *Mediterranean Strategy,* 4.

48. Harold E. Raugh, Jr., *Wavell in the Middle East, 1939–1941: A Study in Generalship* (London: Brassey's, 1993), 204; Smith, *Mussolini,* 259–260.

49. BOH 1, 29–30.

50. GOH 3, 31–32.

51. BOH 1, 35–37; Roderic Owen, *The Desert Air Force: An Authoritative History Published in Aid of the Royal Air Force Benevolent Fund* (London: Hutchinson, 1948), 90.

52. BOH 1, 38–40.

53. Raymond Collishaw to Peter Drummond, "Memorandum on the Italian Air Force," 13 September 1939, AIR 23/755, National Archives; *DLM,* 32–33.

54. This is the term British personnel used to describe special modifications to equipment used in harsh environmental conditions, including deserts and jungles.

55. HQ RAFME, "Minutes of Conference Held at Headquarters, Middle East, on Tuesday, 18 July, 1939," and "Conference on Aircraft Depot, Aboukir, Thursday, 29th June, 1939," both in AIR 23/763, National Archives.

56. "Normal Italian Air Force in Libya," September 1939, AIR 23/781, National Archives; GHQME, "Combined Plan for the Defence of Egypt 1939," February 1939, 6–7, AIR 23/770, National Archives.

57. General Headquarters Middle East, "Combined Plan for the Defence of Egypt, 1939," February 1939, AIR 23/770, National Archives, 11–12.

58. Ibid.

59. Ibid., 16.

60. Peter Drummond, "R.A.F. Middle East Operation Plan: Directive to Air Officer Commanding, Egypt Group," 11 May 1939, AIR 23/781, National Archives, Appendix C, 1–4.

61. BOH 1, 41–45, 56.

62. Ibid., 56–61.

63. Ibid., 65.

64. Ibid., 71–73; Owen, *Desert Air Force,* 112.

65. BOH 1, 74, 81–87.

66. Arthur Longmore, "Despatch No. 1 on Middle East Air Operations, 13–5–40 to 31–12–40," 1 February 1941, AIR 23/808, National Archives, 1–3. Longmore also had 100 percent reserves of Gladiators, Blenheims, and Lysanders available to call up from workshops, or operational training units.

67. GOH 3, 62–98.

68. Ibid., 63.

69. See O'Hara, *Struggle for the Middle Sea,* 31.

70. Brian R. Sullivan, "Downfall of the Regia Aeronautica, 1933–1943," in *Why Air Forces Fail: The Anatomy of Defeat,* ed. Robin Higham and Stephen J. Harris (Lexington: University Press of Kentucky, 2006).

71. GOH 3, 75–85.

72. MacGregor Knox, *Hitler's Italian Allies: Royal Armed Forces, Fascist Regime, and the War of 1940–1943* (Cambridge, UK: Cambridge University Press,

2000), 65–67, 163–167; Lucio Ceva and Andra Curami, "Air Army and Aircraft Industry in Italy, 1936–1943," in *The Conduct of the Air War in the Second World War,* ed. Horst Boog (New York: Berg, 1992), 106, 111; GOH 3, 78–85.

73. Hans Werner Neulen, *In the Skies of Europe: Air Forces Allied to the Luftwaffe, 1939–1945* (Wiltshire, UK: Crowood, 2000), 23–24, 26, 37; *DLM,* 30–32.

74. GOH 3, 81.

75. Ibid., 69, 76.

76. GOH 3, 86–91; O'Hara, *Struggle for the Middle Sea.* In 1940, the Italian Merchant Marine had 786 ships, each weighing more than 500 gross registered tons (GRT), for a total tonnage of 3,318,129 GRT. Of these, 532 totalling 1,947,307 GRT were in the Mediterranean. The rest, 254 totalling 1,370,822 GRT, were outside the Mediterranean and nearly all seized or sunk. The 54 German merchant vessels (MVs) in the Mediterranean when the war began gave the Axis a combined total of 2,135,651 GRT. See GOH 3, 91.

77. BOH 1, 88–97.

78. Arthur Murray Longmore, *From Sea to Sky: 1910–1945—Memoirs* (London: Bles, 1946), 220.

79. Ibid., 97–99, 102–103; *DLM,* 22.

80. John Terraine, "Land/Air Cooperation," in *The End of the Beginning: Bracknell Paper No. 3, A Symposium on the Land/Air Cooperation in the Mediterranean War, 1940–43,* ed. Derek Wood (Bracknell, UK: Royal Air Force Historical Society), 20 March 1992, 8–13.

81. David Ian Hall, *Strategy for Victory: The Development of British Tactical Air Power, 1919–1943* (London: Praeger, 2008), xi–xii, 7. Hall's work is an excellent analysis not only of the ways in which RAF tactical airpower developed but also of the larger operational and military-strategic effects this had in the Middle East theater and later in northwestern Europe.

Chapter Two. War Comes to the Mediterranean

1. Gerhard Schreiber, Bernd Stegemann, and Detlef Vogel, *Germany and the Second World War,* vol. 3, *The Mediterranean, Southeast Europe, and North Africa, 1939–1941,* trans. Dean S. McMurry, Ewald Osers, and Louise Willmot (Oxford, UK: Clarendon, 1995), 55, 122. This work, the German official history, is referred to as "GOH" in subsequent notes.

2. Ibid., 50–62, 99–126.

3. Ibid., 111, 119, 125.

4. Ibid., 186–187.

5. Ibid., 301. The German official historians say that "victory on Europe's Mediterranean flank did not offer new opportunities for decision-seeking operations. For that the Axis powers lacked the geo-strategic position, adequate weapons, and an appropriate infrastructure." This is one of their few questionable statements. Given the grand- and military-strategic benefits that would have accrued from closing the Mediterranean, seizing oil and other resources, opening a back door into the Soviet Union, gaining the active support of the Arabic populace, closing the Red Sea, engaging in intensive U-boat operations against Allied merchant shipping in the Indian Ocean, and perhaps even linking up with the Japanese in India, it is impossible to rule out the possibility that a victory in the Mediterranean theater might have resulted in the British Empire's de facto defeat. At the very least, it would have given

the Axis such a huge wealth of raw materials, allies, and strategic depth that the Allies would have been in an extremely precarious position just to survive, not to mention attempt to reconquer what they had lost.

6. Ibid., 181.

7. Ibid., 148, 151, 184, 187, 190–194. See 197 for Hitler's real plans regarding the British Empire after he had conquered the Soviet Union, which involved an "inheritance of all of Europe and the British Empire" after Nazi Germany had become too powerful to resist. This idea was also embodied in the "Jodl Memorandum" of 30 June 1940, which advocated a "war on the periphery" and even more importantly on the sea lanes to bring the British to the negotiating table. "The struggle against the British Empire," he said, "can be waged only by, or by way of, countries with an interest in the disintegration of the British Empire and with hopes of an abundant inheritance." This was one of several views that pushed Hitler toward alliances with Spain and France (198–199).

8. Ibid., 189–190, 205–206.

9. Ibid., 207.

10. Karl Gundelach, *Die deutsche Luftwaffe im Mittelmeer,* vol. 1 (Frankfurt am Main: Lang, 1981), 34–36, 71–73 (referred to in subsequent citations as *DLM*).

11. Ibid., 212–217.

12. *Fuehrer Conferences on Naval Affairs, 1939–1945* (London: Chatham, 2005), 155. Abbreviated as "*FCNA*" in subsequent citations.

13. Ibid., 132–135; GOH 3, 219.

14. *FCNA,* 132–135, 141–146; GOH 3, 219–221.

15. GOH 3, 247, 252, 254, 259–273. Even the German official historians, no great fans of counterfactuals, commented, "It is intriguing, in this situation, to reflect on the German offer, made in the summer of 1940, to transfer armoured units to North Africa. It seems highly probable that in that case the British would have found themselves in serious difficulties, to put it mildly, at a time when all they had in the desert was 85 tanks."

16. Arthur Murray Longmore, *From Sea to Sky, 1910–1945—Memoirs* (London: Bles, 1946), 221–222; Patrick Dunn, quoted in "Digest of the Group Discussions," in Royal Air Force Historical Society, *The End of the Beginning: Bracknell Paper no. 3, A Symposium on the Land/Air Co-operation in the Mediterranean War, 1940–43,* ed. Derek Wood (Bracknell, UK: Royal Air Force Historical Society), 20 March 1992, 68.

17. I. S. O. Playfair, *The Mediterranean and Middle East,* vol. 1, *The Early Successes against Italy [to May 1941], History of the Second World War: United Kingdom Military Series* (Uckfield, UK: Naval and Military Press, 2004), 113–115. This and all subsequent volumes of the British official history are referred to as "BOH" with volume number.

18. Vincent P. O'Hara, *Struggle for the Middle Sea: The Great Navies at War in the Mediterranean, 1940–1945* (Annapolis, MD: Naval Institute Press, 2009), 35, 44.

19. Ibid., 45.

20. Ibid., 97.

21. Headquarters RAFME, "Royal Air Force Middle East Operational Plan," 14 September 1940, WO 201/335, National Archives. Italics added.

22. Ibid. Italics added. See also Denis Richards, *Royal Air Force, 1939–1945,* vol. 1, *The Fight at Odds* (London: Her Majesty's Stationery Office, 1953), 245. He

observed that the "greatest achievement of No. 202 Group in these early days was that by its aggressive tactics it established a defensive mentality in the opposing air force."

23. Arthur Longmore, "Despatch No. 1 on Middle East Air Operations, 13–5–40 to 31–12–40," 1 February 1941, AIR 23/808, National Archives, 6, 9–12. Collishaw commanded one squadron of Gladiators, three of Blenheim Is, and one of Lysanders. An additional squadron of Gladiators, and one of Blenheims, defended Alexandria. The aircraft-replacement problem was serious until summer 1942—long after the threat of German invasion of the United Kingdom had subsided. Even the Japanese threat to India did not create an absolute shortage. The British government chose to keep many aircraft at home. As will become clear later, this was a major error.

24. Ibid., 13. The first diversion to Greece was a Blenheim squadron, followed by two more of Blenheims and another of Gladiators. The first aircraft to arrive at Takoradi on 5 September 1940 included six Blenheim IVs and six Hurricane IIs. See Richards, *Royal Air Force*, vol. 1, 247–249.

25. Ibid., 9–12.

26. Longmore, *Sea to Sky*, 119–121; *DLM*, 23. To take Malta in 1940, the Italians would have needed their entire navy, supported by most of the IAF, to transport, provide naval gunfire for, and ensure resupply for several Italian divisions. Although the effort was not unrealistic with effective C2 and combined-arms capabilities, the Italians lacked both.

27. Longmore, *Sea to Sky*, 125–128, 141–143. The French had 112 warships (70 of which were submarines) in the Mediterranean. Despite their outrage at the attack, Vichy leaders generally did not cooperate extensively with the Germans. Hitler's distrust of the French was also an issue.

28. Ibid., 145–155, 163.

29. BOH 1, 165–167; MacGregor Knox, *Mussolini Unleashed: Politics and Strategy in Fascist Italy's Last War* (Cambridge, UK: Cambridge University Press, 1982), 150–155.

30. BOH 1, 165–167.

31. Ibid., 113–114.

32. General Headquarters Middle East Forces, "Despatch on East African Operations," 21 May 1942, WO 201/311, National Archives.

33. Ibid.

34. F. H. Hinsley et al., *British Intelligence in the Second World War*, vol. 1 (London: Her Majesty's Stationery Office, 1979), 380–381.

35. BOH 1, 167–168, 180–182.

36. Longmore, "Despatch No. 1 on Middle East Air Operations," 4–5.

37. Ibid., 6. No. 203 Group totaled seventy-two aircraft in one mixed Hurricane/Gladiator fighter squadron, one of Rhodesian Hardys, two of Wellesleys, and one of Blenheim IVs.

38. Longmore, "Despatch No. 1 on Middle East Air Operations," 6–9; Norman MacMillan, *The Royal Air Force in the World War*, vol. 3 (London: Harrap, 1949), 52. Reid's group was even smaller than Slatter's with three mixed squadron of Blenheims, Vincents, and Gladiators.

39. Ibid., 10–11. The SAAF had a squadron of Battles, one of Ju 86 bombers, and one of Gladiators. The mixed squadrons of Hurricanes, Gladiators, Furies, Battles, Ju 86s, Hartebeests, and Ansons in IEA created serious maintenance challenges, but ground crews maintained a high operations tempo.

40. BOH 1, 415–417, 420–423, 431–432.

41. BOH 1, 441–442; MacMillan, *Royal Air Force in the World War,* 62; O'Hara, *Struggle for the Middle Sea,* 106–107.

42. Arthur Longmore, "Despatch No. 2 on Middle East Air Operations, 1–1–41 to 3–5–41," 1 November 1941, AIR 23/808, National Archives, 35–40.

43. BOH 1, 185–187.

44. Ibid., 192–194.

45. Humphrey Wynn, "The RAF in the Mediterranean Theatre," in Royal Air Force Historical Society, *The End of the Beginning: Bracknell Paper no. 3, A Symposium on the Land/Air Co-operation in the Mediterranean War, 1940–43,* ed. Derek Wood (Bracknell, UK: Royal Air Force Historical Society), 20 March 1992, 21.

46. Wynn, "The RAF in the Mediterranean Theatre," 22–23; Roderic Owen, *The Desert Air Force: An Authoritative History Published in Aid of the Royal Air Force Benevolent Fund* (London: Hutchinson, 1948), 59. By August 1941, RAFME had three RSUs and three AMUs in the Western Desert, up from one of each in June 1940.

47. BOH 1, 206–207.

48. Ibid., 208–209.

49. BOH 1, 228; Longmore, *From Sea to Sky,* 224; Hans Werner Neulen, *In the Skies of Europe: Air Forces Allied to the Luftwaffe, 1939–1945* (Wiltshire, UK: Crowood, 2000), 46–47.

50. BOH 1, 217–218.

51. Longmore, "Despatch No. 2 on Middle East Air Operations," 44–46. The planes included one squadron of Hurricanes, one of Wellingtons, another of Sunderland flying boats, and a flight of Martin Maryland reconnaissance aircraft.

52. *FCNA,* 185.

53. BOH 1, 213–218.

54. Ibid., 227–233.

55. Ibid., 235–238.

56. Ibid., 240; Franz Halder, Charles Burton Burdick, and Hans Adolf Jacobsen, *The Halder War Diary, 1939–1942* (Novato, CA: Presidio, 1988), 274. Halder, chief of the general staff, noted in his diary on 1 November, "Fuehrer very much annoyed at Italian maneuvers in Greece. Right now he is not in a mood to send anything to Libya or Albania. Let the Italians do it by themselves!"

57. Halder, Burdick, and Jacobsen, *Halder War Diary,* 278.

58. BOH 1, 252–254.

59. Arthur Tedder, *With Prejudice: The World War II Memoirs of Marshal of the Royal Air Force Lord Tedder, Deputy Supreme Commander of the Allied Expeditionary Force* (Boston: Little, Brown, 1966), 34–35.

60. Ibid., 35–38.

61. BOH 1, 252–256. Forty additional Hurricanes arrived in late December. By 1 January 1941, following the initial stages of Operation Compass, the RAF had forty-one Wellingtons, eighty-seven Hurricanes, and eighty-five Blenheim IVs in Egypt. Another 126,000 troops arrived between August and November, improving odds on the ground.

62. Douglas Porch, *The Path to Victory: The Mediterranean Theater in World War II* (New York: Farrar, Straus, and Giroux, 2004), 101, 107–111; Brian Bond, ed., *Chief of Staff: The Diaries of Lt. Gen. Sir Henry Pownall,* vol. 2 (London: Cooper, 1974), 95.

63. BOH 1, 261–264.

64. Raymond Collishaw, "Brief Report on Royal Air Force Operations in the Western Desert from the Outbreak of War with Italy: The Capture of Cyrenaica to the Time of the Enemy Counter Offensive," 19 April 1941, AIR 23/6475, National Archives, 1–3.

65. Ibid., 3.

66. Royal Air Force Middle East to Deputy Chief of Air Staff, February 1938, AIR 75/101, National Archives.

67. Cyril Newall, "Memorandum by the Chief of the Air Staff to the War Cabinet, CoS Committee, Production Programmes," 30 May 1940, AIR 75/22, National Archives, 1–2.

68. John Slessor, "Air Force: Middle East," June 1940, AIR 75/22, National Archives.

69. John Slessor, "Expansion of the Air Forces of the Empire after June, 1941," 31 July 1940, AIR 75/22, National Archives; Air Ministry, Expansion, and Re-equipment Policy Committee, "Minutes of Meeting Held on Thursday, July 25th and Saturday, July 27th," 28 July 1940, AIR 75/22, National Archives. Slessor relates his efforts in the United States in *The Central Blue: The Autobiography of Sir John Slessor, Marshal of the RAF* (New York: Praeger, 1957), 314–338.

70. Air Ministry to Headquarters Royal Air Force Middle East (HQ RAFME), 14 August 1940, AIR 23/1402, National Archives; Wilfrid Freeman to Arthur Longmore, 15 August 1940, AIR 23/1402, National Archives; HQ RAFME, "Notes by Air Officer Commanding-in-Chief on Conference Held on 2nd and 3rd September, 1940—Western Desert Problems," 3 September 1940, AIR 23/1402, National Archives.

71. John Slessor to Secretary of State for Air Archibald Sinclair, copy to Portal, 19 October 1940, AIR 75/63, National Archives. Portal noted, in handwritten script below Slessor's signal, that it was a "very restrained account" of Beaverbrook's behavior. Portal knew Slessor's diplomatic acumen and trusted him to secure generous aircraft deliveries.

72. Richards, *Royal Air Force*, vol. 1, 266.

73. Winston Churchill to Arthur Longmore, 13 November 1940, AIR 23/1402, National Archives; Arthur Longmore to Charles Portal, 13 November 1940, AIR 23/1402, National Archives; Charles Portal to Arthur Longmore, 14 November 1940, AIR 23/1402, National Archives; Arthur Longmore to Winston Churchill via Charles Portal, 15 November 1940, AIR 23/1402, National Archives.

74. John Slessor, "The U.S. Air Production Programme: Note by Air Commodore Slessor," 27 November 1940, AIR 75/63, National Archives.

75. John Slessor to Charles Portal, 3 December 1940; John Slessor to Charles Portal, 4 December 1940, AIR 75/63, National Archives, 1–3. During 1941, US aircraft production reached 1,539 a month. The British produced 1,675, the Germans 981, and the Italians 291. See Sebastian Ritchie, Table 46, *Industry and Air Power: The Expansion of British Aircraft Production, 1935–1941* (London: Cass, 1997), 262; Brian R. Sullivan, "Downfall of the Regia Aeronautica, 1933–1943," in *Why Air Forces Fail: The Anatomy of Defeat*, ed. Robin Higham and Stephen J. Harris (Lexington: University Press of Kentucky, 2006), 151. Regardless, the Axis had a numerical advantage until summer 1942 based on its head start and central position.

76. Arthur Longmore to Charles Portal, 3 January 1941, AIR 23/1403, National Archives.

77. Loben Maund to Air Ministry, 13 January 1941, AIR 23/1605, National Archives; Loben Maund to Air Ministry (r) Henry Thorold, 1 February 1941, AIR 23/1061, National Archives.

78. John Slessor to Charles Portal, 3 February 1941, AIR 75/63, National Archives.

Chapter Three. Triumph and Tragedy

1. I. S. O. Playfair, *The Mediterranean and Middle East,* vol. 1, *The Early Successes against Italy [to May 1941], History of the Second World War: United Kingdom Military Series* (Uckfield, UK: Naval and Military Press, 2004), 125–130 (this and all subsequent volumes of the British official history are referred to as "BOH" with volume number); Gerhard Schreiber, Bernd Stegemann, and Detlef Vogel, *Germany and the Second World War,* vol. 3, *The Mediterranean, Southeast Europe, and North Africa, 1939–1941,* trans. Dean S. McMurry, Ewald Osers, and Louise Willmot (Oxford, UK: Clarendon, 1995), 127, 132–135, 153–162, 195. The latter work, the German official history, is referred to as "GOH" in subsequent notes.

2. GOH 3, 199–200.

3. Winston Churchill, *The Second World War,* vol. 2 (Boston: Houghton Mifflin, 1948–1953), 91.

4. Ibid., 489.

5. GOH 3, 559–564; James Ramsay Montagu Butler, *Grand Strategy, II, September 1939 to June 1941: History of the Second World War, UK Military Series* (London: Her Majesty's Stationery Office, 1957), 424; Mark Stoler, *Allies and Adversaries: The Joint Chiefs of Staff, the Grand Alliance, and U.S. Strategy in World War II* (Chapel Hill: University of North Carolina Press, 2000), 25–26.

6. Stoler, *Allies and Adversaries,* 26–36; quoted passage on 32; GOH 3, 564–568.

7. Stoler, *Allies and Adversaries,* 37–40. John Slessor was Charles Portal's representative to the talks and provides a useful perspective in *The Central Blue: The Autobiography of Sir John Slessor, Marshal of the RAF* (New York: Praeger, 1957), 339–365. He said of these vital conversations, "The general strategic concept then agreed—while it became at times a bit frayed at the edges—did continue to govern our combined action throughout the war" (343).

8. GOH 3, 226, 228, 237, 241–243, 248; Mussolini's slogan quoted on 248. Regarding Italy's change from ally to vassal, the process was a gradual one, but as the German official historians note, "The 'parallel war' concept had become an absurdity" by 1941. See GOH 3, 243. A detailed record of the confusion leading up to the attack on Greece is located on 401–418. See also Karl Gundelach, *Die deutsche Luftwaffe im Mittelmeer,* vol. 1 (Frankfurt am Main: Lang, 1981), 23 (referred to in subsequent citations as *DLM*).

9. GOH 3, 259–266.

10. Ibid., 176–179.

11. Ibid., 177.

12. Ibid.

13. Ibid., 177–178.

14. Ibid., 177.

15. Ibid., 178–179.

16. *Fuehrer Conferences on Naval Affairs, 1939–1945* (London: Chatham,

2005), 165–166 (abbreviated as "*FCNA*" in subsequent citations); GOH 3, 196–197, 225–231.

17. *FCNA,* 154–156, 160–161; GOH 3, 232–235.

18. *FCNA,* 160–161; GOH 3, 237–242 (quote on 240).

19. BOH 1, 261–272.

20. Arthur Longmore, "Despatch No. 1 on Middle East Air Operations, 13–5–40 to 31–12–40," 1 February 1941, AIR 23/808, National Archives, 14–15.

21. Ibid., 15–16.

22. Raymond Collishaw, "Brief Report on Royal Air Force Operations in the Western Desert from the Outbreak of War with Italy: The Capture of Cyrenaica to the Time of the Enemy Counter Offensive," 19 April 1941, AIR 23/6475, National Archives, 11–12.

23. Ibid., 13–14.

24. Longmore, "Despatch No. 1 on Middle East Air Operations," 16; BOH 1, 277–278, 281–282.

25. Collishaw, "Brief Report on Royal Air Force Operations," 4.

26. BOH 1, 288.

27. Ibid., 292–294.

28. Longmore, "Despatch No. 1 on Middle East Air Operations," 5–6.

29. Ibid., 7–8.

30. Ibid., 20–21.

31. Ibid., Appendices A and D.

32. BOH 1, 299–312.

33. Ibid., 312–314. Malta's air contingent at this time included twelve Swordfish, twenty Hurricanes, twenty Wellingtons, six Sunderland flying boats, and five Martin Maryland bombers.

34. Longmore, "Despatch No. 1 on Middle East Air Operations," 6–7.

35. Collishaw, "Brief Report on Royal Air Force Operations," 14–16; Headquarters Middle East (HQ ME), "Report on Army Co-operation Carried Out in Connection or with the Land Operations in the Western Desert and Libya," 31 January 1941, WO 201/348, National Archives, 1–3; BOH 1, 353–358.

36. HQ ME, "Report on Army Co-operation," 2–3.

37. Ibid., 4–5.

38. Ibid., 6–10.

39. BOH 1, 362–366; HQ ME, "Report on Army Co-operation," Appendix A, 2.

40. Hellmuth Felmy, "The German Air Force in the Mediterranean Theater of War," K113.107-161, Air Force Historical Research Agency, 28–31.

41. BOH 1, 316–318; Felmy, "German Air Force in the Mediterranean Theater," 39; Cajus Bekker, *The Luftwaffe War Diaries: The German Air Force in World War II* (New York: Da Capo, 1994), 200–201.

42. BOH 1, 367.

43. Oberkommando der Wehrmacht to Oberkommando des Heeres, 6 February 1941, and Oberkommando der Wehrmacht to Oberkommando des Heeres, 19 February 1941: "High Level Reports and Directives Dealing with the North African Campaign, 1941," trans. 7/81, A.H.B.6, 19 November 1948, Air Historical Branch, 12; Ritter von Rintelen, *Mussolini als Bundesgenosse: Errinerungen des Deutschen Militärattachés in Rom, 1936–1943* (Tübingen: Wunderlich, 1951), 55–56; Felmy, "German Air Force in the Mediterranean Theater," 39–41; *DLM,* 81.

44. Felmy, "German Air Force in the Mediterranean Theater," 39–41. Harling-hausen's antishipping skills soon got him reassigned to the Atlantic theater, where he did superb work with limited air assets. His departure was a loss to efforts in the Mediterranean. See Roderic Owen, *The Desert Air Force: An Authoritative History Published in Aid of the Royal Air Force Benevolent Fund* (London: Hutchinson, 1948), 66. See also *DLM*, 83, 86–87, 94.

45. *DLM*, 119, 123.

46. Felmy, "German Air Force in the Mediterranean Theater," 52–53; German Air Ministry, "The Activities of Fliegerkorps X in the Mediterranean, January–February 1941," 26 February 1941, trans. 7/54, Air Ministry A.H.B.6, 15 November 1947, Air Historical Branch, 1–3; *DLM*, 104–105, 123.

47. German Air Ministry, "Activities of Fliegerkorps X in the Mediterranean," 1–3. Italics added. Luftwaffe directives were consistently long on providing army support and short on gaining air superiority. Ironically, the only way to achieve the former with persistence was to first achieve the latter.

48. Felmy, "German Air Force in the Mediterranean Theater," 54–56; "Situation Report of the Air Force High Command Intelligence Officer," 156–292; Erwin Rommel and Fritz Bayerlein, *The Rommel Papers,* ed. B. H. Liddell Hart (London: Collins, 1953), 98. Stephan Fröhlich's command consisted of four Ju 87 Stuka squadrons, one Me. 109 squadron, and one-third of a Ju 88 squadron. It also included several signals units, three flak battalions, flak transport vehicles, petroleum, oil, and lubricants (POL) trucking units, and medical units. Fröhlich also had one squadron of Henschel 126 reconnaissance aircraft and a motorized flak repair shop. See also *DLM*, 96–98, 107–109, 125.

49. For an excellent account of the differences between Luftwaffe and RAF employment of fighters, see Christopher Shores and Hans Ring, *Fighters over the Desert: The Air Battles in the Western Desert, June 1940–December 1942* (New York: Arco, 1969). For their overall conclusions, see 217–221.

50. Felmy, "German Air Force in the Mediterranean Theater," 54–56; *DLM*, 108, 119. In the existing literature, especially Hellmuth Felmy's account, there is an almost constant criticism of Erwin Rommel, including his lack of understanding and imagination regarding airpower and his constant bullying of Stephan Fröhlich. Rommel's writings, on the other hand, portray the working relationship between army and air officers as relatively effective if hindered by poor Oberkommando der Wehrmacht, Oberkommando des Heeres, and Oberkommando der Luftwaffe decisions and the insufficient numbers of aircraft to support his army (see Rommel and Bayerlein, *Rommel Papers*). Albert Kesselring's memoirs tend to back Rommel's view, but there is no question that Felmy's criticisms have merit. Karl Gundelach emphasizes the strained and often ineffective relationship, in effect backing Felmy's view. Both were Luftwaffe officers and bitter about Rommel's interactions with airmen, so the reader must approach their assertions with caution. Primary sources indicate a marginal working relationship between Rommel and Fröhlich.

51. Felmy, "German Air Force in the Mediterranean Theater," 51–52, 60–61; *DLM*, 126–127, 131–133.

52. "Axis Operations in the Mediterranean, 31st March–5th April, 1941, Situation Reports Issued by Luftwaffe Fuehrungsstab IC," trans. 7/113, Air Ministry A.H.B.6, July 1952, Air Historical Branch, 19–20; "Axis Operations in the Mediterranean, 20–27 April, 1941, Situation Reports Issued by Luftwaffe Fuehrungsstab

IC," trans. 7/115, Air Ministry A.H.B.6, July 1952, Air Historical Branch, 1, 9, 12–13, 18–19, 35; *DLM*, 128–129.

53. Gerhard Weinberg, *A World at Arms: A Global History of World War II* (Cambridge, UK: Cambridge University Press, 1994), 215–216; MacGregor Knox, *Mussolini Unleashed: Politics and Strategy in Fascist Italy's Last War* (Cambridge, UK: Cambridge University Press, 1982), 279–282; BOH 1, 366–368; Robin Higham, *Diary of a Disaster: British Aid to Greece, 1940–1941* (Lexington: University Press of Kentucky, 1986), 26–27, 34; *DLM*, 24.

54. Norman Rich, *Hitler's War Aims: Ideology, the Nazi State, and the Course of Expansion* (New York: Norton, 1992), 177.

55. I. S. O. Playfair et al., *The Mediterranean and Middle East*, vol. 2, *The Germans Come to the Aid of Their Ally*, History of the Second World War: United Kingdom Military Series (Uckfield, UK: Naval and Military Press, 2004), 14–15 (hereafter cited as BOH 2); Stephen Bungay, *The Most Dangerous Enemy: A History of the Battle of Britain* (London: Aurum, 2000).

56. Quoted in Philip Guedalla, *Middle East, 1940–1942: A Study in Air Power* (London: Hodder and Stoughton, 1944), 89.

57. Weinberg, *World at Arms*, 212–214.

58. Franz Halder, Charles Burton Burdick, and Hans Adolf Jacobsen, *The Halder War Diary, 1939–1942* (Novato, CA: Presidio, 1988), 355; Weinberg, *World at Arms*, 172–173.

59. *DLM*, 110–111, 116–117, 120–122.

60. GOH 3, 693; Horst Boog, Werner Rahn, and Reinhard Stumpf, *Germany and the Second World War*, vol. 6: *The Global War* (Oxford, UK: Oxford University Press, 2001), 121, 131 (hereafter referred to as GOH 6); Geoffrey P. Megargee, *Inside Hitler's High Command* (Lawrence: University Press of Kansas, 2000), 233; Jurgen Förster, "The Dynamics of *Volksgemeinschaft*: The Effectiveness of the German Military Establishment in the Second World War," in *Military Effectiveness*, vol. 3, *The Second World War*, ed. Williamson Murray and Allan R. Millett (Boston: Unwin Hyman, 1988), 182, 204–205.

61. Rommel and Bayerlein, *Rommel Papers*, 101, 103. In a letter to his wife dated 5 March 1941, Rommel said, "Just back from a two-day journey—or rather flight—to the front, which is now 450 miles away to the east" (104).

62. Ibid., 102, 109, 112–113, 116.

63. Ibid., 105–106. Rommel's claim that he knew nothing of Operation Marita appears to be true. See *DLM*, 135.

64. Rommel and Bayerlein, *Rommel Papers*, 119–120, 139, 148. Rommel's later reflections on these possibilities, and the reasons for the Axis defeat in North Africa, are interesting. He returned often to the need for more and better-organized and -employed airpower, even if this recognition sank in too late (see 511–514, 519). However, his comments about the importance of capturing Malta ring hollow because Rommel convinced Hitler to scrap Operation Hercules in favor of what he said would be a ten-day campaign to capture Cairo. More on this later.

65. Ibid., 148, 152.

66. Ibid., 191.

67. Rommel and Bayerlein, *Rommel Papers*, 134. For an excellent discussion of the Luftwaffe's army-support capabilities and the problems bringing them to maturity in the same way as did the RAF and later the USAAF, see Richard Muller, *The*

German Air War in Russia (Baltimore, MD: Nautical and Aviation, 1992). Although Muller's focus is on the Eastern Front, he records many of the same coordination issues that hampered operations in the Mediterranean theater.

68. Robert Citino, *Death of the Wehrmacht: The German Campaigns of 1942* (Lawrence: University Press of Kansas, 2007), 117.

69. Ibid., 214.

70. Longmore, "Despatch No. 1 on Middle East Air Operations," 9–10.

71. Ibid., 8–9.

72. Ibid., 10–13.

73. Ibid., 14–15.

74. Felmy, "German Air Force in the Mediterranean Theater," 48–49; *DLM*, 116, 123; BOH 2, 48. Between April and June, 229 Hurricanes arrived, with 109 staying and the rest continuing to Egypt. This proved timely as German raids peaked in April.

75. BOH 2, 58–59.

76. BOH 1, 318–322; Felmy, "German Air Force in the Mediterranean Theater," 44–47; *DLM*, 99.

77. Felmy, "German Air Force in the Mediterranean Theater," 46–47; German Air Historical Branch (Abteilung 8), "German Air Force Activities in the Mediterranean: Tactics and Lessons Learned, 1941–1943," A.H.B.6, trans. 7/2, 8 October 1946, Air Historical Branch, 2–3.

78. BOH 1, 323–326.

79. Ibid., 327–328.

80. Felmy, "German Air Force in the Mediterranean Theater," 62–65.

81. Ibid., 65–67. Air base commands included a motorized field repair battalion, a vehicle repair platoon, and a transport staff with six supply columns and two supply companies.

82. Felmy, "German Air Force in the Mediterranean Theater," 63–65; German Liaison Staff and Italian Air Staff, "The Mediterranean Campaign: Review of the Situation in the Central and South-Eastern Mediterranean for the Period July 11–August 31, 1941," trans. 7/65, A.H.B.6, March 1948, Air Historical Branch, 7; Rommel and Bayerlein, *Rommel Papers,* 138.

83. Felmy, "German Air Force in the Mediterranean Theater," 67–70.

84. Felmy, "German Air Force in the Mediterranean Theater," 73–74, 79–80; Rommel and Bayerlein, *Rommel Papers,* 121; *DLM,* 135, 146; Halder, Burdick, and Jacobsen, *Halder War Diary,* 348.

85. Felmy, "German Air Force in the Mediterranean Theater," 83–84.

86. Ibid., 84.

87. Ibid., 84–85. Once again, the effects of seams in the command structure are evident. Fliegerkorps X, now stationed in Greece and with most of its bombers and long-range fighters on Crete, was intermittently responsive to Stephan Fröhlich's requests for bomber support. This unit's assets constituted roughly half of total Luftwaffe assets in theater but did not engage in concerted mining or other indirect-support missions.

88. Rommel and Bayerlein, *Rommel Papers,* 128–129.

89. BOH 1, 377.

90. BOH 1, 385–389; Churchill, *Second World War,* vol. 3, 9, 13–14.

91. Felmy, "German Air Force in the Mediterranean Theater," 35–36. For German actions in Romania, see GOH 3, 452–453.

92. Longmore, "Despatch No. 1 on Middle East Air Operations," 17–24; Guedalla, *Middle East,* 124. The RAF destroyed ninety-three enemy aircraft and probably destroyed another twenty-six, mostly fighters, for the loss of eight fighters, from which six pilots were rescued.

93. J. H. D'Albiac to Arthur Tedder, "Report on the Operations Carried Out by the Royal Air Force in Greece: November 1940 to April 1941," 15 August 1941, AIR 23/6370, National Archives, 3–12; F. H. Hinsley et al., *British Intelligence in the Second World War,* vol. 1 (London: Her Majesty's Stationery Office, 1979), 407–409; BOH 2, 78–82.

94. German Air Historical Branch (Abteilung 8), "German Air Force Activities in the Mediterranean," 1–2.

95. BOH 2, 67–69. The formation and accomplishments of No. 201 Group receive detailed attention later.

96. BOH 2, 86–89; Chris McNab, *Order of Battle: German Luftwaffe in World War II* (London: Amber, 2009), 80–109, 173. Luftflotte 4 had 576 aircraft with 168 more on call from Fliegerkorps X, which played a minor role in Operation Marita when they could have been giving direct support to Stephan Fröhlich and Erwin Rommel during a crucial window of opportunity in North Africa. Fliegerkorps VIII added another 414 aircraft and had twice the normal complement of vehicles, ensuring maximum mobility. It is ironic that while Wolfram von Richthofen had double the normal vehicle complement, Fröhlich had half in a huge and mobility-intensive theater. The best work on von Richthofen is James Corum, *Wolfram von Richthofen: Master of the German Air War* (Lawrence: University Press of Kansas, 2008). His chapter on the Balkans campaign (236–256) closes with a very important note that even with von Richthofen's large complement of vehicles and Ju 52s, the campaign in Greece taxed the Luftwaffe's logistics over 150-mile advances. This did not bode well for later logistical efforts in Russia and Africa, over much longer distances and in harsh environments with poor lines of communication, but nobody put any improvements in place.

97. Felmy, "German Air Force in the Mediterranean Theater," 98–99; BOH 2, 90–93.

98. D'Albiac to Tedder, "Report on the Operations Carried Out by the Royal Air Force in Greece," 13–19.

99. BOH 2, 90–93.

100. BOH 2, 94–98, 104n; Felmy, "German Air Force in the Mediterranean Theater," 112–118; Denis Richards, *Royal Air Force, 1939–1945,* vol. 1, *The Fight at Odds* (London: Her Majesty's Stationery Office, 1953), 301–302.

101. Felmy, "German Air Force in the Mediterranean Theater," 114, 117–118.

102. Ibid., 118–120.

103. Longmore, "Despatch No. 2 on Middle East Air Operations, 1-1-41 to 3-5-41," 1 November 1941, AIR 23/808, National Archives, 31–32.

104. BOH 2, 128.

105. Felmy, "German Air Force in the Mediterranean Theater," 120, 125–127.

106. BOH 2, 123–125; GOH 3, 528; Weinberg, *World at Arms,* 228–229; Walter Warlimont, *Inside Hitler's Headquarters, 1939–45* (New York: Praeger, 1966), 131–134.

107. BOH 2, 126–128; Hinsley et al., *British Intelligence in the Second World War,* vol. 1, 413, 415–417.

108. General Headquarters Fliegerkorps XI Section 1a Br.B 5626/41, "Order for

the Cooperation of Ground Troops with the Fliegerkorps VIII," 14 May 1941, Appendix B; Headquarters Royal Air Force Middle East Intelligence Branch, "German Air-Borne Attack on Crete," 1 November 1941, AIR 23/6110, National Archives, 10–12.

109. Ibid.

110. Arthur Tedder, *Report by Air Marshal Tedder on His Tenure in the Post of Air Officer Commanding-in-Chief* (London: MLRS, 2010), 23–27. Original report also available under the title "Marshal of the Royal Air Force Lord Tedder's Despatch on Middle East Operations, May 1941–January 1943," AIR 40/1817, National Archives.

111. BOH 2, 129–147; Tedder, *With Prejudice: The World War II Memoirs of Marshal of the Royal Air Force Lord Tedder, Deputy Supreme Commander of the Allied Expeditionary Force* (Boston: Little, Brown, 1966), 105; David A. Thomas, *Crete 1941: The Battle at Sea* (London: Cassell, 2003). Unfortunately, the admiralty's account of this battle and the larger effort in the Mediterranean through the fall of Tunis downplays the role of airpower in its many combined-arms operations with the navy. See Lewis Anthem Ritchie, *East of Malta, West of Suez: The Official Admiralty Account of the Mediterranean Fleet, 1939–1943* (Boston: Little, Brown, 1944).

112. Tedder, *With Prejudice*, 75–76. Lord Beaverbrook's full name was William Maxwell ("Max") Aitken.

113. Ibid., 77.

114. Ibid.

115. Ibid., 105–107; Tedder, *Report*, 24.

116. Norman MacMillan, *The Royal Air Force in the World War*, vol. 3 (London: Harrap, 1949), 103.

117. Douglas Porch, *The Path to Victory: The Mediterranean Theater in World War II* (New York: Farrar, Straus, and Giroux, 2004), 174; Felmy, "German Air Force in the Mediterranean Theater," 128.

Chapter Four. The RAF Holds—and Learns

1. Mark Stoler, *Allies and Adversaries: The Joint Chiefs of Staff, the Grand Alliance, and U.S. Strategy in World War II* (Chapel Hill: University of North Carolina Press, 2000), 44–47; Gerhard Schreiber, Bernd Stegemann, and Detlef Vogel, *Germany and the Second World War*, vol. 3, *The Mediterranean, Southeast Europe, and North Africa, 1939–1941*, trans. Dean S. McMurry, Ewald Ossers, and Louise Willmot (Oxford, UK: Clarendon, 1995), 569–572. The latter work, the German official history, is referred to as "GOH" in subsequent notes.

2. GOH 3, 566–567.

3. Ibid., 573–585.

4. Ibid., 448.

5. Arthur Tedder to Wilfrid Freeman, 1 May 1941, AIR 23/1386, National Archives.

6. Tedder to Freeman, 1 May 1941; Arthur Tedder to Wilfrid Freeman, 25 April 1941, AIR 23/1386, National Archives.

7. Tedder to Freeman, 25 April 1941. At this point, Tedder had fourteen fighters on air-defense duties in the Nile Delta, nineteen in the Western Desert, eighteen coming out of maintenance units in the next week, and twelve about to leave Takoradi.

8. "The Middle East Campaigns," vol. 1, AIR 41/44, National Archives, 6, 25, 27–28; "Middle East: Strategy for Operations (August 1940–February 1941)," AIR 8/514, "Air Support," AIR 10/5547, National Archives, 47; "Army Air Support," WO 277/34, National Archives, 40.

9. Brad Gladman, *Intelligence and Anglo-American Air Support in World War Two: The Western Desert and Tunisia, 1940–43* (New York: Palgrave MacMillan, 2009), 42.

10. "Air Support," 47–48; "Army Air Support," 39–40; "The Middle East Campaigns," vol. 1, 7.

11. H. W. Wynter, *The History of the Long Range Desert Group, June 1940 to March 1943* (Kew, UK: National Archives, 2008), 13–17, 25.

12. Notes on Combined Services Detailed Intelligence Centre (CSDIC) Mediterranean, pt. 1 (1941), WO 208/3248, National Archives, 1–2, 7; F. H. Hinsley et al., *British Intelligence in the Second World War*, vol. 1 (London: Her Majesty's Stationery Office, 1979), 205.

13. I. S. O. Playfair et al., *The Mediterranean and Middle East*, vol. 2, *The Germans Come to the Aid of Their Ally, History of the Second World War: United Kingdom Military Series* (Uckfield, UK: Naval and Military Press, 2004), 110–113. This and all subsequent volumes of the British official history are referred to as "BOH" with volume number.

14. Arthur Tedder, *With Prejudice: The World War II Memoirs of Marshal of the Royal Air Force Lord Tedder, Deputy Supreme Commander of the Allied Expeditionary Force* (Boston: Little, Brown, 1966), 108.

15. Arthur Longmore, *From Sea to Sky: 1910–1945—Memoirs* (London: Bles, 1946), 282–283; Tedder, *With Prejudice*, 86.

16. Tedder, *With Prejudice*, 81.

17. Ibid., 82.

18. Ibid.

19. Longmore, *From Sea to Sky*, 284; Ralph Bennett, *Ultra and Mediterranean Strategy* (New York: Morrow, 1989), 34–36.

20. Arthur Longmore, "Despatch No. 2 on Middle East Air Operations, 1-1-41 to 3-5-41," 1 November 1941, AIR 23/808, National Archives, 50–51. Axis aircraft losses on the ground included 450 during Operation Crusader, more than 1,000 in the pursuit after El Alamein, and 633 in Torch. Details and sources are provided at those points in the narrative. When combined with the 1,100 IAF aircraft lost during Operation Compass, total Axis aircraft lost on the ground was 3,200. The RAF lost 27 planes on the ground leading up to and during Operation Compass, bringing the total for the entire campaign to 59.

21. Longmore, "Despatch No. 2 on Middle East Air Operations," Appendix F: Own and Enemy Aircraft Casualties during Period 1st January, 1941–30th April, 1941.

22. Arthur Tedder to Charles Portal, 5 May 1941, AIR 23/1395, National Archives.

23. C. V. Whitney [Maj. Gen. Lewis Brereton's director of intelligence], interview with US Army Air Forces Director of Intelligence, "Air Marshall Tedder's Use of Air against Rommel," 6 April 1943, 612.624-1, Air Force Historical Research Agency, 1.

24. Charles Portal to Arthur Tedder, 4 May 1941, AIR 23/1395, National Archives.

25. Tedder, *With Prejudice*, 3–4, 7.

26. Ibid., 7–8.

27. Ibid., 3–8. For the best biography of Tedder, see Vincent Orange, *Tedder: Quietly in Command* (London: Cass, 2004). Orange's close relationship with Tedder's children and extended family gave him unique access to materials not available in the archives, and these give the work great detail, personality, and utility. Tedder's near loss of his command receives detailed attention later.

28. Tedder, *With Prejudice*, 83.

29. Arthur Tedder, *Report by Air Marshal Tedder on His Tenure in the Post of Air Officer Commanding-in-Chief* (London: MLRS, 2010), 1–2.

30. Ibid., 3–5.

31. Ibid., 5.

32. Ibid., 7–9.

33. William Dobbie, quoted in James Douglas-Hamilton, *The Air Battle for Malta: The Diaries of a Spitfire Pilot* (Barnsley, UK: Pen and Sword Aviation, 2006), 15.

34. Arthur Tedder, letter to Rosalinde Tedder, 27 May 1941, Tedder Box 4, Air Historical Branch.

35. Tedder, *Report*, 76–81. At the end of March 1941, there were 30 aircraft on Malta. By the end of June, there were 133, mostly modern types such as the Hurricane 2, Blenheim IV, and Martin Maryland.

36. Ibid., 52.

37. Ibid., 55–59. Arthur Cunningham was a New Zealander. His nickname, "Mary," was a corruption of "Maori," although he did not have any Maori blood.

38. Vincent Orange, "The Commanders and the Command System," in Royal Air Force Historical Society, *The End of the Beginning: Bracknell Paper No. 3, A Symposium on the Land/Air Co-operation in the Mediterranean War, 1940–43*, ed. Derek Wood (Bracknell, UK: Royal Air Force Historical Society), 20 March 1992, 34.

39. Chaz Bowyer, *Men of the Desert Air Force* (London: Kimber, 1984), 144.

40. Bennett, *Ultra and Mediterranean Strategy*, 16–17, 27; Hinsley et al., *British Intelligence in the Second World War*, vol. 1, 19, 570–571; Hinsley et al., *British Intelligence in the Second World War*, vol. 2, 283–287.

41. Patrick Dunn, quoted in "Digest of the Group Discussions," in Royal Air Force Historical Society, *The End of the Beginning: Bracknell Paper no. 3, A Symposium on the Land/Air Co-operation in the Mediterranean War, 1940–43* (Bracknell, UK: Royal Air Force Historical Society), 20 March 1992, 68; Kenneth Cross with Vincent Orange, *Straight and Level* (London: Bugg Street, 1993), 260–261.

42. Phillip Neame, "Report on Operations in Cyrenaica," 7 April 1941, CAB 106/676, National Archives, 5; "Some Signals Lessons of the Libyan Campaign, November 1941 to February 1942," pt. 1, WO 201/369, National Archives, 1; "Lord Tedder's Despatch on Middle East Operations, May 1941–January 1943," AIR 40/1817, National Archives, Signals Appendix.

43. Gladman, *Intelligence and Anglo-American Air Support in World War Two*, 9, 58–59, 61–62; "Some Signals Lessons of the Libyan Campaign, November 1941 to February 1942," pt. 2, "Detailed Technical Lessons," WO 201/369, National Archives, paragraph 16 (i); GS, 30 Corps War Diary 1941, "Notes on First Phase Operations in Libya, 18 November–10 December, 1940," WO 169/1123, National Archives; Eighth Army Main Headquarters Signals, December 1941, WO 169/3904, National Archives; Tedder, *Report*, Signals Appendix, 9.

44. Gladman, *Intelligence and Anglo-American Air Support in World War Two,* 2–3; Military Intelligence Service, War Office, "Notes and Lessons on Operations in the Middle East," 30 January 1943, WO 169/6638, National Archives, 14.

45. Gladman, *Intelligence and Anglo-American Air Support in World War Two,* 4–8; "The Middle East Campaigns," vol. 4, "Operations in Libya, the Western Desert and Tunisia, July 1942–May 1943," AIR 41/50, National Archives, 20.

46. Cross, *Straight and Level,* 157.

47. Gladman, *Intelligence and Anglo-American Air Support in World War Two,* 8–9; *Field Service Regulations,* vol. 2, *Operations—General* (London: Her Majesty's Stationery Office, 1935), 26–28.

48. Headquarters Royal Air Force Middle East, "Paper on a Suggested Inter-Allied Inter-Service Communications Scheme," 30 January 1942, AIR 23/1293, National Archives, 1–3.

49. Mediterranean Air Y, 18 September 1943, AIR 40/2252, National Archives, 2; Military Y Mideast, GSI (s), 3 February 1941, WO 208/5021, National Archives; Aileen Clayton, *The Enemy Is Listening: The Story of the Y Service* (London: Crécy, 1993), 150–151.

50. Intelligence Organization Mobile Fighter Group, 31 May 1943, AIR 23/1209, National Archives; Dunn, "Digest of the Group Discussions," 75, 82, 88–89.

51. Dunn, "Digest of the Group Discussions," 51–52, 92–93.

52. Gladman, *Intelligence and Anglo-American Air Support in World War Two,* 45; "Army Air Support and Photographic Interpretation, 1939–1945: Photographic Interpretation in the Middle East," WO 277/34, National Archives, 1.

53. "Report on Visit to the United Kingdom by Chief Signals Officer, R.A.F. M.E., Chief Signals Officer, Air Formation Signals, and Chief Radio Officer, R.A.F. M.E.," 27 October 1941, AIR 23/1293, National Archives, 1–5.

54. Ibid., 5–11. In addition to the major signals infrastructure upgrade efforts already agreed upon, this additional effort would include sending a Beaufighter squadron equipped with air-intercept (AI) radar and ten ground-controlled-intercept (GCI) radar sets. Each GCI set provided area coverage equivalent to 158 searchlights crewed by 3,667 men. See also Denis Richards and Hilary St. George Saunders, *Royal Air Force, 1939–1945,* vol. 2, *The Fight Avails* (London: Her Majesty's Stationery Office, 1954), 161–162.

55. Raymond Collishaw, "Brief Report on the Royal Air Force Operations from the Time of Our Retreat through Cyrenaica Including the Operation Battleaxe," 12 August 1941, AIR 23/6474, National Archives, 1–2.

56. Erwin Rommel and Fritz Bayerlein, *The Rommel Papers,* ed. B. H. Liddell Hart (London: Collins, 1953), 186.

57. Collishaw, "Brief Report," 2–3.

58. Norman MacMillan, *The Royal Air Force in the World War,* vol. 3 (London: Harrap, 1949), 111.

59. Collishaw, "Brief Report," 5–6.

60. Karl Gundelach, *Die deutsche Luftwaffe im Mittelmeer,* vol. 1 (Frankfurt am Main: Lang, 1981), 136–142 (referred to from this point forward as *DLM*). Halder's diary entry the day before he sent word to Rommel emphasized the importance of improving the Mediterranean situation and stated (regarding Rommel's tactical predilections) the importance of "head[ing] off this soldier gone stark mad." See Franz Halder, Charles Burton Burdick, and Hans Adolf Jacobsen, *The Halder War Diary, 1939–1942* (Novato, CA: Presidio, 1988), 373–374.

61. Arthur Tedder to Wilfrid Freeman, 11 March 1941, AIR 23/1386, National Archives, 1.

62. Ibid., 1–2.

63. Arthur Tedder to Wilfrid Freeman, 23 March 1941, AIR 23/1386, National Archives; MacMillan, *Royal Air Force in the World War,* 137.

64. Arthur Tedder to Wilfrid Freeman, 16 April 1941, AIR 23/1386, National Archives.

65. Troopers (Churchill) to Mideast, 6 May 1941, AIR 23/1386, National Archives; Wilfrid Freeman to Arthur Tedder, 8 May 1941, AIR 23/1386, National Archives.

66. Royal Air Force Middle East to Troopers (r) Arthur Freeman and Charles Portal, 11 May 1941, AIR 23/1386, National Archives.

67. Arthur Tedder to Wilfrid Freeman, 15 May 1941, AIR 23/1386, National Archives; *DLM,* vol. 2, *Anlage 25* (Illustration 25).

68. Charles Portal, "Army Air Requirements: Memorandum by the Chief of the Air Staff," June 1941, AIR 75/43, National Archives. The ninety-eight squadrons included thirty-six night and long-range fighter squadrons, twelve of light-bombers, thirty-seven of mediums, and thirteen of heavies.

69. Winston Churchill to Archibald Wavell, 9 June 1941, AIR 23/1395, National Archives.

70. Charles Portal to Arthur Tedder, 17 June 1941, AIR 23/1395, National Archives.

71. BOH 2, 161–171; *DLM,* 143–144; Halder, Burdick, and Jacobsen, *Halder War Diary,* 407.

72. Charles Portal to Arthur Tedder, 19 June 1941, AIR 23/1395, National Archives; Arthur Tedder to Charles Portal, 20 June 1941, AIR 23/1395, National Archives. Even the Germans noted the RAF's air superiority. See Halder, Burdick, and Jacobsen, *Halder War Diary,* 406.

73. Arthur Tedder to Charles Portal, 20 June 1941, AIR 23/1395, National Archives; Arthur Tedder to Charles Portal, 21 June 1941, AIR 23/1395, National Archives; Arthur Tedder to Charles Portal, 23 June 1941, AIR 23/1395, National Archives.

74. Collishaw, "Brief Report," 10–11.

75. Ibid., 12–13.

76. Ibid., 14–15.

77. Portal to Tedder, 19 June 1941.

78. Norman Franks, ed., *The War Diaries of Neville Duke, 1941–1944* (London: Grub Street, 1995), 76, 134.

79. BOH 1, 232–234.

80. Maintenance, Supply, and Salvage Organization, Western Desert, 1942, AIR 23/6493, National Archives.

81. John Slessor to Charles Portal, 1 March 1941, AIR 75/63, National Archives.

82. Henry Self to John Slessor, 1 March 1941, AIR 75/63, National Archives.

83. John Slessor to Charles Portal, 6 March 1941, AIR 75/67, National Archives.

84. John Slessor, "Note for Discussion with Secretaries Stimson and Knox: British Air Policy," 22 March 1941, AIR 75/66, National Archives.

85. Charles Portal to Arthur Longmore, 7 March 1941, AIR 23/1061, National Archives.

86. John Slessor to William Donovan, 1 April 1941, AIR 75/63, National

Archives; Chiefs of Staff (CoS), "British–United States Staff Conversations, Main Report, Air Agreement," 20 April 1941, AIR 75/65, National Archives.

87. Loben Maund to Arthur Tedder, 21 April 1941, Appendix A: Aircraft Position in Greece and Western Desert, AIR 23/1061, National Archives.

88. Charles Portal to Arthur Tedder, 17 May 1941, AIR 23/1061, National Archives; Charles Portal to Arthur Tedder, 17 May 1941, AIR 23/1395, National Archives.

89. Arthur Tedder to Charles Portal, 20 May 1941, AIR 23/1395, National Archives.

90. British Air Commission (BAC) to Arthur Tedder, 16 July 1941, AIR 23/1315, National Archives; Charles Portal to Arthur Tedder, 2 September 1941, AIR 23/1315, National Archives; Charles Portal to Arthur Tedder, 25 September 1941, AIR 23/1315, National Archives.

91. Headquarters Royal Air Force Middle East, "Details of American Aid Required," August 1941, AIR 23/1315, National Archives.

92. *DLM,* 224–233.

Chapter Five. The Back Door Secured and the Front Door Strengthened

1. Karl Gundelach, *Die deutsche Luftwaffe im Mittelmeer,* vol. 1 (Frankfurt am Main: Lang, 1981), 256–259, 263 (referred to from this point forward as *DLM*).

2. I. S. O. Playfair et al., *The Mediterranean and Middle East,* vol. 2, *The Germans Come to the Aid of Their Ally, History of the Second World War: United Kingdom Military Series* (Uckfield, UK: Naval and Military Press, 2004), 177–178 (this and all subsequent volumes of the British official history are referred to as "BOH" with volume number); Daniel Silverfarb, *Britain's Informal Empire in the Middle East: A Case Study of Iraq, 1929–1941* (New York: Oxford University Press, 1986), 123–140; Lukasz Hirszowicz, *The Third Reich and the Arab East* (London: Routledge and Kegan Paul, 1966), Chapters 7–8.

3. BOH 2, 177–178; *DLM,* 235–237.

4. *DLM,* 238.

5. BOH 2, 177–188.

6. Walter Ansel, *Hitler and the Middle Sea* (Durham, NC: Duke University Press, 1972), 422–424; *DLM,* 239, 241.

7. Hellmuth Felmy, "The German Air Force in the Mediterranean Theater of War," K113.107–161, Air Force Historical Research Agency, 131–133.

8. Felmy, "The German Air Force in the Mediterranean Theater of War," 133–134, 193–197; *DLM,* 243; *Fuehrer Conferences on Naval Affairs, 1939–1945* (London: Chatham, 2005), 194–195 (abbreviated as *FCNA* in subsequent citations). Alex Blomberg was Field Marshal Werner von Blomberg's son and the only serving Luftwaffe officer with prior experience in Iraq.

9. *DLM,* 244.

10. Felmy, "The German Air Force in the Mediterranean Theater of War," 135–136.

11. Ibid., 134–135.

12. BOH 2, 197, 204–205; "The Middle East Campaigns, vol. 9, The Campaign in Syria June 1941," Royal Air Force Narrative (First Draft), Air Historical Branch, 7–9; A. B. Gaunson, *The Anglo-French Collision in Lebanon and Syria, 1940–1945* (London: MacMillan, 1986), 33–34.

13. Arthur Tedder, *Report by Air Marshal Tedder on His Tenure in the Post of Air Officer Commanding-in-Chief* (London: MLRS, 2010), 29–32; Hans Werner Neulen, *In the Skies of Europe: Air Forces Allied to the Luftwaffe, 1939–1945* (Wiltshire, UK: Crowood, 2000), 228–230. They had 103 aircraft, including 60 fighters (with 25 capable D.520s) and 20 bombers. The 30 German fighters expected to intervene never materialized.

14. Arthur Tedder to Charles Portal, 19 May 1941, AIR 23/1395, National Archives; Arthur Tedder to Charles Portal, 31 May 1941, AIR 23/1395, National Archives.

15. Portal to Tedder, 19 May 1941.

16. Tedder, *Report*, 33–36; Neulen, *In the Skies of Europe*, 230–231. In a discussion with his RAF counterparts after the campaign in Syria, General Jeannequin, the Vichy air commander, said that the RAF's early and constant attacks on airfields neutralized his air units from the outset. For the cost of three planes, RAF aircrews had destroyed fifty-five on the ground. The RAF lost another ten in air combat but destroyed thirty more Vichy machines. See Denis Richards, *Royal Air Force, 1939–1945*, vol. 1, *The Fight at Odds* (London: Her Majesty's Stationery Office, 1953), 341–342.

17. Tedder, *Report*, 37–39.

18. BOH 2, 227–228.

19. US Army Air Force Director of Intelligence, "Signal Intelligence Service of the German Luftwaffe," vol. 13, "Cryptanalysis within the Luftwaffe SIS," 1945, 519.6314-10 V.13, Air Force Historical Research Agency, 3, 8–11, 25; Franz Halder, Charles Burton Burdick, and Hans Adolf Jacobsen, *The Halder War Diary, 1939–1942* (Novato, CA: Presidio, 1988), 404.

20. Studiengruppe Geschichte des Luftkrieges (Karlsruhe, Germany)/Andreas L. Nielson, "The Collection and Evaluation of Intelligence for the German Air Force High Command," 1955, K113.107-171, Air Force Historical Research Agency, 154–156.

21. Ibid., 156–159.

22. Ibid., 159–164.

23. Ibid., 21–22.

24. Ibid., 24.

25. "Minutes of Joint Meetings of the Air Support Committee," August 1941, AIR 20/2996, National Archives, Appendix C; David Ian Hall, *Strategy for Victory: The Development of British Tactical Air Power, 1919–1943* (London: Praeger, 2008), 104–105; Arthur Tedder, *With Prejudice: The World War II Memoirs of Marshal of the Royal Air Force Lord Tedder, Deputy Supreme Commander of the Allied Expeditionary Force* (Boston: Little, Brown, 1966), 163; BOH 2, 14–15, 294–295, 414–419. The German total does not include Fliegerkorps X assets stationed in Greece and on Crete.

26. "The Middle East Campaigns," vol. 1, AIR 41/44, National Archives, 174–175; "Air Support," 52–54, AIR 10/5547, National Archives; Tedder, *With Prejudice*, 124–128, 138–143.

27. "Army Training Instruction No. 6," 31 October 1941, reprinted in "Air Support," AIR 10/5547, National Archives, Appendix 2, 194–199.

28. "The Middle East Campaigns," vol. 2, AIR 41/25, National Archives, 40, 213–214; "Army Air Support," WO 277/34, National Archives, 46–48; Richard Townshend Bickers, *The Desert Air War, 1939–1945* (London: Cooper, 1991), 72;

I. S. O. Playfair, *The Mediterranean and Middle East,* vol. 3, *British Fortunes Reach Their Lowest Ebb (September 1941 to September 1942), History of the Second World War: United Kingdom Military Series* (Uckfield, UK: Naval and Military Press, 2009), 43; Sebastian Cox, quoted in "Digest of the Group Discussions," in Royal Air Force Historical Society, *The End of the Beginning: Bracknell Paper No. 3, A Symposium on the Land/Air Co-operation in the Mediterranean War, 1940–43,* ed. Derek Wood (Bracknell, UK: Royal Air Force Historical Society), 20 March 1992, 81; Brad Gladman, "The Development of Tactical Air Doctrine in North Africa, 1940–1943," in *Air Power History: Turning Points from Kitty Hawk to Kosovo,* ed. Sebastian Cox and Peter Gray (London: Routledge, 2002), 188–206.

29. Brad Gladman, *Intelligence and Anglo-American Air Support in World War Two: The Western Desert and Tunisia, 1940–43* (New York: Palgrave MacMillan, 2009), 47; "No. 202 Group Index," Air 25, National Archives; Arthur Longmore, "Air Operations in the Middle East, 1 January to 3 May 1941," CAB 106/626, National Archives, 3.

30. "Middle East (Army and RAF) Training Pamphlet No. 3—Direct Air Support," in "The Middle East Campaigns," vol. 2, AIR 41/25, National Archives, 57–61; "Air Support," AIR 10/5547, National Archives, 56–59; "Army-Air Co-operation," CAB 101/136, National Archives, 29–32; BOH 2, 295.

31. General Headquarters Middle East Forces and Headquarters Royal Air Force Middle East, "Revised Draft of M.E. Training Pamphlet No. 3—Direct Air Support," AIR 23/1762, National Archives.

32. Air Headquarters Egypt, "Standing Operating Instruction No. 20 of 1942 (Replaced No. 7 of 1942)—G.C.I. Control of Interceptions," 10 December 1942, AIR 23/808, National Archives, 1.

33. Ibid., 2.

34. Basil Embry, *Mission Completed* (London: Quality Book Club, 1957), 209–216.

35. Arthur Tedder to Charles Portal, 25 October 1941, Tedder, Box 3, Air Historical Branch; Portal to Tedder, 25 October 1941, Tedder, Box 3, Air Historical Branch; Tedder to Portal, 30 October 1941, Tedder, Box 3, Air Historical Branch; Arthur Tedder, journal, 21 November 1941, Tedder Box 3, Air Historical Branch. Cross said of Embry, "He was, in my opinion, the most outstanding leader in the air up to the rank of Air Vice-Marshal that the Royal Air Force produced in the war." See Kenneth Cross with Vincent Orange, *Straight and Level* (London: Bugg Street, 1993), 151.

36. Arthur Tedder, journal, 21 November 1941.

37. Robin Higham, *Bases of Air Strategy: Building Airfields for the RAF, 1914–1945* (London: Airlife, 1998), 15.

38. Ibid., 19.

39. Ibid., 27.

40. Headquarters Royal Air Force Middle East, "Airfield Construction M. E. F.," n.d. (ca. 1945), WO 201/2807, National Archives, 1–5.

41. Ibid., Appendix A: Some Technical Notes on Airfield Works in M.E., 1–4.

42. Higham, *Bases of Air Strategy,* 151–153.

43. Ibid., 111, 145.

44. H. W. Wynter, *The History of the Long Range Desert Group, June 1940 to March 1943* (Kew, UK: National Archives, 2008), 65, 73, 97–99, 114–115, 122.

45. Gerhard Weinberg, *A World at Arms: A Global History of World War II* (Cambridge, UK: Cambridge University Press, 1994), 233–234; DLM, 265, 267.

46. *DLM*, 269–272.

47. "Appreciation of the Air Situation in the Mediterranean Theatre during the Period 11/7–31/8/1941," trans. 7/41, A.H.B.6, 1 September 1947, Air Historical Branch. Raids on Alexandria involved 72 planes and those on Suez 157—tiny numbers given these targets' importance. Cross emphasized the Luftwaffe's dissipation of effort in *Straight and Level,* 142.

48. *DLM*, 268, 272–278; Vincent P. O'Hara, *Struggle for the Middle Sea: The Great Navies at War in the Mediterranean, 1940–1945* (Annapolis, MD: Naval Institute Press, 2009), 141.

49. *DLM*, 279–283.

50. Arthur Tedder to Wilfrid Freeman, 29 May 1941, AIR 23/1386, National Archives.

51. Arthur Tedder to Wilfrid Freeman, 3 June 1941, AIR 23/1386, National Archives, 1. Cross noted that while he commanded No. 252 Wing at Alexandria, the fleet often moved without Tedder's knowledge and thus ability to provide air cover. See *Straight and Level,* 145.

52. Tedder to Freeman, 3 June 1941, 2.

53. Arthur Tedder to Charles Portal, 7 June 1941 (two memoranda), AIR 23/1395, National Archives.

54. Andrew Cunningham to Arthur Tedder, 11 June 1941, Tedder Box 3, Air Historical Branch; Tedder to Cunningham 27 June 1941, Tedder Box 3, Air Historical Branch; Charles Portal to Arthur Tedder, 28 September 1941, Tedder Box 3, Air Historical Branch; Portal to Tedder, 4 October 1941, Tedder Box 3, Air Historical Branch; Portal to Tedder, 6 October 1941, Tedder Box 3, Air Historical Branch.

55. BOH 2, 274–280.

56. Headquarters Royal Air Force Middle East Tactics Assessment Office, "Air Tactics and Operational Notes on 201 Naval Co-operation Group, R.A.F.," n.d. (ca. June 1943), AIR 23/1282, National Archives, Foreword and Chapter 1, 1; *DLM*, 298. In October 1941, Malta had one squadron of Fleet Air Arm Albacores. No. 201 Group also had a squadron of Blenheim IVs, a Sunderland squadron, a mixed Maryland/Beaufort reconnaissance and torpedo-bombing squadron, and a Beaufighter squadron detached for strafing operations in the Western Desert. This force was far too small to control the SLOCs.

57. Headquarters Royal Air Force Middle East Tactics Assessment Office, METM No. 1, "Practice Low Level Attacks by Marylands on H.M.S. *Carlisle* (A.A.) Suez Eighth September 1941," October 1941, AIR 23/1281, National Archives.

58. US Army Air Forces Director of Intelligence, "Signal Intelligence Service of the German Luftwaffe," vol. 10, "Technical Operations in the South," 1945, 519.6314–10 V.9, Air Force Historical Research Agency, 71–72.

59. BOH 2, 280–282.

60. Ibid., 282.

61. Ibid.

62. Ibid., 283.

63. O'Hara, *Struggle for the Middle Sea,* 143–147.

64. BOH 2, 292–293. For further insights regarding the circumstances surrounding Graham Dawson's assignment to Royal Air Force Middle East, and the superb work he did there, see Denis Richards and Hilary St. George Saunders, *Royal Air Force, 1939–1945,* vol. 2, *The Fight Avails* (London: Her Majesty's Stationery Office, 1954), 162–167.

65. Tedder, *With Prejudice,* 108–109.

66. Anthea M. Lewis, "A Tribute to Air Vice-Marshal GG Dawson, 1895–1944, CMSO ME & N Africa 1941–1944," Air Historical Branch, 3.

67. Ibid., 10–11.

68. Ibid., 12–15.

69. Ibid., 16–18.

70. Ibid., 19.

71. Lewis, "Tribute to Air Vice Marshal GG Dawson," 19–20. Cross, commanding No. 219 Group, was responsible for the successful operational employment of Dawson's modified fighters. See *Straight and Level,* 218. For additional insights, see Richards and Saunders, *Royal Air Force,* vol. 2, 227–228.

72. Ibid., 21–23.

73. Ibid., 23–24.

74. Ibid., 24.

75. Ibid.

76. Lewis, "Tribute to Air Vice Marshal GG Dawson," 14–15, 24–25.

77. Ibid., 25–26.

78. Royal Air Force Middle East, "Maintenance Diagrams," June 1943, AIR 23/3417, National Archives, Charts 1, 3, 14.

79. Quoted in Norman MacMillan, *The Royal Air Force in the World War,* vol. 3 (London: Harrap, 1949), 143.

80. Air Ministry, "The Mediterranean Campaign: Signals and Directives Concerning Problems of Supply and German-Italian Co-operation, April–June 1941," trans. 7/128, A.H.B.6, November 1953, Air Historical Branch, 1–3.

81. Ibid., 4–5; Air Ministry, *The Supply Organisation of the German Air Force, 1935–1945,* reprint ed. (Uckfield, UK: Naval and Military Press, 2009), 8, 43, 191–193. The original 1946 report is available at the Air Historical Branch.

82. Air Ministry, "The Mediterranean Campaign . . . April–June 1941,"6–11.

83. Ibid., 14–16.

84. Ibid.

85. Ibid., 22–23.

86. Johannes Steinhoff, *The Straits of Messina* (London: Deutsch, 1971), 105.

87. Air Ministry, "The Mediterranean Campaign . . . April–June 1941," 24–26.

88. Ibid., 28–31.

89. Ibid.

90. Richards and Saunders, *Royal Air Force,* vol. 2, 168.

91. Air Ministry, "The Mediterranean Campaign: Signals and Directives Concerning Problems of Supply and German-Italian Co-operation, June–October 1941," trans. 7/129, A.H.B.6, February 1954, Air Historical Branch, 11–12.

92. Ibid., 22–23.

93. Ibid., 41–43.

94. Ibid.

95. Ibid., 44–45.

96. *FCNA,* 240.

97. Ibid., 241.

98. Studiengruppe Geschichte des Luftkrieges (Karlsruhe, Germany), "The German Air Force: Aircraft Procurement," 1955, K113.107–170, Air Force Historical Research Agency, secs. 456–457; Air Ministry, *Supply Organisation of the GAF,*

198; Adam Tooze, *The Wages of Destruction: The Making and Breaking of the Nazi Economy* (New York: Viking, 2007), 598.

99. Christopher Shores and Hans Ring, *Fighters over the Desert: The Air Battles in the Western Desert, June 1940–December 1942* (New York: Arco, 1969), 228.

100. Studiengruppe Geschichte des Luftkrieges, "The German Air Force," secs. 466–468, 471.

101. Ibid., secs. 472–473; *DLM*, 507–511. On a related note, see Tooze, *The Wages of Destruction*, 582–585, for the pernicious effects of continued large-scale production of increasingly obsolete combat types.

102. Studiengruppe Geschichte des Luftkrieges, "The German Air Force," secs. 477, 482.

103. Ibid., sec. 481.

104. Ibid.

105. Ibid., secs. 486–487.

106. Ibid., sec. 488.

107. Ibid., sec. 489.

108. Air Ministry A.D.I. (K), "Interrogation of German Officers from the Reichsluftfahrt Ministerium Conducted by Air Ministry A.D.I. (K) on Friday, 15th June 1945," 553.619, Air Force Historical Research Agency, 7.

109. Ibid.; Air Ministry, *Supply Organisation of the GAF*, 231–232, 237; Arthur Tedder to Charles Portal, 23 January 1942, AIR 23/1396, National Archives.

110. Air Ministry, *Supply Organisation of the GAF*, 48–51, 57, 59, 63, 78–81, 199.

111. Ibid., 63–64.

112. Ibid., 68–71.

113. Ibid., 71.

114. Ibid., 208–209.

115. Air Ministry, "The Mediterranean Campaign . . . June–October 1941," 5–6.

116. Ibid.

117. Ibid., 6–7.

118. Ibid., 9–10.

119. Ibid., 10–11.

Chapter Six. Preparations for and the Conduct of Operation Crusader

1. Mark Stoler, *Allies and Adversaries: The Joint Chiefs of Staff, the Grand Alliance, and U.S. Strategy in World War II* (Chapel Hill: University of North Carolina Press, 2000), 50–56.

2. Ibid., 57.

3. Ibid., 58.

4. Ibid., 64–67; Gerhard Weinberg, *A World at Arms: A Global History of World War II* (Cambridge, UK: Cambridge University Press, 1994), 306.

5. Charles Portal to Arthur Tedder, 5 September 1941, AIR 23/1395, National Archives.

6. I. S. O. Playfair et al., *The Mediterranean and Middle East*, vol. 2, *The Germans Come to the Aid of Their Ally, History of the Second World War: United Kingdom Military Series* (Uckfield, UK: Naval and Military Press, 2004), 290–292 (this

and all subsequent volumes of the British official history are referred to as "BOH" with volume number); Wilfrid Oulton, quoted in "Digest of the Group Discussions," in Royal Air Force Historical Society, *The End of the Beginning: Bracknell Paper No. 3, A Symposium on the Land/Air Co-operation in the Mediterranean War, 1940–43,* ed. Derek Wood (Bracknell, UK: Royal Air Force Historical Society), 20 March 1992, 73; Norman Franks, ed., *The War Diaries of Neville Duke, 1941–1944* (London: Grub Street, 1995), 74–88, 132–143.

7. Vincent P. Orange, *Coningham: A Biography of Air Marshal Sir Arthur Coningham* (Washington, DC: Center for Air Force History, 1992), 66–67.

8. BOH 2, 294–295; Roderic Owen, *The Desert Air Force: An Authoritative History Published in Aid of the Royal Air Force Benevolent Fund* (London: Hutchinson, 1948), 61.

9. Vincent Orange, "The Commanders and the Command System," in Royal Air Force Historical Society, *The End of the Beginning: Bracknell Paper No. 3, A Symposium on the Land/Air Co-operation in the Mediterranean War, 1940–43,* ed. Derek Wood (Bracknell, UK: Royal Air Force Historical Society), 20 March 1992, 36–38.

10. Ibid., 39.

11. Ibid., 39–40.

12. Geoffrey Morley-Mower, *Messerschmitt Roulette: The Western Desert, 1941–42* (Shrewsbury, UK: Airlife, 2003), 119.

13. Philip Guedalla, *Middle East, 1940–1942: A Study in Air Power* (London: Hodder and Stoughton, 1944), 85.

14. Franks, *War Diaries of Neville Duke,* 50–51.

15. Werner Held and Ernst Obermaier, *The Luftwaffe in the North African Campaign, 1941–1943* (West Chester, PA: Schiffer, 1992), 44.

16. Arthur Tedder to Ludlow-Hewitt, 6 September 1941, Tedder Box 4, Air Historical Branch; BOH 3, 11–13.

17. BOH 3, 14–15; Arthur Tedder, *With Prejudice: The World War II Memoirs of Marshal of the Royal Air Force Lord Tedder, Deputy Supreme Commander of the Allied Expeditionary Force* (Boston: Little, Brown, 1966), 171. Air Ministry intelligence estimates were 600 Axis aircraft to the RAF's 540. In fact, the British had more than 650 aircraft, of which 540 were serviceable. This included 74 at Malta (66 serviceable). The Axis had 536 aircraft, of which 342 were serviceable. It had another 750 serviceable aircraft (excluding transports) in Tripolitania, Sicily, Sardinia, Greece, and Crete. Additionally, the Italian Metropolitan Air Force and the Italian Army and Navy controlled several hundred.

18. Arthur Tedder, *Report by Air Marshal Tedder on His Tenure in the Post of Air Officer Commanding-in-Chief* (London: MLRS, 2010), 60.

19. Charles Portal to Arthur Tedder, 14 October 1941 (two signals), AIR 23/1395, National Archives.

20. Arthur Tedder to Charles Portal, 15 October 1941, AIR 23/1395, National Archives.

21. Charles Portal to Arthur Tedder, 15 October 1941, AIR 23/1395, National Archives.

22. Arthur Tedder to Charles Portal, 19 October 1941, AIR 23/1395, National Archives; Winston Churchill to Claude Auchinleck, 14 October 1941, AIR 23/1395,

National Archives; Churchill to Auchinleck, 16 October 1941, AIR 23/1395, National Archives; Auchinleck to Churchill, 21 October 1941, AIR 23/1395, National Archives.

23. Winston Churchill to Arthur Tedder, 24 October 1941, AIR 23/1395, National Archives; Tedder to Churchill, 24 October 1941, AIR 23/1395, National Archives.

24. Wilfrid Freeman to Charles Portal, 30 October 1941, AIR 23/1395, National Archives; Sebastian Cox, "'The Difference between White and Black': Churchill, Imperial Politics, and Intelligence before the 1941 Crusader Offensive," *Intelligence and National Security* 9, no. 3 (1994). Freeman and Tedder arrived at the following final figures: the RAF had 600 (528 serviceable) and the Axis 642 (385 serviceable). Of the Axis planes, 435 were Italian and 207 German. The Luftwaffe had another 156 aircraft of all types in the Aegean and Crete, and the RAF had 64 bombers (No. 205 Group) in Palestine and a similar number of fighters in the Nile Delta. The RAF had 50 percent reserves in workshops and other elements of the aircraft "pipe-line" and the enemy almost none.

25. Claude Auchinleck to Charles Portal, 11 December 1941, Tedder Box 3, Air Historical Branch.

26. Winston Churchill to Claude Auchinleck, 25 November 1941, Tedder Box 3, Air Historical Branch.

27. BOH 3, 15.

28. Tedder, *With Prejudice*, 184–190.

29. BOH 3, 16–17.

30. Erwin Rommel and Fritz Bayerlein, *The Rommel Papers*, ed. B. H. Liddell Hart (London: Collins, 1953), 158.

31. BOH 3, 18; Karl Gundelach, *Die deutsche Luftwaffe im Mittelmeer*, vol. 1 (Frankfurt am Main: Lang, 1981), 308–310 (referred to from this point forward as *DLM*).

32. Tedder, *With Prejudice*, 191–192.

33. BOH 3, 18–19.

34. Richard J. Overy, *The Air War, 1939–1945* (New York: Stein and Day, 1980), 66; F. H. Hinsley et al., *British Intelligence in the Second World War*, vol. 2 (London: Her Majesty's Stationery Office, 1979), 291, 322–325; Owen, *Desert Air Force*, 66; *DLM*, 311; BOH 3, 20–22; *Fuehrer Conferences on Naval Affairs, 1939–1945* (London: Chatham, 2005), 243–244 (abbreviated as "*FCNA*" in subsequent citations). Although Hitler originally ordered the move of a *Fliegerkorps*, he soon expanded this to include Luftflotte 2 as well as Fliegerkorps II. Halder stated on 19 May 1941, "All the Fuehrer cares about is that Rommel should not be hampered by any superior Hq. put over him." Rommel acted accordingly. See Franz Halder, Charles Burton Burdick, and Hans Adolf Jacobsen, *The Halder War Diary, 1939–1942* (Novato, CA: Presidio, 1988), 389.

35. Air Ministry, "RAF Summaries Nos. 328–834, 12 June 1940–31 October 1941," AIR 22/10-27, National Archives.

36. General Headquarters Middle East Forces and Headquarters Royal Air Force Middle East, "Middle East (Army and RAF) Directive on Direct Air Support," 30 September 1941, WO 201/386, National Archives, 1–8.

37. Headquarters Western Desert Air Force, "Operation Crusader Air Plan, Air Appreciation and Plan," 2 November 1941, AIR 23/6476, National Archives.

38. Hellmuth Felmy, "The German Air Force in the Mediterranean Theater of War," K113.107–161, Air Force Historical Research Agency, 152–155.

39. Ibid., 155–157.

40. Ibid., 149–150; BOH 2, 223, 281.

41. Tedder, *Report*, 65–66.

42. Ibid., 67.

43. Felmy, "The German Air Force in the Mediterranean Theater of War," 172–181; Kenneth Cross with Vincent Orange, *Straight and Level* (London: Bugg Street, 1993), 157.

44. Ibid., 185–191.

45. Arthur Tedder to Charles Portal, 3 November 1941, AIR 23/1396, National Archives; Tedder to Portal, 17 November 1941, AIR 23/1396, National Archives; Tedder to Portal, 21 November 1941, AIR 23/1396, National Archives.; Tedder to Portal, 24 November 1941, AIR 23/1396, National Archives.

46. Arthur Coningham to Arthur Tedder, 10 December 1941, Tedder Box 3, Air Historical Branch.

47. Tedder, *With Prejudice*, 205.

48. Arthur Coningham to Arthur Tedder, 5 January 1942, in Tedder, "Note on Air Operations in Support of Crusader, January–December 1942," Tedder Box 6, Air Historical Branch, 10.

49. Arthur Tedder to Charles Portal, 7 December 1941, AIR 23/1396, National Archives.

50. Lewis Anthem Ritchie to Arthur Tedder, 12 December 1941, AIR 23/1396, National Archives.

51. *DLM*, 318–321.

52. Tedder, *With Prejudice*, 206–210.

53. Arthur Tedder to Wilfrid Freeman, 16 December 1941, Tedder Box 3, Air Historical Branch; Freeman to Tedder, 30 December 1941, Tedder Box 3, Air Historical Branch. The refusal to send heavy-bombers must be placed within the context of Bomber Command's continuing struggle to build a capable bomber force, and the US Army Air Forces' (USAAF's) shortage of B-24s and B-17s. Tedder would not receive nearly as many heavy-bombers as he requested, and they arrived relatively late in the campaign despite their clear value in a theater where range and payload were exceptionally important.

54. Arthur Coningham to Arthur Tedder, 14 January 1942, in Tedder, "Note on Air Operations in Support of Crusader," 18; *DLM*, 336–337, 351.

55. Felmy, "The German Air Force in the Mediterranean Theater of War," 204–205.

56. Rommel and Bayerlein, *Rommel Papers*, 172.

57. Ibid., 175.

58. Ibid., 177.

59. "Operations in the Middle East, 5 July 1941–31 October 1941," CAB 106/535, National Archives; "Army Air Support and Photographic Interpretation, 1939–1945, Organization and Training of Air Liaison Officers," WO 277/34, National Archives, 35; "Organization and Application of Air Intelligence in a Tactical Air Force," n.d., AIR 23/1209, National Archives.

60. "War Diary of No. 2 Army Air Support Control," January and March 1942, WO 169/6638, National Archives; "The Middle East Campaigns, June 1941–January

1942," AIR 41/25, National Archives, 60–62, 129; "Army Air Support and Photographic Interpretation, 1939–1945, 4.

61. GS, 30 Corps War Diary 1941, "Air Support and Signals Exercise," October 1941, and "30 Corps Operational Instruction No. 1, Ground-Air Recognition," 7 November 1941, WO 169/1123, National Archives, Appendix KK.

62. John Ferris, "The Usual Source: Signals Intelligence and Planning for the 'Crusader' Offensive, 1941," *Intelligence and National Security* 14, no. 1 (1999): 112; "Lessons from Operations—Cyrenaica No. 1, November 1941," WO 106/2255, National Archives, 2.

63. Frederick Rosier, "How the Joint System Worked (1)," in Royal Air Force Historical Society, *The End of the Beginning: Bracknell Paper No. 3, A Symposium on the Land/Air Co-operation in the Mediterranean War, 1940–43,* ed. Derek Wood (Bracknell, UK: Royal Air Force Historical Society), 20 March 1992, 26–30. See also Patrick Dunn, quoted in "Digest of the Group Discussions," in *The End of the Beginning,* 68, 70; Tedder, *Report,* 68. Cross emphasized RAF–army communications problems in *Straight and Level*: "W/T communications from forward Army HQ and vice-versa simply did not work. . . . The location of friendly formations was often unknown and this made planning bomber operations difficult and sometimes impossible" (158–159).

64. Tedder, "Note on Air Operations in Support of Crusader," 28.

65. Arthur Coningham to Arthur Tedder, 2 January 1942, AIR 23/1391, National Archives; Coningham to Tedder, 5 January 1942, AIR 23/1391, National Archives.

66. Tedder, *Report,* 69–72.

67. BOH 3, 69–70; Hinsley et al., *British Intelligence in the Second World War,* vol. 2, 326–328.

68. Arthur Tedder to Sinclair, 17 February 1942, AIR 23/1396, National Archives.

69. Franks, *War Diaries of Neville Duke,* 47.

70. BOH 3, 76.

71. Ibid., 77–78.

72. Arthur Tedder to Sinclair, 12 February 1942, AIR 23/1396, National Archives. In every campaign, gaining air superiority at the outset and maintaining it had been the key to success. From May 1941 to January 1942, the RAF destroyed 1,155 German and 947 Italian aircraft for a loss of 1,009 in combat.

73. Felmy, "The German Air Force in the Mediterranean Theater of War," 208–209.

74. Ibid., 209–210.

75. Albert Kesselring, "The War in the Mediterranean," pt. I, written for the U.S. Historical Division, May 1948, trans. 7/104, A.H.B.6, February 1951, Air Historical Branch, 1.

76. Ibid., 2–5.

77. Halder, Burdick, and Jacobsen, *Halder War Diary,* 402.

78. Kesselring, "The War in the Mediterranean," pt. I, 6–12 (quote on 8).

79. Ibid., 14–17.

80. Ibid., 18–22 (quote on 21).

81. Ibid., 23–25.

82. Felmy, "The German Air Force in the Mediterranean Theater of War," 210–213; *DLM,* 356; *FCNA,* 245.

83. *FCNA,* 263.

84. Ibid., 266.

85. Ibid., 267.

86. BOH 3, 89–92; Tedder, *With Prejudice,* 210–212.

87. Rommel and Bayerlein, *Rommel Papers,* 179.

88. Ibid.

89. Arthur Tedder to Douglas Evill, 13 March 1942, AIR 23/1315, National Archives.

90. Tedder, *With Prejudice,* 226.

91. Ibid., 227.

92. Arthur Tedder to Douglas Evill, 13 March 1942, in Tedder, "Note on Air Operations in Support of Crusader," 68.

93. Douglas Evill to Arthur Tedder, 13 March 1942, AIR 23/1315, National Archives; Tedder to Evill, 14 March 1942, AIR 23/1315, National Archives.

94. BOH 3, 93–94, 99.

95. Ibid., 126–127. The RAF immediately sent six Blenheim squadrons, an air stores park, and a repair-and-salvage unit to India. The Royal Navy redeployed eleven ships, including two submarines, leaving the Mediterranean Fleet a shadow of its former self. The carrier HMS *Indomitable* also ferried forty-seven Hurricanes to the Far East followed by another sixty Hurricanes, while twenty-three more flew directly to India. In total, RAFME sent five Blenheim and seven Hurricane squadrons (two from the theater and five diverted en route).

96. Headquarters Royal Air Force Middle East, "Some Notes on Administrative and Organisational Difficulties Affecting the Royal Air Force in the Western Desert and Cyrenaica during the Period of Mobile Warfare—November 1941–February 1942," 30 March 1942, AIR 23/1048, National Archives, 1–3.

97. Ibid., 4–10.

98. Headquarters Royal Air Force Middle East to All Subordinate Headquarters, "Mobility of Squadrons and Units," 6 April 1942, AIR 23/1048, National Archives.

Chapter Seven. Rommel Strikes Again

1. Oberkommando der Wehrmacht (OKW), Foreign Armies (Western) Intelligence Department, "The Middle East Situation in 1942," A.H.B.6, trans. 7/15, 30 December 1946, Air Historical Branch.

2. War Cabinet Joint Planning Staff, "Middle East and Mediterranean," 29 March 1942, AIR 23/1204, National Archives, Appendix C, 1–7. The Axis had three battleships, nine cruisers, fifty-five destroyers and torpedo boats, and sixty-four submarines. The Royal Navy had zero, four, fifteen, and fifteen, respectively.

3. Chiefs of Staff to Commanders in Chief Middle East, 1 April 1942, AIR 23/1204, National Archives, 1–5.

4. Mark Stoler, *Allies and Adversaries: The Joint Chiefs of Staff, the Grand Alliance, and U.S. Strategy in World War II* (Chapel Hill: University of North Carolina Press, 2000), 64–72.

5. Hellmuth Felmy, "The German Air Force in the Mediterranean Theater of War," K113.107–161, Air Force Historical Research Agency, 213–214.

6. Erwin Rommel and Fritz Bayerlein, *The Rommel Papers,* ed. B. H. Liddell Hart (London: Collins, 1953), 192.

7. I. S. O. Playfair, *The Mediterranean and Middle East,* vol. 3, *British Forces*

Reach Their Lowest Ebb, History of the Second World War: United Kingdom Military Series (Uckfield, UK: Naval and Military Press, 2004), 140–141 (this and all subsequent volumes of the British Official History are referred to as "BOH" with volume number).

8. Arthur Tedder, *Report by Air Marshal Tedder on His Tenure in the Post of Air Officer Commanding-in-Chief* (London: MLRS, 2010), 72–76; Arthur Coningham to Arthur Tedder, 24 January 1942, in Tedder, "Note on Air Operations in Support of Crusader, January–December 1942," Tedder Box 6, Air Historical Branch, 31.

9. "War Diary of Panzer Army, 21 January–6 February 1942," trans. 7/118, A.H.B.6, September 1952, Air Historical Branch, 24–27.

10. Felmy, "The German Air Force in the Mediterranean Theater of War," 216–224; Karl Gundelach, *Die deutsche Luftwaffe im Mittelmeer,* vol. 1 (Frankfurt am Main: Lang, 1981), 348–349 (referred to from this point forward as *DLM*).

11. Arthur Coningham to Arthur Tedder, 24 January 1942, AIR 23/1391, National Archives.

12. Arthur Tedder, *With Prejudice: The World War II Memoirs of Marshal of the Royal Air Force Lord Tedder, Deputy Supreme Commander of the Allied Expeditionary Force* (Boston: Little, Brown, 1966), 242–243. Cross noted that the Luftwaffe missed yet another opportunity to savage RAF formations. Aircraft were crowded onto several small airfields. "Fortunately," he said, "the Luftwaffe had been left far behind by the Wehrmacht and were out of range for we were certainly an excellent target." See Kenneth Cross with Vincent Orange, *Straight and Level* (London: Bugg Street, 1993), 174. Neville Duke emphasized the damage Luftwaffe bombers did even with their small, sporadic, and short-lived raids on RAF landing grounds from February to March 1942. With better ground mobility to get within steady range, and with a dedicated effort to destroy remaining RAF fighter units, the Luftwaffe could almost certainly have done so. See Norman Franks, ed., *The War Diaries of Neville Duke, 1941–1944* (London: Grub Street, 1995), 67–71.

13. BOH 3, 156–158.

14. Ibid., 159–162.

15. BOH 3, 163–175, 193; *DLM,* 356–357. No. 201 Group assets included three bomber, four torpedo-bomber, six reconnaissance, and three long-range fighter squadrons. The effectiveness the Axis achieved by combining the Luftwaffe, Italian Navy, and good intelligence and reconnaissance to keep Malta isolated prompts us to consider why this was not the norm. Persistence may well have led to decisive results in combination with an invasion of Malta. The Royal Navy would have sortied to stop such an invasion, but without constant air cover, it would have been savaged by the Luftwaffe, which controlled the central Mediterranean at this point and could also have covered the invasion fleet and troops.

16. BOH 3, 176–178.

17. Franks, *War Diaries of Neville Duke,* 63.

18. "War Diary of Panzer Army, 21 January–6 February 1942," 24–27; BOH 4, 52–53.

19. Ibid., 63.

20. Tedder, *Report,* 82–83.

21. Cajus Bekker, *The Luftwaffe War Diaries: The German Air Force in World War II* (New York: Da Capo, 1994), 239. From 20 March to 28 April, Luftwaffe raids comprised 5,807 bomber sorties with another 5,667 by fighters and 345 by reconnaissance aircraft—11,819 in total.

22. Tedder, *Report*, 84–85; Douglas-Hamilton gives a detailed account of the first *Wasp* delivery on 45–54.

23. Ibid., 87–89.

24. Ibid., 90–92. Tedder sent 300 fighters to India during this period, leaving 64 in the Western Desert. Further, 139 Blenheims transferred, leaving 32. He also gave up 12 Beaufort torpedo-bombers during a critical phase in the battle for the SLOCs, and a squadron of B-24 heavy bombers—the only bombers he had that could reach Tripoli. By the time the diversions and transfers ended, Tedder had sent 530 aircraft.

25. Tedder, *Report*, 93–94; Roderic Owen, *The Desert Air Force: An Authoritative History Published in Aid of the Royal Air Force Benevolent Fund* (London: Hutchinson, 1948), 88.

26. Tedder, *Report*, 74–75.

27. Operations Record Book of the Forward Desert Air Force, February 1942, AIR 24/443, National Archives; "War Diary of No. 2 Army Air Support Control," 14 March and 9 May 1942, WO 169/6638, National Archives; Arthur Tedder, "Despatch on Middle East Operations, May 1941–January 1943," AIR 40/1817, National Archives, Signals Appendix.

28. Tedder, "Despatch on Middle East Operations," Signals Appendix; John Ferris, "The British Army, Signals and Security," in *Intelligence and Military Operations,* ed. Michael I. Handel (London: Routledge, 1990), 278; "Some Signal Lessons of the Libyan Campaign November 1941 to February 1942," WO 201/369, National Archives; Cross, *Straight and Level,* 156–157.

29. Military "Y" Middle East, GSI (s), 2 March 1942, WO 208/5021, National Archives; "Y Daily Reports Middle East for 20 May 1942," AIR 40/2345, National Archives; Panzer Army Headquarters Ia, memorandum, 5 September 1942, in "War Diary of Panzer Army Africa, 28 July–23 October," AIR 20/7706, National Archives; Aileen Clayton, *The Enemy Is Listening: The Story of the Y Service* (London: Crécy, 1993), 151.

30. Brad Gladman, *Intelligence and Anglo-American Air Support in World War Two: The Western Desert and Tunisia, 1940–43* (New York: Palgrave MacMillan, 2009), 80–81.

31. "Instructions for the Collection and Quick Dissemination of Information, 9 May 1942," WO 201/539, National Archives; Major McNeill to BGS Eighth Army, memorandum discussing employment of the AASC, 1 April 1942, in "War Diary of No. 2 Army Air Support Control," WO 169/6638, National Archives.

32. H. W. Wynter, *The History of the Long Range Desert Group, June 1940 to March 1943* (Kew, UK: National Archives, 2008), 135–136.

33. "Army Air Support and Photographic Interpretation, 1939–1945," WO 277/34, National Archives, 44; Vincent P. Orange, *Coningham: A Biography of Air Marshal Sir Arthur Coningham* (Washington, DC: Center for Air Force History, 1992), 214; "The Middle East Campaigns II, June 1941–January 1942," AIR 41/25, National Archives, 60; "Report by Air Marshal Sir T. Leigh-Mallory on His Visit to North Africa, April 1943: The Application of Direct Support, December 1941," WO 201/488, National Archives, 14.

34. BOH 3, 250–252.

35. "Minutes of Middle East Defence Committee, 20 March 1942," in Tedder, "Note on Air Operations in Support of Crusader," 79–81.

36. Stafford Cripps to Winston Churchill, 20 March 1942, AIR 23/1204, National Archives.

37. BOH 3, 263–266.

38. BOH 3, 271, 276–279; Arthur Tedder to Noel Beresford-Peirse, 15 March 1942, in Tedder, "Note on Air Operations in Support of Crusader."

39. Tedder, *With Prejudice*, 236–239; Gladman, *Intelligence and Anglo-American Air Support in World War Two*, 94–95.

40. Graham Dawson to Arthur Tedder, "Summary of Wing Commander Messiter's Signals on American Projects," 30 October 1941, AIR 23/1315, National Archives; Lyttelton to Headquarters Middle East, 10 December 1941, AIR 23/1315, National Archives.

41. Arthur Tedder to Wilfrid Freeman, 16 December 1941, AIR 23/1386, National Archives; Arthur Tedder to Charles Portal, 12 December 1941, AIR 23/1396, National Archives.

42. Arthur Coningham to Arthur Tedder, 14 January 1942, AIR 23/1391, National Archives.

43. Arthur Tedder to Charles Portal, 23 January 1942, AIR 23/1396, National Archives; Tedder to Portal (r) Noel Beresford-Peirse, 4 February 1942, AIR 23/1396, National Archives.

44. Charles Portal to Arthur Tedder, 6 February 1942, AIR 23/1396, National Archives; Portal to Tedder, 26 February 1942, AIR 23/1396, National Archives.

45. Peter Drummond to Arthur Coningham, 4 March 1942, AIR 23/1396, National Archives; Arthur Tedder to Peter Drummond, 4 March 1942, AIR 23/1396, National Archives.

46. Arthur Tedder to Peter Drummond, 4 March 1942, AIR 23/1396, National Archives.

47. Peter Drummond to Charles Portal, 5 March 1942, AIR 23/1396, National Archives.

48. Arthur Tedder to Charles Portal, 17 March 1942, AIR 23/1396, National Archives, 1–2.

49. Arthur Tedder to Douglas Evill, 23 March 1942, AIR 23/1315, National Archives; Graham Dawson to Royal Air Force Delegation, 3 April 1942, AIR 23/1315, National Archives; Arthur Tedder to Wilfrid Freeman, 6 April 1942, AIR 23/1386, National Archives.

50. Douglas Evill to Arthur Tedder, 13 April 1942, AIR 23/1315, National Archives; Evill to Tedder, 18 April 1942, AIR 23/1315, National Archives; Arthur Tedder to Charles Portal, 6 May 1942, AIR 23/1315, National Archives.

51. Air Ministry to Headquarters Middle East, 12 May 1942, AIR 23/1359, National Archives.

52. US Army Air Forces Director of Intelligence, "Signal Intelligence Service of the German Luftwaffe," vol. 10, "Technical Operations in the South," 1945, 519.6314-10 V.9, Air Force Historical Research Agency, 24–28, 30.

53. John Slessor, "Anglo-American Visit, Draft Agenda and Report," 2 May 1942, AIR 75/10, National Archives. The first Arnold-Towers-Portal agreement was signed on 13 January 1942. The document listed by categories and by months the specific number of planes to be made available to Great Britain during 1942 from US production. The totals were 589 heavy-bombers, 1,744 medium-bombers, 2,745 light-bombers, 4,050 fighters, 402 observation planes, and 852 transports. See Wesley Frank Craven and James Lea Cate, *The Army Air Forces in World War II*, vol. 1 (Washington, DC: Office of Air Force History, 1983), 248–249. By this time, Slessor was the assistant chief of Air Staff for planning, or ACAS(P), and in this capacity

made two more lengthy visits to the United States to continue facilitating the transfer of US-built aircraft to the RAF. He addressed the Arnold-Towers-Portal meetings in *The Central Blue: The Autobiography of Sir John Slessor, Marshal of the RAF* (New York: Praeger, 1957), 404–417.

54. John Slessor, "Portal's Opening Notes to Arnold, Towers, and American Delegation," second Arnold-Towers-Portal conference, 17 May 1942, AIR 75/10, National Archives, 1–8.

55. Ibid.

56. John Slessor, "Opening Speech," Arnold-Towers-Portal meeting, June 1942, AIR 75/10, National Archives.

57. Charles Portal to Arthur Tedder, 26 June 1942, AIR 23/1397, National Archives, 2–4; Portal to Tedder, 20 June 1942, AIR 23/1359, National Archives; Douglas Evill to Arthur Tedder (r) Charles Portal, 22 June 1942, AIR 23/1359, National Archives.

58. Henry Arnold to John Slessor, 4 December 1942, AIR 75/10, National Archives.

59. BOH 3, 178–181.

60. Arthur Tedder to Charles Portal, 23 April 1942, AIR 23/1397, National Archives, 1–3.

61. Arthur Tedder to Charles Portal, 25 April 1942, AIR 23/1397, National Archives.

62. Tedder to Portal, 23 April 1942, AIR 23/1397, National Archives, 3.

63. Arthur Tedder to Charles Portal, 22 April 1942, AIR 23/1397, National Archives.

64. BOH 3, 182–187.

65. Norman MacMillan, *The Royal Air Force in the World War,* vol. 3 (London: Harrap, 1949), 169.

66. BOH 3, 187; Douglas-Hamilton provides a detailed account of this delivery on 62, 74–79.

67. Ibid., 188.

68. Ibid., 189–193.

69. Air Ministry to Wilfrid Freeman, 10 May 1942, AIR 23/1204, National Archives. The Axis lost 233 aircraft with another 78 probable and 330 damaged. The RAF lost 204 and had another 74 damaged.

70. Felmy, "The German Air Force in the Mediterranean Theater of War," 230–231.

71. Ibid., 232–235; Walter Warlimont, *Inside Hitler's Headquarters, 1939–45* (New York: Praeger, 1966), 235–237, 241.

72. Air Ministry, "Kesselring's Plan for the Invasion of Malta," trans. 7/141, A.H.B.6, May 1955, Air Historical Branch, 1–2; *DLM,* 358–359.

73. BOH 3, 195.

74. Air Ministry, "Kesselring's Plan for the Invasion of Malta," 2–4.

75. Ibid., 4–5.

76. Italian Air Staff, "The Air and Naval Bases on Malta, June 1940 to October 1942," trans. 7/43, A.H.B.6, 1 September 1947, Air Historical Branch, 2–4; Italian Air Ministry, "The Plan for the Invasion of Malta," trans. 7/43, A.H.B.6, 10 October 1947, Air Historical Branch, 1–4.

77. Italian Air Staff, "Memorandum on the Malta Situation, 10th May 1942," trans. 7/57, A.H.B.6, December 1947, Air Historical Branch, 1–4; *DLM,* 362–363.

78. Felmy, "The German Air Force in the Mediterranean Theater of War," 246–247.

79. Italian Air Staff, "Memorandum on the Malta Situation, 10th May 1942," 246–254.

80. BOH 3, 194–195; Felmy, "The German Air Force in the Mediterranean Theater of War," 255. Citino provides a succinct and informative discussion of the Malta agreement and Rommel's permission from Hitler to scrap it. He also reminds us that Rommel's impetus to advance and win a decisive victory was, like many things Rommel did, the result of the influence of "300 years of German military history." See Robert Citino, *Death of the Wehrmacht: The German Campaigns of 1942* (Lawrence: University Press of Kansas, 2007), 193–195, 205. See also the concurring opinion in Gerhard Schreiber, Bernd Stegemann, and Detlef Vogel, *Germany and the Second World War*, vol. 6, *The Global War*, trans. Dean S. McMurry, Ewald Ossers, and Louise Willmot (Oxford, UK: Clarendon, 1995), 664.

81. *Fuehrer Conferences on Naval Affairs, 1939–1945* (London: Chatham, 2005), 279.

82. BOH 3, 195. Orange argued that "Britain's army, navy, and RAF commanders in the Mediterranean could hardly have prevented a conquest of Malta had it been attempted with all the Axis power available before August 1942." Based on an assessment of the available evidence, his statement appears compelling. See Vincent Orange, "Getting Together: Tedder, Coningham, and Americans in the Desert and Tunisia, 1940–43," in *Airpower and Ground Armies: Essays on the Evolution of Anglo-American Air Doctrine, 1940–1943*, ed. Daniel R. Mortensen (Maxwell Air Force Base, AL: Air University Press, 1998), 18.

Chapter Eight. Axis High Tide

1. Gerhard Weinberg, *A World at Arms: A Global History of World War II* (Cambridge, UK: Cambridge University Press, 1994), 348.

2. H. R. Trevor-Roper, ed., *Blitzkrieg to Defeat: Hitler's War Directives, 1939–1945* (New York: Holt, Rinehart, and Winston, 1964), 117, 129–130.

3. Weinberg, *World at Arms*, 355–356.

4. Richard Overy, *Russia's War: A History of the Soviet War Effort, 1941–1945* (New York: Penguin, 1998), 193–198.

5. Weinberg, *World at Arms*, 357–360; Mark Stoler, *Allies and Adversaries: The Joint Chiefs of Staff, the Grand Alliance, and U.S. Strategy in World War II* (Chapel Hill: University of North Carolina Press, 2000), 73–90.

6. I. O. S. Playfair, *The Mediterranean and Middle East*, vol. 3, *British Fortunes Reach Their Lowest Ebb (September 1941 to September 1942)*, *History of the Second World War: United Kingdom Military Series* (Uckfield, UK: Naval and Military Press, 2009), 17, 201–202 (this and all subsequent volumes of the British official oistory are referred to as "BOH" with volume number).

7. Ibid., 205–206. In December 1941, the Air Ministry, in view of Arthur Tedder's stated requirements, had increased the target figure to eighty-five and one-half squadrons by August 1942. Modifications led to plans for larger squadrons (sixteen as opposed to twelve in most units) and a consequent decrease to eighty squadrons. Of these, three were heavy-bomber units and thirty-five short-range fighter squadrons, including fifteen of Spitfires. This did not come together until spring 1943.

However, Tedder had his heavy-bomber squadrons—one of Liberators and two of Halifaxes—by summer 1942.

8. Charles Portal to Defence Committee, memorandum, 1 April 1942, CAB 69/4 and AIR 8/989 (DO[42]34), National Archives. Of course, the RAF needed airfields only the army could occupy through offensive action, so the interdependence deepened.

9. Arthur Bryant, *Turn of the Tide: A History of the War Years Based on the Diaries of Field-Marshal Lord Alan Brooke, Chief of the Imperial General Staff* (New York: Doubleday, 1957), 386; Denis Richards, *Portal of Hungerford: The Life of Marshal of the Royal Air Force, Viscount Portal of Hungerford, KG, GCB, OM, DSO, MC* (London: Heinemann, 1977), 202–203.

10. "Air Support," AIR 10/5547, CAB 80/36, 31; "Defence Committee Meeting," 20 May 1942, COS(42)271, National Archives.

11. "Command and Planning," 21 May 1942, AIR 8/1063, JP(42)517; "Continental Operations 1943: Operational Organisation and System of Command of the RAF," 21 July 1942, WO 277/34, Appendix K, "Army Air Support."

12. Secretary of State for War to CIGS, memorandum, 14 September 1942, WO 216/217, National Archives.

13. COS(42)138 (o), 5 October 1942; Prime Minister to Secretaries of State for War and Air, 7 October 1942, PREM 3/8; "Air Support," AIR 10/5547, National Archives, 36, 59.

14. Maurice Dean, *Royal Air Force and Two World Wars* (Worthing, UK: Littlehampton, 1979), 215.

15. David Ian Hall, *Strategy for Victory: The Development of British Tactical Air Power, 1919–1943* (London: Praeger, 2008), 128.

16. Ibid., 128–129; Arthur Tedder, *With Prejudice: The World War II Memoirs of Marshal of the Royal Air Force Lord Tedder, Deputy Supreme Commander of the Allied Expeditionary Force* (Boston: Little, Brown, 1966), 211–212.

17. Hall, *Strategy for Victory*, 129–130; Tedder, *With Prejudice*, 163–164. See also Arthur Tedder, *Air Power in War* (London: Hodder and Stoughton, 1948).

18. James Corum, "The Luftwaffe's Army Support Doctrine, 1918–1941," *Journal of Military Affairs* 59 (January 1995): 53–76; Vincent Orange, "The Commanders and the Command System," in Royal Air Force Historical Society, *The End of the Beginning: Bracknell Paper No. 3, A Symposium on the Land/Air Co-operation in the Mediterranean War, 1940–43*, ed. Derek Wood (Bracknell, UK: Royal Air Force Historical Society), 20 March 1992, 35–36. See also Richard Muller, *The German Air War in Russia* (Baltimore, MD: Nautical and Aviation, 1992).

19. Various German senior officers, "High Level Reports and Directives Dealing with the North African Campaign, 1942," trans. 7/80, A.H.B.6, 30 October 1948, Air Historical Branch.

20. "War Diary of Panzer Army Africa, 24 April 1942 to 25 May 1942," trans. 7/110, A.H.B.6, April 1952, Air Historical Branch, 6–7. The German aircraft earmarked for the Malta invasion included one fighter *Geschwader* with four *Gruppen*, a dive-bomber *Geschwader* with three *Gruppen*, a twin-engine fighter *Gruppe*, and a bomber *Geschwader* from Fliegerkorps X.

21. Ibid., 1–7. This reinforcement included a *Gruppe* each of single-engine fighters, twin-engine fighters, and dive-bombers. The IAF contingent would consist of a Macchi 202 Stormo and two more of mixed single-engine fighters. Regarding Ju 88

deployments to Derna, see Karl Gundelach, *Die deutsche Luftwaffe im Mittelmeer*, vol. 1 (Frankfurt am Main: Lang, 1981), 368 (referred to from this point forward as *DLM*).

22. "War Diary of Panzer Army Africa, 24 April 1942 to 25 May 1942," 7–8; *DLM*, 368.

23. Arthur Tedder, *Report by Air Marshal Tedder on His Tenure in the Post of Air Officer Commanding-in-Chief* (London: MLRS, 2010), 96; Hall, *Strategy for Victory*, 133.

24. Tedder, *Report*, 96.

25. Arthur Tedder to Charles Portal, 27 May 1942, AIR 23/1397, National Archives.

26. BOH 3, 229–230.

27. Arthur Tedder to Charles Portal, 1 June 1942, AIR 23/1397, National Archives.

28. Hellmuth Felmy, "The German Air Force in the Mediterranean Theater of War," K113.107–161, Air Force Historical Research Agency, 257–259.

29. Ibid., 260–262. Rommel still did not recognize that air attacks on tanks were largely ineffective.

30. Ibid., 264–266.

31. Norman MacMillan, *The Royal Air Force in the World War*, vol. 3 (London: Harrap, 1949), 151–152.

32. Tedder, *Report*, 101–103.

33. Felmy, "The German Air Force in the Mediterranean Theater of War," 271; *DLM*, 369.

34. Felmy, "The German Air Force in the Mediterranean Theater of War," 272–273.

35. Ibid., 274–275. Predictably, Rommel's view of the Bir Hacheim assault is different. See Erwin Rommel and Fritz Bayerlein, *The Rommel Papers*, ed. B. H. Liddell Hart (London: Collins, 1953), 214, 218. The 368 aircraft that made the final raids on Bir Hacheim represented nearly all remaining air strength.

36. H. W. Wynter, *The History of the Long Range Desert Group, June 1940 to March 1943* (Kew, UK: National Archives, 2008), 152–153, 160–162, 171, 303–304, 320–321, 328.

37. BOH 3, 223–230n1.

38. Ibid., 233–235; Rommel and Bayerlein, *Rommel Papers*, 212, 222; *DLM*, 366, 373.

39. BOH 3, 254–256; *DLM*, 370–371.

40. Rommel and Bayerlein, *Rommel Papers*, 227.

41. BOH 3, 257–259.

42. Felmy, "The German Air Force in the Mediterranean Theater of War," 285–286; *DLM*, 372–373.

43. Felmy, "The German Air Force in the Mediterranean Theater of War," 292–296.

44. Ibid., 278–280.

45. Arthur Tedder, "Note on Air Operations in Support of Crusader, January–December 1942," Tedder Box 6, Air Historical Branch, 130–131.

46. Ibid., 133–135; Tedder, *With Prejudice*, 290–291.

47. Tedder, *With Prejudice*, 298–300; *DLM*, 377–378.

48. *DLM*, 379. Gundelach said an attack on Malta was of "life-or-death" importance to the campaign.

49. Tedder, *With Prejudice*, 308–309.

50. Cajus Bekker, *The Luftwaffe War Diaries: The German Air Force in World War II* (New York: Da Capo, 1994), 251.

51. Arthur Tedder to Charles Portal, 29 June 1942, AIR 23/1397, National Archives, 4.

52. Felmy, "The German Air Force in the Mediterranean Theater of War," 281–283.

53. Ibid., 298–302.

54. BOH 3, 311–313; *DLM*, 374–375.

55. Arthur Tedder to Charles Portal, 16 June 1942, in Tedder, "Note on Air Operations in Support of Crusader," 127–129; Tedder to Portal, 16 June 1942, AIR 23/1397, National Archives.

56. Tedder, *Report*, 121–123; *DLM*, 410.

57. Felmy, "The German Air Force in the Mediterranean Theater of War," 296–298.

58. Arthur Tedder to Charles Portal, 29 June 1942, AIR 23/1397, National Archives, 1–2.

59. Tedder, *Report*, 108–111.

60. Tedder to Portal, 29 June 1942, 1–2; Orange, "Commanders and the Command System," 40.

61. Arthur Tedder to Charles Portal, 21 June 1942, AIR 23/1397, National Archives.

62. Tedder to Portal, 29 June 1942, 3–4.

63. Advanced Air Headquarters, Western Desert, "Report on Operations during the Withdrawal from Cyrenaica," pt. 2, 15 September 1942, AIR 23/6480, National Archives, 1–10.

64. Ibid., 4–10, pt. 3.

65. BOH 3, 301, 306, 309–310, 314–317; Tedder, *Report*, 110–111.

66. Army Battle Headquarters to Oberkommando der Wehrmacht Operational Staff, memorandum, 1 August 1942, in "War Diary of Panzer Army Africa," AIR 20/7706, National Archives; Army Battle Headquarters to German General at Headquarters Italian Armed Forces Rome, memorandum, 2 August 1942, in "War Diary of Panzer Army Africa," AIR 20/7706, National Archives; Army Battle Headquarters to Army General, Staff Operations Department, memorandum, 8 August 1942, in "War Diary of Panzer Army Africa," AIR 20/7706, National Archives.

67. War Office Military Intelligence Service, "Notes and Lessons on Operations in the Middle East," 30 January 1943, WO 106/2270, National Archives, 14.

68. "War Diary of the German Africa Corps, June 1942," trans. 7/88, A.H.B.6, 19 July 1949, Air Historical Branch, 21–23, 43–45, 47, 52; *DLM*, 384–385.

69. "War Diary of the German Africa Corps, June 1942," 52.

70. Various German senior officers, "High Level Reports and Directives Dealing with the North African Campaign, 1942," 8–11.

71. Felmy, "The German Air Force in the Mediterranean Theater of War," 298–300.

72. Ibid., 340.

73. Ibid., 340–341, 345; Kurt Student, *Der Deutsche Fallschirmjaeger*, no. 11, 1956, 3–4.

74. Felmy, "The German Air Force in the Mediterranean Theater of War," 334, 340–341.

75. Ibid., 346–350.

76. Rommel and Bayerlein, *Rommel Papers*, 193.

77. Tedder, *Report*, 301–305; *DLM*, 388.

78. Tedder, *Report*, 305–306, 310–311.

79. BOH 3, 281–283.

80. Rommel and Bayerlein, *Rommel Papers*, 237.

81. BOH 3, 289–290, 295–297. For Rommel's references to RAF air attacks, see Rommel and Bayerlein, *Rommel Papers*, 236. See *DLM*, 390, for air strengths.

82. Tedder, *Report*, 124.

83. BOH 3, 315–317, 320–324. From Italy's entry into the war until January 1941, 21 MVs delivered 160,000 tons of cargo to Malta without loss. After this, the cost was high. From the Luftwaffe's arrival in January 1941 until it ceased to be a serious threat to convoys in August 1942, 46 merchant ships managed to discharge 320,000 tons of cargo at Malta. Of the 82 setting out, 23 were sunk, 10 turned back, and 3 had unproductive cargoes as a result of flooding. However, the 46 productive arrivals kept the island going. Arrivals increased as the fighter garrison grew. From August 1940 to September 1942, 670 Hurricanes and Spitfires were successfully flown from carriers to Malta in 19 operations—a huge but wise commitment.

84. Ibid., 327–329.

85. "War Diary of the German Africa Corps, July 1942," 1–16.

86. Ibid., 26, 30–35.

87. "War Diary of Panzer Army Africa, 28 July–23 October 1942," trans. 7/105, A.H.B.6, 12 May 1951, Air Historical Branch, 1–2. One consumption unit (*Versorgungzsetze*) gave an army unit enough fuel to move 100 kilometers (62 miles) over roads or firm and relatively flat terrain. See *DLM*, 396, for Fliegerführer Afrika aircraft losses.

88. Rommel and Bayerlein, *Rommel Papers*, 305.

89. Ibid., 245.

Chapter Nine. The Tide Turns

1. Mark Stoler, *Allies and Adversaries: The Joint Chiefs of Staff, the Grand Alliance, and U.S. Strategy in World War II* (Chapel Hill: University of North Carolina Press, 2000), 90–97.

2. Gerhard Weinberg, *A World at Arms: A Global History of World War II* (Cambridge, UK: Cambridge University Press, 1994), 353–355.

3. Arthur Coningham to Arthur Tedder, 31 May 1942, AIR 23/1391, National Archives.

4. Douglas Evill to Arthur Tedder, 3 June 1942, AIR 23/1315, National Archives; Arthur Tedder to Douglas Evill, 12 June 1942, AIR 23/1315, National Archives.

5. Air Ministry to Arthur Tedder, 24 June 1942, AIR 23/1359, National Archives; Douglas Evill to Arthur Tedder (r) Charles Portal, 25 June 1942, AIR 23/1359, National Archives; Headquarters US Middle East Air Force, General Order No. 1, 28 June 1942, AIR 23/1359, National Archives.

6. Roderic Owen, *Tedder* (London: Collins, 1952), 166.

7. Ibid., 190.

8. Air Ministry, "The Middle East Campaigns," vol. 4, "Operations in Libya

and the Western Desert: July 1942–May 1943," 512.041-12, Air Force Historical Research Association, 185, 198; Alan J. Levine, *The War against Rommel's Supply Lines, 1942–43* (Mechanicsburg, PA: Stackpole, 2008), 27; Wesley Frank Craven and James Lea Cate, *The Army Air Forces in World War II,* vol. 2 (Washington, DC: Office of Air Force History, 1983), 25; Lewis H. Brereton, *The Brereton Diaries* (New York: Morrow, 1946), 142.

9. "Direct Support in the Libyan Desert," 533.04, Air Force Historical Research Association; Lewis Brereton to Henry Arnold, 22 August 1943, in "Ninth U.S. Air Force, Middle East," 533.04, Air Force Historical Research Association; "Interview with Major P. R. Chandler" by Assistant Chief of Air Staff, Intelligence, 17 June 1943, 142.052, Air Force Historical Research Association.

10. Christopher M. Rein, *The North African Air Campaign: U.S. Army Air Forces from El Alamein to Salerno* (Lawrence: University Press of Kansas, 2012), 62; quoted in Karl Gundelach, *Die deutsche Luftwaffe im Mittelmeer,* vol. 1 (Frankfurt am Main: Lang, 1981), 442 (referred to from this point forward as *DLM*).

11. Douglas Evill to Arthur Tedder, 4 November 1942, AIR 23/1359, National Archives; Arthur Tedder to Douglas Evill, 13 November 1942, AIR 23/1359, National Archives; Graham Dawson to Arthur Tedder, 21 November 1942, AIR 23/1359, National Archives.

12. Douglas Evill to Arthur Tedder, 13 November 1942, AIR 23/1359, National Archives; Arthur Tedder to Charles Portal, 14 November 1942, AIR 23/1359, National Archives.

13. William Foster to Arthur Tedder, 24 December 1942, AIR 23/1315, National Archives. This agreement included a "qualified promise" of 600 C-47 transports, 600 P-40s, 600 Baltimores, 398 B-24s, 200 B-25s, and 276 B-26s, along with spare parts and manuals required both for new arrivals and aircraft already in theater. The majority of these aircraft went to the Mediterranean. One of the astounding things about RAFME is how heterogeneous it was and yet how cohesive. Of ninety-six total squadrons, sixty were British, thirteen American, thirteen South African, five Australian, one Rhodesian, two Greek, one Free French, and one Yugoslav. Many of the aircrews in the British squadrons were from the dominions. The entire force totaled more than 1,500 first-line aircraft. Ironically, the Axis had more than 3,000 aircraft in the Mediterranean at this time, but only 689 in Africa and most of the rest on Sicily attacking Malta or on Crete doing far less than was possible and necessary. See Denis Richards and Hilary St. George Saunders, *Royal Air Force, 1939–1945,* vol. 2, *The Fight Avails* (London: Her Majesty's Stationery Office, 1954), 233.

14. Charles Portal, "Appreciation on Air Forces Required in the Eastern Mediterranean in Order to Free Naval Units," 10 December 1941, AIR 23/1204, National Archives.

15. Wilfrid Freeman to Arthur Tedder, 25 December 1941, AIR 23/1457, National Archives.

16. Headquarters Royal Air Force Middle East, "Discussion on 27th December 1941 on Employment of Air Forces in Control of Sea Communications—Eastern Mediterranean," 7 January 1942, AIR 23/1457, National Archives, 1–6.

17. Assistant Senior Air Staff Officer to Arthur Tedder (forwarded to Charles Portal), "Control of Sea Communications in the Central Mediterranean," 4 February 1942, AIR 23/1457, National Archives, 1–2.

18. Ibid.

19. Leonard Slatter to Arthur Tedder, "Operation of Sea Reconnaissance and Striking Forces," 22 February 1942, AIR 23/1457, National Archives.

20. Leonard Slatter to Arthur Tedder, "Air Blockade of the Central Mediterranean," 6 April 1942, AIR 23/1457, National Archives.

21. Leonard Slatter to Arthur Tedder, 24 June 1942, AIR 23/1457, National Archives. Emphasis in original.

22. Slatter to Tedder, 24 June 1942, AIR 23/1457, National Archives, emphasis in original; *DLM,* 411.

23. Headquarters Royal Air Force Middle East (HQ RAFME) Tactics Assessment Office, "Air Tactics and Operational Notes on 201 Naval Co-operation Group, R.A.F.," n.d. (probably June 1943), AIR 23/1282, National Archives, Chapter 34.

24. Ibid.

25. *DLM,* 399–401, 405, 407–409.

26. HQ RAFME Tactics Assessment Office, "Air Tactics and Operational Notes on 201 Naval Co-operation Group, R.A.F.," Appendix 12: Enemy Merchant Shipping Losses; Admiralty Trade Division, "Summary of Enemy Merchant Shipping Losses," 24 September 1945, AIR 23/1175, National Archives.

27. Arthur Tedder, *Report by Air Marshal Tedder on His Tenure in the Post of Air Officer Commanding-in-Chief* (London: MLRS, 2010), 115–116; Norman MacMillan, *The Royal Air Force in the World War,* vol. 3 (London: Harrap, 1949), 160.

28. Arthur Tedder to Charles Portal, 12 July 1942, AIR 23/1398, National Archives, 3–4, 7–8.

29. Ibid., Appendix A.

30. Tedder, *Report,* 119.

31. Erwin Rommel and Fritz Bayerlein, *The Rommel Papers,* ed. B. H. Liddell Hart (London: Collins, 1953), 255, 257.

32. Ibid., 256, 259.

33. Military Intelligence Service, War Office, "Notes and Lessons on Operations in the Middle East," 30 January 1943, WO 106/2270, National Archives, 14; Tedder, *Report,* 119–120; Rommel and Bayerlein, *Rommel Papers,* 263, 265.

34. Arthur Tedder, *With Prejudice: The World War II Memoirs of Marshal of the Royal Air Force Lord Tedder, Deputy Supreme Commander of the Allied Expeditionary Force* (Boston: Little, Brown, 1966), 317, 341, 348, 354–355; Arthur Tedder to Charles Portal, 22 October 1942, AIR 23/1398, National Archives.

35. Tedder, *With Prejudice,* 318–327, 347; Vincent P. Orange, *Coningham: A Biography of Air Marshal Sir Arthur Coningham* (Washington, DC: Center for Air Force History, 1992), 106–107; Corelli Barnett, *The Desert Generals* (Edison, NJ: Castle, 2001), 231–239; Bernard Law Montgomery, *The Memoirs of Field Marshal the Viscount Montgomery of Alamein, K.G.* (London: World, 1958), 102. Cross relates that he watched Churchill stride past all the other dignitaries arrayed for his arrival, going straight to Tedder and shaking his hand. Kenneth Cross with Vincent Orange, *Straight and Level* (London: Bugg Street, 1993), 218.

36. Tedder, *With Prejudice,* 347–348; Frederick Rosier, quoted in "Digest of the Group Discussions," in Royal Air Force Historical Society, *The End of the Beginning: Bracknell Paper No. 3, A Symposium on the Land/Air Co-operation in the Mediterranean War, 1940–43,* ed. Derek Wood (Bracknell, UK: Royal Air Force Historical Society), 20 March 1992, 65–66.

37. Arthur Tedder to Charles Portal, 7 September 1942, in Tedder, "Note on Air

Operations in Support of Crusader, January–December 1942," Tedder Box 6, Air Historical Branch, 206.

38. Ibid.

39. Ibid., 208.

40. Tedder, *With Prejudice,* 356–361. Wynn highlights the intensity of RAF operations: one Kittyhawk pilot commented that "at four shows a day, our ground crews have to work damned hard." Humphrey Wynn, *Desert Eagles* (Shrewsbury, UK: Airlife, 1993), 63.

41. Arthur Coningham and Bernard Law Montgomery, "Air Power in the Land Battle," published January 1943 but delivered verbally before El Alamein, AIR 23/1299, National Archives.

42. Ibid.

43. Nigel Hamilton, *Master of the Battlefield: Monty's War Years, 1942–1944* (New York: McGraw-Hill, 1983), 12; Montgomery, *Memoirs,* 137–138; "Evaluation of the British and American Commands and Troops in North Africa," 1943, CAB 146/27, National Archives.

44. Ralph Bennett, *Ultra and Mediterranean Strategy* (New York: Morrow, 1989), 126, 142, 157; Aileen Clayton, *The Enemy Is Listening: The Story of the Y Service* (London: Crécy, 1993), 151, 200; Military "Y" Middle East, GSI (s), 2 March 1941, WO 208/5021, National Archives; Mediterranean Air "Y," 18 September 1942, AIR 40/2252, National Archives, 2; No. 285 Reconnaissance Wing, "Report on Organisation and Operational Methods from Formation until the Conclusion of the European War," 1945, AIR 23/6472, National Archives; "Army Air Support and Photographic Interpretation, 1939–1945," WO 277/34, National Archives, 4.

45. General Headquarters Middle East Forces and Headquarters Royal Air Force Middle East, "Middle East Training Pamphlet (Army and R.A.F.) No. 3A—Direct Air Support," March 1942, AIR 23/1763, National Archives, 1–2.

46. Ibid., 1–5.

47. Commander, No. 1 Air Support Control, "Report on Air Support Control—R.A.F. Signals Organisation," 23 July 1942, AIR 23/1764, National Archives, 1–3, 7–10.

48. Headquarters Royal Air Force Middle East to Advanced Air Headquarters Western Desert, "Air Support," 10 August 1942, AIR 23/1764, National Archives; "Selection of Air Liaison Officers," 3 August 1942, AIR 23/1764, National Archives.

49. M. T. Judd to OC, 239 Wing, "Report on Possible Tactical Employment of Fighter Bomber," 25 August 1942, AIR 23/1764, National Archives, 1–4.

50. "War Diary of No. 2 Army Air Support Control," 28 August 1942, WO 169/6638, National Archives; "Ground to Air Recognition Signals," 23 August 1942, WO 169/6638, National Archives; "War Diary of No. 5 Army Air Support Control," 21 July 1942, WO 169/6640, National Archives.

51. Headquarters Royal Air Force Middle East, "Suggestions for Direct Air Support by Night, Air Tactics," February 1942, AIR 39/141, National Archives; Panzer Army Headquarters 1a to Panzer Army Africa, "Memorandum on RAF Operations," 8 September, in "War Diary of Panzer Army Africa, 28 July–23 October 1942," AIR 20/7706, National Archives; Advanced Air Headquarters, "Western Desert Operation Instruction No. 5," 21 October 1942, AIR 41/50, National Archives, Appendix 10.

52. "War Diary of Panzer Army Africa, 28 July–23 October," 7–11.

53. Ibid., 12–14.

54. Ibid., 34, 38–39, 41–42. The German shipping backlog included 4,000 men, 1,250 trucks, more than 100 tanks, 230 half-tracks, and 101 guns. For Allied shipping tonnage, see John North, ed., *The Memoirs of Field-Marshal Earl Alexander of Tunis, 1940–1945* (London: Cassell, 1962), 18.

55. "War Diary of Panzer Army Africa, 28 July–23 October," 41–42.

56. Ibid., 54, 59, 66–68.

57. F. H. Hinsley et al., *British Intelligence in the Second World War,* vol. 2 (London: Her Majesty's Stationery Office, 1979), 412; "War Diary of the German Africa Corps, 3rd August–22nd November 1942," trans. 7/101, A.H.B.6, November 1950, Air Historical Branch, 14, 19.

58. Hans von Luck, *Panzer Commander: The Memoirs of Colonel Hans von Luck* (New York: Praeger, 1989), 97; *DLM*, 419–420.

59. Rommel and Bayerlein, *Rommel Papers,* 276.

60. "War Diary of the German Africa Corps, 3rd August–22nd November 1942," 24–28, 39; Various German senior officers, "High Level Reports and Directives Dealing with the North African Campaign, 1942," trans. 7/80, A.H.B.6, 30 October 1948, Air Historical Branch, 69.

61. "War Diary of the German Africa Corps, 3rd August–22nd November 1942," 24–28, 39; Various German Senior Officers, "High Level Reports and Directives Dealing with the North African Campaign, 1942," 69–70; *DLM*, 417, 420–421.

62. Rommel and Bayerlein, *Rommel Papers,* 279.

63. Ibid. Rommel referred to the "party rally" formations as an indication that the formation flying was as precise as that of Luftwaffe squadrons flying over the Nuremberg party rallies. On 3 September, these formations of eighteen to twenty-four bombers attacked his troops twelve times, while fighter-bombers engaged them incessantly.

64. "War Diary of the German Africa Corps, 3rd August–22nd November 1942," 24–28, 39; Various German senior officers, "High Level Reports and Directives Dealing with the North African Campaign, 1942," 69–70.

65. Various German senior officers, "High Level Reports and Directives Dealing with the North African Campaign, 1942," 71–72.

66. Ibid., 72–73.

67. Rommel and Bayerlein, *Rommel Papers,* 284.

68. Ibid.

69. Ibid.

70. Various German senior officers, "High Level Reports and Directives Dealing with the North African Campaign, 1942," 87–89.

71. Ibid., 90–91.

72. Rommel and Bayerlein, *Rommel Papers,* 288–289.

73. Ibid., 290, 294.

74. Brad Gladman, *Intelligence and Anglo-American Air Support in World War Two: The Western Desert and Tunisia, 1940–43* (New York: Palgrave MacMillan, 2009), 111; Headquarters Royal Air Force Middle East, "Fourth Conference of Air Officers Commanding," 18 September 1942, AIR 23/1292, National Archives; Roderic Owen, *The Desert Air Force: An Authoritative History Published in Aid of the Royal Air Force Benevolent Fund* (London: Hutchinson, 1948), 78.

75. Various German senior officers, "High Level Reports and Directives Dealing with the North African Campaign, 1942," 33–34.

76. Tedder, *Report,* 128–129; Arthur Tedder to Charles Portal, 22 October 1942, in Tedder, "Note on Air Operations in Support of Crusader, January–December 1942," Tedder Box 6, Air Historical Branch, 229–230.

77. Tedder, *Report,* 130–132. The Axis lost 118 aircraft, 37 probable, and had 110 damaged compared with 19 Spitfires for the British. Regarding Axis aircraft requirements for this effort, see Richards and Saunders, *Royal Air Force,* vol. 2, 232.

78. Ibid., 132–133.

Chapter Ten. El Alamein and Operation Torch

1. I. S. O. Playfair, *The Mediterranean and Middle East,* vol. 4, *The Destruction of the Axis Forces in Africa, History of the Second World War: United Kingdom Military Series* (Uckfield, UK: Naval and Military Press, 2004), 13–14 (this and all subsequent volumes of the British official history are referred to as "BOH" with volume number). It is important to remember that Bomber Command was going through a period of crisis and growth at this point, while the Eighth Air Force was beginning its buildup. The imperative for heavy-bomber units in the United Kingdom posed a constant problem for their provision to RAFME and to the USAAF Twelfth Air Force after Operation Torch.

2. Ibid., 18–20; Kenneth Cross with Vincent Orange, *Straight and Level* (London: Bugg Street, 1993), 218.

3. BOH 4, 32–33.

4. Ibid., 33–34.

5. Ibid., 36, 44.

6. "War Diary of the German Africa Corps, 3rd August–22nd November 1942," trans. 7/101, A.H.B.6, November 1950, Air Historical Branch, 44–49.

7. Hans von Luck, *Panzer Commander: The Memoirs of Colonel Hans von Luck* (New York: Praeger, 1989), 104.

8. Arthur Tedder, *Report by Air Marshal Tedder on His Tenure in the Post of Air Officer Commanding-in-Chief* (London: MLRS, 2010), 134–136.

9. Fifteenth Panzer Division, "Report on the Battle of Alamein and the Retreat to Marsa El Brega," AIR 20/7706, National Archives, Appendix to the Report for the Period 23 October–20 November 1942 (see esp. 2 November 1942); Twenty-first Panzer Division, "Report on the Battle of Alamein and the Retreat to Marsa El Brega, 2 November 1942," AIR 20/7706, National Archives; Erwin Rommel and Fritz Bayerlein, *The Rommel Papers,* ed. B. H. Liddell Hart (London: Collins, 1953), 298, 305.

10. Twenty-first Panzer Division, "Report on the Battle of Alamein and the Retreat to Marsa El Brega"; "The Middle East Campaigns," vol. 4, "Operations in Libya and the Western Desert: July 1942–May 1943," 512.041-12, Air Force Historical Research Association, 280, 320; "RAF Bomb Weights Dropped by Desert Air Force, 3 November 1942," AIR 41/50, National Archives; Rommel and Bayerlein, *Rommel Papers,* 307.

11. Rommel and Bayerlein, *Rommel Papers,* 311.

12. Ibid., 315.

13. BOH 4, 59, 63–64, 68.

14. Rommel and Bayerlein, *Rommel Papers*, 317.

15. BOH 4, 72–75.

16. Ibid., 83–88.

17. Arthur Tedder to Charles Portal, 5 November 1942, AIR 23/1398, National Archives, 1–2.

18. Ibid. Barnett argued that Bernard Law Montgomery's slow pursuit robbed the Eighth Army and the WDAF of a knockout blow. He also credited Claude Auchinleck with stopping Erwin Rommel at Alam Halfa ridge, giving Harold Alexander and Montgomery time to prepare and execute their counterattack. Corelli Barnett, *The Desert Generals* (Edison, NJ: Castle, 2001), 9.

19. BOH 4, 88–89; Arthur Tedder to Charles Portal, 22 October 1942, AIR 23/1398, National Archives. The Luftwaffe managed 242 sorties on its peak day during El Alamein. The average daily sortie count from 14 October through 2 November was 162, about 13.5 percent of the 1,200 British sorties. German figures came from Hans Seidemann. See Rommel and Bayerlein, *Rommel Papers*, 335.

20. BOH 4, 88–89; Von Luck, *Panzer Commander*, 117–119.

21. BOH 4, 367.

22. Ibid., 369–370; "Interview with General Ritter von Thoma" by F/Lt. Henry of Royal Air Force Section, Combined Services Detailed Interrogation Center, November 1942, in Arthur Tedder, "Note on Air Operations in Support of Crusader, January–December 1942," Tedder Box 6, Air Historical Branch, 252.

23. Robert Citino, *Death of the Wehrmacht: The German Campaigns of 1942* (Lawrence: University Press of Kansas, 2007), 222.

24. Advanced Air Headquarters, "Western Desert Operation Buster," 11 October 1942, AIR 23/1776, National Archives; "The Middle East Campaigns," vol. 4, AIR 41/50, National Archives, 361, 365; Vincent P. Orange, *Coningham: A Biography of Air Marshal Sir Arthur Coningham* (Washington, DC: Center for Air Force History, 1992), 114.

25. "Operation Record Book for No. 205 Group," 4–5 November and 5–6 November 1942, AIR 25/816, National Archives; Rommel and Bayerlein, *Rommel Papers*, 339, 345; James J. Sadkovich, *The Italian Navy in World War II* (Westport, CT: Greenwood Press, 1994), 344; "Account of Operations of Eighth Army Formations under Comd First Army in the Final Phase of the North Africa Campaign," Appendix 1: Axis Losses in the African Campaigns, CAB 106/572, National Archives.

26. Humphrey Wynn, *Desert Eagles* (Shrewsbury, UK: Airlife, 1993), 105–106.

27. Arthur Tedder to Charles Portal, 5 November 1942; Tedder to Portal, 14 November 1942, in Tedder, "Note on Air Operations in Support of Crusader, January–December 1942," Tedder Box 6, Air Historical Branch, 236–238, 241.

28. Robin Higham, *Bases of Air Strategy: Building Airfields for the RAF, 1914–1945* (London: Airlife, 1998), 162–164.

29. Ibid., 101–106.

30. Tedder, *Report*, 136–139.

31. BOH 4, 111–112.

32. Arthur Tedder, *With Prejudice: The World War II Memoirs of Marshal of the Royal Air Force Lord Tedder, Deputy Supreme Commander of the Allied Expeditionary Force* (Boston: Little, Brown, 1966), 392; Tedder, "Air Command in the Mediterranean," December 1945, Tedder Box 8, Air Historical Branch, 29, 29.

33. "Minutes of the 58th Meeting, 16 January 1943, Combined Chiefs of Staff," Folder 3, Box 169, RG 218, US National Archives and Records Administration; for

German casualties, see Percy Schramm, *Kriegstagebuch des Oberkommandos der Wehrmacht: 1940–1945,* 8 vols. (Bonn: Bernard and Graefe, 1961), 1508–1511.

34. Hellmuth Felmy, "The German Air Force in the Mediterranean Theater of War," K113.107–161, Air Force Historical Research Agency, 544–545.

35. Albert Kesselring, "The War in the Mediterranean," pt. 2, trans. Air Ministry A.H.B.6, September 1951, Air Historical Branch, 3–7.

36. Ibid., 9–11.

37. Daniel Mortensen, "The Legend of Laurence Kuter: Agent for Airpower Doctrine," in *Airpower and Ground Armies: Essays on the Evolution of Anglo-American Air Doctrine, 1940–1943,* ed. Daniel Mortensen (Maxwell Air Force Base, AL: Air University Press, 1998), 97.

38. BOH 4, 113–114; Richard G. Davis, *Carl A. Spaatz and the Air War in Europe* (Washington, DC: Smithsonian Institution Press, 1992), 145; Mortensen, *Airpower and Ground Armies,* xv. In 1982, William Momyer, commander of the USAAF Thirty-third Fighter Group during Operation Torch, said Coningham was the man "who brought the thing together" in terms of operational airpower effectiveness. See Richard H. Kohn and Joseph P. Harahan, eds., *Air Superiority in World War II and Korea: An Interview with Gen. James Ferguson, Gen. Robert M. Lee, Gen. William Momyer, and Lt. Gen. Elwood R. Quesada* (Washington, DC: Office of Air Force History, 1983), 30–31.

39. Vincent Orange, "Getting Together: Tedder, Coningham, and Americans in the Desert and Tunisia, 1940–43," in *Airpower and Ground Armies: Essays on the Evolution of Anglo-American Air Doctrine, 1940–1943,* ed. Daniel R. Mortensen (Maxwell Air Force Base, AL: Air University Press, 1998), 11.

40. BOH 4, 128–130.

41. Ibid., 131–132.

42. Ibid., 136.

43. Ibid., 116–120.

44. Felmy, "The German Air Force in the Mediterranean Theater of War," 547–550; Walter Warlimont, *Inside Hitler's Headquarters, 1939–45* (New York: Praeger, 1966), 270–272.

45. Felmy, "The German Air Force in the Mediterranean Theater of War," 560–573; James Doolittle to Henry Arnold, 19 November 1942, Henry Harley Arnold Papers, Box 13, Library of Congress. One of the most succinct discussions of airborne operations during Operation Torch is in Monro MacCloskey, *Torch and the Twelfth Air Force* (New York: Richards Rosen, 1971), 57–58, 68, 135–139.

46. Felmy, "The German Air Force in the Mediterranean Theater of War," 573–581.

47. Albert Kesselring, "The War in the Mediterranean," pt. 2, written for the US Historical Division, May 1948, trans. 7/106, A.H.B.6, September 1951, Air Historical Branch, 8.

48. Abteilung 8, "German Air Force Activities in the Mediterranean Tactics and Lessons Learned, 1941–1943," trans. 7/2, A.H.B.6, 8 October 1946, Air Historical Branch, 6; MacCloskey, *Torch and the Twelfth Air Force,* 104–105, 132; Karl Gundelach, *Die deutsche Luftwaffe im Mittelmeer,* vol. 1 (Frankfurt am Main: Lang, 1981), 453–454, 460–461 (referred to from this point forward as *DLM*). See *DLM,* 407–408, for KG 26's earlier exploits.

49. Felmy, "The German Air Force in the Mediterranean Theater of War," 582–583.

50. Ibid., 583–584; David R. Mets, "A Glider in the Propwash of the Royal Air Force? Gen Carl A. Spaatz, the RAF, and the Foundations of American Tactical Air Doctrine." In *Airpower and Ground Armies: Essays on the Evolution of Anglo-American Air Doctrine, 1940–1943,* ed. Daniel R. Mortensen (Maxwell Air Force Base, AL: Air University Press, 1998), 48; Daniel R. Mortensen, *A Pattern for Joint Operations: World War II Close Air Support North Africa* (Washington, DC: Office of Air Force History, 1989); H. H. Arnold, *Global Mission* (New York: Harper and Brothers, 1949), 379. For an indication of the frequency with which Allied forces endured Luftwaffe raids until March 1943, see David Rolf, *The Bloody Road to Tunis: Destruction of the Axis Forces in North Africa, November 1942–May 1943* (London: Greenhill, 2001).

51. Christopher M. Rein, *The North African Air Campaign: U.S. Army Air Forces from El Alamein to Salerno* (Lawrence: University Press of Kansas, 2012), 6, 20; Richard J. Overy, *The Air War, 1939–1945* (New York: Stein and Day, 1980); Mortensen, "The Legend of Laurence Kuter," 99.

52. US War Department, *Field Manual 31-35: Aviation in Support of Ground Forces* (Washington, DC: Government Printing Office, 1942), para. 1–6.

53. Ibid.

54. Rein, *The North African Air Campaign,* 22; Mets, "A Glider in the Propwash of the Royal Air Force?" 36, 52–53, 75. Mets makes a very strong case that US airmen had many of the same insights and qualities as their Commonwealth counterparts but had to build the experience and de facto equality with ground commanders before they could maximize the effectiveness of their air assets.

55. Rein, *The North African Air Campaign,* 24–25; Lee Kennett, "Developments to 1939," in *Case Studies in the Development of Close Air Support,* ed. Benjamin Franklin Cooling (Washington, DC: Office of Air Force History, 1990), 157.

56. David Syrett, "The Tunisian Campaign, 1942–1943," in *Case Studies in the Development of Close Air Support,* ed. Benjamin Franklin Cooling (Washington, DC: Office of Air Force History, 1990), 169.

57. Syrett, "The Tunisian Campaign," 169; Mets, "A Glider in the Propwash of the RAF?" 79–80.

58. Carl Spaatz to Henry Arnold, 7 March 1943, 168.49-1, Air Force Historical Research Agency, Frame 968.

59. Rein, *The North African Air Campaign,* 35–36.

60. B. Michael Bechtold, "A Question of Success: Tactical Air Doctrine and Practice in North Africa, 1942–1943," *Journal of Military History* 68, no. 3 (July 2004): 824, 841, 846.

61. Ibid.

62. Rein, *The North African Air Campaign,* 76.

63. Davis, *Carl A. Spaatz,* 124; Ira C. Eaker, diary, 4 August 1942, MS10, US Air Force Academy.

64. "History, 51st Troop Carrier Wing," Henry Harley Arnold Papers, Library of Congress.

65. Rein, *The North African Air Campaign,* 91–92, 95; Arnold, *Global Mission,* 352; Carl Spaatz to Henry Arnold, 23 November 1942, MS10, US Air Force Academy.

66. Rein, *The North African Air Campaign,* 97, 124–126, 132–134; Douglas Porch, *The Path to Victory: The Mediterranean Theater in World War II* (New York: Farrar, Straus, and Giroux, 2004), 412; MacCloskey, *Torch and the Twelfth Air Force,* 145–149.

67. "Army Cooperation Plan, Operation Torch," 5 October 1942, AIR 23/6560, National Archives; Plans—Op "Torch" (Eastern Air Command and Air Force Headquarters), September–October 1942, Army Cooperation Plan "Torch," AIR 23/6575, National Archives; George F. Howe, *Northwest Africa: Seizing the Initiative in the West,* vol. 2, pt. 1 of *The United States Army in World War II* (Washington, DC: Department of the Army, 2002), 40–43; Dwight D. Eisenhower, *Crusade in Europe* (Garden City, NY: Doubleday, 1948), 91.

68. "Axis Air Operations North Africa and Mediterranean: The Last Phase in North Africa, 1st January 1943–12th May 1943," AIR 40/2358, National Archives.

69. William L. Welsh, "Operation Torch (Plan B) Air Appreciation," 5 September 1942, AIR 47/13, National Archives, 5; Laurence Kuter, "Memorandum on the Organization of American Air Forces," 12 May 1943, Spaatz Box 12, Library of Congress; Carl Spaatz, memorandum, 17 January 1943, Spaatz Box 10, Library of Congress.

70. Felmy, "The German Air Force in the Mediterranean Theater of War," 584–585.

71. Ibid., 590–592.

72. Ibid., 595–596. Philip Jordan, a British journalist who witnessed the entire campaign, referred repeatedly to the Luftwaffe's frequent raids and freedom of action up to March 1943 and to the Luftwaffe's initial superiority. However, Jordan also remarked, "These thorough Germans are as regular as the Greenwich time signal, and come over to be knocked down at regular hours of day." See Philip Jordan, *Jordan's Tunis Diary* (London: Collins, 1943), 5, 60.

73. Dwight D. Eisenhower, *Crusade in Europe,* rep. ed. (Baltimore, MD: Johns Hopkins University Press, 1997), 120.

74. Felmy, "The German Air Force in the Mediterranean Theater of War," 601–603, 608.

75. "The North African Campaign, November 1942–May 1943," AIR 41/33, National Archives, 76–77.

76. Kuter, "Memorandum on the Organization of American Air Forces."

77. German Air Historical Branch, "The Operational Use of the Luftwaffe in the War at Sea, 1939–43," January 1944, trans. 7/102, A.H.B.6, October 1950, Air Historical Branch, 14; Hans Werner Neulen, *In the Skies of Europe: Air Forces Allied to the Luftwaffe, 1939–1945* (Wiltshire, UK: Crowood, 2000), 51–52.

78. BOH 4, 166–168.

79. Ibid., 173–179.

80. Ibid., 179–186.

81. Rein, *The North African Air Campaign,* 105–106.

82. Henry Arnold to Carl Spaatz, 15 November 1942, Spaatz Box 8, Library of Congress.

83. James Doolittle to Carl Spaatz, 25 December 1942, 1943 Ops File, Doolittle Box 19, Library of Congress.

84. Harry Coles, *Ninth Air Force in the Western Desert Campaign to 23 January 1943* (Washington, DC: US Army Air Forces Historical Division, 1945), 90, 123–124. Cross marveled at the damage heavy-bombers did to Axis airfields. See Cross, *Straight and Level,* 266.

85. See Rein, *The North African Air Campaign,* 132 and 132n4. This calculation is based on the number of sorties divided by plane-months available in each command. A plane-month equals one plane assigned for an entire month. Using this

methodology, the Eighth Air Force had 6,114 plane-months and the Twelfth Air Force 2,396—figures that make the discrepancy in total and effective sorties all the more telling.

Chapter Eleven. Building a New Air Command and Clearing North Africa

1. Mark Stoler, *Allies and Adversaries: The Joint Chiefs of Staff, the Grand Alliance, and U.S. Strategy in World War II* (Chapel Hill: University of North Carolina Press, 2000), 97–98.

2. Ibid., 100–102.

3. Ibid., 103–104; Gerhard Weinberg, *A World at Arms: A Global History of World War II* (Cambridge, UK: Cambridge University Press, 1994), 380–381, 437.

4. Arthur Tedder, *With Prejudice: The World War II Memoirs of Marshal of the Royal Air Force Lord Tedder, Deputy Supreme Commander of the Allied Expeditionary Force* (Boston: Little, Brown, 1966), 369.

5. Ibid., 370–371.

6. Ibid., 374.

7. Charles Portal to Arthur Tedder, 4 December 1942, AIR 23/1398, National Archives.

8. Arthur Tedder, *Report by Air Marshal Tedder on His Tenure in the Post of Air Officer Commanding-in-Chief* (London: MLRS, 2010), 143–145; Dwight Eisenhower to Chiefs of Staff (CoS), 30 November 1942, Tedder Box 7, Air Historical Branch; Tedder, "Air Command in the Mediterranean," December 1945, Tedder Box 8, Air Historical Branch, 12; Portal to Tedder, 4 December 1942, AIR 23/1398, National Archives; Tedder to Portal, 5 December 1942, AIR 23/1398, National Archives.

9. Tedder, *With Prejudice*, 375, 378, 380.

10. Ibid., 385–386.

11. Arthur Tedder to Middle East Defence Committee, "Command in the Mediterranean," 21 November 1942, in Tedder, "Note on Air Operations in Support of Crusader, January–December 1942," Tedder Box 6, Air Historical Branch, 247–248; Tedder to Charles Portal, 28 November 1942, "Note on Air Operations in Support of Crusader, January–December 1942," Tedder Box 6, Air Historical Branch, 256–257; Portal to Tedder, 29 November 1942, "Note on Air Operations in Support of Crusader, January–December 1942," Tedder Box 6, Air Historical Branch, 256–257; Portal to Tedder, 9 December 1942, "Note on Air Operations in Support of Crusader, January–December 1942," Tedder Box 6, Air Historical Branch, 273.

12. Meeting of Middle East Commanders in Chiefs, 10 December 1942, in Tedder, "Note on Air Operations in Support of Crusader, January–December 1942," Tedder Box 6, Air Historical Branch, 278; Arthur Tedder to Charles Portal, 13 December 1942, in Tedder, "Note on Air Operations in Support of Crusader, January–December 1942," Tedder Box 6, Air Historical Branch, 284; Tedder to Portal, 17 December 1942, in Tedder, "Note on Air Operations in Support of Crusader, January–December 1942," Tedder Box 6, Air Historical Branch, 286.

13. Arthur Tedder to Charles Portal, 9 December 1942, AIR 23/1398, National Archives; Portal to Tedder, 9 December 1942, AIR 23/1398, National Archives.

14. Arthur Tedder to Charles Portal, 13 December 1942, AIR 23/1398, National Archives, 1–3.

15. Tedder, *Report*, 145–146; Tedder, "Air Command in the Mediterranean," 18–23.

16. Tedder, *Report*, 147–149; Tedder, "Air Command in the Mediterranean," 25.

17. Tedder, *Report*, 150.

18. Tedder, *With Prejudice*, 393–396; Tedder, "Air Command in the Mediterranean," 29–31.

19. Tedder, "Air Command in the Mediterranean," 31–33.

20. Ibid., 34–37.

21. Ibid., 37–41.

22. Anthea M. Lewis, "A Tribute to Air Vice-Marshal GG Dawson, 1895–1944, CMSO ME & N Africa 1941–1944," Air Historical Branch, 35–37.

23. Ibid., 46.

24. Arthur Tedder to Charles Portal, 18 February 1943, AIR 23/7772, National Archives.

25. Tedder, *With Prejudice*, 398.

26. Vincent Orange, "The Commanders and the Command System," in Royal Air Force Historical Society, *The End of the Beginning: Bracknell Paper No. 3, A Symposium on the Land/Air Co-operation in the Mediterranean War, 1940–43*, ed. Derek Wood (Bracknell, UK: Royal Air Force Historical Society), 20 March 1992, 43.

27. Ibid., 44.

28. I. S. O. Playfair, *The Mediterranean and Middle East*, vol. 4, *The Destruction of the Axis Forces in Africa, History of the Second World War: United Kingdom Military Series* (Uckfield, UK: Naval and Military Press, 2004), 197–201 (this and all subsequent volumes of the British official history are referred to as "BOH" with volume number).

29. Ibid., 202–206.

30. Ibid., 216–217.

31. Ibid., 221–222, 225.

32. Erwin Rommel and Fritz Bayerlein, *The Rommel Papers*, ed. B. H. Liddell Hart (London: Collins, 1953), 350–351, 355, 359, 363, 370, 373, 384, 389.

33. Headquarters Western Desert Air Force to Headquarters Allied Air Forces, Northwest Africa, "Notes on the Move of No. 239 Wing, Consisting of Four Kittyhawk-Bomber Squadrons, by Air Transport from BELANDAH to MARBLE ARCH on 18 December 1942," 24 December 1942, AIR 23/1292, National Archives.

34. BOH 4, 230–233.

35. Ibid., 234–236.

36. Ibid., 237–238.

37. Ibid., 239–245; Karl Gundelach, *Die deutsche Luftwaffe im Mittelmeer*, vol. 1 (Frankfurt am Main: Lang, 1981), 480–483 (referred to from this point forward as *DLM*).

38. Abteilung 8, "The Luftwaffe in the Battle for Tunis: A Strategical Survey," trans. 7/5, A.H.B.6, October 1946, National Archives, 2–4.

39. Abteilung 8, "A Tactical Appreciation of the Air War in Tunisia," trans. 7/6, A.H.B.6, 21 October 1946, Air Historical Branch, 1–2.

40. Ibid., 4–6.

41. Ibid., 6.

42. Ibid., 7.

43. Ibid., 7–11.

44. BOH 4, 252–253.

45. Ibid., 279–280.

46. Ibid., 281–284.

47. Hellmuth Felmy, "The German Air Force in the Mediterranean Theater of War," K113.107–161, Air Force Historical Research Agency, 627–628; Walter Warlimont, *Inside Hitler's Headquarters, 1939–45* (New York: Praeger, 1966), 311.

48. Felmy, "The German Air Force in the Mediterranean Theater of War," 629–632; Warlimont, *Inside Hitler's Headquarters,* 311.

49. Felmy, "The German Air Force in the Mediterranean Theater of War," 634–636.

50. Ibid., 638–640.

51. Ibid., 640–641; *DLM,* 505.

52. BOH 4, 291, 294, 297–298, 301–302.

53. Rommel and Bayerlein, *Rommel Papers,* 404.

54. Felmy, "The German Air Force in the Mediterranean Theater of War," 667; BOH 4, 301–302; Operations Record Book of 242 Group RAF, 23 February 1943, AIR 24/1041, National Archives, 16–17, 18–19; Carl Spaatz, memorandum, 23 February 1943, Spaatz Box 10, Library of Congress; "Air Power in Peace and War, North Africa," n.d., Spaatz Box 270, Library of Congress.

55. Rommel and Bayerlein, *Rommel Papers,* 408.

56. Arthur Tedder to Charles Portal, 28 February 1943, AIR 23/7772, National Archives.

57. Ibid.

58. "XII Air Support Command in the Tunisian Campaign, January–May 1943," 655.01-2, Air Force Historical Research Agency, 7.

59. Ultra reports of 6 March 1943, located at DEFE 3, National Archives; "The Axis in Tunisia: The Battle of Medenine," March 1943, CAB 146/26, National Archives, 81; Ralph Bennett, *Ultra and Mediterranean Strategy* (New York: Morrow, 1989), 379; Vincent P. Orange, *Coningham: A Biography of Air Marshal Sir Arthur Coningham* (Washington, DC: Center for Air Force History, 1992), 142.

60. BOH 4, 334–336.

61. Ibid., 345–346.

62. Ibid., 347–348. Cross related that Air Vice-Marshal Harry Broadhurst, who arrived just after El Alamein to instruct fighter pilots on the latest tactics in the west, did such an exceptional job that he replaced Arthur Coningham as WDAF commander in time to plan this superb air effort to help break the Mareth line. Kenneth Cross with Vincent Orange, *Straight and Level* (London: Bugg Street, 1993), 225.

63. BOH 4, 348–349.

64. Felmy, "The German Air Force in the Mediterranean Theater of War," 683–687.

65. Northwest African Tactical Air Forces to All Units, "General Operational Directive," 2 March 1943, AIR 23/1299, National Archives. Arthur Coningham wrote this document before Bernard Law Montgomery and Harry Broadhurst won at the Mareth line but did so in anticipation, with Carl Spaatz, of the major ground offensive that would follow.

66. Tedder, *With Prejudice,* 409.

67. Ibid., 359–362, 368–369.

68. Arthur Tedder to Charles Portal, 5 March 1943, AIR 23/7772, National Archives.

69. Felmy, "The German Air Force in the Mediterranean Theater of War," 690–691.

70. Ibid., 694.

71. BOH 4, 374–379.

72. Felmy, "The German Air Force in the Mediterranean Theater of War," 723–726.

73. BOH 4, 249.

74. Norman Franks, ed., *The War Diaries of Neville Duke, 1941–1944* (London: Grub Street, 1995), 124.

75. "The Axis in Tunisia: The End in Africa, April–May 1943," CAB 146/27, Appendix 30, 15; "The North African Campaign, November 1942–May 1943," AIR 41/33, National Archives, 185–186; "The Battle Story of Flax," Spaatz Box 19, Library of Congress; Bennett, *Ultra and Mediterranean Strategy,* 215; James J. Sadkovich, *The Italian Navy in World War II* (Westport, CT: Greenwood, 1994), 343; Tedder, *With Prejudice,* 414–415; F. H. Hinsley et al., *British Intelligence in the Second World War,* vol. 2 (London: Her Majesty's Stationery Office, 1979), 611–614. By the time Johannes Steinhoff's unit received orders to evacuate Sicily, Adolf Galland told him, "There'll be no transport aircraft—Air Corps haven't a Ju left."

76. Alan Moorehead, *Desert War: The North African Campaign, 1940–1943* (New York: Penguin, 2001), 603.

77. BOH 4, 450–451, 454–455, 459; Hans von Luck, *Panzer Commander: The Memoirs of Colonel Hans von Luck* (New York: Praeger, 1989), 138; Johannes Steinhoff, *The Straits of Messina* (London: Deutsch, 1971), 24.

78. BOH 4, 460; Norman MacMillan, *The Royal Air Force in the World War,* vol. 3 (London: Harrap, 1949), 213.

79. Headquarters Royal Air Force Middle East Intelligence Officer to Headquarters Royal Air Force Middle East Records, "Enemy Aircraft Casualties," 30 January 1945, AIR 23/1207, National Archives; Headquarters Fifteenth Air Force, "The Statistical Story of Fifteenth Air Force," 1945, SMS1136, Package, US Air Force Academy, 1–11; Headquarters Mediterranean Allied Air Forces, "Air Power in the Mediterranean, November 1942–February 1945,"AIR 23/1478, National Archives, 32; Samuel W. Mitcham, Jr., and Friedrich von Stauffenberg, *The Battle for Sicily: How the Allies Lost Their Chance for Total Victory* (New York: Orion, 1991), 61; Christopher Shores and Hans Ring, *Fighters over the Desert: The Air Battles in the Western Desert, June 1940–December 1942* (New York: Arco, 1969), 217.

80. Shores and Ring, *Fighters over the Desert,* 217–221. The testimonies of thirteen prominent fighter pilots from both sides return again and again to their different approaches (221–236).

81. Headquarters Royal Air Force Middle East Tactics Assessment Office, "Air Tactics and Operational Notes on 201 Naval Co-operation Group, R.A.F.," n.d. (probably June 1943), AIR 23/1282, National Archives, Appendix 12: Enemy Merchant Shipping Losses; Admiralty Trade Division, "Summary of Enemy Merchant Shipping Losses," 24 September 1945, AIR 23/1175, National Archives.

82. See Vincent P. O'Hara, *Struggle for the Middle Sea: The Great Navies at War in the Mediterranean, 1940–1945* (Annapolis, MD: Naval Institute Press, 2009), 258–259.

83. Rommel's detailed appraisal of the reasons for Axis defeat in the Mediterranean, and the lost strategic opportunities resulting from it, are well worth reading, especially because he wrote them shortly before his death in 1944 and had thought

long and hard about this issue. Rommel and Bayerlein, *Rommel Papers,* 511–514, 519.

84. The definitive work on Quesada's efforts in North Africa and later in northwestern Europe is Thomas Alexander Hughes, *Over Lord: General Pete Quesada and the Triumph of Tactical Air Power in World War II* (New York: Free Press, 1995).

85. Roderic Owen, *Tedder* (London: Collins, 1952), 150–151.

Chapter Twelve. Operations Husky, Avalanche, and Baytown

1. Mark Stoler, *Allies and Adversaries: The Joint Chiefs of Staff, the Grand Alliance, and U.S. Strategy in World War II* (Chapel Hill: University of North Carolina Press, 2000), 120–121; Gerhard Weinberg, *A World at Arms: A Global History of World War II* (Cambridge, UK: Cambridge University Press, 1994), 439–441.

2. US Army Air Forces Historical Office, "Air Phase of the Italian Campaign to 1 January 1944," June 1946, Air Force Historical Research Agency, 17–18.

3. C. J. C. Molony et al., *The Mediterranean and Middle East,* vol. 5, pt. 1, *The Campaign in Sicily 1943 and the Campaign in Italy, 3rd September 1943 to 31st March 1944, History of the Second World War: United Kingdom Military Series* (Uckfield, UK: Naval and Military Press, 2004), 206–207, 216–217 (henceforth referred to as "BOH 5" with the part number.

4. Ibid., 197–202.

5. Ibid., 14.

6. Ibid., 17–19, 22.

7. Arthur Tedder, *With Prejudice: The World War II Memoirs of Marshal of the Royal Air Force Lord Tedder, Deputy Supreme Commander of the Allied Expeditionary Force* (Boston: Little, Brown, 1966), 431–436; Arthur Tedder to Charles Portal, 1 May 1943, AIR 23/7772, National Archives, 1–2.

8. Tedder, *With Prejudice,* 438.

9. BOH 5, pt. 1, 26–31.

10. Ibid., 32–33.

11. Ibid., 33–34.

12. Christopher M. Rein, *The North African Air Campaign: U.S. Army Air Forces from El Alamein to Salerno* (Lawrence: University Press of Kansas, 2012), 45.

13. BOH 5, pt. 1, 37–40, 44. For additional insights regarding the often overlooked but vital role of flak in German defensive efforts, see Edward B. Westermann, *Flak: German Anti-Aircraft Defenses, 1914–1945* (Lawrence: University Press of Kansas, 2001), with specific reference to the Straits of Messina on 232.

14. Ibid., 46.

15. Adolf Galland, "Report" (Luftwaffe Intelligence Summaries), 27 August–19 September 1943, 520.056-289, Box 23, Air Force Historical Research Agency; Johannes Steinhoff, *The Straits of Messina* (London: Deutsch, 1971), 78.

16. Paul Deichmann, "High Level Luftwaffe Policy in the Mediterranean, May–June 1943," extract from "The Campaign in Italy," pt. 1, written for the US Historical Division, trans. 7/94, A.H.B.6, February 1950, Air Historical Branch, 1.

17. Ibid., 2–4.

18. Deichmann, "High Level Luftwaffe Policy in the Mediterranean," 5.

19. James Corum, *Wolfram von Richthofen, Master of the German Air War* (Lawrence: University Press of Kansas, 2008), 318–340.

20. Deichmann, "High Level Luftwaffe Policy in the Mediterranean," 5–6.

21. Heinrich von Vietinghoff, "The Campaign in Italy," pt. 1, written for the US Historical Division, trans. 7/94, A.H.B.6, , February 1950, Air Historical Branch; Walter Warlimont, *Inside Hitler's Headquarters, 1939–45* (New York: Praeger, 1966), 322.

22. BOH 5, pt. 1, 47–48, 51.

23. Eduard Mark, *Aerial Interdiction in Three Wars* (Washington, DC: Center for Air Force History, 1994), 61.

24. Keith Park, "Report on Operations Carried Out from Malta Covering the Period from the Fall of Tunisia until July 20th, 1943," 26 July 1943, AIR 23/6630, National Archives.

25. BOH 5, pt. 1, 49; Tedder, *With Prejudice*, 426–428, 441.

26. BOH 5, pt. 1, 50.

27. Ibid., 52–53, 62.

28. Ibid., 63–67; Steinhoff, *Straits of Messina*, 145–146, 151, 169; Headquarters Mediterranean Air Command Air Plans, "Notes on Discussion Held with Air Commander-in-Chief on Air Operations—Operation 'Husky,'" 23 July 1943, AIR 23/6630, National Archives.

29. BOH 5, 1, 94, 97–98; Samuel W. Mitcham, Jr., and Friedrich von Stauffenberg, *The Battle for Sicily: How the Allies Lost Their Chance for Total Victory* (New York: Orion, 1991), 64.

30. BOH 5, pt. 1, 93–97.

31. Ibid., 98–99.

32. Ibid., 127–130.

33. Steinhoff, *Straits of Messina*, 185. The irony of this statement must have struck Steinhoff hard. As one of the older veterans who had flown in the Polish campaign and every subsequent one, he no doubt remembered when the Luftwaffe had adhered at least in part to Wever's seminal doctrine document, *Luftkrieg Führung* (Luftwaffe Regulation 16, Conduct of the Air War). It was published in 1935, the official year of the Luftwaffe's birth.

34. BOH 5, pt. 1, 153–157.

35. Ibid., 161–162.

36. Mark, *Aerial Interdiction in Three Wars*, 68–74, 77–78; BOH 5, pt. 1, 164–168.

37. Mark, *Aerial Interdiction in Three Wars*, 78–79; Rein, *The North African Air Campaign*, 158.

38. Rein, *The North African Air Campaign*, 162–167; Wesley Frank Craven and James Lea Cate, *The Army Air Forces in World War II*, vol. 2 (Washington, DC: Office of Air Force History, 1983), 474–476; Harry Coles, *Ninth Air Force in the Western Desert Campaign to 23 January 1943* (Washington, DC: USAAF Historical Division, 1945), 157, 170, 175; Mark, *Aerial Interdiction in Three Wars*, 70.

39. Rein, *The North African Air Campaign*, 168–183.

40. Samuel W. Mitcham, Jr., *Blitzkrieg No Longer: The German Wehrmacht in Battle, 1943* (Mechanicsburg, PA: Stackpole, 2010), 213; Von Vietinghoff, "The Campaign in Italy," Chapter 6, 2. For an excellent assessment of the Allied failure to stop the Axis evacuation of Sicily, see Richard G. Davis, *Bombing the European Axis Powers: A Historical Digest of the Combined Bomber Offensive, 1939–1945* (Maxwell Air Force Base, AL: Air University Press, 2006), 162–171.

41. BOH 5, pt. 1, 169; Tedder, *With Prejudice*, 445.

42. Headquarters Allied Air Forces, Northwest Africa to Henry Arnold, "Air Support in Tunisia," 15 September 1943, 248.211-135, Air Force Historical Research Agency; Headquarters V Corps, "Adjustment of Artillery Fire by P-51 Airplanes (Salerno)," 26 September 1943, 248.211-135, Air Force Historical Research Agency.

43. Allied Forces Headquarters, "Lessons from the Sicilian Campaign," 20 November 1943, AIR 23/6630, National Archives, 41, 44, 65–66.

44. BOH 5, pt. 1, 169–171.

45. Ibid., 171–172. See Richard Overy, *The Bombers and the Bombed: Allied Air War over Europe, 1940–1945* (New York: Viking, 2013), 336–340, for a discussion of the role of Allied bombing in Italy's surrender. See also Davis, *Bombing the European Axis Powers*, 148.

46. BOH 5, pt. 1, 173–174.

47. Headquarters Mediterranean Air Command (MAC), "Lessons Learnt in Operation 'Husky,'" 23 August 1943, AIR 23/6630, National Archives, 14–19, 23–24, and Appendix A: Lessons Learned, 1–2.

48. Headquarters MAC, "Lessons Learnt in Operation 'Husky,'" Appendix A, 2, 6–8.

49. Ibid., 1, 28, 30.

50. Douglas Porch, *The Path to Victory: The Mediterranean Theater in World War II* (New York: Farrar, Straus, and Giroux, 2004), 679. The Allies inflicted 536,000 casualties on the Germans in Italy, including about 100,000 prisoners taken at the end of the war, whereas they suffered 312,000. A German accounting arrived at 434,636, including 48,067 killed, 172,531 wounded, and 214,048 missing but not the 100,000 prisoners taken at the end of the war. See Dominick Graham and Shelford Bidwell, *Tug of War: The Battle for Italy, 1943–1945* (New York: St. Martin's, 1986), 403.

51. BOH 5, pt. 1, 218, 224–225. MAC had 3,280 aircraft (2,460 serviceable) for initial support of the two operations. Axis air forces totaled 612 aircraft (310 serviceable).

52. Headquarters Allied Air Forces, Northwest Africa, "Salerno Plan: Task of the Air Forces," Spaatz Box 140, Library of Congress.

53. Charles Portal to Arthur Tedder and William Welsh, 13 July 1943, Tedder Box 7, Air Historical Branch.

54. Charles Portal to Arthur Tedder and Wigglesworth, 16 July 1943, Tedder Box 7, Air Historical Branch.

55. Arthur Tedder to Charles Portal, 14 August 1943, Tedder Box 7, Air Historical Branch; Robert S. Ehlers, Jr., *Targeting the Third Reich: Air Intelligence and the Allied Bombing Campaigns* (Lawrence: University Press of Kansas, 2009), Chapters 9–13.

56. Arthur Tedder to Chief of Air Staff (CAS), 28 July 1943, Tedder Box 7, Air Historical Branch; Roderic Owen, *The Desert Air Force: An Authoritative History Published in Aid of the Royal Air Force Benevolent Fund* (London: Hutchinson, 1948), 164–165.

57. Allied Forces Headquarters, "Avalanche Report," October 1943, AIR 23/6635, National Archives.

58. Mark, *Aerial Interdiction in Three Wars*, 99–100, 107–108.

59. US Army Air Force (USAAF) Historical Office, "Air Phase of the Italian

Campaign to 1 January 1944," June 1946, Record 101-115, Air Force Historical Research Agency, 61–69, 74–75.

60. BOH 5, pt. 1, 227–229; Kenneth Cross with Vincent Orange, *Straight and Level* (London: Bugg Street, 1993), 267–268.

61. USAAF Historical Office, "Air Phase of the Italian Campaign to 1 January 1944," 84–90.

62. Ibid., 80–83, 93–94, 98, 104.

63. Ibid., 112–113, 117.

64. BOH 5, pt. 1, 287–289.

65. USAAF Historical Office, "Air Phase of the Italian Campaign to 1 January 1944," 121–122, 127–128; Allied Forces Headquarters, "Avalanche Report," October 1943, AIR 23/6635, National Archives, 4, 5.

66. BOH 5, pt. 1, 302.

67. Ibid., 304–307; Allied Forces Headquarters, "Avalanche Report," 6–8.

68. BOH 5, pt. 1, 311.

69. USAAF Historical Office, "Air Phase of the Italian Campaign to 1 January 1944," 131–133, 137–146; Mark, *Aerial Interdiction in Three Wars*, 105.

70. USAAF Historical Office, "Air Phase of the Italian Campaign to 1 January 1944," 131–133, 137–146.

71. Arthur Tedder to Charles Portal, 16 September 1943, Tedder Box 7, Air Historical Branch.

72. Heinrich von Vietinghoff, diary, quoted in Denis Richards and Hilary St. George Saunders, *Royal Air Force 1939–1945*, vol. 2, *The Fight Avails* (London: Her Majesty's Stationery Office, 1954), 339.

73. USAAF Historical Office, "Air Phase of the Italian Campaign to 1 January 1944," 179.

74. Dwight Eisenhower to Combined Chiefs of Staff (CCS), 21 September 1943, Tedder Box 7, Air Historical Branch. Eisenhower's comments about the "so-called Strategic Air Force" indicate his clear understanding of the heavy-bombers' versatility.

75. Arthur Tedder to Charles Portal, 22 September 1943, Tedder Box 7, Air Historical Branch; Mark, *Aerial Interdiction in Three Wars*, 104–108.

76. USAAF Historical Office, "Air Phase of the Italian Campaign to 1 January 1944," 131–133, 137–146.

77. Ibid., 153–154.

78. Ibid., 156–159, 170, 174–175, 177.

79. Ibid., 180–182.

80. BOH 5, pt. 1, 333n1.

81. Ibid., 334–340, 344–345.

82. Ibid., 344, 348, 353, 355.

83. USAAF Historical Office, "Air Phase of the Italian Campaign to 1 January 1944," 183–185, 188.

84. Ibid., 187–189, 200–202, 209–211, 214. For the role of RAF Bomber Command raids in the dispersal of German factories and the dislocation and quality-control crises this created, see Ehlers, *Targeting the Third Reich*, 151–159, 174–176.

85. USAAF Historical Office, "Air Phase of the Italian Campaign to 1 January 1944," 191, 194–195.

86. BOH 5, pt. 1, 131, 356–360.

87. Ibid., 349, 352–354. The Luftwaffe had 125 torpedo-bombers, 9 long-range reconnaissance planes, and 18 long-range fighters.

Chapter Thirteen. The Italian Campaign and the Invasion of Southern France

1. C. J. C. Molony et al., *The Mediterranean and Middle East*, vol. 5, pt. 1, *The Campaign in Sicily 1943 and the Campaign in Italy 3rd September 1943 to 31st March 1944, History of the Second World War: United Kingdom Military Series* (Uckfield, UK: Naval and Military Press, 2004), 378–383 (henceforth referred to as "BOH 5" with the part number). Details of ground actions in Italy are recorded in Martin Blumenson, *Salerno to Cassino: The U.S. Army in World War II—The Mediterranean Theater of Operations* (Washington, DC: Government Printing Office, 1969).

2. BOH 5, pt. 1, 416, 432–433, 440–441.

3. Ibid., 446–462.

4. Ibid., 462–469.

5. Ibid., 491–493.

6. Ibid., 516–518, 523.

7. Ibid., pt. 2, 593–594, 603–605.

8. Headquarters Mediterranean Allied Air Forces (HQ MAAF) Historian, "The History of M.A.A.F., 10 December 1943–1 September 1944," April 1945, 622.01-1, Box 7, Air Force Historical Research Agency, 155–157.

9. Ibid., 162–165.

10. Ibid., 165–167.

11. BOH 5, pt. 2, 652–653.

12. Ibid., 654–657.

13. Various German senior officers, "High Level Reports and Directives Dealing with the Italian Campaign in 1944," 30 November 1948, trans. 7/82, A.H.B.6, Air Historical Branch, 1–4.

14. BOH 5, pt. 2, 664–665, 668–669, 754–764.

15. Eduard Mark, *Aerial Interdiction in Three Wars* (Washington, DC: Center for Air Force History, 1994), 109, 135–140. See esp. ammunition shortage chart on 137.

16. HQ MAAF Historian, "History of M.A.A.F., 10 December 1943–1 September 1944," 169–171.

17. BOH 5, pt. 2, 681–683.

18. Ibid., 779.

19. Ibid., 780–782.

20. Ibid., 785–787.

21. Ibid., 789.

22. Ibid., 703, 708–709.

23. Mark, *Aerial Interdiction in Three Wars*, 143.

24. William Jackson and T. P. Gleave, *The Mediterranean and Middle East*, vol. 6, *Victory in the Mediterranean*, pt. 1, *1st April to 4th June 1944, History of the Second World War: United Kingdom Military Series* (Uckfield, UK: Naval and Military Press, 2004), 7, 12. Referred to subsequently as "BOH 6" with part number.

25. Allied Forces Headquarters (AFHQ) Supreme Allied Commander's (SACMED) Despatch, "Italian Campaign, 10 May 1944–12 August 1944," AIR 75/138, National Archives, 1–4.

26. John Slessor, *The Central Blue: The Autobiography of Sir John Slessor, Marshal of the Royal Air Force* (New York: Praeger, 1957), 578–579.

27. HQ MAAF Historian, "History of M.A.A.F., 10 December 1943–1 September 1944," 190.

28. Ira Eaker to Henry Arnold, 6 March 1944, Eaker Box 22, vol. 1, Library of Congress. Emphasis added to highlight the clear understanding that the air phase would set the conditions for success during the combined-arms phase—Operation Diadem.

29. AFHQ SACMED Despatch, "Italian Campaign," 11.

30. James Parton, *Air Force Spoken Here: General Ira Eaker and the Command of the Air* (Maxwell Air Force Base, AL: Air University Press, 2000), 381–383.

31. Lauris Norstad, Headquarters Mediterranean Allied Air Forces (HQ MAAF), "Air Force Participation in 'Diadem,'" 31 July 1944, AIR 23/8572, National Archives, 1.

32. Ibid., 2.

33. HQ MAAF Historian, "History of M.A.A.F., 10 December 1943–1 September 1944," 191–194.

34. Henry Arnold to Ira Eaker, 27 March 1944, MS57, Box 2, USAFA.

35. Norstad, HQ MAAF, "Air Force Participation in 'Diadem,'" 2.

36. Ibid., 3.

37. AFHQ SACMED Despatch, "Italian Campaign," 10.

38. Ibid.

39. Ibid., 11.

40. Ibid.

41. BOH 6, pt. 1, 38–42. See also Headquarters Mediterranean Allied Air Forces (HQ MAAF), "Air Power in the Mediterranean, November 1942–February 1945," AIR 23/1478, National Archives, 10.

42. John Slessor to Charles Portal, 16 April 1944. Quoted in full in Slessor, *The Central Blue,* 571–572. Emphasis in original.

43. HQ MAAF Historian, "History of M.A.A.F., 10 December 1943–1 September 1944," 196–197; HQ MAAF, "Air Power in the Mediterranean, November 1942–February 1945," 12–14.

44. Slessor, *The Central Blue,* 579.

45. BOH 6, pt. 1, 64–65.

46. BOH 5, pt. 2, 813.

47. Mark, *Aerial Interdiction in Three Wars,* 172–178; BOH 6, pt. 1, 43–44.

48. BOH 6, pt. 1, 67–68, 79.

49. US Army Air Forces (USAAF) Director of Intelligence, "Signal Intelligence Service of the German Luftwaffe," vol. 10, "Technical Operations in the South," 1945, 519.6314-10 V.9, Air Force Historical Research Agency, 17–18.

50. BOH 6, pt. 1, 142, 149, 150–156.

51. HQ MAAF Historian, "History of M.A.A.F., 10 December 1943–1 September 1944," 199–200.

52. BOH 6, pt. 1, 94, 99, 112; AFHQ SACMED Despatch, "Italian Campaign," 15–17.

53. AFHQ SACMED Despatch, "Italian Campaign," 27–28.

54. Ibid., 26–28, 36.

55. BOH 6, pt. 1, 162–165.

56. Ibid., 203–205.

57. Ibid., 205–211.

58. Ibid., 232–234.

59. AFHQ SACMED Despatch, "Italian Campaign," 22–23.

60. HQ MAAF Historian, "History of M.A.A.F., 10 December 1943–1 September 1944," 186–187.

61. Ibid., 201.

62. Ibid., 205.

63. Ibid., 205–206.

64. Ibid., 209–211.

65. Henry Arnold, "Trip to England, June 8, 1944–June 21, 1944," MS33, Box 88, US Air Force Academy, 16. The title of this memo is misleading because Arnold also visited Algeria and Italy.

66. Ibid., 21.

67. Norstad, HQ MAAF, "Air Force Participation in 'Diadem,'" 6. Emphasis in original.

68. Mark, *Aerial Interdiction in Three Wars,* 206–208; Norstad, HQ MAAF, "Air Force Participation in 'Diadem,'" Appendix: The Effect of Air Power in a Land Offensive, 1–2.

69. Norstad, HQ MAAF, "Air Force Participation in 'Diadem,'" Appendix, 2.

70. Ibid., 3.

71. Mark, *Aerial Interdiction in Three Wars,* 208.

72. BOH 6, pt. 1, 284.

73. Ibid., 313.

74. John Slessor to Trenchard, 8 May 1944, AIR 75/115, National Archives.

75. John Slessor to Charles Portal, 4 March 1944, in AIR 75/115, National Archives.

76. John Slessor to Charles Portal, 20 January 1944, AIR 75/115, National Archives, 1–2.

77. Ibid.

78. John Slessor, "The Effect of Air Power in a Land Offensive," 18 June 1944, AIR 75/42, National Archives, 1–2.

79. Ibid., 2–3.

80. Ibid., 3–4.

81. Ibid., 5.

82. Ibid.

83. John Slessor to Thomas Gleave, 9 October 1973, AIR 75/137, National Archives. Emphasis in original.

84. Ibid.

85. Ibid.

86. Headquarters Eighth Army, "P.W. Reports Concerning Destruction of M.T., a/c and Shipping by Allied Aircraft," 21 June 1944, WO 204/7930, National Archives, 1–2.

87. AFHQ SACMED Despatch, "Italian Campaign," 45–58.

88. Ibid., 65–69.

89. Ibid., 76.

90. Ibid., 77.

91. HQ MAAF, "Air Power in the Mediterranean, November 1942–February 1945," 77.

92. Ibid.

93. Ibid., 78.

94. Ibid.

95. Ibid.

96. Headquarters Mediterranean Allied Air Forces (HQ MAAF) Director of Intelligence, "What Is the German Saying?" 1945, 622.619-2, Air Force Historical Research Agency.

97. Ibid.

98. Mark, *Aerial Interdiction in Three Wars*, 204–206.

99. HQ MAAF Director of Intelligence, "What Is the German Saying?"

100. Ibid.

101. Steve Weiss, *Allies in Conflict: Anglo-American Strategic Negotiations, 1938–1944* (New York: Palgrave MacMillan, 1997), 83.

102. Mark Stoler, *Allies and Adversaries: The Joint Chiefs of Staff, the Grand Alliance, and U.S. Strategy in World War II* (Chapel Hill: University of North Carolina Press, 2000), 169–173; BOH 5, pt. 2, 840–845.

103. BOH 6, pt. 1, 332–333, 335.

104. HQ MAAF Historian, "History of M.A.A.F., 10 December 1943–1 September 1944," 123–125, 220.

105. BOH 6, pt. 2, 51, 61.

106. HQ MAAF, "Air Power in the Mediterranean, November 1942–February 1945"; BOH 5, pt. 2, 817–818; BOH 6, pt. 2, 98–101.

107. Headquarters Seventh Army, "Air Support Plan: Section 1 Air Support Control," 19 July 1944, 685.327-1, Air Force Historical Research Agency, 1.

108. Ibid.

109. BOH 6, pt. 2, 173–179.

110. Headquarters Mediterranean Allied Tactical Air Forces (HQ MATAF), "Report of Operation Dragoon: Mediterranean Allied Tactical Air Force," 1 November 1944, 168.7044-12, Air Force Historical Research Agency, 10; BOH 6, pt. 2, 179–183.

111. BOH 6, pt. 2, 183.

112. HQ MAAF Historian, "History of M.A.A.F., 10 December 1943–1 September 1944," 216–217.

113. HQ MATAF, "Report of Operation Dragoon," 17–18; BOH 6, pt. 2, 228–232.

114. HQ MATAF, "Report of Operation Dragoon," 52.

115. Samuel W. Mitcham, Jr., *Retreat to the Reich: The German Defeat in France, 1944* (New York: Praeger, 2000), 163–166.

116. HQ MATAF, "Report on Operation Dragoon," Air Force Historical Research Agency, Reel 626.43009, 5.

117. Alan F. Wilt, *The Riviera Campaign of August 1944* (Carbondale: Southern Illinois University Press, 1981), 45.

118. Mitcham, *Retreat to the Reich*, 169.

119. Wilt, *Riviera Campaign*, 71.

120. HQ MAAF Historian, "History of M.A.A.F., 10 December 1943–1 September 1944," 229, 233, 235–239, 246.

121. Ira Eaker to Henry Arnold, 21 August 1944, 622.161-2, Box 11, Air Force Historical Research Agency, 1–6.

122. Jeffrey J. Clarke and Robert Ross Smith, *Riviera to the Rhine* (Washington, DC: US Army Center of Military History, 1993), 200.

123. Ibid., 210–214.

124. Ibid., 215.

125. Ibid.

126. HQ MATAF, "Report of Operation Dragoon," 57.

127. Combined Operations Headquarters, "Bulletin Y/42: Operation DRA-GOON, December 1944," Air Force Historical Research Agency, Reel 626. 430–439, 21.

128. Thomas Alexander Hughes, *Over Lord: General Pete Quesada and the Triumph of Tactical Air Power in World War II* (New York: Free Press, 1995), 128–139.

129. Ibid., 205–226.

130. Clark and Smith, *Riviera to the Rhine*, 197–200.

131. Eaker to Arnold, 21 August 1944, 5–7.

Chapter Fourteen. The Second Aerial Front

1. Mark Stoler, *Allies and Adversaries: The Joint Chiefs of Staff, the Grand Alliance, and U.S. Strategy in World War II* (Chapel Hill: University of North Carolina Press, 2000), 167–168; Gerhard Weinberg, *A World at Arms: A Global History of World War II* (Cambridge, UK: Cambridge University Press, 1994), 628–630.

2. C. J. C. Molony et al., *The Mediterranean and Middle East*, vol. 5, *The Campaign in Sicily 1943 and the Campaign in Italy 3rd September to 31st March 1944*, *History of the Second World War: United Kingdom Military Series* (Uckfield, UK: Naval and Military Press, 2004), 571–576 (this volume of the British official history is referred to subsequently as "BOH 5" with part number).

3. Arthur Tedder to Charles Portal, 31 October 1943, AIR 23/7773, National Archives; Combined Chiefs of Staff (CCS), "106th Meeting Minutes," 14 August 1943, Arnold Box 214, Library of Congress; Ira Eaker to Haywood Hansell, 25 August 1943, Spaatz Box 323, Library of Congress.

4. USAAF Historical Office, "Air Phase of the Italian Campaign to 1 January 1944," June 1946, Air Force Historical Research Agency, 239–244.

5. Ibid., 217–219.

6. Charles Portal to Arthur Tedder, 13 October 1943, Tedder Box 7, Air Historical Branch; Tedder to Portal, 14 October 1943, Tedder Box 7, Air Historical Branch; Combined Chiefs of Staff (CCS) to Dwight Eisenhower, 23 October 1943, Tedder Box 7, Air Historical Branch.

7. USAAF Historical Office, "Air Phase of the Italian Campaign to 1 January 1944," 239–244.

8. Combined Chiefs of Staff (CCS) 400/2, "Control of Strategic Air Forces in Northwest Europe and in the Mediterranean," 4 December 1943, Tedder Box 7, Air Historical Branch; Carl Spaatz to Mediterranean Allied Air Forces, 5 January 1944, Tedder Box 7, Air Historical Branch; Spaatz to Ira Eaker, "Operational Directive," 11 January 1944, Tedder Box 7, Air Historical Branch; Dwight Eisenhower to Headquarters Mediterranean Air Command, "General Orders No. 67," 20 December 1943, AIR 23/3278, National Archives.

9. USAAF Historical Office, "Air Phase of the Italian Campaign to 1 January 1944," 220.

10. Ibid., 220–222.

11. Headquarters Fifteenth Air Force, "The Statistical Story of Fifteenth Air Force," 1945, SMS1136, Package, US Air Force Academy, 1–11.

12. Ibid.

13. USAAF Historical Office, "Air Phase of the Italian Campaign to 1 January 1944," 245–246, 264, 269–274.

14. Ibid., 234–235.

15. Ibid., 236–238.

16. Headquarters Mediterranean Allied Air Forces (HQ MAAF) Historian, "The History of M.A.A.F., 10 December 1943–1 September 1944," April 1945, 622.01-1, Box 7, Air Force Historical Research Agency, 1–4.

17. Ibid., 4–7, 14–20, 24.

18. Ibid., 27–28.

19. Ibid., 42, 48.

20. Ibid., 145–149.

21. Ira Eaker to Henry Arnold, 21 March 1944, MS57, Box 2, US Air Force Academy, 16–18.

22. Ibid., 134–138. For the effects of the transportation campaign in France to isolate the Normandy battle area, see Robert S. Ehlers, Jr., *Targeting the Third Reich: Air Intelligence and the Allied Bombing Campaigns* (Lawrence: University Press of Kansas, 2009), Chapter 9.

23. Ira Eaker to Henry Arnold, 7 May 1944, MS57, Box 2, US Air Force Academy, 1–4, 7–10.

24. US Army Air Forces Director of Intelligence, "Signal Intelligence Service of the German Luftwaffe," vol. 10, "Technical Operations in the South," 1945, 519.6314-10 V.9, Air Force Historical Research Agency, 39–46.

25. Ibid.

26. Ira Eaker to Henry Arnold, 1 June 1944, MS57, Box 2, US Air Force Academy, 6–7.

27. HQ MAAF Historian, "The History of M.A.A.F., 10 December 1943–1 September 1944," 84–87, 111–114.

28. Ibid., 114–115; HQ MAAF Director of Intelligence to HQ MAAF, "Proposed Amalgamation of M.A.P.R.C. and M.E.I.U.," 25 February 1944, AIR 23/6692, National Archives; HQ MAAF to Headquarters Royal Air Force Middle East, "Amalgamation of Middle East Interpretation Unit with Mediterranean Allied Photographic Interpretation Command," 13 March 1944, AIR 23/6692, National Archives.

29. Ibid., 117–122.

30. "The Diary of USSAFE Mission to Russia in Connection with Frantic Project and Other Matters," May 1944, Spaatz Box 316, Library of Congress, 4; John Deane to Carl Spaatz, cable, 10 May 1944, MS16/S3/B4/F1, US Air Force Academy. Deane said the "Russians considered the oil in Ploesti as being a primary objective."

31. "Report on Visit to Russia by Mission of USSAFE Officers," 21 May 1944, Spaatz Box 17, Library of Congress, Appendix B: Conference with Gen. Grendal, 2.

32. John R. Deane, *The Strange Alliance* (New York: Viking, 1947), 207.

33. Ibid. It is interesting to note that the Red Army received a double advantage from the Allies. The oil offensive created severe fuel shortages for the Germans, whereas nearly 500,000 US lend-lease vehicles gave the Russians great mobility.

34. R. Cooke and Roy Conyers Nesbit, *Target: Hitler's Oil—Allied Attacks on German Oil Supplies, 1939–45* (New York: HarperCollins, 1985), 78–79; HQ MAAF, "Air Power in the Mediterranean: November 1942–February 1945," 13 February 1945, Spaatz Box 95, Library of Congress, 39.

35. Aleksanar Milutinovic, "First Foray: The United States Army Air Forces'

Strategic Strike at Ploesti, 1 August, 1943," unpublished M.A. thesis, The Ohio State University, 2001, RG341/E217/B62 and RG341/E217/B71, National Archives and Records Administration, 155–156.

36. Cooke and Nesbit, *Target: Hitler's Oil*, 86–87.

37. Milutinovic, "First Foray," 158.

38. HQ MAAF, "Air Power in the Mediterranean: November 1942–February 1945," 39–40. The Romanians were only producing 4.8 million tons of petroleum, oil, and lubricants (POL) annually when raids began.

39. US Office of Special Services (OSS), Report No. SQ. 590, "Rumania: Economic, Oil Production, Transport, etc.," 12 January 1944, Spaatz Box 187, Library of Congress, 2.

40. See AIR 8/1332, PRO; COSMED 117, 30 May 1944; COSMED 124, 6 June 1944, PRO. Eaker's comments are in Spaatz Box 17, in the cover memo for a report, "Result of Attacks on Axis Oil Installations," 15 May 1944. Emphasis added.

41. Supreme Headquarters Allied Expeditionary Force (SHAEF), "Commanders Weekly Intelligence Review No. 4," 30 April 1944, MS16/S5/B12/F4, US Air Force Academy, 1.

42. Denis Richards and Hilary St. G. Saunders, *Royal Air Force, 1939–1945*, vol. 3 (London: Her Majesty's Stationery Office, 1953–1954), 226–227. For an excellent account of No. 205 Group's mining efforts, see Maurice G. Lihou, *Out of the Italian Night: Wellington Bomber Operations, 1944–45* (Shrewsbury, UK: Airlife, 2000), 52–56.

43. Lihou, *Out of the Italian Night*, 131.

44. William Jackson and T. P. Gleave, *The Mediterranean and Middle East*, vol. 6, *Victory in the Mediterranean*, pt. 2, *June to October 1944*, *History of the Second World War: United Kingdom Military Series* (Uckfield, UK: Naval and Military Press, 2004), 215. Referred to subsequently as "BOH 6" with part number.

45. US Strategic Bombing Survey, "USSBS Interview No. 11, Subject: Reichsminister Albert Speer, 31 May 1945, Minutes of a 19 May 1945 Interrogation," RG243/E31/B1, National Archives and Records Administration, 9.

46. Joint Oil Target Committee (JOTC), Working Committee, "Weekly Bulletin No. 2," 11 July 1944, MS16/S5/B8/F7, US Air Force Academy, 4–8; HQ MAAF to US Strategic Air Forces in Europe (USSAFE) and Armed Forces Headquarters (AFHQ), "Priorities for Bombing Effort in South-East Europe," 25 July 1944, MS16/S3/B4/F1, US Air Force Academy.

47. "Draft History of Oil-versus-Transportation Debate," 2 November 1944, Spaatz Box 202, Library of Congress; Lihou, *Out of the Italian Night*, 65–67.

48. Joint Oil Target Committee (JOTC), "Weekly Bulletin No. 9," 29 August 1944, AIR 2/8011, National Archives, 5–8. Emphasis in original.

49. Joint Intelligence Centre (JIC) (44) 285 (O) (Final), 4 July 1944, AIR 8/1018, National Archives, 3–4.

50. HQ MAAF Director of Intelligence (A2), "What Is the German Saying?" 1945, 622.619-2, Air Force Historical Research Agency.

51. AI3(e), "M.S.S. References," 30 December 1944, 6 September intercept, AIR 40/2073, National Archives.

52. Ibid., 13 October intercept.

53. Ibid., 17 November intercept.

54. War Department, Captured Personnel and Material Branch, Military

Intelligence Division, "Morale and Conditions in the German Army," 13 December 1944, RG243/E36/B114, National Archives and Records Administration, 2.

55. Ibid., 4.

56. Henry Arnold to Ira Eaker, 8 September 1944, MS57, Box 2, US Air Force Academy.

57. HQ MAAF, "Air Power in the Mediterranean, November 1942–February 1945," 37–41; Ira Eaker to Henry Arnold, 21 November 1944, MS57, Box 2, US Air Force Academy.

58. Combined Strategic Targets Committee (CSTC) Oil Production, "Bulletin No. 25," 19 December 1944, AIR 40/2073, National Archives, 2–3, 7.

59. See Ehlers, *Targeting the Third Reich*, 346–349.

60. US Office of Special Services (OSS) Research and Analysis, "Notes on Certain Effects of US 15th AAF Bombing of Railroad Transportation in Rumania," 16 October 1944, 670.454-2, Box 12, Air Force Historical Research Agency.

61. The first major raids against the capitals of Axis satellite states in the Balkans were designed to prompt their governments to switch sides. The raids did not have any clear or major effectiveness here, in stark contrast to the oil and transportation offensives. See Richard G. Davis, *Bombing the European Axis Powers: A Historical Digest of the Combined Bomber Offensive, 1939–1945* (Maxwell Air Force Base, AL: Air University Press, 2006), 325–326.

62. J. C. E. Luard, Chief of Intelligence Section, MAAF, "The Balkan Situation— Possibilities of Air Attack," 24 April 1944, Spaatz Box 17, Library of Congress, 1.

63. Ibid., 2–3.

64. Ibid., 8.

65. HQ MAAF to AFHQ and USSAFE, "Priorities for Bombing Effort in South-East Europe," 5 September 1944, MS16/S3/B4/F2, US Air Force Academy.

66. Davis, *Bombing the European Axis Powers*, 317.

67. HQ MAAF, "The History of the 15th Air Force," vol. 2, 1945, 670.01-1, Air Force Historical Research Agency, Appendices, 288–289.

68. David M. Glantz, *Red Storm over the Balkans: The Failed Soviet Invasion of Rumania, Spring 1944* (Lawrence: University Press of Kansas, 2007), 18, 29, 38, 45, 52, 66, 199, 202.

69. Ibid., 100–102, 147, 162, 181, 203.

70. Ibid., 215, 232, 246, 247, 261, 267, 317.

71. Ibid., 269.

72. Ibid., 371.

73. Ibid., 372–373. Glantz substantiates these statements further on 373–375.

74. Ibid., 159, 272–273, 366, 379–381.

75. BOH 6, pt. 1, 45–46.

76. Ira Eaker to Henry Arnold, 31 August 1944, MS57, Box 2, US Air Force Academy, 2–3.

77. Ibid; Davis, *Bombing the European Axis Powers*, 429.

78. BOH 6, pt. 2, 318–321.

79. Ibid., 303–304, 307.

80. Ibid., 314–315.

81. Ira Eaker to Henry Arnold, 21 November 1944, MS57, Box 2, US Air Force Academy, 4–5, 8.

82. USAAF A2, "Signal Intelligence Service of the German Luftwaffe," vol. 10, 1.

83. Ibid., 1–2.

84. Ibid., 36.

85. Ibid.

86. BOH 6, pt. 2, 317–318; HQ MAAF A2, "What Is the German Saying?"

87. HQ MAAF, "The History of the 15th Air Force," vol. 2, Appendices, 319–320.

88. Combined Services Detailed Interrogation Center (CSDIC), "A Survey of the Supply System of the German Army, 1939–45," 25 August 1945, RG243/E36/B187, National Archives and Records Administration, 4–7.

89. Ibid., 17–19; Geoffrey P. Megargee, *Inside Hitler's High Command* (Lawrence: University Press of Kansas, 2000), 122. In a postwar interrogation, Albert Speer said, "Naturally we had a supply organization of the Army maintaining large ammunition dumps behind the front in order to satisfy the needs of the front. In fact these dumps were very large, and out of all proportion to what was available at the front itself. It was just this fact which caused the heavy losses . . . during retreats, because then these dumps were lost." See Technical CCG, FIA, Report No. 54, "Problems of Supply in the Armed Forces," 11 October 1945, AIR 20/8780, National Archives, 5, based on a 13 July 1945 interrogation of Albert Speer.

90. US Strategic Bombing Survey (USSBS), Military Analysis Division, *The Impact of Allied Air Effort on German Logistics*, 2nd ed., January 1947, RG243/E2/B59, National Archives and Records Administration, 32. Based on 3 November 1945 interrogation.

91. Ziemke noted the disastrous impact of Hitler's stand-fast orders during the Russian counterattack at Stalingrad and in virtually every major action afterward. See Earl Frederick Ziemke, *Stalingrad to Berlin: The German Defeat in the East* (Washington, DC: US Army, Office of the Chief of Military History, 1968), 193.

92. See "Interrogation of Colonel Otto Eckstein," RG243/E32/B2, National Archives and Records Administration, 10–11. Five consumption units were enough to move a unit 310 miles over roads. In practice, however, it moved a division at most 200 miles in good weather over open terrain and on backroads. When weather and road conditions worsened, the figure dropped to 125 miles.

93. CSDIC, "Survey of the Supply System of the German Army," 43–44.

94. Ziemke, *Stalingrad to Berlin*, 417.

95. "USSBS Interview No. 11, Subject: Reichsminister Albert Speer," 5.

96. USSBS Military Analysis Division, *The Impact of Allied Air Effort on German Logistics*, 8.

97. Ibid., 8–9.

98. Davis, *Bombing the European Axis Powers*, 566–567.

Chapter Fifteen. The Balkan Air Force

1. For a useful overview of the development and effectiveness of the Balkan Air Force, see John Slessor, *The Central Blue: The Autobiography of Sir John Slessor, Marshal of the RAF* (New York: Praeger, 1957), 591–610.

2. Allied Forces Headquarters (AFHQ) Supreme Allied Commander's (SACMED) Despatch, "Yugoslav Campaign, 8 January 1944–12 December 1944," AIR 75/139, National Archives, 2–4.

3. C. J. C. Molony et al., *The Mediterranean and Middle East*, vol. 5, pt. 1, *The Campaign in Sicily 1943 and the Campaign in Italy 3rd September 1943 to 31st March 1944, History of the Second World War: United Kingdom Military Series*

(Uckfield, UK: Naval and Military Press, 2004), 372–373 (this volume of the British official history is henceforth referred to as "BOH 5" with part number).

4. AFHQ SACMED Despatch, "Yugoslav Campaign, 8 January 1944–12 December 1944," 5–9.

5. Ibid., 10–16.

6. US Army Air Forces Director of Intelligence (USAAF A2), "Signal Intelligence Service of the German Luftwaffe," vol. 10, "Technical Operations in the South," 1945, 519.6314-10 V.9, Air Force Historical Research Agency, 58–64.

7. Slessor, *Central Blue*, 558.

8. Headquarters Mediterranean Allied Air Forces (HQ MAAF) Historian, "The History of M.A.A.F., 10 December 1943–1 September 1944," April 1945, 622.01-1, Box 7, Air Force Historical Research Agency, 84–87, 125; John Slessor to Henry Maitland Wilson, 28 April 1944, AIR 75/93, National Archives.

9. Headquarters Mediterranean Allied Air Forces (HQ MAAF), "Assessment of Possible Rate of Supply to the Partisans per Squadron of U.S. C-47s," 19 April 1944, AIR 23/6274, National Archives; HQ MAAF, "Notes on Air Force Aspects of Assistance to the Jugoslav Partisans and Use of Bases in Jugoslavia," 19 April 1944, AIR 23/6274, National Archives, 1–2.

10. Headquarters Mediterranean Allied Air Forces (HQ MAAF), "Notes on a Meeting to Discuss Air Assistance to Jugoslav Partisans," 21 April 1944, AIR 23/6274, National Archives, 1–3.

11. Headquarters Mediterranean Allied Air Forces (HQ MAAF), "A Plan for the Use of Air Power to Take Advantage of the Situation in Yugoslavia in the Early Summer of 1944," 23 April 1944, AIR 23/6274, National Archives, 1–2.

12. Ibid., 2–4.

13. John Slessor to Charles Portal, 31 May 1944, AIR 75/69, National Archives.

14. Ira Eaker to Henry Arnold, 7 May 1944, MS57, Box 2, US Air Force Academy, 12–13.

15. Headquarters Mediterranean Allied Air Forces (HQ MAAF), "Command and Co-ordination of Trans-Adriatic Operations," 11 May 1944, AIR 75/93, National Archives, 1–3.

16. John Slessor, "The Strategic Value of the Yugoslav Partisans," 18 June 1944, AIR 75/93, National Archives. Tito's quote is in Roderic Owen, *The Desert Air Force: An Authoritative History Published in Aid of the Royal Air Force Benevolent Fund* (London: Hutchinson, 1948), 217. For additional insights on the German offensive, the evacuation of Tito, and the following talks at Bari, see Hilary St. George Saunders, *Royal Air Force, 1939–1945*, vol. 3, *The Fight Is Won* (London: Her Majesty's Stationery Office, 1954), 236–239.

17. Ira Eaker to William Elliot, "Draft Directive to A.O.C. Balkan Force from Air C. in C.," June 1944, AIR 75/93, National Archives.

18. AFHQ SACMED Despatch, "Yugoslav Campaign, 8 January 1944–12 December 1944," 11–19; Maurice G. Lihou, *Out of the Italian Night: Wellington Bomber Operations, 1944–45* (Shrewsbury, UK: Airlife, 2000), 141, 147.

19. William Jackson and T. P. Gleave, *The Mediterranean and Middle East,* vol. 6, *Victory in the Mediterranean,* pt. 2, *June to October 1944, History of the Second World War: United Kingdom Military Series* (Uckfield, UK: Naval and Military Press, 2004), 201–202, 216–218, 224. Referred to subsequently as "BOH 6" with part numbers.

20. Allied Forces Headquarters (AFHQ), "Army/Air Cooperation Notes No. 2,"

9 July 1945, WO 204/6797, National Archives, Appendix F: Notes on the Air Liaison Organisation with Balkan Air Force, June 1944–May 1945, 1–2.

21. BOH 6, pt. 3, 173–176.

22. Ira Eaker to Henry Arnold, 21 November 1944, MS57, Box 2, US Air Force Academy.

Chapter Sixteen. Taking Stock

1. William Jackson and T. P. Gleave, *The Mediterranean and Middle East,* vol. 6, *Victory in the Mediterranean,* pt. 2, *June to October 1944, History of the Second World War: United Kingdom Military Series* (Uckfield, UK: Naval and Military Press, 2004), 220 (this volume of the British official history is henceforth referred to as "BOH 6" with part number).

2. Richard Overy, *The Air War, 1939–1945* (New York: Stein and Day, 1980); Overy, *Why the Allies Won* (New York: Norton, 1995), 15–24, 316–325.

3. Gerhard Weinberg, *A World at Arms: A Global History of World War II* (Cambridge, UK: Cambridge University Press, 1994), 233–234, 307–348.

4. Overy, *Air War.*

5. For a discussion of how many Grand Alliance soldiers the air effort likely saved, see Robert S. Ehlers, Jr., *Targeting the Third Reich: Air Intelligence and the Allied Bombing Campaigns* (Lawrence: University Press of Kansas, 2009), 346–347.

6. Douglas Porch, *The Path to Victory: The Mediterranean Theater in World War II* (New York: Farrar, Straus, and Giroux, 2004), 174.

7. Headquarters Royal Air Force Middle East (HQ RAFME) Intelligence Officer to HQ RAFME Records, "Enemy Aircraft Casualties," 30 January 1945, AIR 23/1207, National Archives; Headquarters Fifteenth Air Force, "The Statistical Story of Fifteenth Air Force," 1945, SMS1136, Package, US Air Force Academy, 1–11; Headquarters Mediterranean Allied Air Forces (HQ) MAAF, "Air Power in the Mediterranean, November 1942–February 1945," AIR 23/1478, National Archives, 32. For German aircraft losses at the hands of the USAAF, see *Army Air Forces Statistical Digest,* Table 168, at http://permanent.access.gpo.gov/lps51153 /airforcehistory/usaaf/digest/t166.htm. The total listed for the Mediterranean theater of operations is 9,749, of which 7,849 were destroyed after May 1943. Aircraft loss figures for the Eastern Front are notoriously difficult to confirm, but the average, compiled from a variety of sources, appears to be around 11,000.

8. Headquarters Royal Air Force Middle East (HQ RAFME) Tactics Assessment Office, "Air Tactics and Operational Notes on 201 Naval Co-operation Group, R.A.F.," n.d. (probably June 1943), AIR 23/1282, National Archives, Appendix 12: Enemy Merchant Shipping Losses; Admiralty Trade Division, "Summary of Enemy Merchant Shipping Losses," 24 September 1945, AIR 23/1175, National Archives.

9. For lend-lease figures and an excellent assessment of what these items did to transform the Red Army's war-making ability, see Richard Overy, *Russia's War: A History of the Soviet War Effort, 1941–1945* (New York: Penguin, 1998), 193–198.

10. John Slessor, *The Central Blue: The Autobiography of Sir John Slessor, Marshal of the RAF* (New York: Praeger, 1957), 565.

11. For a succinct assessment of these important questions and their implications for future military effectiveness, see Ervin J. Rokke, Thomas A. Drohan, and Terry C. Pierce, "Combined Effects Power," *Joint Forces Quarterly* 73, no. 2 (2014): 26–31.

SELECTED BIBLIOGRAPHY

Archival Sources (US)

AIR FORCE HISTORICAL RESEARCH AGENCY, MAXWELL AIR FORCE BASE, AL

Records of the Air Corps Tactical School
Records of the Allied Air Offensive against Germany
Records of the Army Air Corps (USAAC), later Army Air Forces (USAAF)
Records of the Committee of Operations Analysts
Records of the Eighth Air Force
Records of the Fifteenth Air Force
Records of Mediterranean Allied Air Forces
Records of the Military Intelligence Division
Records of the US Strategic Air Forces in Europe
Records of the US Strategic Bombing Survey

Library of Congress Manuscript Collection, Washington, DC

Arnold, Henry Harley, Papers
Doolittle, James H., Papers
Eaker, Ira C., Papers
Spaatz, Carl A., Papers

National Archives and Records Administration, Washington, DC

RG18 Records of the Army Air Forces (USAAF)
RG243 Records of the US Strategic Bombing Survey

US Air Force Academy Special Collections, US Air Force Academy, CO

MS 57, John W. Huston Henry ("Hap") Harley Arnold Collection
MS 58, Reade F. Tilley Collection
SMS1136

Archival Sources (British)

NATIONAL ARCHIVES, LONDON

AIR 8	Chiefs of Staff Papers
AIR 22	Royal Air Force (RAF) Operational Summaries
AIR 23	Royal Air Force (RAF) Overseas Commands: Reports and Correspondence
AIR 75	Slessor, John, Papers
WO 201	War Office, Middle East Forces Military Headquarters Papers, WWII

WO 204   War Office, Allied Forces, Mediterranean Theatre Military
          Headquarters, WWII

AIR HISTORICAL BRANCH, ROYAL AIR FORCE (RAF), NORTHOLT, UK
Axis Translations, vol. 7, World War II
Axis Translations, vol. 8, World War II
Axis Translations, vol. 9, World War II
Axis Translations, vol. 10, World War II
Axis Translations, vol. 11, World War II
Axis Translations, vol. 12, World War II
Axis Translations, vol. 13, World War II
Lewis, Anthea M., "A Tribute to Air Vice-Marshal GG Dawson, 1895–1944, CMSO
    ME and N. Africa, 1941–1944"
Tedder, Arthur, Box 3, Correspondence and Papers, 1940–1941
Tedder, Arthur, Box 4, Diaries, December 1940–July 1942
Tedder, Arthur, Box 5, List of Events, 1941–1945; Notes on Greece, 1941–1943;
    Battle of Egypt, October 1942; Moscow Conference, January 1945
Tedder, Arthur, Box 6, Correspondence and Papers, 1942
Tedder, Arthur, Box 7, Correspondence and Papers, 1943
Tedder, Arthur, Box 8, Despatches, Report on Strategic Bombing in Mediterranean

Published Works

Air Ministry. *The Origins and Development of Operational Research in the Royal
    Air Force.* London: Her Majesty's Stationery Office, 1963.
———. *Supply Organisation of the German Air Force, 1935–1945.* Reprint of 1946
    original. East Sussex, UK: Naval and Military Press, 2009.
Ansel, Walter. *Hitler and the Middle Sea.* Durham, NC: Duke University Press, 1972.
Arnold, Henry Harley. *Global Mission.* New York: Harper, 1949.
Austin, Alexander Berry. *Birth of an Army.* London: Gollancz, 1943.
Ball, Edmund F. *Staff Officer with the Fifth Army: Sicily, Salerno, and Anzio.* New
    York: Exposition Press, 1958.
Ball, S. J. *The Bitter Sea.* London: Harper, 2010.
Barnett, Corelli. *The Desert Generals.* New York: Viking, 1961.
———. *Engage the Enemy More Closely: The Royal Navy in the Second World War.*
    London: Hodder and Stoughton, 1991.
Barr, Niall. *Pendulum of War: The Three Battles of El Alamein.* Woodstock, NY:
    Overlook, 2005.
Bartov, Omer. *Hitler's Army: Soldiers, Nazis, and War in the Third Reich.* New York:
    Oxford University Press, 1992.
Bath, Alan Harris. *Tracking the Axis Enemy: The Triumph of Anglo-American Na-
    val Intelligence.* Lawrence: University Press of Kansas, 1998.
Bechtold, B. Michael. "A Question of Success: Tactical Air Doctrine and Practice in
    North Africa, 1942–1943." *Journal of Military History* 68, no. 3 (July 2004).
Behrendt, Hans. *Rommel's Intelligence in the Desert Campaign: 1941–1943.* Lon-
    don: Kimber, 1985.
Bekker, Cajus. *The Luftwaffe War Diaries: The German Air Force in World War II.*
    New York: Da Capo, 1994.

Bennett, Ralph Francis. *Behind the Battle: Intelligence in the War with Germany, 1939–1945*. London: Pimlico, 1999.

———. *Ultra and Mediterranean Strategy*. New York: Morrow, 1989.

Beurling, George Frederick, and Leslie Roberts. *Malta Spitfire: The Story of a Fighter Pilot*. New York: Little and Ives, 1944.

Bickers, Richard Townshend. *The Desert Air War, 1939–1945*. London: Cooper, 1991.

Bierman, John, and Colin Smith. *The Battle of Alamein: Turning Point, World War II*. New York: Viking, 2002.

Blumenson, Martin. *Salerno to Cassino*. Washington, DC: US Army, Office of the Chief of Military History, 2002.

Bond, Brian, ed. *Chief of Staff: The Diaries of Lt. Gen. Sir Henry Pownall*. London: Cooper, 1974.

Boog, Horst. *The Conduct of the Air War in the Second World War: An International Comparison—Proceedings of the International Conference of Historians in Freiburg im Breisgau, from 29 August to 2 September 1988*. New York: Berg, 1992.

Boog, Horst et al. *Germany and the Second World War*. Volume 6. *The Global War*. Edited by the Militärgeschichtliches Forschungsamt. Translated by John Brownjohn, Patricia Crampton, Ewald Ossers, Reinhard Stumpf, and Louise Willmot. Oxford, UK: Oxford University Press, 2001.

Bosworth, R. J. B. *Mussolini's Italy: Life under the Dictatorship, 1915–1945*. New York: Penguin, 2006.

Bowyer, Chaz, and Christopher F. Shores. *Desert Air Force at War*. London: Allan, 1981.

———. *Men of the Desert Air Force, 1940–1943*. London: Kimber, 1984.

Bragadin, Marc. *The Italian Navy in World War II*. Annapolis, MD: US Naval Institute, 1957.

Brereton, Lewis H. *The Brereton Diaries*. New York: Morrow, 1946.

Brookes, Andrew J. *Air War over Italy, 1943–1945*. Shepperton, UK: Allan, 2000.

Bryant, Sir Arthur. *Triumph in the West*. New York: Doubleday, 1959.

———. *Turn of the Tide: A History of the War Years Based on the Diaries of Field-Marshal Lord Alan Brooke, Chief of the Imperial General Staff*. New York: Doubleday, 1957.

Buckley, Christopher. *Greece and Crete, 1941*. Athens: Efstathiadis, 2003.

Bungay, Stephen. *The Most Dangerous Enemy: A History of the Battle of Britain*. London: Aurum, 2000.

Burdick, Charles Burton. *Germany's Military Strategy and Spain in World War II*. Syracuse, NY: Syracuse University Press, 1968.

Butler, James Ramsay Montagu. *Grand Strategy*, vol. 2, *September 1939 to June 1941*, History of the Second World War, UK Military Series. London: Her Majesty's Stationery Office, 1957.

Cabell, Charles P., and Charles P. Cabell, Jr. *A Man of Intelligence: Memoirs of War, Peace, and the CIA*. Colorado Springs, CO: Impavide, 1997.

Caldwell, Donald L., and Richard Muller. *The Luftwaffe over Germany: Defense of the Reich*. London: Greenhill, 2007.

Callahan, Raymond. *Churchill and His Generals*. Lawrence: University Press of Kansas, 2007.

Carver, Michael. *Dilemmas of the Desert War: A New Look at the Libyan Campaign, 1940–1942.* Bloomington: Indiana University Press, 1986.

Center for Air Force History. *The AAF in Northwest Africa: An Account of the Twelfth Air Force in the Northwest Africa Landings and the Battle for Tunisia—an Interim Report.* Washington, DC: CAFH, 1992.

Ceva, Lucio, and Andra Curami. "Air Army and Aircraft Industry in Italy, 1936–1943." In *The Conduct of the Air War in the Second World War,* edited by Horst Boog. New Haven, CT: Berg, 1986.

Chappell, F. R. *Wellington Wings: An RAF Intelligence Officer in the Western Desert.* New ed., Bodmin, UK: Crécy, 1992.

Ciano, Galeazzo, and H. Gibson. *The Ciano Diaries, 1939–1943: The Complete, Unabridged Diaries.* Edited by H. Gibson. Introduction by S. Welles. Garden City, NY: Doubleday, 1946.

Citino, Robert Michael. *Death of the Wehrmacht: The German Campaigns of 1942.* Lawrence: University Press of Kansas, 2007.

———. *The German Way of War: From the Thirty Years' War to the Third Reich.* Lawrence: University Press of Kansas, 2012.

Clark, Alan. *The Fall of Crete.* London: Cassell, 2001.

Clark, Lloyd. *Anzio: Italy and the Battle for Rome—1944.* New York: Grove, 2006.

Clark, Mark W. *Calculated Risk.* New York: Harper, 1950.

Clarke, Jeffrey J., and Robert Ross Smith. *Riviera to the Rhine.* Washington, DC: US Army, Center of Military History, 1993.

Clayton, Aileen. *The Enemy Is Listening.* Manchester, UK: Crécy, 1993.

Clayton, Tim, and Phil Craig. *The End of the Beginning: From the Siege of Malta to the Allied Victory at El Alamein.* New York: Free Press, 2003.

Coakley, Robert W., and Richard M. Leighton. *Global Logistics and Strategy: 1943–1945.* Washington, DC: US Army, Office of the Chief of Military History, 1968.

Coles, Harry. *Ninth Air Force in the Western Desert Campaign to 23 January 1943.* Washington, DC: US Army Air Forces Historical Division, 1945.

Cooke, R., and Roy Conyers Nesbit. *Target: Hitler's Oil—Allied Attacks on German Oil Supplies, 1939–45.* New York: HarperCollins, 1985.

Cooling, B. Franklin. *Case Studies in the Achievement of Air Superiority.* Washington, DC: Center for Air Force History, 1994.

———. *Case Studies in the Development of Close Air Support.* Washington, DC: Office of Air Force History, 1990.

Corum, James. *The Luftwaffe: Creating the Operational Air War, 1918–1940.* Lawrence: University Press of Kansas, 1997.

———. "The Luftwaffe's Army Support Doctrine, 1918–1941." *Journal of Military Affairs* 59 (January 1995).

———. *Wolfram von Richthofen: Master of the German Air War.* Lawrence: University Press of Kansas, 2008.

Corum, James S., and Richard Muller. *The Luftwaffe's Way of War: German Air Force Doctrine, 1911–1945.* Baltimore, MD: Nautical and Aviation, 1998.

Corvaja, Santi. *Hitler and Mussolini: The Secret Meetings.* New York: Enigma, 2008.

Cox, Sebastian, and Peter Gray, eds. *Air Power History: Turning Points from Kitty Hawk to Kosovo.* London: Routledge, 2002.

Craven, Wesley Frank, and James Lea Cate. *The Army Air Forces in World War II.* 7 vols. Washington, DC: Office of Air Force History, 1983.

Croissant, Michael P., and Bülent Aras, eds. *Oil and Geopolitics in the Caspian Sea Region.* Westport, CT: Greenwood, 1999.

Cross, Kenneth, with Vincent Orange. *Straight and Level.* London: Grub Street, 1993.

Daso, Dik A. *Hap Arnold and the Evolution of American Airpower.* Washington, DC: Smithsonian Institution, 2000.

Davis, Richard G. *Bombing the European Axis Powers: A Historical Digest of the Combined Bomber Offensive, 1939–1945.* Maxwell Air Force Base, AL: Air University Press, 2006.

———. *Carl A. Spaatz and the Air War in Europe.* Washington, DC: Smithsonian Institution Press, 1992.

Deakin, F. W. *The Brutal Friendship: Mussolini, Hitler, and the Fall of Italian Fascism.* New York: Harper and Row, 1962.

Dean, Maurice. *Royal Air Force and Two World Wars.* Worthing, UK: Littlehampton, March 22, 1979.

Deane, John R. *The Strange Alliance.* New York: Viking, 1947.

D'Este, Carlo. *Eisenhower: A Soldier's Life.* New York: Henry Holt, 2002.

———. *World War II in the Mediterranean, 1942–1945.* Chapel Hill, NC: Algonquin, 1990.

DiNardo, R. L. *Germany and the Axis Powers from Coalition to Collapse.* Lawrence: University Press of Kansas, 2005.

Douglas, Sholto, and Robert Wright. *Combat and Command: The Story of an Airman in Two World Wars.* New York: Simon and Schuster, 1966.

Duke, Neville, and Norman L. R. Franks. *The War Diaries of Neville Duke: The Journals of Squadron Leader NF Duke DSO, OBE, DFC & 2 Bars, AFC, MC(Cz), 1941–44.* London: Grub Street, 1995.

Dunning, Chris. *Courage Alone: The Italian Air Force, 1940–1943.* Revised and updated. Ottringham, UK: Hikoki, 2009.

———. *Regia Aeronautica: The Italian Air Force, 1923–1945: An Operational History.* Stamford, CT: Classic, 2010.

Ehlers Jr., Robert S. *Targeting the Third Reich: Air Intelligence and the Allied Bombing Campaigns.* Lawrence: University Press of Kansas, 2009.

Eisenhower, Dwight D. *Crusade in Europe.* Reprint edition. Baltimore, MD: Johns Hopkins University Press, 1997.

Ellis, John. *Brute Force: Allied Strategy and Tactics in the Second World War.* New York: Viking, 1990.

———. *Cassino, the Hollow Victory: The Battle for Rome, January–June 1944.* London: Deutsch, 1984.

Embry, Basil Edward. *Mission Completed.* London: Methuen, 1957.

Ferris, John. "The British Army, Signals and Security," in *Intelligence and Military Operations,* edited by Michael I. Handel. London: Routledge, 1990.

———. "The Usual Source: Signals Intelligence and Planning for the 'Crusader' Offensive, 1941," *Intelligence and National Security* 14, no. 1 (1999).

Finney, Robert T. *History of the Air Corps Tactical School, 1920–1940.* Washington, DC: Center for Air Force History, 1992.

Fischer, Louis. *Dawn of Victory.* New York: Duell, Sloan, and Pearce, 1942.

Fisher, Ernest F. *Cassino to the Alps.* Washington, DC: US Army, Center of Military History, 2002.

Förster, Jurgen. "The Dynamics of *Volksgemeinschaft:* The Effectiveness of the German Military Establishment in the Second World War," in *Military Effectiveness,* vol. 3, *The Second World War,* edited by Williamson Murray and Allan R. Millett. Boston: Unwin Hyman, 1988.

Garland, Albert N., and Howard M. Smyth. *United States Army in World War II: The Mediterranean Theater of Operations, Sicily, and the Surrender of Italy.* Washington, DC: US Army, Center of Military History, 2002.

Gaunson, B. *The Anglo-French Collision in Lebanon and Syria, 1940–1945.* London: MacMillan, 1986.

Giziowski, Richard J. *The Enigma of General Blaskowitz.* London: Cooper, 1997.

Gladman, Brad. *Intelligence and Anglo-American Air Support in World War Two: The Western Desert and Tunisia, 1940–43.* Basingstoke, UK: Palgrave Macmillan, 2009.

Glantz, David M. *Red Storm over the Balkans: The Failed Soviet Invasion of Rumania, Spring 1944.* Lawrence: University Press of Kansas, 2007.

Goda, Norman J. W. *Tomorrow the World: Hitler, Northwest Africa, and the Path toward America.* College Station: Texas A & M University Press, 1998.

Goddard, George W., and DeWitt S. Copp. *Overview: A Life-long Adventure in Aerial Photography.* Garden City, NY: Doubleday, 1969.

Gooch, John. *Mussolini and His Generals: The Armed Forces and Fascist Foreign Policy, 1922–1940.* Cambridge, UK: Cambridge University Press, 2007.

Gooderson, Ian. *Air Power at the Battlefront: Allied Close Air Support in Europe, 1943–45.* London: Cass, 1998.

Gordon IV, John. "Operation Crusader: Preview of the Non-linear Battlefield." *Military Review* 71 (1991): 48–61.

Graham, Dominick, and Shelford Bidwell. *Tug of War: The Battle for Italy, 1943–1945.* New York: St. Martin's, 1986.

Greenfield, George. *Chasing the Beast: One Man's War.* London: Cohen, 1998.

Griehl, Manfred, and Joachim Dressel. *Luftwaffe Combat Aircraft, 1935–1945: Development, Production, Operations.* Atglen, PA: Schiffer, 1994.

Guedalla, Philip. *Middle East, 1940–1942: A Study in Air Power.* London: Hodder and Stoughton, 1944.

Gundelach, Karl. *Die deutsche Luftwaffe im Mittelmeer, 1940–1945.* 2 vols. Frankfurt am Mann: Lang, 1981.

Halder, Franz, Charles Burton Burdick, and Hans Adolf Jacobsen. *The Halder War Diary, 1939–1942.* Novato, CA: Presidio, 1988.

Hall, David Ian. *Strategy for Victory: The Development of British Tactical Air Power, 1919–1943.* Westport, CT: Praeger, 2008.

Hall, R. Cargill. *Case Studies in Strategic Bombardment.* Washington, DC: US Air Force History and Museums Program, 1998.

Hamilton, James. *The Air Battle for Malta: The Diaries of a Spitfire Pilot.* Barnsley, UK: Pen and Sword, 2006.

Hamilton, Nigel. *Master of the Battlefield: Monty's War Years, 1942–1944.* New York: McGraw-Hill, 1983.

Handel, Michael. *Intelligence and Military Operations.* London: Cass, 1990.

Hattendorf, John B. *Naval Policy and Strategy in the Mediterranean: Past, Present, and Future.* London: Cass, 2000.

Hauenstein, Ralph W., and Donald E. Markle. *Intelligence Was My Line: Inside Eisenhower's Other Command.* New York: Hippocrene, 2005.

Hehn, Paul N. *The German Struggle against Yugoslav Guerrillas in World War II: German Counter-Insurgency in Yugoslavia.* New York: Columbia University Press, 1979.

Held, Werner, and Ernst Obermaier. *The Luftwaffe in the North African Campaign, 1941–1943: A Photo Chronicle.* West Chester, PA: Schiffer, 1992.

Herington, John. *Air Power over Europe, 1944–1945.* Canberra: Australian War Memorial, 1963.

———. *Air War against Germany and Italy, 1939–1943.* Sydney: Halstead Press, 1954.

Higham, Robin. *Bases of Air Strategy: Building Airfields for the RAF, 1914–1945.* Shrewsbury, UK: Airlife, 1998.

———. *Diary of a Disaster: British Aid to Greece, 1940–1941.* Lexington: University Press of Kentucky, 1986.

Higham, Robin, and Stephen John Harris. *Why Air Forces Fail: The Anatomy of Defeat.* Lexington: University Press of Kentucky, 2006.

Hinsley, F. H. et al. *British Intelligence in the Second World War: Its Influence on Strategy and Operations.* 3 vols. London: Her Majesty's Stationery Office, 1979–1988.

———. *Codebreakers: The Inside Story of Bletchley Park.* Oxford, UK: Oxford University Press, 2001.

Hirszowicz, Lukasz. *The Third Reich and the Arab East.* London: Routledge and Kegan Paul, 1966.

Hitler, Adolf, Helmut Heiber, and David M. Glantz. *Hitler and His Generals: Military Conferences, 1942–1945: The First Complete Stenographic Record of the Military Situation Conferences, from Stalingrad to Berlin.* New York: Enigma, 2003.

Holland, James. *Fortress Malta: An Island under Siege, 1940–1943.* New York: Miramax, 2003.

———. *Together We Stand: America, Britain, and the Forging of an Alliance.* New York: Hyperion, 2005.

Hooton, E. R. *Phoenix Triumphant: The Rise and Rise of the Luftwaffe.* London: Brockhampton, 1999.

Horne, Alistair, and David Montgomery. *Monty: The Lonely Leader, 1944–1945.* New York: HarperCollins, 1994.

House, Jonathan M. *Combined Arms Warfare in the Twentieth Century.* Lawrence: University Press of Kansas, 2001.

Howard, Michael. *Mediterranean Strategy in the Second World War.* London: Greenhill, 1993.

Howard, Michael, Peter Paret, and Rosalie West. *Carl von Clausewitz: On War.* Princeton, NJ: Princeton University Press, 1984.

Howe, George F. *Northwest Africa: Seizing the Initiative in the West.* Washington, DC: US Army, Center of Military History, 2002.

Hughes, Thomas Alexander. *Over Lord: General Pete Quesada and the Triumph of Tactical Air Power in World War II.* New York: Free Press, 1995.

Hyndhope, Andrew Browne Cunningham. *A Sailor's Odyssey; The Autobiography of Admiral of the Fleet, Viscount Cunningham of Hyndhope.* New York: Dutton, 1951.

Ingersoll, Ralph. *The Battle Is the Pay-off.* New York: Harcourt, Brace, 1943.

Ismay, Lord. *The Memoirs of General the Lord Ismay, K.G., P.C., G.C.P., C.H., D.S.O.* London: Heinemann, 1960.

Jackson, W. G. F. *The Battle for North Africa, 1940–43*. New York: Mason/Charter, 1975.

Jackson, William, and T. P. Gleave. *The Mediterranean and Middle East*. Volume 6. *Victory in the Mediterranean, June to October 1944*. Parts 1–3. Reprint of 1954 original. *History of the Second World War: United Kingdom Military History Series—Official Campaign History*. Uckfield, UK: Naval and Military Press, 2004.

Jones, Martyn R. *Desert Flyer: The Log and Journal of Flying Officer William E. Marsh*. Atglen, PA: Schiffer, 1997.

Jones, Matthew. *Britain, the United States and the Mediterranean War, 1942–44*. New York: St. Martin's, 1996.

Jones, Reginald V. *Most Secret War*. London: Hamilton, 1978.

Jordan, Philip. *Jordan's Tunis Diary*. London: Collins, 1943.

Kahn, David. *Hitler's Spies: German Military Intelligence in World War II*. New York: Macmillan, 1978.

Keitel, Wilhelm, and Walter Gorlitz. *The Memoirs of Field-Marshal Wilhelm Keitel*. New York: Cooper Square, 2000.

Kesselring, Albert. *The Memoirs of Field-Marshal Kesselring*. London: Kimber, 1953.

Knox, MacGregor. *Common Destiny: Dictatorship, Foreign Policy, and War in Fascist Italy and Nazi Germany*. Cambridge, UK: Cambridge University Press, 2000.

———. *Hitler's Italian Allies: Royal Armed Forces, Fascist Regime, and the War of 1940–1943*. Cambridge, UK: Cambridge University Press, 2000.

———. *Mussolini Unleashed, 1939–1941: Politics and Strategy in Fascist Italy's Last War*. Cambridge, UK: Cambridge University Press, 1982.

Kreis, John F. *Piercing the Fog: Intelligence and Army Air Forces Operations in World War II*. Honolulu, HI: University Press of the Pacific, 2004.

Larrabee, Eric. *Commander in Chief: Franklin Delano Roosevelt, His Lieutenants, and Their War*. Annapolis, MD: Naval Institute Press, 2004.

Lavigne, J. P. A. Michel, and James F. Edwards. *Kittyhawk Pilot: Wing Commander J. F. (Stocky) Edwards*. Battleford, SK, Canada: Turner-Warwick, 1983.

Leverkuehn, Paul. *German Military Intelligence*. New York: Praeger, 1954.

Levine, Alan J. *The War against Rommel's Supply Lines, 1942–43*. Mechanicsburg, PA: Stackpole, 2008.

Lewin, Ronald. *Churchill as Warlord*. New York: Stein and Day, 1973.

———. *The Chief: Field Marshal Lord Wavell Commander-in-Chief and Viceroy, 1939–1947*. New York: Farrar, Straus, and Giroux, 1980.

———. *Montgomery as Military Commander*. New York: Stein and Day, 1971.

Lihou, Maurice G. *Out of the Italian Night: Wellington Bomber Operations, 1944–45*. Shrewsbury, UK: Airlife, 2003.

Lloyd, Hugh Pughe. *Briefed to Attack: Malta's Part in African Victory*. London: Hodder and Stoughton, 1949.

Longmore, Arthur Murray. *From Sea to Sky, 1910–1945: Memoirs*. London: Bles, 1946.

Lucas, Laddie. *Malta: The Thorn in Rommel's Side—Six Months That Turned the War*. London: Paul, 1992.

Luck, Hans von. *Panzer Commander: The Memoirs of Colonel Hans von Luck*. New York: Praeger, 1989.

MacCloskey, Monro. *Torch and the Twelfth Air Force*. New York: Rosen, 1971.

MacDonald, Charles B., and Sidney T. Mathews. *United States Army in World War II: Three Battles—Arnaville, Altuzzo, and Schmidt.* Washington, DC: US Army, Center of Military History, 1952.

Macksey, Kenneth. *Kesselring: The Making of the Luftwaffe.* London: Batsford, 1978.

Maclean, Fitzroy. *Eastern Approaches.* London: Cape, 1949.

Macmillan, Norman. *The Royal Air Force in the World War.* 3 vols. London: Harrap, 1949.

Mallmann Showell, Jak P. *Fuehrer Conferences on Naval Affairs, 1939–1945.* London: Chatham, 2005.

Mark, Eduard. *Aerial Interdiction in Three Wars.* Washington, DC: Center for Air Force History, 1994.

Marshall, George C., Henry Harley Arnold, and Ernest Joseph King. *The War Reports of General of the Army George C. Marshall, Chief of Staff, General of the Army H. H. Arnold, Commanding General, Army Air Forces [and] Fleet Admiral Ernest J. King, Commander-in-Chief, United States Fleet and Chief of Naval Operations.* Philadelphia, PA: Lippincott, 1947.

Matloff, Maurice. *Strategic Planning for Coalition Warfare, 1943–1944.* Washington, DC: US Army, Office of the Chief of Military History, 1959.

Matloff, Maurice, and Edwin M. Snell. *Strategic Planning for Coalition Warfare, 1941–1942.* Washington, DC: US Army, Center of Military History, 1999.

McCarthy, Michael C. *Air-to-Ground Battle for Italy.* Maxwell Air Force Base, AL: Air University Press, 2004.

McFarland, Keith D., and David L. Roll. *Louis Johnson and the Arming of America: The Roosevelt and Truman Years.* Bloomington: Indiana University Press, 2005.

McNab, Chris. *German Luftwaffe in WWII.* London: Amber, 2009.

Megargee, Geoffrey P. *Inside Hitler's High Command.* Lawrence: University Press of Kansas, 2000.

Mets, David R. "A Glider in the Propwash of the Royal Air Force? Gen Carl A. Spaatz, the RAF, and the Foundations of American Tactical Air Doctrine." In *Airpower and Ground Armies: Essays on the Evolution of Anglo-American Air Doctrine, 1940–1943.* Edited by Daniel R. Mortensen, 45–91. Maxwell Air Force Base, AL: Air University Press, 1998.

———. *Master of Airpower: General Carl A. Spaatz.* Novato, CA: Presidio, 1988.

Mierzejewski, Alfred C. *The Collapse of the German War Economy, 1944–1945: Allied Air Power and the German National Railway.* Chapel Hill: University of North Carolina Press, 1988.

Milutinovic, Aleksanar. "First Foray: The United States Army Air Forces' Strategic Strike at Ploesti, 1 August, 1943," unpublished M.A. thesis, The Ohio State University, 2001.

Mitcham, Jr., Samuel W. *Blitzkrieg No Longer: The German Wehrmacht in Battle, 1943.* Mechanicsburg, PA: Stackpole, 2010.

———. *Retreat to the Reich: The German Defeat in France, 1944.* New York: Praeger, 2000.

Mitcham, Jr., Samuel W., and Friedrich von Stauffenberg. *The Battle of Sicily: How the Allies Lost Their Chance for Total Victory.* New York: Orion, 1991.

Molony, C. J. C., with F. C. Flynn, H. L. Davies, and T. P. Gleave. *The Mediterranean and Middle East.* Vol. 5. *The Campaign in Sicily 1943 and the Campaign in Italy 3rd September 1943 to 31st March 1944.* Parts 1–2. Reprint of 1954 original.

History of the Second World War: United Kingdom Military History Series—Official Campaign History. Uckfield, UK: Naval and Military Press, 2004.

Momyer, William W., and A. J. C. Lavalle. *Airpower in Three Wars.* Reprint ed. Maxwell Air Force Base, AL: Air University Press, 2003.

Montgomery, Bernard Law. *The Memoirs of Field Marshal the Viscount Montgomery of Alamein, K. G.* London: World, 1958.

Moorehead, Alan. *Desert War: The North African Campaign, 1940–1943.* New York: Penguin, 2001.

Morely-Mower, Geoffrey. *Messerschmitt Roulette: The Western Desert, 1941–42.* Shrewsbury, UK: Airlife, 2003.

Morewood, Steven. *The British Defence of Egypt: Conflict and Crisis in the Eastern Mediterranean.* London: Routledge, 2004.

Morgan, Philip. *The Fall of Mussolini: Italy, the Italians, and the Second World War.* Oxford, UK: Oxford University Press, 2007.

Mortensen, Daniel R. "The Legend of Laurence Kuter: Agent for Airpower Doctrine." In *Airpower and Ground Armies: Essays on the Evolution of Anglo-American Air Doctrine, 1940–1943.* Edited by Daniel R. Mortensen, 93–145. Maxwell Air Force Base, AL: Air University Press, 1998.

Motter, T. H. *The Middle East Theater: The Persian Corridor and Aid to Russia.* Washington, DC: US Army, Office of the Chief of Military History, 1952.

Muller, Richard. *The German Air War in Russia.* Baltimore, MD: Nautical and Aviation Press, 1992.

Murray, Williamson. *Luftwaffe.* Baltimore, MD: Nautical and Aviation Press, 1985.

———. *Strategy for Defeat: The Luftwaffe, 1933–1945.* Maxwell Air Force Base, AL: Air University Press, 1983.

Murray, Williamson, and Allan R. Millet, eds. *Military Innovation in the Interwar Period.* Cambridge, UK: Cambridge University Press, 1998.

———. *A War to Be Won: Fighting the Second World War, 1937–1945.* Cambridge, MA: Belknap Press of Harvard University Press, 2000.

Nesbit, Roy Conyers. *Eyes of the RAF: A History of Photo-reconnaissance.* Phoenix Mill, UK: Sutton, 1996.

Neulen, Hans Werner. *In the Skies of Europe: Air Forces Allied to the Luftwaffe, 1939–1945.* Ramsbury, UK: Crowood, 2005.

North, John. *The Alexander Memoirs, 1940–1945: Field-Marshal Earl Alexander of Tunis.* London: Cassell, 1962.

O'Hara, Vincent P. *Struggle for the Middle Sea: The Great Navies at War in the Mediterranean, 1940–1945.* Annapolis, MD: Naval Institute Press, 2009.

Orange, Vincent. *Coningham: A Biography of Air Marshal Sir Arthur Coningham, KCB, KBE, DSO, MC, DFC, AFC.* Washington, DC: Center for Air Force History, 1992.

———. "Getting Together: Tedder, Coningham, and Americans in the Desert and Tunisia, 1940–43." In *Airpower and Ground Armies: Essays on the Evolution of Anglo-American Air Doctrine, 1940–1943.* Edited by Daniel R. Mortensen, 1–44. Maxwell Air Force Base, AL: Air University Press, 1998.

———. *Slessor: Bomber Champion—The Life of Marshal of the Royal Air Force Sir John Slessor, GCB, DSO, MC.* London: Grub Street, 2006.

———. *Tedder: Quietly in Command.* London: Cass, 2004.

Osborne, Richard E. *World War II in Colonial Africa: The Death Knell of Colonialism.* Indianapolis, IN: Riebel-Roque, 2001.

Overy, Richard J. *The Air War, 1939–1945*. New York: Stein and Day, 1980.

———. *The Bombers and the Bombed: Allied Air War over Europe, 1940–1945*. NY: Viking, 2013.

———. *Russia's War: A History of the Soviet Military Effort: 1941–1945*. NY: Penguin, 1998.

———. *Why the Allies Won*. New York: Norton, 1997.

Owen, Roderic. *The Desert Air Force: An Authoritative History Published in Aid of the Royal Air Force Benevolent Fund*. London: Hutchinson, 1948.

———. *Tedder*. London: Collins, 1952.

Paine, Lauran. *German Military Intelligence in World War II: The Abwehr*. New York: Stein and Day, 1984.

Parton, James. *"Air Force Spoken Here": General Ira Eaker and the Command of the Air*. Bethesda, MD: Adler and Adler, 1986.

Paxton, Robert O. *Vichy France: Old Guard and New Order, 1940–1944*. New York: Columbia University Press, 2001.

Payne, Stanley G. *Franco and Hitler: Spain, Germany, and World War II*. New Haven, CT: Yale University Press, 2008.

Phillips, C. E. Lucas. *Alamein*. Boston: Little, Brown, 1962.

Pitt, Barrie. *The Crucible of War*. 2nd ed. London: Macmillan, 1986.

Playfair, I. S. O. *The Mediterranean and Middle East*. Vol. 1. *Early Successes against Italy (to May 1941)*. Reprint of 1954 original. *History of the Second World War: United Kingdom Military History Series—Official Campaign History*. Uckfield, UK: Naval and Military Press, 2004.

———. *The Mediterranean and Middle East*. Vol. 3. *British Fortunes Reach Their Lowest Ebb*. Reprint of 1954 original. *History of the Second World War: United Kingdom Military History Series—Official Campaign History*. Uckfield, UK: Naval and Military Press, 2004.

———. *The Mediterranean and Middle East*. Vol. 4. *The Destruction of the Axis Forces in Africa*. Reprint of 1954 original. *History of the Second World War: United Kingdom Military History Series—Official Campaign History*. Uckfield, UK: Naval and Military Press, 2004.

Playfair, I. S. O., F. C. Flynn, C. J. C. Molony, and S. E. Toomer. *The Mediterranean and Middle East*. Vol. 2. *The Germans Come to the Help of Their Ally (1941)*. Reprint of 1954 original. *History of the Second World War: United Kingdom Military History Series—Official Campaign History*. Uckfield, UK: Naval and Military Press, 2004.

Porch, Douglas. *The Path to Victory: The Mediterranean Theater in World War II*. New York: Farrar, Straus, and Giroux, 2004.

Powell, Lewis F., and Diane T. Putney. *ULTRA and the Army Air Forces in World War II: An Interview with Associate Justice of the U.S. Supreme Court Lewis F. Powell, Jr.* Washington, DC: Office of Air Force History, 1987.

Preda, Gavriil. "German Foreign Policy towards the Romanian Oil during 1938–1940." *International Journal of Social Science and Humanity* 3, no. 3 (May 2013): 326–329.

Prien, Jochen. *Einsatz des Jagdgeschwaders 77 von 1939 bis 1945: ein Kriegstagebuch nach Dokumenten, Berichten und Erinnerungen*. Eutin, Germany: Struve-Druck, 1994.

Raugh, Harold E. *Wavell in the Middle East, 1939–1941: A Study in Generalship*. London: Brassey's, 1993.

Rein, Christopher M. *The North African Air Campaign: U.S. Army Air Forces from El Alamein to Salerno*. Lawrence: University Press of Kansas, 2012.

Reuth, Ralf Georg. *Entscheidung im Mittelmeer die südliche Peripherie Europas in der deutschen Strategie des Zweiten Weltkrieges, 1940–1942*. Edited by Genehmigte Lizenzausg. Erlangen, Germany: K. Muller, 1999.

Rich, Norman. *Hitler's War Aims: Ideology, the Nazi State, and the Course of Expansion*. New York: Norton.

Richards, Denis. *Portal of Hungerford: The Life of Marshal of the Royal Air Force, Viscount Portal of Hungerford, KG, GCB, OM, DSO, MC*. London: Heinemann, 1977.

———. *Royal Air Force, 1939–1945*. Vol. 1. *The Fight at Odds*. London: Her Majesty's Stationery Office, 1953.

Richards, Denis, and Hilary St. George Saunders. *Royal Air Force, 1939–1945*. Vol. 2. *The Fight Avails*. London: Her Majesty's Stationery Office, 1954.

Ring, Hans, and Werner Girbig. *Jagdgeschwader 27; die Dokumentation uber den Einsatz an allen Fronten, 1939–1945*. Stuttgart: Motorbuch Verlag, 1971.

Ritchie, Lewis Anthem. *East of Malta, West of Suez: The Official Admiralty Account of the Mediterranean Fleet, 1939–1943*. Boston: Little, Brown, 1944.

Roberts, Walter N. *Tito, Mihailović, and the Allies, 1941–1945*. New Brunswick, NJ: Rutgers University Press, 1973.

Rolf, David. *The Bloody Road to Tunis: Destruction of the Axis Forces in North Africa, November 1942–May 1943*. London: Greenhill, 2001.

Rommel, Erwin, and Fritz Bayerlein. *The Rommel Papers*. Edited by B. H. Liddell Hart, with the assistance of Lucie-Maria Rommel, Manfred Rommel, and General Fritz Bayerlein. Translated by Paul Findlay. London: Collins, 1953.

Ross, William F., and Charles F. Romanus. *The Quartermaster Corps: Operations in the War against Germany*. Washington, DC: US Army, Office of the Chief of Military History, 1965.

Royal Air Force Historical Society. *Air Intelligence: A Symposium, 22 March 1996*. Bracknell, UK: Royal Air Force Historical Society, 1997.

———. *The End of the Beginning: Bracknell Paper No. 3, A Symposium on the Land/Air Co-operation in the Mediterranean War, 1940–43*. Edited by Derek Wood. Bracknell, UK: Royal Air Force Historical Society, 1992.

———. *Reaping the Whirlwind: A Symposium on the Strategic Bomber Offensive, 1939–45—26 March 1993*. Bracknell, UK: Royal Air Force Historical Society, 1993.

Russell, Edward T. *Africa to the Alps: The Army Air Forces in the Mediterranean Theater*. Washington, DC: Air Force History and Museums Program, 1999.

Sadkovich, James J. *The Italian Navy in World War II*. Westport, CT: Greenwood, 1994.

———. "Of Myths and Men: Rommel and the Italians in North Africa, 1940–1942." *International History Review* 13 (1991): 298–301.

Salerno, Reynolds M. *Vital Crossroads: Mediterranean Origins of the Second World War, 1935–1940*. Ithaca, NY: Cornell University Press, 2002.

Sallager, F. M. *Operation Strangle*. Santa Monica, CA: Rand, 1972.

Sassoon, Joseph. *Economic Policy in Iraq, 1932–1950*. London: Cass, 1987.

Saunders, Hilary St. George. *Royal Air Force, 1939–1945*. Vol. 3. *The Fight Is Won*. London: Her Majesty's Stationery Office, 1954.

Schmider, Klaus. "The Mediterranean in 1940–1941: Crossroads of Lost Opportunities?" *War and Society* 15, no. 2 (October 1997): 19.

Schramm, Percy. *Kriegstagebuch des Oberkommandos der Wehrmacht: 1940–1945.* 8 vols. Bonn: Bernard and Graefe, 1961.

Schreiber, Gerhard, Bernd Stegemann, and Detlef Vogel. *Germany and the Second World War.* Vol. 3. *The Mediterranean, South-East Europe and North Africa, 1939–1941: From Italy's Declaration of Non-Belligerence to the Entry of the United States into the War.* Edited by the Militärgeschichtliches Forschungsamt. Translated by Dean S. McMurry, Ewald Ossers, and Louise Willmot. Oxford, UK: Clarendon, 1995.

Segre, Claudio G. *Italo Balbo: A Fascist Life.* Berkeley: University of California Press, 1987.

Shores, Christopher F., and Brian Cull. *Air War for Yugoslavia, Greece and Crete, 1940–41.* Carrollton, TX: Squadron Signal, 1987.

Shores, Christopher F., and Giovanni Massimello. *A History of the Mediterranean Air War, 1940–1945.* London: Grub Street, 2012.

Shores, Christopher F., and Hans Ring. *Fighters over the Desert: The Air Battles in the Western Desert, June 1940 to December 1942.* New York: Arco, 1969.

Silverfarb, Daniel. *Britain's Informal Empire in the Middle East: A Case Study of Iraq, 1929–1941.* New York: Oxford University Press, 1986.

Simpson, Michael. "Super Highway to the World Wide Web: The Mediterranean in British Imperial Strategy, 1900–45." In *Naval Policy and Strategy in the Mediterranean: Past, Present and Future.* Edited by John B. Hattendorf. London: Cass, 2000.

Slessor, John. *Air Power and Armies.* Tuscaloosa: University of Alabama Press, 2009.

———. *The Central Blue: The Autobiography of Sir John Slessor, Marshal of the RAF.* New York: Praeger, 1957.

Smith, Colin. *England's Last War against France: Fighting Vichy, 1940–42.* London: Weidenfeld and Nicolson, 2009.

Smith, Denis. *Mussolini.* New York: Knopf, 1982.

Speer, Albert. *Inside the Third Reich: Memoirs.* New York: Galahad, 1995.

Stanley, Roy M. *World War II Photo Intelligence.* New York: Scribner, 1981.

Starr, Chester G. *From Salerno to the Alps: A History of the Fifth Army, 1943–1945.* Washington, DC: Infantry Journal Press, 1948.

Steinhoff, Johannes. *The Straits of Messina: Diary of a Fighter Commander.* London: Deutsch, 1971.

Stewart, Richard A. *Sunrise at Abadan: The British and Soviet Invasion of Iran, 1941.* New York: Praeger, 1988.

Stoler, Mark A. *Allies and Adversaries: The Joint Chiefs of Staff, the Grand Alliance, and U.S. Strategy in World War II.* Chapel Hill: University of North Carolina Press, 2000.

———. *George C. Marshall: Soldier-Statesman of the American Century.* Boston: Twayne, 1989.

Stout, Jay A. *Fortress Ploesti: The Campaign to Destroy Hitler's Oil.* Havertown, PA: Casemate, 2003.

Strawson, John. *The Battle for North Africa.* New York: Scribner, 1969.

Strong, Kenneth. *Intelligence at the Top: The Recollections of an Intelligence Officer.* London: Cassell, 1968.

Student, Kurt. *Der Deutsche Fallschirmjaeger,* no. 11. Rothenburg o.d.T.: Scheider, 1956.

Suchenwirth, Richard. *Historical Turning Points in the German Air Force War Effort, etc.* Reprinted ed. New York: Arno, 1968.

Sutherland, Jonathan, and Diane Canwell. *Air War East Africa, 1940–41: The RAF versus the Italian Air Force.* Barnsley, UK: Pen and Sword Aviation, 2009.

Tedder, Arthur. *Air Power in War: The Lees Knowles Lectures.* London: Hodder and Stoughton, 1948.

———. *Report by Air Marshal Tedder on His Tenure in the Post of Air Officer Commanding-in-Chief.* London: MLRS, 2010.

———. *With Prejudice: The War Memoirs of Marshal of the Royal Air Force Lord Tedder G.C.B.* Boston: Little, Brown, 1966.

Terraine, John. *The Right of the Line: The Royal Air Force in the European War, 1939–1945.* London: Hodder and Stoughton, 1985.

———. *A Time for Courage: The Royal Air Force in the European War, 1939–1945.* New York: Macmillan, 1985.

Thomas, David Arthur. *Crete, 1941: The Battle at Sea.* London: Cassell, 2003.

Thompson, H. L. *New Zealanders with the Royal Air Force.* Wellington, NZ: War History Branch Department of International Affairs, 1959.

Thruelsen, Richard, and Elliott Arnold. *Mediterranean Sweep: Air Stories from El Alamein to Rome.* New York: Duell, Sloan, and Pearce, 1944.

Tooze, J. Adam. *The Wages of Destruction: The Making and Breaking of the Nazi Economy.* New York: Viking, 2007.

Tregaskis, Richard. *Invasion Diary.* New York: Random House, 1944.

Trevor-Roper, H. R. *Blitzkrieg to Defeat: Hitler's War Directives, 1939–1945.* New York: Holt, Rinehart, and Winston, 1964.

US Army Air Forces. *ULTRA and the History of the United States Strategic Air Force in Europe versus the German Air Force.* Frederick, MD: University Publications of America, 1980.

US War Department. *Field Manual 31-35, Aviation in Support of Ground Forces.* Washington, DC: Government Printing Office, 1942.

Van Creveld, Martin, Steven L. Canby, and Kenneth S. Brower. *Air Power and Maneuver Warfare.* Maxwell Air Force Base, AL: Air University Press, 1994.

von Rintelen, Ritter. *Mussolini als Bundesgenosse: Errinerungen des Deutschen Militärattachés in Rom, 1936–1943.* Tübingen, Germany: Wunderlich, 1951.

Warlimont, Walter. *Inside Hitler's Headquarters, 1939–45.* Translated by R. H. Barry. New York: Praeger, 1964.

Warner, Oliver. *Admiral of the Fleet: Cunningham of Hyndhope—the Battle for the Mediterranean; a Memoir.* Athens: Ohio University Press, 1967.

Warnock, A. Timothy. *Air Power versus U-boats: Confronting Hitler's Submarine Menace in the European Theater.* Washington, DC: Air Force History and Museums Program, 1999.

Weal, John A. *Ju 88 Kampfgeschwader of North Africa and the Mediterranean.* Oxford, UK: Osprey, 2009.

Weinberg, Gerhard L. *Germany, Hitler, and World War II: Essays in Modern German and World History.* Cambridge, UK: Cambridge University Press, 1996.

———. *Visions of Victory: The Hopes of Eight World War II Leaders.* Cambridge, UK: Cambridge University Press, 2005.

———. *A World at Arms: A Global History of World War II.* Cambridge, UK: Cambridge University Press, 1994.

Weiss, Steve. *Allies in Conflict: Anglo-American Strategic Negotiations, 1938–1944.* New York: Palgrave MacMillan, 1997.

Westermann, Edward B. *Flak: German Anti-Aircraft Defenses, 1914–1945.* Lawrence: University Press of Kansas, 2001.

Wilt, Alan F. *The French Riviera Campaign of August 1944.* Carbondale: Southern Illinois University Press, 1981.

———. *War from the Top: German and British Military Decision Making during World War II.* London: Tauris, 1990.

Woodward, E. L. *British Foreign Policy in the Second World War.* London: Her Majesty's Stationery Office, 1962.

Wynn, Humphrey. *Desert Eagles.* Shrewsbury, UK: Airlife, 1993.

Wynter, H. W. *Special Forces in the Desert War, 1940–1943.* Kew, UK: National Archives, 2008.

Ziemke, Earl Frederick. *Stalingrad to Berlin: The German Defeat in the East.* Washington, DC: US Army, Office of the Chief of Military History, 1968.

INDEX

First Army, 280, 282, 354
First Infantry Division, 304
First Parachute Brigade, 302
First Parachute Division, 328, 336
First Provisional Bomb Group, 225
First Tactical Air Force, 272
II Corps, 280
Second Tactical Air Force (TAF), 272,
   288, 356
Second Ukrainian Army, MAAF and, 381
Fifth Army, 315, 332, 343, 347
   advance of, 323, 324, 325, 327–338
   air support for, 349
   rations for, 282
   Salerno and, 309, 312
Fifth Light Division, 60, 79
Fifth Squadra, 37
Sixth Army, 374
Seventh Armored Division, 28
Seventh Army, 307, 349, 350–351, 355
Eighth Air Force, 250, 251, 263, 357,
   361, 363, 365, 381, 384
   buildup of, 451n1
   coordination with, 360
   Fifteenth Air Force and, 316, 360
   missions by, 258, 264, 358
   plane-months for, 456n85
   radius of action for, 359 (map)
Eighth Army, 89, 123, 156, 178, 197,
   209, 214, 221, 230, 241, 291
   advance of, 124, 164, 174, 243, 245,
   247, 274, 275, 315, 316, 327–328
   air superiority and, 164
   air support for, 161, 180, 258, 294,
   310–311, 349
   DAF and, 323, 324
   defeat of, 200, 212, 217, 228
   delaying, 207
   Foggia and, 318
   Headquarters, 205
   leadership of, 232
   Monte Cassino and, 325
   Operation Crusader and, 161, 163, 164
   Operation Torch and, 251
   RAF and, 103, 105, 168–169,
   174–175, 215
   retreat for, 180, 184, 187, 210, 211,
   215, 216

Straits of Messina and, 309
strength of, 240
survival of, 400
Tedder and, 218
WDAF and, 180, 186, 242, 247,
   248, 282
Eighth Bomber Command, 310
Ninth Air Force, 257, 315, 368
Ninth TAC, 288
X Air Corps, 79
Tenth Army, 316, 329
   casualties for, 336–337
   supply shortage for, 332
Twelfth Air Force, 252, 270, 310, 357, 362
   AASC for, 259
   groups of, 259–260
   Montelimar and, 353
   Operation Torch and, 251
   plane-months for, 456n85
   provisions for, 451n1
   Sicily and, 305
   sorties by, 264
   southern France and, 355
Twelfth Air Support Command (ASC),
   258, 279, 282, 302, 311
   command for, 270
   headquarters for, 315
   objectives for, 280
   VHF radios for, 307
Twelfth Army, 80
Twelfth TAC, 334, 336, 349
Fourteenth Army, 329, 332
Fourteenth Panzer Corps, 336
Fifteenth Air Force, 273, 291, 357,
   382, 403
   air supremacy and, 358
   Eighth Air Force and, 360, 361
   formation of, 358, 360
   mission of, 361, 365, 377, 380–381
   oil/transportation attacks and, 372,
   384–385
   problems for, 361–362
   radius of action for, 359 (map)
   Red Army and, 381
   replacements for, 362
   Southeastern Europe and, 382
   upgrade for, 363–364
Fifteenth Panzer Division, 60, 79, 215

Fifteenth Panzer Grenadier Division, 304, 336
Sixteenth Panzer Division, 322–323
Eighteenth Army Group, 286
Nineteenth Army, 350, 351, 353, 354
Twentieth Corps, 217
Twenty-first Army Group, 288, 357
Twenty-first Panzer Division, 138, 210, 237, 243
Thirty-fourth Division, 259
38 Bomber Squadron, 132
Forty-fourth Infantry Division, 336
Sixty-fourth Fighter Wing, 326
Seventy-eighth Division, 259
Ninetieth Division, 217
148th Division, 351
242nd Division, 351
244th Division, 351
338th Division, 351
505th Parachute Regimental Combat Team, 302
525th Assault Gun Battery, 344
577th Panzer Grenadier Regiment, 344

A-29 Hudsons, 32, 217, 230, 247, 274
A-36s, 309
AAA. *See* Antiaircraft artillery
AASC. *See* Army air support control
Abadan, refinery at, 118
ABC-1. *See* American, British, and Canadian Talks
Abwehr, 119
Abyssinia, 5–6, 7, 137
Acroma, 207, 208
Advanced Air and Rear Air Headquarters, 175, 186, 247
Aegean Sea, 126, 129, 357
AFHQ. *See* Allied Forces Headquarters
Afrika Korps, 67, 78, 215, 238, 243, 245
    communications and, 148
    Luftwaffe and, 79, 91
    resupply for, 140
    transport of, 60
Airborne Task Force, 350
Air Command Libya, 163
Air Corps Tactical School (ACTS), 272
Aircraft
    battle for, 43–45, 110–112, 189–194, 224–226, 231
    delivery of, 112, 129, 130, 135, 415n71
    PR, 43, 56, 58, 95, 97, 100, 120, 158, 200, 205, 227, 300, 381

reconnaissance, 57, 69, 98, 122, 148, 228, 229
reinforcement, 30, 31, 442n7
shortage of, 29, 43–45, 57, 70, 97, 143, 185, 207, 209, 224, 435n54
Tac R, 160, 167, 205, 240
tropicalized, 15
Air District Tunis, 278
Airfield attacks, 42, 53, 76, 95, 160, 167, 169
    Allied, 255, 299, 310, 323, 351
    British, 66, 73, 75, 104, 108–109, 120–121, 127, 140, 159, 182, 195, 205, 207, 208, 215, 259, 275, 286
    German, 62, 63, 66, 76, 159, 207
Airfields
    building, 124–125
    capture of, 229, 294, 312
    controlling, 169–170
    dispersal of, 278
    environmental impacts at, 124–125
    forward, 56, 58, 59
    protecting, 180, 277
    shortage of, 198
Airfields Organization (AO), 124
Air-ground cooperation, 59, 98, 160, 167, 255, 256, 288, 294, 345, 347, 349–352
    importance of, 208, 210, 303, 342, 343
    problems with, 328
    refining, 121–123, 233
Air Headquarters Egypt, 123
Air Headquarters Western Desert, 98
Air liaison officers (ALOs), 117, 166, 187, 225, 233, 235, 307
Air Materiel Command Africa, fuel dumps by, 77
Air Ministry, 135, 152, 156, 175, 271, 366
    Armament Branch, 94
    Army Cooperation Command and, 272
    Establishments Committee, 98
    goals of, 202
Air officer commanding (AOC), 89, 213, 268, 302
Airpower, 60–62, 107, 257, 258, 269, 271, 292, 336, 337, 345, 353, 356
    Allied, 256, 279, 285–286, 303, 398
    British, 71, 93, 101
    combined-arms operations and, 13
    development of, 124, 192, 193
    effective, 138, 345
    importance of, 3–5, 6, 13, 69–71, 87, 178, 238, 285, 318, 397–398

Italian, 7, 19
land-based, 3, 13, 83, 93, 152, 221,
    397–398, 401
at Malta, 38, 68, 76, 87, 103, 120,
    129, 141, 159, 180, 182, 184, 195,
    196, 197, 198, 199, 205,
    255, 259, 272, 296, 299, 302,
    417n33, 433n17, 446n83
role of, 153, 288, 291, 343, 346, 347
at Sicily, 251, 267, 280, 285, 294, 295,
    297, 301, 315, 340, 346
Slessor and, 43, 45, 192, 332–333,
    342–343, 415n71
strategic, 268
tactical, 268, 343
Air Power and Armies (Slessor), 94, 341
Air stores parks (ASPs), 15, 56, 175, 259,
    270, 274, 437n95
development of, 260
logistics and, 136–137
Air superiority
Allied, 153, 271, 279, 283, 285, 294,
    303, 311, 322, 357, 400–401
British, 23, 30, 32, 34, 35, 37, 41,
    42, 53, 56, 68, 73, 75, 89, 107,
    108, 112, 117–118, 120, 125, 127,
    157, 159, 164, 170, 182, 188–189,
    190, 191, 196, 202, 205, 207–210,
    214, 231, 239, 252, 260, 282, 286,
    287, 331
gaining, 118, 120, 156, 157, 158, 159,
    160, 264, 276, 287, 291–292, 309
German, 62, 63, 65, 66, 67, 70, 80,
    81, 82, 104, 106, 112, 127, 170,
    175, 179, 185, 190, 196, 203, 204,
    208, 211, 212, 238, 239, 255, 259,
    260, 276
importance of, 37, 67, 122, 276, 288,
    291, 336, 340
maintaining, 158, 191, 271, 331
US, 257, 258–259, 264, 399
Air support, 167, 187, 256, 259, 261,
    295, 376
Allied, 332–333, 338
British, 122, 130, 193
German, 274, 375
impact of, 121, 122, 379
lack of, 274, 277, 352
Air support controls (ASCs), 59, 98, 160,
    189, 234, 306
chain of command for, 122
impact of, 23
operational, 122–123

Air transport, German, 73, 77, 145,
    146, 148, 161, 166, 204, 230, 247,
    278–279, 282, 283, 284
Air War Plans Division Plan I (AWPD I), 87
Aircraft-maintenance units (AMUs), 36,
    414n46
Aitken, William Maxwell. *See*
    Beaverbrook, Lord
Alam Halfa, 225, 236, 237, 452n18
Albacores, 133, 190, 205, 214, 218, 225,
    226, 243, 273
missions for, 72, 235–236
Albanella Railway Station, 315
Aleppo-Nerab airfield, 118
Alexander, Harold, 294, 328, 329, 332,
    341, 343, 355, 452n18
command for, 232, 270
Eisenhower and, 268
Operation Anvil and, 344, 345
Operation Diadem and, 332
Po River Valley and, 344
Sicily and, 302
on Tunisian campaign, 285
Wilson and, 344
Alexandria, 29, 40, 83, 91, 92, 107,
    182, 196
attacks on, 51, 52, 62, 76, 127, 148,
    174, 420n47
convoys from, 300
defending, 76, 98, 129, 130
reinforcements from, 213
Allied Air Forces, Northwest Africa
    (NAAF), 268, 295, 316, 362
air superiority and, 309
airfield raids by, 317
losses for, 320
MAC and, 269
Allied Expeditionary Force, 248, 271–
    272, 276–277
Allied Forces Headquarters (AFHQ), 248,
    251, 259, 261, 264, 266, 268, 329,
    338, 341
ALOs. *See* Air liaison officers
American, British, and Canadian Talks
    (ABC-1), 48, 49, 179, 201, 224
Ammunition, 176, 196, 376, 402
AAA, 195
consumption rates for, 338
disruption of, 380
shortage of, 166, 282, 326, 331,
    334, 346
stocking, 314
AMUs. *See* Aircraft-maintenance units

Ancona, 323, 344, 345
Antiaircraft artillery (AAA), 19, 20, 21,
    31, 76, 91, 104, 106, 116, 120, 130
  ammunition for, 195
  Italian, 28, 251
  lack of, 182
  suppressing, 210
Antishipping operations, 205, 213,
    229–230, 231, 255, 262, 283, 403
  aircraft for, 227
  ASV-guided, 192
  impact of, 236, 287
Antisubmarine warfare (ASW), 34, 269,
    272, 300, 326
  success for, 321
  U-boats and, 276
Anzio, 110, 291, 292, 329, 347, 401
  air support at, 326
  breakout from, 337–338
  landings at, 324, 327
AOC. *See* Air officer commanding
Aosta, duke of, 31, 35
Arcadia Conference, 152, 178, 179,
    201, 224
Armed Forces Motor Transport Office, 139
Army Air Photo Interpretation Unit, 23
Army air support control (AASC),
    121–122, 123, 160, 166, 187, 235,
    259, 307, 349
Army Cooperation Command, 202–
    203, 272
Army Group C, 329, 336, 337, 341
Army Group Center, 350
Army Group E, 380
Army Group G, 350, 351
Army Group North, 371
Army Group South Ukraine, 274, 368,
    374, 375, 376
  collapse of, 372
Army Support Group Command
    (ASGC), 261
"Army Training Instruction No. 6," 121
Arnold, Henry "Hap," 258, 264, 327,
    349, 358, 363, 365, 381, 466n65
  aircraft for, 112
  bombing technique and, 331
  Brereton and, 225
  Eaker and, 305, 364, 377
  on German supply problems, 338–339
  on Luftwaffe command problems, 339
  Mediterranean theater and, 265
  Operation Torch and, 250
  Ploesti and, 367
  Portal agreement with, 192–193

reorganization and, 263
replacement rate and, 362
Slessor and, 194, 341
Spaatz and, 256, 257, 263
Arnold-Towers-Portal agreement, 192,
    194, 440n41, 440n53
ASCs. *See* Air support controls
ASPs. *See* Air stores parks
ASW. *See* Antisubmarine warfare
Atlantic Conference, 86
Auchinlek, Claude, 152, 170, 452n18
  air-ground cooperation and, 160
  air situation and, 188, 202
  appointment of, 121
  ASC and, 189
  Churchill and, 121, 157, 202
  counterattacks by, 232
  Portal and, 157
  Rommel and, 166

B-17s, 225, 259, 263, 299, 304, 305, 307,
    317, 336
  arrival of, 191
  bombing by, 246, 328
  replacement, 361
  shortage of, 435n53
B-24 Liberators, 217, 246, 263, 272, 365,
    368, 443n7, 447n13
  arrival of, 191
  bombing by, 225, 247, 299, 328
  losses of, 305
  Mandrel-equipped, 351
  photo of, 368
  replacement, 361
  shortage of, 435n53
  transfer of, 439n24
B-25 Mitchells, 217, 225, 242, 307, 336,
    447n13
  ASV-equipped, 282
  bombing by, 328
B-26s, 307, 328, 336, 447n13
Baade, Ernst-Günther, 296, 304
Badoglio, Pietro, 12, 27, 28, 294
BAF. *See* Balkan Air Force
Bagnold, R. A., 89–90
Balbo, Italo, 28
Balfour Declaration, 50
Balkan Air Force (BAF), 292, 347, 394,
    472n1
Balkans, 78, 250, 295, 296, 318, 357, 373
  airpower in, 77, 292
  developments in, 79–82
  Red Army advance through, 378–379
    (map), 381

as strategic liability, 85
withdrawal from, 394
Baltimores, 32, 190, 231, 242, 447n13
Barce, raid on, 187, 208–209
Bardia, 42, 56, 141, 148, 221
fall of, 56, 164
mining at, 63
Bastico, Ettore, 212
Battle of the Atlantic, 46, 62, 63, 78, 86,
133, 179, 195, 202, 401
assets for, 126, 262
changes for, 265, 321
focusing on, 266
Mediterranean theater and, 398
Battle of Britain, 29, 67, 113, 135, 153,
214, 297
Battle of Calabria, 29, 293, 311
Battle of Capo Spado, 29
Battle of Matapan, 29, 80
Battle of Montelimar, 352
described, 353–355
Baubatallinen, 105, 125
Bayerlein, Fritz, 158, 283
photo of, 61
Beamish, George, 271
Beaufighters, 91, 95, 107, 218, 227, 272,
273, 415n54, 430n56
AI, 263, 319
ASV-equipped, 282
attacks by, 99 (photo), 247
delivery of, 129, 130
impact of, 319
oil barges and, 369
photo of, 92
Beauforts, 109, 112, 226, 247, 430n56,
439n24
ASV-equipped, 282
photo of, 229
Beaverbrook, Lord (William Maxwell
Aitken), 43, 44, 84, 134, 135,
422n112
Bechtold, B. Michael, 257
Beda Fomm, 57, 59
Belgrade, attacks on, 364
Benghazi, 29, 65, 76, 103, 109, 128,
133, 139
airfields at, 77, 180, 182
attacks on, 38, 42, 62, 66, 72, 75,
104, 127, 188, 225, 230, 241,
246, 272
convoys at, 161, 170, 227
fall of, 56–57, 58, 141, 165, 180, 248
German occupation of, 75
mining at, 63

Benina, 77, 167, 168, 187
attack on, 104, 209
Beresford-Peirse, Noel, 108
Bergeret, Jean, 118
Bernhard line, 324
Bewegungskrieg, 13, 71, 246, 273, 374,
375, 376
Bir Hacheim, 204, 205, 209, 444n35
attacks on, 207, 208
combined-arms problems at, 212
Bizerte, 140, 255, 259, 263, 283
raids on, 312
Bizerte-Gabes Railway, 75, 140
Blaskowitz, Johannes, 351
Blenheims, 15, 32, 76, 104, 107, 108,
111, 132, 165, 189, 214, 219,
413n23, 413n24, 413n38
attacks by, 72, 103
destruction of, 81
reserves of, 410n66
transfer of, 439n24
Blenheim IVs, 17, 43, 44, 57, 109, 128,
413n37, 414n61, 424n35, 430n56
arrival of, 30, 73, 413n24
photo of, 55
superiority of, 53
Bletchley Park, 34, 89, 102
Blomberg, Alex, 427n8
Bomb damage assessment (BDA), 351, 366
Bône, 252, 258, 260, 263
Bostons, 190, 214, 242, 336
constructing, 112
performance of, 231
Brenner Pass, 295, 363
Brereton, Lewis, 257, 263
command for, 315–316
Tedder and, 224–225
Brett, George, 112
Bridges, attacks on, 323, 326, 331, 332,
333, 334
Brindisi, 141, 318
British Army, 21, 116, 789
Operation Crusader and, 161
RAF and, 66, 71, 257, 267
Rommel and, 70
British Overseas Airways, assets of, 106
British Somaliland, offensive against, 25
Broadhurst, Harry, 281, 302, 319,
458n62, 458n65
Brooke, Alan, 93, 202, 232
Bucharest
attacks on, 364
bottleneck at, 373
Bullitt, William C., 189

C-47s, 110, 270, 273, 447n13
  arrival of, 284
C2. *See* Command and control
Cabell, Charles, 365, 366
Cabrank method, effectiveness of, 335
Calabria, 29, 293, 311
Canary Islands, 8, 10, 26
Cannon, John K. "Joe," 258, 326, 349, 353
  directive by, 334
  Operation Diadem and, 332
  Twelfth Air Force and, 280, 357
Cape Matapan, 29, 80
Cape of Good Hope, 6, 31, 318, 320
Casablanca, 251, 258, 260
  aircraft assembly at, 270
Casablanca Conference, 249, 266, 267
  air command issue at, 268
  Operation Pointblank and, 269–270
Cassino, 292, 332, 336
  air-ground cooperation at, 328
  attacks on, 291, 327, 328
  capture of, 325
  disaster at, 329, 401
Castel Benito, 77, 275
Castelvetrano, 294, 296
Catania, 295, 310
Caucasus, 200, 223, 372, 409n32
Cauldron, disaster at, 209
Cavallero, Ugo, 133, 141, 172, 196–197, 212
CBO. *See* Combined bomber offensive
CCS. *See* Combined Chiefs of Staff
Central Interpretation Unit, 102–103
Central Task Force, 259
Chetniks, 88
Chi-Stelle (signals organizations), 119
Chiefs of Staff (CoS), 37–38, 48, 152, 192, 202, 224
  aircrew shortage and, 153
  Casablanca Conference and, 266
  JCS and, 292
  Mediterranean theater and, 265–266
  Middle East/airpower and, 178
Churchill, Winston, 3, 18, 25, 36, 91, 106, 328
  airpower and, 107, 152, 269
  Auchinlek and, 121, 157, 202
  Battle of the Atlantic and, 202
  Beaverbrook and, 44
  Casablanca and, 266, 267
  Crete and, 82
  Cripps and, 188
  cross-channel invasion and, 266

in Egypt, 232
grand strategy and, 79
Greece and, 93
Italian surrender and, 293, 294
Malta and, 189
Mediterranean theater and, 5, 9, 121–122, 265, 356
Operation Battleaxe and, 107
Operation Compass and, 44
Operation Shingle and, 327
Ploesti and, 367
Portal and, 202
Red Army and, 201
Roosevelt and, 13, 47, 48, 86, 87, 137, 151, 152, 192, 223, 269
Russian aid and, 125–126
Syria and, 117
Tedder and, 156, 157, 448n35
Tunisian campaign and, 285
Ciano, Count, 40, 49, 51
Civil Flying Schools, 94
Clark, Mark, 315, 323, 325, 337
  on air support, 316
  Rome and, 341
  winter offensive by, 324
Clausewitz, Carl von, 402, 404
Collishaw, Raymond, 21, 30, 75, 98, 103, 154
  air-ground cooperation and, 59
  aircraft for, 57
  IAF and, 14, 53
  leadership style of, 105
  mobility and, 58
  O'Connor and, 42
  Operation Battleaxe and, 109
  Tedder and, 105
Comando Supremo, 42, 172, 249
Combined-arms operations, 13, 35, 60, 61, 66, 160, 167, 203, 219, 240, 252, 256–258, 264, 292, 317, 319, 329
  airpower and, 13, 85
  communications-intensive, 307
  effective, 159
  impact of, 101, 149, 238, 330, 332, 333, 340, 341, 342, 346, 398, 400
  lack of, 303
  planning for, 266–267
  push for, 11
  victory and, 398
Combined bomber offensive (CBO), 196, 258, 266, 291, 305, 309, 316, 322, 323, 357, 360

Combined Bureau Middle East, 102
Combined Chiefs of Staff (CCS), 152,
   224, 305, 332, 360
  Casablanca Conference and, 266
  compromise strategy of, 292–293
  coordination and, 192
  Cunningham and, 183
  mining and, 369
  Operation Anvil and, 344
  Tedder and, 268
  Wilson and, 329, 347–348
Combined Operations Headquarters, 353
Combined Services Detailed Intelligence
   Centre (CSDIC), 90
Combined Strategic Targets Committee
   (CSTC), 366
Comiso, 294, 296
Command and control (C2), 6, 13, 42,
   61, 66, 126
  Allied, 259, 261, 267, 302, 303
  British, 89, 90, 153, 186, 232, 233,
   255, 260
  collapse of, 101, 337
  effective, 14, 101
  German, 67, 70–71, 78, 82, 83,
   148–149, 209, 296, 297, 300, 302,
   336, 337, 345, 399
  improving, 123, 124, 267
  problems with, 101, 167, 259, 261,
   297, 299, 307, 315
Commonwealth forces, 35, 36, 75, 82,
   165, 168, 202, 221, 348
  attacks on, 169
  Egypt and, 94
  evacuation of, 84
  resurgence of, 231–236
Communications, 59, 115, 120, 131, 148,
   261, 274
  air-to-air/air-to-ground, 77
  improvement in, 101, 228
  joint, 166
  land, 234
  long-haul, 100
  loss of, 198, 243, 298
  problems with, 20, 57, 119, 176, 213,
   259, 260, 281, 307, 380
  radio, 91, 103, 278
  RAF, 100–103
  sea, 234, 240
  tactical, 326
Coningham, Arthur "Mary," 99, 105,
   123, 155, 156, 163, 164, 168,
   175, 188, 190, 211, 214, 239,

   241, 247, 268, 272, 280, 288,
   302, 356
  air-ground cooperation and, 154, 233
  air offensive and, 281
  aircraft shortage for, 224
  airpower and, 205, 271
  assets for, 166, 227
  command for, 270, 279
  intelligence/communications and, 101
  maintenance and, 180, 208
  Momyer on, 453n38
  Montgomery and, 232–233
  nickname of, 424n37
  Operation Lightfoot and, 242
  photo of, 273
  RAF Regiment and, 180
  relief of, 458n62
  replacements for, 209
  Sicily and, 304–305
  Sidi Barani and, 210
  Tedder and, 154, 255, 256, 304
  Tripoli and, 275
  WDAF and, 98, 154, 165, 237
Consumption units, 221–222, 238, 282,
   446n87, 472n92
Convoys, 77, 221, 248, 262, 276–277,
   300
  attacks on, 161, 166, 173, 219–220,
   219 (photo), 228, 230, 255, 259,
   315, 326, 399
  escorting, 119, 138, 139, 151, 173,
   231, 255
  losses for, 133, 238, 241, 245
  protecting, 129, 133, 141, 142, 161,
   170, 227, 236, 272, 320
Cooperation, 11, 23, 48–49, 192, 277,
   352, 360
  air-ground, 59, 98, 121–123,
   154, 160, 167, 180, 189, 203,
   207–208, 210, 233, 234, 255, 256,
   288, 294, 328, 342, 343, 345, 347,
   349–352
  air-ground-sea, 42, 329
  air-sea, 80, 131, 133, 149, 173, 174,
   196
  Axis, 10–12, 148, 200
  command and, 149
  interservice, 14, 93
  lack of, 57
Corsica, 325, 349, 351, 354
  evacuation of, 317
CoS. *See* Chiefs of Staff
Craig, General, 267

Crete, 5, 51, 88, 98, 107, 126, 130, 141,
    142, 156, 160, 170, 177
  defending, 63, 66, 67, 129
  fall of, 38, 76, 82, 83, 84–85, 91, 93,
    114, 129, 130, 132, 151
  Mediterranean Fleet and, 226
  ship losses at, 80
  strategic importance of, 83 (map)
Cripps, Stafford, 188
Cross, Kenneth, 430n47, 438n12,
    455n84
  Broadhurst and, 458n62
  Churchill and, 448n35
  Dawson and, 431n71
  on Embry, 429n35
  No. 252 Wing and, 430n51
Cross-channel invasion, 223, 224, 265,
    272, 293, 295, 356
Crüwell, Ludwig, 207, 404
CSDIC. *See* Combined Services Detailed
    Intelligence Centre
CSTC. *See* Combined Strategic Targets
    Committee
Cubs, reconnaissance by, 306
Cunningham, Alan, 34, 89, 97
Cunningham, Andrew, 39, 79, 90, 270,
    295, 304
  air support and, 130
  CCS and, 183
  communications and, 131
  Crete and, 129, 130
  Malta and, 21
  strength of, 183
  Tedder and, 129–130, 131, 227
Cyprus, 85, 117, 118, 177
Cyrenaica, 53, 57, 67, 69, 76
  map of, 96

DAF. *See* Desert Air Force
Dakotas, 32
D'Albiac, J. H., 38, 39, 80, 81
Damascus, 117, 118
Danube River, 373
  mining of, 72, 364, 365, 368, 369,
    370, 372
Darlan, François, 88
Dawson, Graham, 36, 138, 188, 189,
    231, 269
  aircraft from, 112, 191
  character of, 134
  as CMO, 135
  Cross and, 431n71
  Harriman and, 137
  logistics problems and, 270

  maintenance and, 134, 135–136, 137,
    154, 185, 205, 217
  RAFME and, 430n64
  salvage and, 154, 225, 270
  Takoradi route and, 135, 137
  Tedder and, 135, 136, 155, 214
  on Tomahawks, 137
  US aircraft shipments and, 135
Deane, John, 367
De Gaulle, Charles, 46–47
Deichmann, Paul, 298
Derna, 42, 141, 148, 149, 167, 168, 180,
    209, 247
  abandonment of, 56
  attacks on, 89, 104, 127
  base at, 77
Desert Air Force (DAF), 6, 310, 334, 349,
    357, 362
  aircrew of, 99
  convoy raids by, 315
  Eighth Army and, 323, 324
  See also Western Desert Air Force
Dickson, William F. "Tommy," 334,
    349, 357
"Direct Air Support in the Libyan
    Desert," 225
Djibouti, offensive against, 25
Djidjelli airfield, 252
Dobbie, William, 97
Dodecanese Islands, 204, 322, 356
Donovan, William, 111
Doolittle, James H. "Jimmy," 266, 279,
    357, 360
  air superiority and, 264
  on Cannon, 258
  command for, 270
  Operation Torch and, 250
  Spaatz and, 270
  on Tunisian War, 264
  Twelfth Air Force and, 250
Dornier Do 217s, 314, 319, 352
  raids by, 325
Douglas, Sholto, 268, 270
Douglas Aircraft, 112
Drummond, Peter, 91, 190
Duke, Neville, 110, 155, 283
  on air attacks, 183–184
  Gazala offensive, 110
Düllberg, Ernst, 143
Dunn, Patrick, 100

EAC. *See* Eastern Air Command
Eaker, Ira, 355, 360, 362, 363, 365,
    470n40

on air offensive, 377
Arnold and, 305, 327, 364, 377
bombing technique and, 331
combined-arms operations and, 333
on communications, 330
Fifteenth Air Force and, 381
MAAF and, 357, 397
MAAF Instruction No. 8 and, 330
mining and, 369
Operation Anvil and, 348
Operation Diadem and, 333, 334
Operation Strangle and, 332
Operation Torch and, 258
Ploesti and, 367, 371–372
Portal and, 310
Slessor and, 341, 356
Eastern Air Command (EAC), 252, 261, 262, 288
AASC of, 259
Operation Torch and, 250, 251
Eastern Front, 3, 249, 269, 291, 312, 322, 327, 329
collapse of, 374
fuel for, 370
German supply on, 373
Mediterranean theater and, 177–178
resources from, 318
Eastern Group Central Provision Office, 36
Eisenhower, Dwight, 263, 315, 349, 355, 360, 367, 397
airfield protection and, 277
air operations and, 258
Alexander and, 268
CCS and, 317
on German air superiority, 260
Italian campaign and, 322
Italian surrender and, 294
Monte Cassino and, 328
on naval/air strength, 316
Operation Anvil and, 348
Operation Dragoon and, 354
Operation Torch and, 248, 250, 251, 258
Sicily and, 302, 305
Spaatz and, 257, 259
Tedder and, 258, 264, 266–267, 268, 269, 288, 356
Tunisia and, 266–267
El Adem, 205, 208, 209–210
El Agheila, 62, 66, 69, 103, 104, 107, 164, 165, 245, 248, 273–274
El Alamein, 63, 68, 71, 101, 105, 110, 205, 217, 219, 220, 232, 241
advance on, 61, 215, 222, 223
beginning of, 225, 243, 245, 246

defeat at, 70, 244 (map)
German losses at, 209, 247, 275, 280
preparing for, 233, 242–243
retreat to, 211
supplies for, 216
El Aouina, 263
photo of, 284
El Daba, 237, 241
Electronic warfare, 242, 243, 295, 308
Elmhirst, Thomas, 155
Embry, Basil, 123, 429n35
Enigma cipher, 102, 121, 186
Ethiopians, IEA and, 32
Evill, Douglas, 224

F4F Wildcats, 252
photo of, 197, 254
FAA. *See* Fleet Air Arm
Fascist Republican Air Force, 293
FASLs. *See* Forward air support liaisons
Fast Carrier Striking Group, 223
Fellers, Bonner, 101
Felmy, Helmuth, 78, 116, 418n50
Ferdinand (deception plan), 350
Field Manual 31–35: 255, 256, 272, 288
Field Manual 100–20, 257
Field Service Regulations, 101
Fieseler Storch, 69, 207, 217
Fleet Air Arm (FAA), 16, 72, 117, 131, 134, 183, 196, 226, 430n56
shipping attacks by, 98, 182, 272, 273
Fliegerführer Afrika, 66, 82, 89, 104, 108, 160, 197, 230, 254
air superiority and, 179
challenges for, 112, 113
command of, 149, 170
cooperation with, 138, 184
El Alamein and, 63, 216
Gazala offensive and, 205
headquarters for, 148
IAF and, 138
issues for, 62
LOCs and, 139
losses for, 221
Luftflotte 2 and, 172
operational control over, 149
priorities for, 65, 207
Rommel and, 166, 236
serviceability rates for, 126
strength of, 70, 78, 183, 204, 212, 218
supplies and, 139, 141
Fliegerkorps, 147, 371, 434n34
challenges for, 119
as organization formation, 146

Fliegerkorps I, fuel shortages and, 371
Fliegerkorps II, 158, 197, 212, 260, 311, 434n34
  convoy escort by, 231
  redeployment of, 150
  strength of, 196
Fliegerkorps VIII, 75, 80, 81, 82, 239, 421n96
  airfields for, 125
  mission of, 82
Fliegerkorps X, 79, 104, 108, 158, 159, 170, 204
  arrival of, 59, 66, 127
  command structure and, 172, 420n87
  convoy escort by, 129, 133, 142, 161
  effectiveness of, 112, 113
  Greece and, 76, 78, 80–81, 100, 128, 131
  headquarters, 75
  IAF and, 138
  intelligence and, 119, 120, 149
  Malta and, 75
  mining by, 63
  missions by, 60, 65, 119, 129, 207
  Operation Marita and, 421n96
  priorities for, 129
  reinforcements from, 163
  relocation of, 139, 141, 168
  serviceability rates for, 126
  Sicily and, 70, 139, 197
  strength of, 160, 212
Fliegerkorps XI, 82, 85, 197
Fliegerkorps Tunisien, 278
Focke-Wulf FW 190s, 277
Focke-Wulf FW-200 Condors, 63, 107, 127, 262
Foggia airfields, 4, 291, 319, 347, 358
  capture of, 318, 401
  raids on, 303, 312
Force K, 134, 273
Force Q, 273
Force V, 311–312
Forward air support liaisons (FASLs), 122, 187, 234, 235
Franco, Francisco, 10, 27, 28, 51, 409n35
Free French, 208, 249, 319, 336, 344, 351
  de Gaulle and, 46
Freeman, Wilfrid, 40, 91, 156, 157, 227, 434n24
  Beaufighters and, 129
  on dive-bombing, 106
  light-bomber shortages and, 191
  Longmore and, 43
  Tedder and, 94–95, 107, 165

French Forces of the Interior (FFI), 348
French National Railway, 364
French Navy, 31, 252
Freyberg, Bernard, 212
Fröhlich, Stephan, 139, 140, 163, 239, 420n87
  army support and, 67
  C2 and, 79
  command of, 418n48
  fuel shortages for, 134
  Göring and, 63
  Libya and, 68, 159
  Luftwaffe and, 70
  Rommel and, 66, 67, 69, 71, 78, 148, 418n50, 421n96
  serviceability rates and, 126
Fuel, 11, 164, 176, 358, 374, 376, 383
  consumption rates for, 338
  delivery of, 215, 228
  destruction of, 209, 314 (photo)
  disrupting, 380
  limits on, 339
  shortage of, 20, 134, 166, 169, 213, 219, 222, 236–237, 245, 274, 282, 285, 287, 306, 322–323, 326, 327, 331, 333, 367, 368, 371, 377, 381, 382, 383, 384
  supplying, 220, 314
  surplus, 370
Fuel dumps, 77, 382
"Führer Directive No. 18," 51, 80
"Führer Directive No. 30, Middle East," 116
"Führer Directive No. 38 of 2 December 1941," 158
"Führer Directive No. 41 of April 1942," 200
"Führer Directive No. 45 of 23 July 1942," 200
Fuqa, 127, 217, 241

Gabês, 259, 263
Gafsa, offensive against, 282
Galland, Adolf, 297, 298, 459n75
Gambut, 79, 109, 212, 272
  airfields at, 77, 210, 217
Gazala, 42, 56, 63, 109, 159, 163, 165, 171, 186
  airfields at, 164
  attacks on, 89, 104, 136, 208, 214, 223, 228
  German forces at, 104
  German victory at, 201, 223
Gazala offensive, 110

beginning of, 205, 207–212
map of, 207
planning of, 204–205
GCCS. *See* Government Code and Cipher School
Geisler, Ferdinand, 60, 159
Gela, landing at, 294
General Headquarters (GHQ) Mobile Division, 15, 89
General Headquarters Middle East (GHQME), 16
Genoa, 251
attacks on, 307, 351
Gerbini, 294, 295, 296
German Air Force Liaison Staff in Italy (Italuft), 77, 142, 149
German Army
assistance from, 59–60, 62–63, 65–69
combined-arms operations and, 340
counterattacks by, 312, 315
logistics problems for, 382–384
losses for, 246, 265, 347
Luftwaffe and, 85, 438n12
Gesternberg, Alfred, 368
GHQME. *See* General Headquarters Middle East
Gibraltar, 5, 8, 9, 26, 87, 88, 120, 195, 216, 218
capturing, 10, 52, 126
convoy to, 252
importance of, 11, 150, 172
meeting at, 360
protecting, 46
Gladiators, 21, 55, 80, 115, 116, 413n23, 413n24, 413n37, 413n38, 413n39
reserves of, 410n66
Göring, Hermann, 28, 70, 216, 342
antishipping operations and, 262
development-and-repair bureaus and, 145
Fliegerkorps X and, 60, 65
Fröhlich and, 63
Galland and, 297
Greece and, 141
joint capabilities and, 78
logistics and, 142
maritime airpower and, 320
Raeder and, 62
Government Code and Cipher School (GCCS), 89, 100, 186
Grand Alliance, 292, 307, 358, 366, 379, 474n5
casualties for, 373, 402

initiative for, 265
reducing losses for, 318
Grand Harbour, 183
photo of, 220
Grand strategy, 9, 95, 114, 287, 398, 399, 401, 403, 405
Allied, 86–87, 292–293, 356–357
British, 46–47, 79, 177–178
German, 144, 177, 231, 339
See also Strategic decisions
Gray, Colin, 404
Graziani, Rodolfo, 27, 28, 37, 49–50, 69
Greece
defeat in, 114, 122, 130, 132, 151
encirclement of, 93
Italian claims in, 25, 88
Italian invasion of, 28, 38–40
Greek Air Force, decline of, 38
Greek Army, RAF and, 80
Grendal, Vladimir, 367
Grigg, P. J., 203
Grossdeutschland Division, 375
Gruppen, 197, 443n20
defined, 62
Gulf of Aden, 7, 14, 34, 35, 111, 150, 172
Gundelach, Karl, 418n50, 445n48
Gura, 112, 189
Gustav Line, 317, 325, 327, 336

H2X Pathfinder Force (PFF), 352, 364
Habbaniya, 115, 116, 117
Halder, Franz, 105, 133, 414n56, 425n60, 434n34
Halifaxes, 194, 247, 443n7
Halverson Detachment, 217, 225
Harlinghausen, Martin, 60, 78, 418n44
Harpoons, arrival of, 213
Harriman, Averell, 137
He 111s, 63, 115, 127, 159
arrival of, 116
He 177s, 352
Headquarters Royal Air Force Middle East (RAFME) Tactics Assessment Office, 132, 231, 287, 403
Heliopolis, 102, 135, 137
Himmler, Heinrich, 351
Hitler, Adolf, 3, 13, 25, 41, 65, 81, 99, 151, 285, 336, 337, 340
Balkans and, 85
convoy escorts and, 255
Crete and, 82, 83, 85, 93
Eastern Front and, 178, 318

Hitler, Adolf, *continued*
  Egypt and, 50
  Greece and, 28, 59, 66
  Italian theater and, 37, 322–324, 326
  Malta and, 38, 196, 198, 204
  maritime air warfare and, 60–62
  Mediterranean theater and, 8–9, 10,
    12, 26, 27, 28, 40, 46, 51–52, 68,
    88, 114, 277, 402, 409n35
  Middle East and, 4, 28, 67, 69,
    116, 200
  Mussolini and, 142, 293
  oil reserves and, 6
  Operation Barbarossa and, 52, 150
  Operation Hercules and, 171, 173,
    199, 211, 212, 419n64
  Pact of Steel and, 11
  Ploesti and, 366–367
  Raeder and, 133, 174, 408n25
  Rashid and, 115
  Rommel and, 61, 68, 211, 212, 216,
    245, 434n34, 452n18
  Sicily and, 293
  Stalingrad and, 472n91
  Student and, 216
  Tobruk and, 199
  Vichy and, 28, 47, 88, 413n27
HMS *Ark Royal*, sinking of, 173–174
HMS *Barnham*, 84, 174
HMS *Coventry*, 130
HMS *Eagle*, 195, 196, 197, 218
HMS *Formidable*, 84, 254
HMS *Furious*, 218
HMS *Illustrious*, 76
HMS *Indomitable*, 437n95
HMS *Liverpool*, 213
HMS *Nelson*, attack on, 127
HMS *Queen Elizabeth*, attack on, 174
HMS *Ranger*, 191
HMS *Southhampton*, 76
HMS *Valiant*, attack on, 174
HMS *Warspite*, 84, 314
Hopkins, Harry, 86
House, Edwin, 311, 315
Howard, Michael, 10, 12, 13
Hube, Hans-Valentin, 304, 306
Hurribombers, 161, 191
Hurricanes, 17, 32, 38, 43, 44, 45, 57,
    81, 84, 107, 112, 130, 136, 185,
    193, 252, 282, 413n37, 413n39,
    414n51, 417n33, 437n95, 446n83
  air superiority and, 108
  arrival of, 30, 40, 68, 73, 75, 194,

    197, 414n61, 420n74
  loss of, 182, 184, 240
  Me. 109s and, 78, 79, 184
  photo of, 55
  raids by, 42
  reconnaissance and, 58
  superiority of, 53, 231
Hurricane IIs, 55, 103, 107, 424n35
  arrival of, 111, 413n24

Iachino, Angelo, 80
IAF. *See* Italian Air Force
Identification system, 166, 167
IEA. *See* Italian East Africa
Indian Ocean, 4, 131, 132, 399, 411n5
  strategy, 150, 177, 200, 223
Intelligence, 6, 119–121, 187, 221, 279,
    288, 333, 374
  air, 160, 319–320
  improvements in, 17, 101
  photographic, 58, 102, 103, 120, 131,
    209, 234, 351, 366, 381
  planning and, 338
  radio, 120
  RAF, 100–103
  signals, 121, 218, 243
  strategic, 100
  See also Ultra intelligence
Intelligence Division (Fliegerführer
    Afrika), 216
Intelligence Division (MAAF), 373
Inter-Service Intelligence Staff
    Conference, 166
Inter-Service Operational Staff
    Conference, 166
Interrogation friend or foe (IFF), 299, 308
Iran, 170
  securing, 400
Iranian Air Arm, 118
Iraq, 170
  revolt in, 114–117
  securing, 400
Iraqi Air Force, 115
Iraqi Army, 115
Irwin, Samuel D., 112
Istres airfield, concentration at, 307
Italia, damage to, 293
Italian Air Force (IAF), 38, 42, 140, 210,
    260, 398, 400
  aircraft problems for, 17, 19, 20, 70
  air superiority and, 37
  Allied control of, 293
  assets of, 52–53, 127–128, 204

attacks by, 15, 16, 35, 37, 127, 138, 207
C2 and, 299, 399
convoy escort and, 139, 338
destruction of, 16, 29, 30, 34, 35, 80,
    287, 293–294, 402, 403
Fliegerkorps X and, 138
fuel shortages for, 20
Italian Army and, 53
logistics and, 142
losses for, 59, 70, 93, 274
Malta and, 30–31, 198
problems for, 14, 20–21, 27, 29,
    37, 57
RAF and, 14–15, 17, 29, 31, 53, 57, 159
shipping schedule and, 236
strength of, 57, 121, 170, 190
Western Desert and, 211
Italian Air Force High Command, 149
Italian Army, 52, 56, 172, 433n17
    destruction of, 287
    Egypt and, 30
    equipment shortage for, 19–20
    IAF and, 53
    manpower shortage for, 20
    shortcomings for, 27
    surrender of, 57, 293
Italian East Africa (IEA), 5, 15, 17, 21,
    41, 42, 76, 95, 97, 287, 318, 400
    campaign in, 28, 33 (map), 34–36, 46,
        111, 130, 132, 170
    forces in, 20, 32
    importance of, 34
    loss of, 88, 93, 172
    maintenance challenges in, 413n39
    reserves in, 46
    securing, 31–32, 34
    supplies for, 50
Italian High Command, 212
Italian Intelligence Service, 119
Italian Merchant Marine, 278, 287, 400,
    411n76
    losses for, 20, 131–132, 142
Italian Metropolitan Air Force, 6, 433n17
Italian Navy, 17, 39, 77, 84, 170, 182,
    249, 287, 293, 318
    aircraft of, 433n17
    attacks on, 31, 34
    convoys and, 183
    Luftwaffe and, 438n15
    Royal Navy and, 7, 20, 92
    shipping schedule and, 236
Italian Navy Supreme Command, 149
Italian Somaliland, 35, 88

Italuft. *See* German Air Force Liaison
    Staff in Italy
Italy
    declaration of war by, 29–31
    decline of, 49–50
    policy aims of, 7
    preparations by, 17, 19–20
    surrender of, 293–294, 401

Jagdeschwader 27 (JG 27), 43, 65, 155,
    164, 212, 297
Jagdeschwader 53 (JG 53), 65, 164, 297
Jagdeschwader 77 (JG 77), 285
JCS. *See* Joint Chiefs of Staff
Jeannequin, General, 428n16
Jodl, Alfred, 27, 282
Jodl Memorandum, war on the periphery
    and, 412n7
Joint Chiefs of Staff (JCS), 152, 192,
    223, 360
    Balkans and, 344
    CoS and, 292
    lend-lease and, 86
    Mediterranean theater and, 265
    Operation Torch and, 224
    Rainbow Plan 5 and, 48
Joint Intelligence Centre (JIC), 177
Joint Photographic Reconnaissance
    Center (JPRC), 366
Joint Planning Staff (JPS), 177
Jordan, Philip, 455n72
Ju 52s, 75, 77, 84, 105, 106, 126, 142,
    145, 168, 169, 173, 246, 285
    attacks on, 283–284, 303
    dependence on, 139
    destruction of, 247
    evacuation by, 377
    photo of, 73, 284
Ju 86s, 36, 413n39
Ju 87 Stukas, 19, 68, 108, 142, 209, 210,
    231, 235, 260, 418n48
    attacks by, 106–107, 169–170, 205,
    212, 214, 237
    destruction of, 107, 214
    Kasserine Pass and, 279
    photo of, 64
    vulnerability of, 221, 239
Ju 88s, 141, 159, 204, 210, 221, 249,
    276, 318, 352, 418n48, 443–444n21
    photo of, 63
    raids by, 325
    torpedo-bombers, 107, 262
Junck, Werner, 116

Kasserine Pass, 267, 277–279, 279–281
Katastrophdienst, 327
Keitel, Wilhelm, 116, 140, 141, 172, 236
Kenney, George, 327
Kesselring, Albrecht (Albert), 239, 252, 326, 337, 342
  air policy and, 246
  air-sea coordination and, 174
  air transport and, 142, 161, 221
  Comando Supremo and, 250
  command for, 67, 150, 158, 159, 171, 172, 238, 249, 296
  convoys and, 173, 230
  Fliegerführer Afrika and, 163, 204
  Fliegerkorps II and, 231
  on German airmen, 277
  IAF and, 238
  logistics and, 252
  Luftflotte 2 and, 70
  Malta and, 172, 173, 174, 184, 196, 198, 199, 212, 216
  Mediterranean theater and, 172
  Operation Hercules and, 196–197, 210
  Operation Torch and, 249–250, 252
  photo of, 171
  Rommel and, 170, 204, 207, 208, 215, 246
  Sicily and, 170, 301
  SLOCs and, 165
  supply shortage and, 221, 307, 332, 333
  von Waldau and, 209
KG 26: 254, 262
KG 30: 254, 262
Kittyhawks, 163, 207, 209, 210, 217, 224, 225–226, 242, 247, 274, 303
  armament for, 183
  arrival of, 191
  assembly of, 112
  attacks by, 168, 283–284
  development of, 137
  maintaining, 193
  and Me. 109Fs compared, 175
  shortage of, 207
Konev, Ivan, 375, 376
Kriegsmarine, 94, 142, 320, 352, 377
  fuel shortage for, 371
  mine clearing by, 369
Kuter, Laurence, 272
  photo of, 273

Lampedusa, 299, 300
La Senia, 252, 258
Lawson, George, 261

L Detachment Special Air Service (SAS), 90
Leese, Oliver, 357
Leghorn, capture of, 318, 345
Leigh-Mallory, Trafford, 272
Lend-lease, 47, 52, 93, 112, 178, 192, 320, 374, 400
  importance of, 201
  increase in, 86
  Red Army and, 403–404, 474n9
  reverse, 251
Lend-Lease Act, 35, 45
Liberty ships, arrival of, 265
Liddell Hart, B. H., 404
Lines of communication (LOCs), 94, 118, 139, 306, 309, 319, 336, 361, 374, 382
  attacks on, 108, 300, 303, 310, 311, 317, 325, 326, 331, 332, 333, 334, 337, 342, 364, 365, 381
  defending, 323
  disrupting, 295, 330, 334
  lengthening, 240
Liri, bombing at, 337
Littoro Division, 217
Lloyd, Hugh, 188, 270, 326
  Malta and, 30, 213
  Sicily and, 195
  Spitfires and, 195
LOCs. *See* Lines of communication
Loerzer, Bruno, 158, 297
Logistics, 6, 58, 155, 158, 288, 358
  advantage in, 132–134
  Allied, 204, 261, 349
  ASPs and, 136–137
  Axis, 77–79, 88, 97, 142, 168, 182–183, 187, 219, 221, 230, 255, 278, 300, 331
  battle for, 132, 170, 180, 230
  British, 56, 102, 124, 154, 193, 215, 216, 270
  German, 67, 70–71, 75, 78, 102, 104, 113, 116, 132, 138–142, 146–147, 149, 163, 168, 171, 174, 179, 182–183, 216, 217, 230, 231, 232, 236, 239, 250, 252, 261, 262, 277, 303, 326, 329, 330–331, 338, 340, 346, 374–375, 382–384
  mobile warfare and, 379
  operational/supply responsibilities and, 382
  problems with, 126, 138–149, 204, 216, 217, 230, 231, 232, 236, 261, 262, 263, 270, 300, 303, 326, 329, 330–331, 332, 338, 340, 352

SLOCs and, 69
US, 87, 193, 256, 258, 263
Longmore, Arthur, 17, 34, 36, 90, 288, 400, 410n66
  aircraft reinforcement and, 30
  air plan and, 42
  air superiority and, 53
  army cooperation squadrons and, 43
  challenges for, 20, 76, 111
  character of, 84
  command for, 397
  Freeman and, 43
  IAF and, 29, 30
  Malta and, 21, 37, 38
  mobility and, 58
  Operation Compass and, 40, 41, 44
  photo of, 18
  Portal and, 84–85
  RAFME and, 92–93
  resignation of, 91–95, 97–99
  SAAF and, 35
  Tedder and, 40, 41, 91
Long Range Desert Group (LRDG), 89–90, 101, 187
LRDG. *See* Long Range Desert Group
Luard, J. C. E., 373, 374
Luftflotte, 37, 119, 145
Luftflotte 2: 70, 159, 165, 168, 195, 196, 198, 199, 296, 297, 311, 398
  aircraft for, 255
  arrival of, 163, 170, 183
  command structure and, 172
  engagements of, 399–400
  Fliegerführer Afrika and, 172
  Headquarters, 158
  intervention by, 174
  Malta and, 179
  mission of, 173
  redeployment of, 142, 150
Luftflotte 4: 80, 255, 371
Luftflotte Reich, 297, 309, 357, 361, 363, 403
Luftgau VII, 371
Luftministerium, 40, 43, 119, 131, 134, 146, 433n17, 442n7
Luftwaffe, 13, 16, 21, 27, 28, 36, 58, 61
  Afrika Korps and, 79, 91
  air superiority and, 63, 70, 158, 160, 164, 175, 185, 207, 211, 259, 260, 276, 398, 418n47, 455n72
  buildup of, 144
  command relationships of, 66, 67, 339, 398–399

destruction of, 33, 187, 249, 276, 284 (photo), 285, 292, 303, 309, 310–311, 314 (photo), 319–320, 325, 326, 350, 360–361, 400, 402, 403
effectiveness of, 68, 271
flexibility of, 67
German Army and, 85, 438n12
Italian Navy and, 438n15
logistics and, 78, 138–139, 147, 163, 168, 174
losses for, 196, 204, 224, 254, 265, 274, 282, 286, 308, 320, 352
maintenance and, 141, 143
Malta and, 30, 125–126, 239
missions of, 58, 76, 127, 250, 275, 276, 454n50
Operation Crusader and, 166
overhauls by, 65–66
problems for, 75, 113, 246
and RAF compared, 67, 68, 81, 89, 90
reinforcements for, 168
reorganization of, 133, 278
Royal Navy and, 84
shipping schedule and, 236
shortages for, 70, 167, 209, 320
strength of, 31, 163, 199, 245, 279, 352, 434n24
transport capability of, 73, 139
window of opportunity for, 72–73, 75–77
Luftwaffenkommando Südost, 297
Luqua airfield, 13
Lysanders, 43, 58, 410n66, 413n23

MAAF. *See* Mediterranean Allied Air Forces
MAAF Instruction No. 8: 330
MAC. *See* Mediterranean Air Command
MACAF. *See* Mediterranean Allied Coastal Air Force
Maintenance, 16, 155, 226, 232
  British, 154, 180, 185, 189, 208
  challenges of, 303, 413n39
  German, 116, 141, 143–148
  Luftgaue and, 146–147
  personnel/shortage of, 146, 153
  RAF, 134–138
Maison Blanche, 252, 260, 263, 270
Maleme airfield, loss of, 83
Malta, 5, 9, 11, 26, 51, 57, 69, 82, 85, 90, 91
  aircraft ranges from, 39 (map)

Malta, *continued*
　airfields on, 125
　airpower at, 38, 68, 76, 87, 103,
　　120, 129, 141, 159, 180, 182, 184,
　　195, 196, 197, 198, 199, 205, 255,
　　259, 272, 296, 299, 302, 417n33,
　　433n17, 446n83
　attacks on, 30–31, 37, 38, 60, 68, 75,
　　76, 83, 125–126, 138, 139, 173,
　　182, 185, 194, 196, 198, 199, 200,
　　229, 297, 399, 443n20, 445n48,
　　447n13
　defending, 61, 92, 93, 97, 114, 188,
　　189, 194–199, 213
　importance of, 12, 13, 21, 30, 37–38,
　　150, 239, 419n64
　invasion plans for, 198–199, 200, 204,
　　212, 216, 285
　neutralizing, 60, 119, 126, 133, 158,
　　168, 170, 171, 172, 179, 183, 184,
　　195, 198, 241
　photo of, 220
　relief for, 76, 141, 182–183, 212, 213,
　　218, 220, 241, 248, 272–273
　Tedder at, 184–185, 188–189,
　　194–195
Mann, William, 103, 271
MAPRW. *See* Mediterranean Allied
　Photographic Reconnaissance Wing
Marble Arch, 142, 169, 274
Mareth Line, 244, 247, 250, 277–279,
　458n62, 458n65
Marseilles, 292, 348, 354
Marshaling yards, 304, 361
　attacks on, 299, 303, 310, 323, 326,
　　331, 332, 334, 351, 364, 365, 373,
　　379, 380 (photo), 382
　repairing, 327
　targeting, 358
Marshall, George C., 151, 224, 317,
　349, 357
　airpower and, 193
　Balkans and, 344
　cross-channel invasion and, 265
　lend-lease and, 86
　Operation Torch and, 248
　US ground/airpower and, 179
Martlets (F4Fs), 252
　photo of, 254
Martuba, 56, 180, 184, 209
Marylands, 38, 111, 132, 134, 205,
　414n51, 417n33, 424n35, 430n56

MASAF. *See* Mediterranean Allied
　Strategic Air Forces
Massawa, 29, 35, 112, 189
MATAF. *See* Mediterranean Allied
　Tactical Air Forces
Maten Bagush, raid on, 209
Maynard, F. H. M., 30
Me. 109s, 104, 140, 143, 158, 159, 163,
　164, 168, 169, 191, 228, 352
　arrival of, 112, 189
　escorting by, 237
　Hurricanes and, 78, 79
　increase in, 127, 182
　losses of, 361
　and Spitfires compared, 188, 214
　superiority of, 107, 188
Me. 109Es, photo of, 62
Me. 109Fs, 163, 164, 170
　attacks by, 175
　and Hurricanes compared, 184
　and Kittyhawks compared, 175
　photo of, 169
　superiority of, 182, 190, 191, 242
Me. 109Gs, 245
　superiority of, 242
Me. 110s, 108, 115, 116, 204
MEAF. *See* Middle East Air Force
Mechanical transport light repair units
　(MTLRU), 176
Mecuzzi, Amadeo, 19
Mediterranean Air Command (MAC),
　188, 267, 270, 462n51
　commanders of, 268
　formation of, 268–269, 271
　MAAF and, 325–326, 357, 360
　Malta and, 272
　NAAF and, 269, 362
　objectives for, 269
　Operation Avalanche and, 309
Mediterranean Allied Air Forces
　(MAAF), 330, 333, 337, 345, 368,
　397, 403
　aircraft of, 352
　Eaker and, 357, 397
　formation of, 326, 357–358, 360–366
　Headquarters, 365–366
　missions of, 332, 334, 360
　Operation Anvil and, 348
　Operation Diadem and, 332
　Operation Shingle and, 325
　Ploesti and, 372
　Second Ukrainian Army and, 381

Strategic Targets Committee of, 369
  Wilson and, 356
Mediterranean Allied Coastal Air Force
  (MACAF), 326, 362
Mediterranean Allied Photographic
  Reconnaissance Wing (MAPRW),
  311, 365, 366
Mediterranean Allied Strategic Air Forces
  (MASAF), 291, 358, 360, 362, 363,
  364, 365
  missions of, 192, 333, 350, 377,
  394, 401
  Operation Diadem and, 335
  Red Army and, 384
Mediterranean Allied Tactical Air Forces
  (MATAF), 326, 349, 360, 362
  cover/support from, 343
  missions of, 333, 334, 336, 337
  pursuit and, 341
Mediterranean Fleet, 90, 132, 226
Mediterraneanists, position of, 6
Mediterranean Photographic
  Interpretation Center (MPIC), 366
Mediterranean theater
  Eastern Front and, 177–178
  grand-strategic views of, 3–5, 46
  impact of, 26, 277, 397–404
  as laboratory, 405
  map of, 2
  Operation Barbarossa and, 10
  as strategist's nightmare, 4–5
Merchant vessels (MVs), 27, 35, 36, 63,
  75, 82, 87
  arrival of, 265
  attacks on, 183, 247, 262
  increase in, 318
  Italian, 99 (photo)
  protecting, 140, 272
  shortage of, 6
  sinking of, 39, 76, 99 (photo), 127,
  128, 131, 134, 182, 196, 230, 231,
  261, 262, 273, 275, 320
Mers al Kebir, 31, 46
Mersa Matruh, 40, 83, 160, 211, 214,
  215, 217
  advance on, 42
  Italian troops and, 59
  raids on, 127, 209
Messerschmitt Plant, 306–307, 310
Messina, 294, 297
Middle East, 9, 47, 70, 114, 115, 200,
  223

airpower and, 178
British in, 5–6, 46
defending, 178
German-Japanese linkup in, 200
Hitler and, 4, 28, 67, 69, 116
importance of, 6, 49
map of, 2
Operation Barbarossa and, 10
as sideshow, 403
Middle East Air Force (MEAF), 175, 224,
  315–316
Middle East Combined Signals Board,
  101–102
Middle East Command, 283
Middle East Defence Committee, 188,
  202, 267
Middle East Interpretation Unit, 23
"Middle East Training Pamphlet (Army
  and R.A.F.) No. 3A—Direct Air
  Support," 123, 154, 234, 235
Milan, 251, 307
Milch, Erhard, 144, 145, 146, 147
Mining, 51, 62, 63, 118, 132, 133,
  272, 287
  of Danube, 72, 364, 365, 368, 369,
  370, 372
Mitchell, William, 13, 15, 17, 43
Mobile warfare, 20, 285, 326
  developing, 124–125
  equipment for, 147–148
  logistics and, 379
Mobility, 16, 58, 155, 175, 176, 246,
  358, 402
  airfield, 56, 124–125
  disruption of, 380
  importance of, 58, 155
  operational, 382
  strategic, 382
Mohawks, 111
Momyer, William: on Coningham,
  453n38
Monckton, Walter, 194
Monte Cassino, 325, 328
Montecorvino airfield, 314, 315
Montgomery, Bernard Law, 213, 280,
  295, 458n65
  advance by, 218, 274
  air-ground cooperation and, 233, 240
  air superiority and, 294
  air support and, 281, 288
  C2 and, 233
  Coningham and, 232–233

Montgomery, Bernard Law, *continued*
  deception scheme by, 242
  Eighth Army and, 232, 247
  Mareth Line and, 282
  methodological style of, 241
  Operation Lightfoot and, 242–243
  Operation Supercharge and, 243, 245
  Ponte Olivo and, 294
  Tedder and, 240, 241
  Tripoli and, 275
  Twenty-First Army Group and, 357
  Ultra and, 233
Morgenthau, Henry, 43, 44
Morocco, 8, 10, 26
  landing at, 251
Mosquitos, 165, 273
  PR by, 158, 188
Msus, 164, 180
Muhammad Amin al-Husayni, 50
Mulberries, destruction of, 348
Mussolini, Benito, 3, 10, 37, 80, 171,
    408n22, 409n35
  Arab world and, 51
  cooperation and, 59, 88
  Cyrenaica and, 198
  fall of, 294
  foreign policy of, 25
  Greece and, 28, 49
  Hitler and, 142, 293
  Mediterranean theater and, 8
  military efforts of, 7, 17, 20, 25, 28,
    47, 49, 416n8
  oil reserves and, 6
  Operation Hercules and, 173, 199,
    211, 212
  Pact of Steel and, 11
  resources for, 27, 49
  Rommel and, 216
  sphere of interest of, 12
  strategic posture of, 12

NAAF. *See* Allied Air Forces, Northwest
    Africa
NACAF. *See* Northwest African Coastal
    Air Force
Naples, 307, 310, 325, 342
  attacks on, 261, 305
  invasion of, 294, 318, 319
  occupation of, 323
NASAF. *See* Northwest African Strategic
    Air Force
NATAF. *See* Northwest African Tactical
    Air Force

Nehring, Walther, 404
Neumann, Eduard, 212
Neutrality Act, 33, 35
New Zealand Division, 212, 281
Newall, Cyril, 43
Nile Delta, 27, 40, 59, 69, 95, 117, 130,
    171, 213, 215
  advance on, 228, 239
  defending, 205, 219
  railroad from, 248
  reconnaissance of, 136
  reinforcements from, 214
No. 201 Naval Cooperation Group, 34,
    98, 173, 205, 213, 220, 221, 227,
    241, 430n56
  air-sea cooperation and, 132–133
  antishipping operations of, 229,
    230, 231
  ASV for, 230
  attacks by, 399
  formation of, 226
  mission of, 80, 129–132
  needs of, 228
  photo of, 229
  RAF-Royal Navy Combined
    Operations Room at, 272
  Strength of, 183
No. 202 Group, 30
No. 203 Group, 34, 413n37
No. 204 Group, 98
No. 205 Group, 72, 98, 270, 291, 295,
    358, 362, 369, 434n24
  aircraft in, 242
  bombing by, 127
  excellence of, 191–192, 365
  mining by, 364, 368
  Sicily and, 305
No. 206 Group, 135
No. 211 Group, 185, 234
No. 216 Group, 274
No. 219 Group, 431n71
No. 242 Group, 270, 282
No. 247 Wing, 132
No. 252 Wing, 430n51
No. 253 Army Co-operation Wing, 98
No. 258 Wing, 101
No. 285 Air Reconnaissance Wing, 23
No. 333 Group, 251
Normandy, 4, 302, 309, 344, 347, 350,
    354, 356, 364
Norstad, Lauris, 331, 339, 340
Northwest African Air Service
    Command, 270

Northwest African Coastal Air Force
(NACAF), 270, 288, 317
Northwest African Photographic
Reconnaissance Wing (NAPRW), 270
Northwest African Strategic Air Force
(NASAF), 283, 316, 319
command for, 270
missions of, 299, 310, 312, 317, 320
Northwest African Tactical Air Force
(NATAF), 279, 281, 323
command for, 270
missions of, 317
Northwest African Training Command, 270
Northwest African Troop Carrier
Command, 270
Numbered air forces (NAFs), 360

Oberkommando der Luftwaffe (OKL),
14, 62, 67, 139, 249, 296
challenges for, 119–120
micromanaging by, 70
organization structure of, 144
production priorities of, 144
RAF and, 71
Technical Office and, 144
Oberkommando der Wehrmacht (OKW),
27, 40, 59, 78, 115, 141, 142, 177,
212, 221
aircraft dispersal and, 298
convoy protection and, 161
fuel quotas and, 371
logistics and, 230
Mediterranean theater and, 70
micromanaging by, 70
Middle East and, 70
Operation Barbarossa and, 150
supply system and, 383
transfers by, 254
O'Connor, Richard, 42, 52, 57, 59
Office of Special Services (OSS), 366
Oil offensive, 303, 347, 349, 358, 367,
371–372, 374, 394, 402, 469n33,
471n61
impact of, 376, 382–383, 384–385
Oil production, 368, 369, 370, 372
attacks on, 358, 361, 371
synthetic, 372
Oil resources, 47, 69, 87, 88, 150, 201,
215, 399
attacks on, 292, 318, 361, 367, 371,
372, 376, 384
British, 178
defending, 114, 118

German, 9, 177, 178, 223
Japanese, 177
Middle Eastern, 6, 9, 10, 11, 398,
409n31
seizing, 172, 200, 411n5
Soviet, 9, 11
See also Petroleum, oil, and lubricants
assets
OKL. *See* Oberkommando der Luftwaffe
OKW. *See* Oberkommando der
Wehrmacht
Omaha Beach, 352
Operation Accolade, 356, 401
Operation Achse, 296, 394
Operation Acrobat, 174, 180
Operation Anvil, 329, 344, 347
planning, 348–349
support for, 348
See also Operation Dragoon
Operation Avalanche, 132, 291, 293, 325
beginning of, 312–317
map of, 313
naval/air strength and, 316
planning for, 294, 309–310
support for, 311–312
Operation Barbarossa, 26, 60, 62, 67, 69,
78, 87, 88, 91, 144, 402
aircraft for, 97
demands of, 51, 52
importance of, 150
Mediterranean/Middle East and, 9, 10
priority for, 52, 125, 126
timing of, 12
Operation Battleaxe, 121, 154, 170
described, 105–110
failure of, 108, 114
Operation Baytown, 110, 291, 293, 305,
401
beginning of, 312–317
map of, 313
planning for, 309–310
Operation Big Week, 319, 358, 360, 363
Operation Brevity, 170
described, 105–110
Operation Cobra, 354
Operation Compass, 21, 44, 46, 55, 105,
110, 400, 414n61
aircraft losses during, 423n20
air operations during, 89
air superiority and, 41
assets for, 45
beginning of, 15, 52–53, 56–59
as combined-arms operation, 56

Operation Compass, *continued*
  forward airfields for, 59
  lessons of, 106
  logistics and, 58
  map of, 54
  planning for, 40–42
  RAF and, 105, 159
  RT units and, 102
Operation Crusader, 63, 110, 114, 121,
    122–123, 125, 127, 132, 133, 134,
    137, 138, 149, 187, 201, 205, 235
  air preparations for, 152–156, 157
  air support for, 155, 157, 159, 160, 203
  ASCs during, 59
  beginning of, 157–161, 163–165
  cipher operations during, 101
  end of, 169–170, 176
  innovation during, 166–169
  intelligence and, 90
  logistics and, 154, 168
  losses during, 185, 423n20
  Luftwaffe and, 126
  maintenance and, 154
  map of, 162
  PR/PI and, 102–103
  reinforcements for, 120
  Rommel on, 184
  supplies for, 98
Operation Diadem, 292, 324, 332, 401,
    465n28
  air action during, 333, 334, 339–
    340, 341
  combined-arms efforts in, 329
  described, 335–347
  impact of, 292, 330, 332, 339, 343,
    350
  LOC strikes before, 342
  losses in, 338, 345
  planning for, 329–331
  reconnaissance during, 345–346
Operation Dragoon, 329, 344, 351, 355,
    356, 401
  air-ground focus and, 347
  air-sea-ground effort of, 292
  combat squadrons for, 349
  described, 347–348
  focus on, 292
  impact of, 350
  plan for, 352
  Red Army and, 309
  support for, 345
  tactical success of, 354
  See also Operation Anvil

Operation Eilbote, 277
Operation Excess, 76
Operation Felix, 10, 52
Operation Flax, 283, 284
Operation Hercules, 61, 83, 165, 171,
    173, 199, 204, 210, 211, 239,
    419n64
  Malta and, 195, 196
  plan for, 197
  plea for, 212
Operation Husky, 291, 293, 307, 308,
    325, 401
  air-ground cooperation in, 294
  beginning of, 299–300, 300–304
  countering, 312
  map of, 301
  planning for, 294–296
Operation Konstantin, 296
Operation Lehrgang, 304, 306
Operation Lightfoot
  beginning of, 243, 245
  preparing for, 242–243
Operation Marita, 70, 80, 419n63,
    421n96
Operation Overlord, 341, 347, 348, 354,
    357, 405
  air supremacy for, 358
  buildup for, 318, 321, 398
  combat squadrons for, 349
  impact of, 350
  postponement of, 266
  priority for, 329
  support for, 356
Operation Pedestal, 213, 218, 220
Operation Pointblank, 269–270, 291,
    305, 307, 309, 357, 360–361, 363
  air assets and, 316
  objective of, 361
Operation Shingle, 291, 292, 324–325, 328
  air support for, 325–326
  described, 327
Operation Sonnenblume, airpower and,
    60–62
Operation Stonage Convoy, 272
Operation Strangle, 324
  air effort in, 292, 334, 341
  beginning of, 332
  combined-arms efforts in, 329
  described, 330, 331–335
  impact of, 330, 332–333, 340
  losses in, 338
  planning for, 329–331, 333
  shortcomings of, 343

Operation Supercharge, 242, 243, 245
Operation Thesus, 110
    beginning of, 205, 207–212
    planning of, 204–205
Operation Tidalwave, 305, 368
Operation Torch, 152, 179, 196, 223,
    224, 246, 303, 423n20, 451n1,
    453n38, 453n45
    air operations of, 250–252, 258, 401
    Allied advances in, 255–256
    beginning of, 225, 252
    combined-arms operations in, 252,
    256–257
    German response to, 252, 254–258
    map of, 253
    shipping for, 254, 262
    strategic decisions regarding, 248–250
Operational training units (OTUs), 98,
    110, 153, 189, 235
Operations Division (Naval Staff), 240
Oran, 251, 252
Orange, Vincent, 99, 424n27, 442n82
OSS. *See* Office of Special Services
OTUs. *See* Operational training units
Overy, Richard, 256, 398

P-38 Lightnings, 280, 307, 309, 311, 319,
    361, 362, 364
P-40 Tomahawks, 105, 110, 225, 315,
    447n13
    performance of, 44, 137
    reinforcement with, 111
    replacement of, 175
P-47 Thunderbolts, 325, 362, 364
P-51 Mustangs, 306, 309, 364
Pact of Steel (1939), 11
Palermo, 261, 294, 295, 296
Palm Sunday Massacre, 283
Pankhurst, Leonard, 348, 388, 389
Pantelleria, attacks on, 285, 299, 300
Panzerarmee Afrika, 61, 204, 215, 217, 400
    fuel shortages and, 236–237, 245, 371
    logistical pressure on, 248
    Luftwaffe and, 184, 210
    mission of, 228, 234
    retreat of, 247
    supplying, 166, 199
Panzerarmee Afrika Headquarters, 184,
    221
Park, Keith, 30, 99, 205, 213, 299
    air assets for, 302
    MAC and, 268
    RAF Malta and, 270

Partisans, 88, 318, 382
    aid for, 48, 292, 309, 347, 400
    combat power of, 394
    movements of, 48, 380
Patch, Alexander M., 349, 355
Patton, George, 258
Pedestal Convoy, 218, 220
Penney, William "Ronald," 103
"Penny-packet" patrols, 259, 271, 280
Pétain, Philippe, 51, 88, 409n35
    de Gaulle and, 46–47
    Hitler and, 10, 27, 28
Petroleum, oil, and lubricants (POL)
    assets, 81, 173, 409n32, 418n48,
    470n38
    attacks on, 367–373
    German, 282, 367–368
    movement of, 72
    seizure of, 11
    shortage of, 160
    See also Oil resources
Photographic intelligence (PI), 58, 102,
    103, 120, 131, 209, 234, 351,
    366, 381
Photoreconnaissance (PR), 23, 127, 131,
    186, 214, 254, 351, 366
    aircraft, 43, 56, 58, 95, 97, 100, 120,
    158, 205, 227, 300, 325, 381
    resources, 185
    value of, 102, 334
Photoreconnaissance (PR) Detachment, 234
Photoreconnaissance Unit (PRU), 263, 275
Pierced steel planking (PSP), 124, 255, 268
Pisa-Rimini line, 330, 344, 345
Plan Dog, 48, 86
Platt, William, 34
Ploesti, 9, 72, 79, 177, 358, 360, 409n32,
    469n30
    attacks on, 305, 306, 309–310,
    318, 364, 367, 360 (photo), 369,
    371–372, 373, 379
    aviation fuel from, 375
    bottleneck at, 373
    capturing, 274, 370
    defending, 80, 82, 83, 363, 368
    importance of, 366–367
Ploesti campaign, 265, 293
    beginning of, 367–373
    described, 366–367
POL assets. *See* Petroleum, oil, and
    lubricants assets
Ponte Grande Bridge, seizing, 302
Ponte Olivo, airfields at, 294

Porch, Douglas, 4, 402
Po River Valley, 318, 329, 344, 348
Porro, Felip, 37
Port Said, attacks on, 76
Port Sudan, 34, 112, 118, 134
Portal, Charles, 44, 175, 266, 272, 332,
    355, 374
    air situation and, 108, 111, 188–
        189, 193
    airpower and, 192, 193
    army air support group and, 203
    Arnold agreement with, 192–193
    ASV and, 228
    Auchinlek and, 157
    Churchill and, 202
    combined-arms operations and, 333
    communications and, 109
    Eaker and, 310
    Longmore and, 84–85
    on Luftwaffe/IAF, 94
    Ploesti and, 367
    Portal and, 158
    Pound and, 131
    Slessor and, 43, 415n71
    Tedder and, 93, 94, 95, 107, 108,
        109–110, 111, 117, 123, 130, 131,
        134, 156, 157, 163, 189, 190, 194,
        195, 202, 214, 227, 228, 232, 233,
        267–268, 271
Ports
    attacks on, 127, 169, 294, 299
    bottlenecks at, 17
    protecting, 236
Pound, Dudley, 130–131
POWs. *See* Prisoners of war
Pozallo, landing at, 294
PR. *See* Photoreconnaissance
Prendergast, Guy, 90
Pricolo, Francesco, 19, 140, 142, 207
Primosole Bridge, seizing, 302
Prisoners of war (POWs), 59, 168, 230,
    346, 347, 367, 381
    interrogation of, 90, 119, 120, 186,
        231–232, 330, 338, 343, 371
PRU. *See* Photoreconnaissance Unit

Qattara depression, 211
Quadrant Conference, 292
Quebec Conference, 329
Quesada, Pete, 288, 354

Radar, 13, 102, 104, 141, 274, 365
    AI, 103, 263, 326, 425n54

ASV, 21, 132, 133, 165, 190, 192,
    205, 227, 228, 230, 282, 296
    Freya, 237, 278, 283
    GCI, 109, 123, 263, 281–282, 296,
        308, 325, 425n54
    H2X, 352, 360, 361, 364
    HF/DF, 103, 227, 296
    jamming, 198
    lack of, 20
    SCI, 308
    Würzburg, 278
Radio
    AASC, 187
    communications by, 278
    HF, 100, 335
    intercepts, 119
    Rover VHF, 307
    VHF, 100, 186, 187, 272, 296, 299,
        307, 308, 335
Radio direction-finding (RDF), 281
Radio-telephony (RT) interception units,
    102
Raeder, Erich, 26, 63, 114
    antishipping operations and, 262
    combined-arms cooperation and, 11
    Fliegerkorps X and, 141
    Göring and, 62
    Hitler and, 133, 174, 408n25
    Malta and, 38, 196
    Mediterranean theater and, 12, 51–52
    OKW and, 59, 142
    Operation Barbarossa and, 52
    recommendations by, 27–28
    resignation of, 276–277
    on sea communications, 240
    on transports, 142
    vision of, 8
RAF. *See* Royal Air Force
RAF Bomber Command, 47, 86, 169,
    242, 251, 295, 307, 361, 384
    raids by, 143
    Ruhr Offensive of, 320
RAF Coastal Command, 131, 228,
    251, 268
RAF Delegation (RAFDEL), 43
RAF Malta, 230, 270
RAF Regiment, 180, 263
RAF-Royal Navy Operation Room, 272
RAF Staff College, 94
RAFME. *See* Royal Air Force Middle East
Railroads
    attacks on, 323, 326, 331, 332, 333,
        334, 350, 374, 377, 383

collapse of, 382
shortage of, 381
Rainbow Plans, 48, 49
Rashid Ali al Gailani, 50, 51, 88, 115
Rear air support links (RASLs), 122
Reconnaissance, 21, 68, 157
  aerial, 17, 59, 80, 119, 160, 174,
    179, 180, 185, 186, 187, 198, 207,
    211, 217, 218, 226, 228, 234, 243,
    249, 252, 274–275, 282, 300, 306,
    345–346, 360
  artillery, 58, 346
  ASV, 296
  British, 58, 226, 236
  conducting, 159
  effective, 221
  German, 136, 211
  long-range, 132, 205
  strategic, 157, 263
  tactical, 160, 167, 205, 234, 240, 279,
    306, 337
Red Army, 179, 248, 292, 347, 401, 402
  advance of, 249, 291, 306, 309, 350,
    356, 376, 377, 378–379 (map), 379,
    380, 383
  air support for, 361, 373–374, 381
  equipment for, 201
  Fifteenth Air Force and, 381
  lend-lease and, 403–404, 474n9
  losses for, 150, 375
  MASAF and, 384
  Ploesti and, 367
  transformation of, 201
Red Sea, 15, 26, 27, 34, 35, 111, 127,
    128, 411n5
  defending, 93, 400
  importance of, 150
  Italy and, 7, 25, 408n25
  SLOCs in, 33
Regensburg, raids on, 309, 310, 360
Reichsbahn, 173, 310, 382
Reid, G. R. M., 34–35, 413n38
Repair-and-salvage units (RSUs), 15,
    16, 36, 56, 137, 270, 322, 414n46,
    437n95
  development of, 260
  forward, 134
  work of, 136, 175, 180, 188, 211,
    214, 225
Ribbentrop, Joachim von, 49
Ritchie, Lewis Anthem, 123
Robb, James, 259
Roma, sinking of, 293

Romanian Air Force, 79
Rome, 305, 307, 337
  advance on, 324, 329, 330
  defending, 330
  fall of, 338, 341, 344, 346, 347
Rommel, Erwin, 73, 79, 84, 88, 98, 99,
    107, 139, 150, 161, 163, 178, 186,
    197, 202, 219, 225
  Afrika Korps and, 238
  air attacks and, 279–280, 444n28
  air intelligence and, 160
  air policy and, 246
  airpower and, 69–71, 126–127
  on air superiority, 238, 239
  air support for, 65, 68, 70, 71, 79,
    221, 274
  Alam Halfa and, 237
  antishipping operations and, 236
  arrival of, 60, 61
  Auchinlek and, 166
  Benghazi and, 75
  Bir Hacheim and, 208, 444n35
  British Army and, 70
  combined-arms effort and, 60, 66
  convoy protection and, 236
  counterattack by, 72, 137, 175, 176,
    180, 201, 227, 232, 401
  defeat of, 152, 178, 400
  defensive posture for, 237
  Eighth Army and, 213
  El Alamein and, 211
  encirclement of, 165
  Enigma decrypts on, 102
  equipment losses for, 218
  Fröhlich and, 66, 67, 69, 71, 78, 148,
    418n50, 421n96
  fuel shortages for, 134, 222, 274
  Gambut and, 210
  Gazala and, 110, 186, 201, 204, 223
  Halder and, 105
  Hitler and, 61, 68, 211, 212, 216, 245,
    434n34, 442n80
  Kesselring and, 170, 204, 207,
    215, 246
  last gamble for, 236–241
  logistics and, 71, 104, 132, 138, 165,
    171, 174, 179, 182, 216, 217, 230,
    232, 236, 248
  Malta and, 69, 199, 239
  Mareth Line and, 244, 250
  Mediterranean theater and, 459–
    460n83
  Mussolini and, 216

Rommel, Erwin, *continued*
offensive/first by, 63, 66, 68, 74 (map), 75, 90–91, 101, 103–105, 110, 135, 187
offensive/second by, 177, 179–180, 181 (map), 182, 198, 205, 206 (map), 213–215, 217, 224, 228, 231, 236, 237, 278, 279, 280
OKW and, 70, 245
Operation Crusader and, 184
Operation Hercules and, 211, 212, 419n64
Operation Marita and, 419n63
Operation Theseus and, 204, 207
photo of, 61
promotion for, 211
reconnaissance and, 160, 165–166
reinforcing, 250
shipping and, 221
shortages for, 166
signals intelligence and, 121
SLOCs and, 83, 133
Tedder and, 164, 213–214
Tobruk and, 78, 105, 204, 210, 215
transportation and, 209
von Arnim and, 279
von Waldau and, 207, 208, 212, 217, 221
weaknesses for, 61, 70–71
Weichold and, 236–237
withdrawal of, 163, 164, 165, 166, 241, 245–248, 274, 280, 281
Roosevelt, Elliott, 270
Roosevelt, Franklin D., 110, 151, 194, 201, 356
airpower and, 197, 269
Battle of the Atlantic and, 202
Casablanca and, 266, 267
Churchill and, 13, 47, 48, 86, 87, 137, 151, 152, 192, 223, 269
convoy escort and, 151
cross-channel invasion and, 265, 293
Italian surrender and, 293, 294
Mediterranean theater and, 48, 265
Middle East and, 47
military aid and, 13
Operation Torch and, 179, 223
Red Army and, 201
shipping restrictions and, 35
Rosier, Frederick, 232
Rover teams, 307, 334–335
Royal Air Force (RAF)
aircraft shortage for, 43–45

air superiority and, 30, 32, 34, 35, 37, 41, 42, 53, 56, 68, 73, 75, 112, 117–118, 125, 127, 153, 157, 159, 164, 170, 175, 182, 188, 196, 246, 399
air support by, 122, 167
assets of, 106, 120
attacks by, 42, 53, 65, 108, 138, 139, 143, 147, 168, 232
British Army and, 66, 71, 72, 257, 267
challenges for, 20, 72
Crete and, 82, 84
direct-support mission by, 109
effectiveness of, 231–232
Eighth Army and, 103, 105, 168–169, 174–175, 215
Greece and, 82, 93, 113
Greek Army and, 80
IAF and, 14–15, 17, 29, 31, 53, 57, 159
improvements for, 88–90, 125
interservice cooperation and, 14, 34
and Luftwaffe compared, 67, 68, 81, 89, 90
maintenance by, 16, 66, 134–138
priorities for, 75
rebuilding by, 36, 202–204
replacements for, 30, 31
Royal Navy and, 75, 129, 131, 173, 287, 400
shipping attacks and, 98
strength of, 19, 120–121, 178, 218, 434n24
tactical advantages for, 42, 212, 411n81
TAF and, 42
US aircraft for, 193–194, 226
USAAF and, 194, 225
warfare approach of, 6, 64, 71
WDAF and, 247
Royal Air Force delegation to the United States (RAFDEL), 224
Royal Air Force Middle East (RAFME), 56, 103, 107, 154, 155, 159, 175, 188, 268, 362
achievements of, 92–93
aircraft for, 36, 43, 70, 91, 190
air superiority and, 32
area of responsibility for, 97
challenges for, 20
command of, 99, 397
components of, 98
land/sea operations and, 16
logistics and, 58, 132
maintenance and, 134, 185, 226

Mitchell and, 13
Operation Compass and, 45
priorities for, 15
reorganization of, 98
replacements for, 73, 111
strength of, 44, 240
transport and, 176
Royal Navy, 5, 27, 31, 59, 83, 116, 118, 122, 218, 226, 229
advantages for, 20
air support for, 130
attacks by, 34, 76
attacks on, 126, 198
FAA and, 117
intelligence for, 100
Italian Navy and, 7, 20, 29, 92
Luftwaffe and, 84
RAF and, 75, 129, 131, 173, 287, 400
retreat of, 107
RSUs. *See* Repair-and-salvage units
Ruhr Offensive, 320

SAAF. *See* South African Air Force
Salerno, 110, 291, 293, 304, 311, 316
air buildup around, 319
counterattack at, 317
landing at, 308, 309, 312
losses at, 326
problems at, 313–314
reinforcement of, 315
victory at, 318
Salvage assets, 138, 154, 155, 175–176, 188
Salvage crews, 143, 153, 176
Sardinia, 317, 325
airfields in, 283
airpower for, 251, 297
attacks on, 251
defending, 298
SAS. *See* Special Air Services
Saville, Gordon P., 334, 352, 354
Schweinfurt, raids on, 310
Scorpion, breaking of, 233–234
Seafires, 254, 311, 312
Sea Gladiators, 21
Sea lines of communication (SLOCs), 15, 16, 33, 64, 70, 72, 75, 86, 88, 95, 182
attacks on, 60, 82, 157, 220, 261
battle for, 11, 62, 63, 83, 114, 129, 132, 170, 226, 231, 262, 398, 399, 401
controlling, 65, 78, 112, 165, 175, 218, 321

logistics and, 69
protecting, 34, 37, 40, 97, 98, 127, 133, 142, 170, 196
securing, 179, 198, 248, 357
Seidemann, Hans, 71, 230, 239, 243, 274, 452n19
Selassie, Haile, 32
Self, Henry, 110
Serviceability rates, 126, 129, 146, 212, 259, 262, 299, 362
improvements in, 134, 231, 269, 280
problems with, 142, 278–279, 283
Sfax, 259, 263
SHEAF. *See* Supreme Headquarters Allied Expeditionary Force
Shipping, 11, 35, 147, 221, 236, 237, 254, 262, 374
attacks on, 76, 98, 182, 272, 273
losses of, 287
protecting, 132, 295
shortage of, 178, 191, 223, 294
See also Antishipping operations
Shortages, 147, 174, 314, 362
aircraft, 29, 43–45, 57, 70, 97, 143, 165, 169, 175, 185, 207, 209, 224, 320, 435n53
ammunition, 166, 282, 326, 331, 334, 346
communications, 259
equipment, 14, 19–20, 136, 173
flak gun, 126–127
food, 239
fuel, 20, 134, 160, 166, 169, 213, 219, 222, 228, 236–237, 245, 274, 285, 287, 306, 322–323, 326, 327, 331, 333, 367, 368, 371, 377, 381, 382, 383, 384, 469n33
landing craft, 294
maintenance personnel, 146
manpower, 20, 153, 154
shipping, 178, 191, 223, 294
spare parts, 143, 146, 167
supply, 174, 202, 240, 279, 332, 402
transport, 105–106, 125
vehicle, 126, 140, 184, 260, 286, 333
weapons, 14
Sicily, 60, 110, 142, 261, 307, 308, 318, 326, 349, 350
aircraft losses at, 402
airfields on, 170, 198, 283, 294–295, 299
airpower on, 251, 267, 280, 285, 294, 295, 297, 301, 315, 340, 346
assets in, 139

Sicily, *continued*
 attacks on, 195, 231, 295, 310
 cooperation roles in, 309
 defending, 231, 298
 evacuation of, 205, 305, 306
 invasion of, 249, 269, 293, 294, 296,
  300, 302, 306, 312, 346
 Luftwaffe in, 77, 91, 128
 map of, 301
 victory in, 317
Sidi Barani, 42, 210
Siebel ferries, 142, 173, 230, 261, 277
Signal Intelligence Service (SIS), 101, 119,
  132, 192, 335, 364, 365, 381
 RAF strength and, 120–121
 support from, 259
Signals, 101, 261, 335, 362, 366
 capabilities/improving, 185, 425n54
Sirte, 72, 125
SIS. *See* Signal Intelligence Service
Slatter, Leonard, 205, 227, 229, 413n38
 convoy attacks and, 228
 No. 201 Naval Cooperation Group
  and, 34, 132
Slessor, John "Jack," 94, 110, 194, 203,
  355, 362, 363, 440–441n53
 airpower and, 43, 45, 192, 332–333,
  342–343, 415n71
 appointment for, 356
 Arnold and, 341
 on Beaverbrook, 44
 combined-arms operations and, 333,
  341, 343
 Eaker and, 341, 356
 interservice rivalries and, 405
 Morgenthau and, 44
 obsolete aircraft and, 111
 Operation Anvil and, 348
 Operation Diadem and, 333
 Operation Strangle and, 332
 Portal and, 43, 415n71
 Tedder and, 192
 Tomahawks and, 44
 trainers/flying schools and, 110–111
 Zuckerman plan and, 334
SLOCs. *See* Sea lines of communication
SM.79 torpedo-bombers, 19, 127–128,
  170, 247
SM.82 transports, 247
Smart, H. G., 115
Smuts, Jan, 35
Sollum, 77, 164, 165, 401
 mining at, 63

Sollum offensive, 78, 79, 110, 132, 163, 187
Sommerveld Track, 124
Sousse, 259, 263
South African Air Force (SAAF), 35,
  413n39
Spaatz, Carl "Tooey," 264, 327, 360,
  458n65
 on airpower, 256, 257
 Arnold and, 256, 257, 263
 command for, 259, 263, 268, 269
 Doolittle and, 270
 Eisenhower and, 257, 259
 MAC and, 268, 271
 mining and, 369
 NAAF and, 269
 Operation Torch and, 250
 photo of, 273
 Ploesti and, 367
 Tedder and, 267, 309
 USSAFE and, 356, 357, 358
Spare parts, 176, 189, 225
 problems with, 144, 145
 shortage of, 143, 146, 167
Special Air Services (SAS), 125, 147, 168,
  187, 208
Speer, Albert, 370, 383, 384, 472n89
Speidel, Hans, 79
Sperrverband, 60, 138
Spitfires, 87, 110, 185, 193, 205, 286,
  299, 309, 315, 446n83, 451n77
 ambush tactics with, 173
 arrival of, 68, 136, 157–158, 182,
  184, 190, 195–196, 197, 198, 199,
  209, 213, 214, 218, 242
 assembling, 32
 escort by, 252
 Me. 109s and, 188, 214
 need for, 188, 189–190, 195
 operational, 194
 patrols by, 281
 performance of, 231
 photo of, 197
 PR, 205, 209, 275
 reverse lend-lease with, 251
Spitfire Vs, 62, 165, 311
 Me. 109s and, 188
Spitfire IXs, 277
Staff Conference Report (1941), priorities
 of, 48
Stalin, Josef, 201, 347, 375, 403, 404
 Allied bombing and, 376
 Balkans strategy of, 376
 frustrations of, 265

sea/airpower and, 356
unconditional surrender and, 266
Stalingrad, 73, 146, 196, 246, 255, 283,
  284, 403, 472n91
  air transport at, 148
  Tunisia and, 286
Stark, Harold, 48
Stavka (High Command), 375, 376
Steinhoff, Johannes, 140, 285, 303,
  459n75, 461n33
  on air fighting, 301
  Galland and, 297
Stimson, Henry, 45
Stirling, David, 90, 125, 168, 187, 209
Straits of Messina, 297, 304, 305, 309,
  460n13
  map of, 301
Strategic decisions
  Allied, 151–152, 201–202, 248–250,
    265–266
  Axis, 150, 200, 223–224
  See also Grand strategy
Strategic Targets Committee (MAAF), 356
Student, Kurt, 216
Stumme, Georg, 243
Suez Bay, 118, 236
Suez Canal, 7, 9, 26, 27, 65, 83, 92, 159,
  174, 215, 216, 246
  attacks on, 28, 50, 51, 52, 59, 60, 76,
    127, 128, 133, 148, 161, 430n47
  closing, 50, 161
  defending, 76, 98, 114, 117, 398, 400
  importance of, 4, 5, 6, 70, 150, 172, 398
  mining of, 51, 62, 63, 118, 133, 287
  reinforcements at, 211
  shipping in, 11, 147, 237
Sunderland flying boats, 68, 414n51,
  417n33, 430n56
Superaereo, 148–149
Supermarina, 172
Supplies, 19, 50, 98, 139, 155, 215, 227,
  230, 236, 284, 348, 351, 383
  battle for, 185
  handling, 11, 102, 270
  problems with, 146–147, 166, 260,
    277–278, 338–339, 342, 343,
    373, 400
  shortage of, 7, 174, 279, 332, 340, 402
Supply dumps, 145, 381
Supreme Headquarters Allied
  Expeditionary Force (SHEAF), 355
Swordfish, 39, 190, 226, 417n33
Syracuse, 294, 295, 310

Syria, 40, 170
  British occupation of, 117–119

Tactical reconnaissance (Tac R), 160, 167,
  205, 234, 240, 279, 306, 337
Tafaraoui, 258, 260
Takoradi, repair work at, 137, 175
Takoradi air route, 76, 95, 110, 134, 137,
  288, 422n7
  aircraft reinforcement by, 30, 31, 40,
    97, 413n24
  German discovery of, 120
  map of, 32
  US aircraft on, 135
Tanker truck, destruction of, 219 (photo)
Taranto, 141, 318
  raid on, 29, 39–40
Tebessa, 258, 277
Tedder, Arthur, 21, 36, 98, 106, 126, 129,
  152, 170, 202, 225, 235, 263, 270,
  315, 360
  on AAA, 182
  aircraft for, 30, 92, 95, 97, 110,
    111–112, 207, 214–215, 224, 226,
    435n53, 442n7
  air-defense duties and, 422n7
  air-ground cooperation and, 154, 157,
    160, 189
  air situation and, 161, 188–189,
    204–205, 207, 212, 268, 271,
    310, 357
  air superiority and, 89, 106, 108, 131,
    158, 160, 190, 191, 214, 231
  on American consignment, 174–175
  antishipping operations and, 213,
    229–230
  ASC and, 189
  ASV and, 228
  character of, 93–94, 95, 99
  Churchill and, 156, 157, 448n35
  combined-arms efforts and, 203, 240,
    264, 267
  command for, 91–95, 97–99, 231,
    268–269
  communication and, 100, 101, 109, 121
  Coningham and, 154, 255, 256, 280, 304
  Crete and, 82, 84, 85, 88
  Cunningham and, 129–130, 131, 227
  Dawson and, 135, 136, 155, 214
  Eisenhower and, 258, 264, 266–267,
    268, 269, 288, 356
  on fighter-pilot losses, 195
  Freeman and, 94–95, 107, 165

Tedder, Arthur, *continued*
ground-attack and, 106, 107
intelligence and, 100, 101
on Iraq/Syria, 118
Kittyhawks and, 191, 226
logistics and, 180, 182, 247
Longmore and, 40, 41, 91
MAC and, 188, 268–269, 271
Malta and, 184–185, 188–189, 194–195
on MEAF, 175
Montgomery and, 240, 241
NASAF and, 310
Ninth Air Force and, 315
Operation Acrobat and, 174, 180
Operation Baytown and, 309
Operation Compass and, 40, 44, 89, 105, 106
Operation Crusader and, 164, 167
Operation Husky and, 294
Operation Lightfoot and, 242
photo of, 41, 273
pilot shortages and, 154
Ploesti and, 367
Portal and, 93, 94, 95, 107, 108, 109–110, 111, 117, 123, 130, 131, 134, 156, 157, 158, 163, 189, 190, 194, 195, 202, 214, 227, 228, 232, 233, 267–268, 271
priorities of, 357–358
problems for, 156–157
RAF Regiment and, 180
Rommel and, 164, 213–214
shipping attacks and, 98
Sicily and, 195, 295, 296, 304–305
on SLOCs, 214
on supply battle, 185
Takoradi and, 40
transport and, 105–106
WDAF and, 154, 165, 189, 245, 267
Zuckerman report and, 310
Teheran Conference, 356
Telecommunications Centre Middle East, 186
Terraine, John, 21
Thorold, Henry, 40, 134, 135
Tito, Josef Broz, 322, 394, 473n16
assistance to, 292, 309, 347, 400
Tmimi, 56, 77, 180
Tobruk, 69, 83, 85, 88, 139, 163, 221, 226
air/ground losses at, 78
air superiority at, 104
attacks on, 42, 56, 78, 126, 127, 132, 141, 161, 177, 204, 210, 225, 230, 241, 246, 247, 272

defending, 75, 106, 133, 170
fall of, 126, 141, 205, 210, 211, 214, 223
importance of, 75
mining at, 63
railroad to, 248
retreat from, 164
siege of, 105, 129, 163
supplies to, 98, 174, 215, 227, 236
Todt Organization, 327
Toulon, 252, 354
attacks on, 31, 350, 352
combined-arms operations and, 292
occupation of, 348, 350
Towers, John, 192
Trans-Iranian Railway, 118
Transportation, 139, 401
attacks on, 169, 292, 318, 357, 366, 368, 374, 376, 377, 380–381, 383, 384
oil/fuel, 373
shortage of, 125
Transportation offensives, 303, 310, 347, 349, 358, 361, 364, 367, 371–372, 373, 374, 394
impact of, 376, 382–385
intensification of, 377, 380–382
photo of, 380
Trenchard, 342
Trident Conference, 292, 347
Tripartite Pact, joining, 79–80
Tripoli, 56, 60, 104, 133, 204, 215, 233
advance on, 274
attacks on, 56, 72, 76, 90, 188, 230, 275
convoys at, 161, 170, 173, 176, 227
depots at, 143
fuel dump at, 77
port at, 128 (photo)
repair work at, 138
Tripolitania, 59, 139, 152, 174
aircraft at, 179
airfield attacks at, 95
defending, 60, 72
loss of, 249
Tunis, 73, 146, 252, 254, 259
advance on, 255, 263, 264
air transport at, 148
bridgehead in, 250
capture of, 397, 422n111
Tunisia, 60, 196, 258
cooperation roles in, 309
evacuation of, 278
German assets in, 282
lodgment in, 250

ports in, 259
Stalingrad and, 286
Tunisian Railway, 140
Turkish Air Force, 111
Twining, Nathan, 332, 357, 358, 360

U-boats, 173–174, 218, 254, 320, 350, 398, 399
ASW patrols and, 276
deployment of, 126
killing, 321
Udet, Ernst, 147
Ultra intelligence, 82, 98, 134, 233, 237, 280, 371
impact of, 100, 186, 246
US Army Air Corps (USAAC), 44, 48, 189
US Army Air Force (USAAF), 86, 119, 191, 217, 224, 225, 226, 240, 250, 251
aircraft for, 194
air superiority for, 399
CBO and, 258
challenges for, 255–256
combined-arms operations and, 257
expansion of, 48
as learning organization, 257
radius of action for, 359 (map)
RAF and, 194, 225
signal units and, 335
Spitfires for, 251
TAF and, 42
US Army and, 267
US Navy
ABC-1/Arcadia Conference and, 179
convoy escort by, 151
US Strategic Air Force, 381
US Strategic Bombing Survey (USSBS), 357, 360, 363, 366, 384
Spaatz and, 356, 358
USAAC. *See* US Army Air Corps
USAAF. *See* US Army Air Force
USS *Ancon*, 311
USS *Hilary*, 311
USS *Ohio*, 218
photo of, 220
USS *Savannah*, damage to, 314
USS *Wasp*, 182, 185, 195
delivery by, 439n22
photo of, 197
USSBS. *See* US Strategic Bombing Survey
Ustashi, 88
Utah Beach, 352

Valle, Giuseppe, 19
Vehicles, 286
lend-lease, 374
shortage of, 184, 260, 333
tropicalized, 15
Via Balbia, 56, 58, 77
Vichy Air Force, 116, 118, 252
Vichy France, 31, 172, 248, 249, 251
resistance from, 118
Syria and, 115, 117
Vigorous Convoy, 210, 213
Von Arnim, Hans-Jürgen, 279, 280, 282, 285
Von Bismarck, Georg, 404
Von Blomberg, Alex, 116, 427n8
Von Blomberg, Werner, 427n8
Von Boetticher, Friedrich, 87
Von Brauchitsch, Walther, 27
Von Luck, Hans, 237
Von Manteuffel, Hasso, 375, 376
Von Pohl, Ritter, 14, 19, 27, 172
Von Ravenstein, Johann, 138
Von Richthofen, Wolfram, 80, 81, 89, 298, 421n96
airfields by, 125
Fliegerkorps VIII and, 75, 239
OKL and, 296
Von Rintelen, Enno, 12, 139, 168, 239
air superiority and, 37
logistics and, 230
resupply and, 140
Von Senger und Etterlin, Fridolin, 336
Von Thoma, Wilhelm Ritter, 51, 230, 246, 404
Von Vietinghoff, Heinrich, 306, 316, 336–337
Von Waldau, Hoffman, 71, 104, 138, 139, 230, 239
air support and, 212
arrival of, 207
award for, 211
on El Alamein, 225
Kesselring and, 209
missions for, 236
Pricolo and, 207
Rommel and, 207, 208, 212, 217, 221
Von Weichs, Maximilian, 377
Von Weizäcker, Ernst, 50

WAC. *See* Western Air Command
Waffen SS, 375
War Cabinet, 30, 35, 108, 118, 152
Mediterranean/Eastern Front and, 177–178

War Cabinet, *continued*
 Middle East and, 135
 strategic guidance from, 178
Warlimont, Walter, 277–278
Wavell, Archibald, 36, 41, 44, 69, 85,
 97, 115
 air superiority and, 107, 108
 Operation Battleaxe and, 107
 Operation Compass and, 40, 52
 replacement of, 121
 resources and, 31
WDAF. *See* Western Desert Air Force
Weichold, Eberhard, 27, 236–237
Weinberg, Gerhard, 398
Wellingtons, 45, 75, 95, 111, 115, 128,
 189, 202, 205, 214, 225, 272
 ASV-equipped, 132, 133, 190, 205, 282
 bombing by, 38, 39, 72, 76, 80, 104,
 195, 251, 273, 304, 307
 deployment of, 68, 84
 electronic warfare, 242, 243
 fueling/bombing up, 72 (photo)
 missions for, 235–236
 torpedo-equipped, 165
Welsh, William, 250, 259, 266
Wenninger, Rudolf, 334
Western Air Command (WAC), 251
Western Desert, 13, 14, 16, 21, 40, 44,
 55, 61, 75, 76, 84, 86
 air support in, 5, 111, 122, 124, 130, 156
 combined-arms operations in,
 257–258
 RAF strength in, 110, 113, 269
 struggle in, 114, 160
 success in, 57
Western Desert Air Force (WDAF), 30,
 42, 69, 98, 107–108, 163, 164, 165,
 166, 185, 187, 191, 220, 225, 232,
 263, 267
 aircraft for, 152, 179, 274
 air superiority and, 156, 209, 210
 attacks by, 140, 143, 180, 186, 215,
 217–218, 237, 245, 246, 274, 275

 command for, 270
 convoy protection and, 272
 coordination with, 227
 doctrinal/procedural tenets of, 288
 effectiveness of, 246, 274–275
 Eighth Army and, 180, 186, 242, 247,
 248, 282
 innovations by, 273–277
 lessons for, 251
 losses for, 168, 211
 mobility of, 175
 as model, 203
 Operation Crusader and, 170
 Operation Supercharge and, 245
 rebuilding, 189
 strength of, 168, 240
 support from, 280
 See also Desert Air Force
Westphal, Siegfried, 179, 336
Wever, General, 303, 461n33
Wiener Neustadt, 309, 319, 360, 363
 attacks on, 307, 310, 316, 320
Wiese, Friedrich, 351
Willkie, Wendell, 48
Wilson, Henry Maitland, 329, 355, 374
 on airpower, 336, 337, 345
 CCS and, 329, 347–348
 MAAF and, 356
 Operation Anvil and, 344, 348
 Operation Strangle and, 332
Wilson, J. R., 235
With Prejudice (Tedder), 93
World at Arms, A (Weinberg), 398
Wynn, Humphrey, 6, 449n40

Youk les Bains, 252, 263
Y Service, 101, 123, 186, 234, 280, 283,
 298, 308, 326
 improvements for, 102, 119
Yugoslav Air Force, destruction of, 81

Zhukov, Georgi: on lend-lease, 201
Zuckerman, Solly: plan by, 310, 331, 334

$34.24   940. 54 EH   1337973
                     2015